T0321455

Problems of Point-Blast Theory

Problems of Point-Blast Theory

V. P. Korobeinikov

Translated by
George Adashko

© 1985 by Nauka Publishers, Moscow, USSR

© 1991 by Springer New York
All rights reserved
Printed in the United States of America

Library of Congress Cataloging-in-Publication Data

Korobeĭnikov, V. P. (Viktor Pavlovich)
[Zadachi teorii tochechnogo vzryva. English]
Problems of point-blast theory/V. Korobeinikov.
p. cm.--(Translation series)
Translation of: Zadachi teorii tochechnogo vzryva.
Includes bibliographical references.
ISBN 0-88318-674-8
1. Blast effect. 2. Explosions. I. Title. II. Series:
Translation series (New York, N.Y.).
TA748.K6713 1990
662'.2--dc20

90-45115
CIP

Contents

Chapter 1. Fundamental equations and formulation of problems

Chapter 2. Strong blast in a gas

Chapter 3. Linearized non-self-similar one-dimensional problems

Chapter 4. Spherical, cylindrical, and planar blasts with allowance for back pressure at a constant initial density

Chapter 5. Problems of point-blast theory in inhomogeneous media and with asymmetric energy release

Chapter 6. Blast in a combustible gas mixture

Chapter 7. Point blast in an electrically conducting gas with allowance for the influence of a magnetic field

Chapter 8. Propagation of perturbations in solar flares

Preface

The theory of point-blast explosion was developed about 35 years ago by L. I. Sedov as well as by J. Taylor, K. P. Stanyukovich, and J. von Neumann. Many problems of this theory were solved by a number of Soviet and foreign scientists. Principal attention is paid of late to the influence of various physical processes on the propagation of blast waves, and to two-dimensional nonstationary problems.

The present book deals with the development of models of the motion induced in a medium by a blast. Mathematical problems are formulated for these models and methods for their solution are studied.

The emphasis is on theoretical topics treated by the author himself or in collaboration with others. To make the book useful to specialists in various fields without having to refer to many other sources, the first chapter contains the basic information on gasdynamics, magnetohydrodynamics, explosion theory, dimensionality theory, and mathematical physics.

This book is an enlarged revision of an earlier limited-edition monograph (V. P. Korobeinikov, Topics in the theory of point blasts in gases, in *Proceedings of the V. A. Steklov Mathematics Institute of the USSR Academy of Sciences, Vol. 119,* Nauka, 1973). Since all the chapters have been substantially revised and expanded, the present edition is in fact a new book. Its aim is to acquaint the reader with the advances in the theory and to point out the yet-unsolved problems.

The book's eight chapters and three appendices cover all the main trends of modern theory, except for applications to nova and supernova outbursts and to concentrated blasts in two-phase media and in deformable solids.

The first chapter is an introduction. The remaining chapters are expositions of various aspects of the theory. Besides the traditional questions of point-blast theory, considerable attention is paid to modern problems of gasdynamics of radiating media, to nonequilibrium phenomena and kinetics of chemical reactions, to the interaction of high-temperature gases with electromagnetic fields, and to two-dimensional spatial effects. The use of numerical methods is covered quite comprehensively and specific examples of numerical procedures are given. The appendices contain concise information on numerical integral-equation and finite-difference methods. Each chapter is accompanied by a list of references published in the USSR and abroad. They include as a rule only work related to the subject matter or relevant cited results.

Space limitations prevent inclusion of detailed tables of gasdynamic functions, but their sources are appropriately referred to. All the pertinent tables could fill a separate volume. The appendices contain brief descriptions of some calculation results.

The book embodies many years' work by the author and his colleagues on the subject. It summarizes, in particular, the experience gained from courses taught at the Mechanical–Mathematical Department of the Moscow State

University, and from numerous papers delivered at various seminars, symposia, and conferences.

The author is deeply grateful to L. I. Sedov for suggesting many topics, for numerous discussions of the results, and for continuous attention and encouragement.

Thanks are due also to A. G. Kulikovskii, V. V. Markov, L. D. Kondrat'eva, and R. P. Agafonov for help with the preparation of the book. The author is also indebted to his colleagues and comrades, particularly V. P. Karlikov, P. I. Chushkin, E. V. Ryazanov, V. A. Levin, and D. N. Brushlinskii for participating in the development of the theory expounded here. Graduate students L. V. Shidlovskaya, B. V. Putyatin, and I. S. Men'shov offered their results and comments, for which the author is most grateful. The author thanks also S. S. Grigoryan for a thorough review of the book.

V. P. Korobeinikov

Introduction

Point-blast theory (PBT) evolved to meet the need for describing phenomena produced in continuous media by exploding charges of small size and weight but with high specific energy. This theory yields, with sufficient accuracy, many needed data on the nonstationary motion and on the shock waves produced by explosions. For gases, PBT deals with phenomena such as strong explosions in the atmosphere, explosive processes in outer space, motion of thin blunted bodies (including meteorites) in the earth's atmosphere, propagation of shock waves produced by powerful laser radiation and by electric discharges in gases, electric explosions of conductors, and others.

We consider briefly the gist of the point-blast phenomenon. Let a small explosive mass be concentrated in a volume much smaller than the ambient medium, and let an explosion be produced at some instant of time and release energy of density (per unit volume) much higher than in the ambient medium. The pressure and energy of the medium surrounding the explosion will then increase almost instantaneously, but will attenuate as the explosion propagates away from the blast site. Allowance for all the factors influencing the blast-wave propagation under real conditions would make a theoretical description of the phenomenon most complicated. Here, just as in other complex phenomena, a successful solution of the problem requires a certain degree of idealization, wherein account is taken only of the factors that predominate in the development of the entire process. When it comes to explosions, the principal idealization is the assumption that the energy is released instantaneously, while the volume occupied by the explosive and the mass of the charge are zero. In the case of spherical, cylindrical, and planar symmetry, the blast is produced in a point, along a straight line, or over a plane, respectively. This idealized energy-release process is called a point blast. Various idealized schemes of describing the evolution of the blast, and models describing the motion of the medium, are also developed.

To account for various physical processes, we study point-blast-initiated gas motion by using known models as well as new ones, for which mathematical problems are formulated and solution methods are investigated.

In the theoretical study of the point-blast phenomenon, use is made of dimensional and dimensionless variables. Dimensional quantities are more frequently used to derive the fundamental equations and to analyze physical problems, to compare the numerical results of the theory with experimental data, and also in certain other cases. Mathematical formulation of the problems, investigation and integration of the equations, and the graphical representation of the calculation results are more conveniently effected in dimensionless form. This was the author's main tendency. By introducing dimensionless quantities it is frequently possible to obtain simpler analytic solutions and more universal numerical solutions, as well as to use analytic solutions obtained for certain initial parameters to determine other parameters.

In the study of individual problems, the dimensionless quantities are usually ratios of corresponding dimensional ones of like dimensionality. These characteristic quantities can be different in different problems. As a rule, results of other workers are cited in their original physical units and notation.

Chapter 1
Fundamental equations and formulation of problems

1. Gasdynamics equations

1.1. Equations in integral form

We shall follow Refs. 1–6 in the discussion of the known fundamental equations of gasdynamics. In the investigations of a compressible liquid or gas, we regard the gas as a continuous medium, and use hydrodynamic models to describe the motions of a material medium.

To study the motion of a compressible continuous medium it is necessary to introduce scalar, vector, and tensor quantities, such as the temperature T, the velocity $\mathbf{v}$, the stress tensor $\hat{P}$, and others. This motion can be considered in various coordinate frames. If we are primarily interested in phenomena occurring at points in the observer's reference frame, we approach the study of the motion of a continuous medium from Euler's point of view. We investigate then the variation of the velocity, density, temperature, and other quantities at a given point of space. On the other hand, to follow the variation of the parameters of individual points of the medium, we use Lagrange's approach. A certain coordinate frame (x^1, x^2, x^3) that is fixed in space is introduced. Its triad of coordinates is identified with a point of space and with a position vector $\mathbf{r}$. The mathematical concept corresponding to the physical notion of liquid flow is that of a continuous transformation of a three-dimensional Euclidean space $\mathscr{E}_3$ into itself. The parameter t describing this transformation is identified with time. The value $t = 0$ can be taken to be the initial instant, and the range of variation of t can be represented by the entire real axis.

We introduce a Lagrangian or comoving coordinate frame (ξ^1, ξ^2, ξ^3), and denote by ξ a Lagrangian point. Let a moving liquid particle be located in a point $\xi = (\xi^1, \xi^2, \xi^3)$ at the instant $t = 0$ and in a point $\mathbf{r}(x^1, x^2, x^3)$ at the instant t. It can be assumed then that $\mathbf{r}$ is defined as a function of ξ and t, and from the kinematic standpoint the flow of the medium can be regarded as the transformation.

$$x^i = \varphi^i(\xi^i, t). \tag{1.1}$$

If ξ is fixed and t is varied, Eq. (1.1) defines the trajectory of a particle M initially located at the point ξ. If, on the other hand, t is fixed, Eq. (1.1) defines a transformation of the region initially occupied by the liquid into the region occupied by the liquid at the instant t.

If initially different points remain different during the entire time of motion, we can introduce the inverse transformation

$$\xi^j = \Phi^j(\mathbf{r}, t). \tag{1.2}$$

The functions φ^i and Φ^j are assumed to have continuous derivatives up to third order with respect to all the variables, except for some singular surfaces, curves, or points.

We use hereafter mainly Euler variables, although in some of the problems we shall consider also Lagrange variables.

The fundamental laws of the mechanics of a continuous medium are described by the equations of gasdynamics. Neglecting effects due to disequilibrium, heat conduction, and viscosity, the integral forms of these equations for a certain liquid volume Ω^* are

$$\frac{d}{dt}\int_{\Omega^*(t)} \rho \, d\Omega = 0 \tag{1.3}$$

or the mass-conservation law,

$$\frac{d}{dt}\int_{\Omega^*(t)} \rho \mathbf{v} \, d\Omega = \int_{\Omega^*(t)} \mathbf{F}\rho \, d\Omega - \oint_{\Sigma(t)} p\mathbf{n} \, d\Sigma \tag{1.4}$$

the momentum-conservation law, and

$$\frac{d}{dt}\int_{\Omega^*(t)} \rho\left(\varepsilon + \frac{v^2}{2}\right) d\Omega = -\oint_{\Sigma(t)} p\mathbf{v}\mathbf{n} \, d\Sigma + \int \rho(\mathbf{F}\mathbf{v}) \, d\Omega \tag{1.5}$$

the energy-conservation law.

Here $\Omega^*(t)$ is an individual liquid volume bounded by a surface Σ, p is the pressure, ε is the internal energy, $\mathbf{F}$ is the mass force (the field vector of the external forces referred to a unit mass), and $\mathbf{n}$ is the outward normal to the surface Σ.

To account for heat conduction with a heat-influx vector $\mathbf{q}$ specified by the Fourier heat-conduction law

$$\mathbf{q} = -\varkappa \operatorname{grad} T,$$

where $\varkappa$ is the thermal conductivity, it is necessary to add to the right-hand side of Eq. (1.5) a term in the form

$$-\oint_{\Sigma(t)} (\varkappa \operatorname{grad} T)\mathbf{n} \, d\Sigma. \tag{1.6}$$

Equations (1.3)–(1.6) hold for arbitrary volumes, particularly also when Ω^* contains a surface Σ_i on which the hydrodynamic functions become discontinuous.

In Euler variables, a gas as a continuous medium is described by the following fundamental quantities: velocity vector $\mathbf{v}(\mathbf{r}, t)$, density $\rho(\mathbf{r}, t)$, and pressure $p(\mathbf{r}, t)$. The internal energy is assumed as a rule to be specified by the thermodynamic function $\varepsilon = \varepsilon(p, \rho)$ or $\varepsilon = \varepsilon(T, \rho)$.

1.2. Gasdynamics differential equations and their properties

We introduce a four-dimensional Euclidean space-time $\mathscr{E}_4$. The motion of a gas will be called continuous in a region $\Omega \subset \mathscr{E}_4$ if the functions $\mathbf{v}$, ρ, and p are continuous together with their derivatives everywhere in Ω.

Continuous motion satisfies a system of differential equations that are the counterparts of the integral equations (1.3)–(1.5). For gas motion without allowance for mass forces and dissipative processes, this system takes the form

$$\frac{\partial \rho}{\partial t} + \operatorname{div}(\rho \mathbf{v}) = 0, \tag{1.7}$$

$$\frac{\partial (\rho \mathbf{v})}{\partial t} + \operatorname{div} \hat{P} = 0, \tag{1.8}$$

$$\frac{\partial}{\partial t}\left(\rho\varepsilon + \frac{\rho v^2}{2}\right) + \operatorname{div} \mathbf{v}\left(\rho\varepsilon + p + \frac{\rho v^2}{2}\right) = 0, \tag{1.9}$$

where $\varepsilon = (p, \rho)$, $\hat{P}$ is a tensor with components $P_{ik} = p\delta_{ik} + \rho v^i v^k$, and δ_{ik} is a unit tensor. We use the systems (1.3)–(1.5) and (1.7)–(1.9) as the basis for further study. The function $\varepsilon(p, \rho)$ depends on the properties of the gas and is assumed to be continuous and differentiable with respect to p and ρ. A simple yet important example of an ideal compressible medium is an ideal gas with a constant specific-heat ratio $\gamma = c_p/c_v$.

The function $\varepsilon(p, \rho)$ for an ideal gas is of the form

$$\varepsilon = \frac{1}{\gamma - 1}\frac{p}{\rho} + \text{const}, \tag{1.10}$$

and the connection between the density, pressure, and temperature is given by the thermal equation of state

$$p = R\rho T, \tag{1.11}$$

where R is the gas constant.

For adiabatic reversible gas motion, Eq. (1.9) can be transformed with the aid of Eqs. (1.7) and (1.8) into

$$\frac{d\varepsilon}{dt} - \frac{p}{\rho^2}\frac{d\rho}{dt} = 0, \tag{1.12}$$

where $d/dt = \partial/\partial t + \mathbf{v} \cdot \nabla$ is the substantive or total derivative. Introducing the entropy S and using the first law of thermodynamics, we get

$$T\,dS = d\varepsilon + p\,d\frac{1}{\rho}. \tag{1.13}$$

This yields, taking Eq. (1.12) into account,

$$\frac{dS}{dt} = 0. \tag{1.14}$$

The energy equation for adiabatic continuous motions of an inviscid medium can thus be chosen in the form (1.14). The entropy S is a function of the thermodynamic parameters of the medium, say p and ρ or ρ and T. For an ideal gas we have

$$S = c_v \ln \frac{p}{\rho^\gamma} + \text{const}, \tag{1.15}$$

where c_v is the specific heat of the gas at constant volume. If S is constant everywhere in the region in which the gas moves, its motion is called isentropic.

Note some important gas motions characterized by special geometric trajectories in $\mathscr{E}_3$ space, as well as by the geometric properties of the level surfaces of the fundamental quantities.

Motion is called two-dimensional if the fundamental quantities can be described by functions that depend, apart from the time t, on only two spatial variables.

One form of two-dimensional motion is plane-parallel or planar flow, for which the particle trajectories in $\mathscr{E}_3$ are planar curves, the planes are parallel to a fixed plane Π, and the fundamental quantities do not vary in directions perpendicular to Π and can depend only on t. If a Cartesian frame is chosen such that the plane Π is the coordinate plane (Ox^1x^2), the velocity vector $\mathbf{v}$ in plane-parallel motion has two nonzero components $\mathbf{v}(v^1, v^2, 0)$, while v^1 and v^2 are functions of only the variables x^1, x^2, and t. The differential equations of plane-parallel flow in scalar form can be easily obtained from the system (1.7)–(1.9).

Another important example of two-dimensional motion is axisymmetric flow. Motion (or flow) is called asymmetric if the particle trajectories in $\mathscr{E}_3$ are planar curves whose planes pass through some fixed straight line—a symmetry axis. The fundamental functions sought can depend only on t on any circle whose center is on the symmetry axis and whose plane is perpendicular to this axis.

Axisymmetric motions can be conveniently studied in cylindrical or spherical (polar) coordinates. If a cylindrical frame is chosen such that its axis is the symmetry axis (which we designate by Ox^1), the fundamental quantities can be regarded as dependent on the coordinate x^1 and on the radius r. We denote by u the velocity component along Ox^1 and by v the modulus of the velocity projection on a plane perpendicular to the symmetry axis. For Cartesian velocity components, a density ρ, and a pressure p we have $v^1 = u(x, r, t)$, $v^2 = v(x, r, t) \cos \varphi$, $v^3 = v(x, r, t) \sin \varphi$, $\rho = \rho(x, r, t)$, and $x = x^1$, where φ is the polar angle. For a chosen coordinate frame, the differential equations of axisymmetric motion differ from those of plane-parallel motion in that the components v^1 and v^2 are replaced by u and v, the role of the variable x^2 is assumed by r, and the divergence is given by

$$\operatorname{div}(\psi \mathbf{v}) = \frac{\partial(\psi u)}{\partial x} + \frac{\partial(\psi v)}{\partial r} + \frac{\psi v}{r},$$

where ψ is some scalar function.

In a number of cases it is expedient to study axisymmetric motions in a polar spherical system with coordinates r, θ, and φ and with the center on the symmetry axis. Now θ is the angle between the symmetry axis and the position vector of the point in space. If the functions sought are the velocity component v_r along the position vector and the component of v_θ in a direction tangent to a circle with center at the origin and passing through the symmetry axis and the given point of space, then the fundamental quantities sought are independent of the angle φ. The gasdynamics equations with allowance for the mass forces can be expressed here as

$$r^2 \frac{\partial \rho}{\partial t} + \frac{\partial}{\partial r}(r^2 \rho v_r) + \frac{r}{\cos\theta}\frac{\partial}{\partial \theta}(\rho v_\theta \cos\theta) = 0,$$

$$\frac{dv_r}{dt} - \frac{v_\theta^2}{r} + \frac{1}{\rho}\frac{\partial p}{\partial r} = F_r, \qquad \frac{dS}{dt} = 0,$$

$$\frac{dv_\theta}{dt} - \frac{v_r v_\theta}{r} + \frac{1}{r\rho}\frac{\partial p}{\partial \theta} = F_\theta, \tag{1.16}$$

$$\frac{d}{dt} = \frac{\partial}{\partial t} + v_r \frac{\partial}{\partial r} + \frac{v_\theta}{r}\frac{\partial}{\partial \theta},$$

where F_r and F_θ are the components of the acceleration due to the mass forces.

In the presence of gravitational force along the flow symmetry axis we have $F_r = g\cos\theta$ and $F_\theta = g\sin\theta$, where g is the acceleration of gravity.

We present also, for the flow of an ideal gas in a gravitational field, equations in the cylindrical coordinates r and z (the z axis is directed counter to the force of gravity and is the symmetry axis for the axisymmetric flow):

$$\frac{1}{\rho}\frac{d\rho}{dt} + \frac{\partial v_z}{\partial z} + \frac{\partial v_r}{\partial r} + \frac{(\nu - 2)v_r}{r} = 0,$$

$$\frac{d}{dt} = \frac{\partial}{\partial t} + v_r \frac{\partial}{\partial r} + v_z \frac{\partial}{\partial z},$$

$$\frac{dv_r}{dt} + \frac{1}{\rho}\frac{\partial p}{\partial r} = 0, \qquad \rho\frac{dv_r}{dt} + \frac{\partial p}{\partial z} = -\rho g, \tag{1.17}$$

$$\frac{\partial}{\partial t}\left[\rho\left(\varepsilon + \frac{v^2}{2}\right)\right] + \frac{\partial}{\partial z}\left\{v_z\left[\rho\left(\varepsilon + \frac{v^2}{2}\right) + p\right]\right\} + \frac{\partial}{\partial r}\left\{v_r\left[\rho\left(\varepsilon + \frac{v^2}{2}\right) + p\right]\right\}$$
$$+ \frac{(\nu - 2)v_r(\rho\varepsilon + p + \rho v^2/2)}{r} + \rho v_z g = 0.$$

Here $\nu = 2$ for flow in the (r, z) plane and $\nu = 3$ for axisymmetric flow.

Motion is called one-dimensional if the fundamental quantities can be expressed in terms of functions that depend on only one spatial variable, in addition to the time t.

We shall distinguish between one-dimensional motions with planar, cylindrical, and spherical symmetry.

In one-dimensional motion with planar symmetry (or with plane waves) the particle trajectories form in $\mathscr{E}_3$ space a family of straight lines perpendicular to a certain fixed plane Π, and the fundamental quantities along any plane parallel to Π can depend only on the time t, for any point on the plane. If the Cartesian axis Ox^1 (or Or in the other notation) is chosen perpendicular to the plane Π, the velocity vector $\mathbf{v}$ has only one nonzero component $v^1 = v$, and $v = v(r, t)$, $p = p(r, t)$, and $\rho = \rho(r, t)$. The differential equations for this case are easily obtained from the basic system (1.7)–(1.9).

One-dimensional motion with cylindrical symmetry is axisymmetric motion in which the particle trajectories form in $\mathscr{E}_3$ a family of straight lines that cross perpendicularly the symmetry axis, and the fundamental quantities sought depend only on the time and are constant on the surfaces of circular cylinders whose axes coincide with the symmetry axis.

It is convenient here to use the cylindrical coordinates x, r, and φ and choose the fundamental quantities to be the radial velocity component $v_r = v$, the density ρ, and the pressure p, in which case we have

$$v^1 = 0, \quad v^2 = v(r, t)\cos\varphi, \quad v^3 = v(r, t)\sin\varphi,$$

$$p = p(r, t), \quad \rho = \rho(r, t).$$

The system of differential equations can be obtained from the basic system (1.7)–(1.9) in the manner used above for the axisymmetric case.

In one-dimensional motion with spherical symmetry (or with spherical waves) the particle trajectories form in $\mathscr{E}_3$ a family of straight lines drawn from a fixed point O; the fundamental quantities remain constant on any sphere with center at O and can depend only on the time t. For an analytic description of these flows it is convenient to use a spherical frame (r, θ, φ) with origin at the point O. We denote by v the radial component of the velocity $\mathbf{v}$ ($v_r = v$), so that for this flow v, p, and ρ are all functions of r and t.

The set of differential equations can be obtained here easily from the system (1.16), in which we need put only $v_\theta = 0$ and $v_r = v$, and all the functions must be regarded as independent of θ. We shall frequently find it convenient to consider all three types of one-dimensional motion jointly. To this end we introduce a symmetry parameter ν such that $\nu = 1$, 2, and 3 for planar, cylindrical, and spherical symmetry, respectively. The parameter ν identifies the dimensionality of that $\mathscr{E}_3$ subspace which suffices to describe the coordinate dependences of the hydrodynamic functions.

In the notation introduced, the unified system of gasdynamic differential equations takes the form

$$\frac{\partial}{\partial t}(r^{\nu-1}\rho) + \frac{\partial}{\partial r}(r^{\nu-1}v\rho) = 0, \tag{1.18}$$

$$\frac{\partial}{\partial t}(\rho v) + \frac{\partial}{\partial r}(p + \rho v^2) + \frac{(\nu-1)\rho v^2}{r} = 0, \tag{1.19}$$

$$\frac{\partial}{\partial t}\left[\rho r^{\nu-1}\left(\varepsilon + \frac{v^2}{2}\right)\right] + \frac{\partial}{\partial r}\left[r^{\nu-1}\rho v\left(\varepsilon + \frac{p}{\rho} + \frac{v^2}{2}\right)\right] = 0. \tag{1.20}$$

The considered one-dimensional motions with planar, cylindrical, and spherical symmetry are examples of motions that are exactly compatible with the system (1.7)–(1.9).

In some applications, use is made of one-dimensional motions of another form, corresponding to quasi-one-dimensional flows of a compressible medium. Examples are flows in pipes of variable cross section, flows describable in a one-dimensional hydraulic approximation. We shall not dwell on them here in detail. A derivation of the relevant differential equations in Euler variables is given, for example, in Ref. 2.

The gasdynamic equations take in Lagrange variables the form

$$\rho J = \rho_0 J_0, \tag{1.21}$$

$$\nabla \mathbf{r}\,\frac{d^2\mathbf{r}}{dt^2} + \frac{1}{\rho}\nabla p = 0, \tag{1.22}$$

$$\frac{dS}{dt} = 0, \tag{1.23}$$

where $\rho_0 = \rho_0(\mathbf{r})$ is the initial density distribution, J is the Jacobian of the transformation (1.1):

$$J = \frac{\partial(x^1, x^2, x^3)}{\partial(\xi^1, \xi^2, \xi^3)} = \det\left(\frac{\partial x^i}{\partial \xi^j}\right),$$

and the gradient is taken in the space of the Lagrangian coordinates ξ^j. In terms of the coordinates, Eq. (1.22) takes the form

$$\frac{\partial x^i}{\partial \xi^j}\frac{d^2 x^i}{dt^2} = -\frac{1}{\rho}\frac{\partial p}{\partial \xi^j},$$

where summation over i is implied.

For one-dimensional motions, the system (1.21)–(1.23) can be rewritten in the form

$$\rho\frac{\partial r}{\partial \xi} = \rho_0\frac{\xi^{\nu-1}}{r^{\nu-1}}, \qquad v - \frac{\partial r}{\partial t} = 0,$$
$$\frac{\partial p}{\partial \xi} = -\rho\frac{\xi^{\nu-1}}{r^{\nu-1}}\frac{\partial v}{\partial t}, \qquad \frac{\partial S}{\partial t} = 0. \tag{1.24}$$

Here ξ is the initial coordinate of the particle, i.e., the distance from the particle to the zero plane, to the symmetry axis, or to the origin in the planar, cylindrical, and spherical cases, respectively.

Note certain properties of the gasdynamic equations. Let φ, ψ, u, f, $\mathbf{v}$, and Φ be certain continuously differentiable functions. An equation of the type

$$\frac{\partial\varphi(u, f, \mathbf{r}, t)}{\partial t} + \operatorname{div}\,[\psi(u, f, \mathbf{r}, t)\mathbf{v}] = \Phi(u, \mathbf{v}, \mathbf{r}, t)$$

will be called an equation in conservation-law form or in divergence form.

Property 1. The gasdynamic equations (1.7)–(1.9) and (1.18)–(1.20) are expressed in divergence form. The system (1.24) in Lagrangian coordinates can also be reduced to the divergence form. Note that Eqs. (1.8)–(1.20) in divergence form were given in Ref. 4.

Property 2. The systems (1.7)–(1.9), (1.18)–(1.20), (1.21), (1.22), and (1.24) are hyperbolic partial-differential equation systems in Euclidean space–time coordinates.

Property 3. The system (1.7)–(1.9) is invariant to the Galileo–Newton transformations $p_* = p$, $\rho_* = \rho$, $t_* = t + \tau$, $\mathbf{r}_* = \mathbf{r} + \mathbf{a}t + \mathbf{r}_0$, and $\mathbf{v}_* = \mathbf{v} + \mathbf{a}$, where $\mathbf{a}$ and $\mathbf{r}_0$ are constant vectors, τ is a constant scalar, and asterisks mark quantities pertaining to the state of motion after the transformation. The truth of property 3 can be verified directly.

In many gasdynamic problems the fundamental quantities or the fundamental gasdynamic functions are chosen to be the velocity $\mathbf{v}$, the pressure p, and the density ρ. In many cases (particularly for numerical solutions), the fundamental quantities are taken to be ρ, $\mathbf{v}$, and the internal energy ε, while the pressure is obtained from the caloric equation of state

$$p = p(\rho, \varepsilon). \tag{1.25}$$

It is sometimes convenient to use as a basis the velocity $\mathbf{v}$, the density ρ, and the total energy per unit volume E, which is given in gasdynamics by

$$E = \rho(\varepsilon + v^2/2). \tag{1.26}$$

The system (1.7)–(1.9) takes in these variables the form

$$\frac{\partial\rho}{\partial t} + \operatorname{div}\,(\rho\mathbf{v}) = 0, \qquad \frac{\partial(\rho\mathbf{v})}{\partial t} + \operatorname{div}\,(\rho\mathbf{v}\cdot\mathbf{v} + \delta_{ik}p) = 0,$$
$$\frac{\partial E}{\partial t} + \operatorname{div}\,(E\mathbf{v}) + \operatorname{div}\,(p\mathbf{v}) = 0, \tag{1.27}$$

where p, in accordance with Eqs. (1.25) and (1.26), is a specified function of E, ρ, and v^2.

Note also the following simple property of the system (1.7)–(1.9):

$$\mathbf{v} = 0, \quad \rho = \rho_0(\mathbf{r}), \quad p = p_0 = \text{const}. \tag{1.28}$$

We shall call this solution the equilibrium state or the quiescent state.

1.3. Certain simplified and particular forms of the equations of motion of fluids

We introduce in the gasdynamic equations certain simplifications that yield approximate solutions of gas-motion problems. One of the common simplifications is linearization, which consists of the following. Assume a certain basic motion, i.e., an exact solution of the nonlinear gasdynamic equations. If these equations are, for example, Eqs. (1.7), (1.8), and (1.14), this solution takes the form

$$\mathbf{v}_0(\mathbf{r},t), \quad \rho_0(\mathbf{r},t), \quad p_0(\mathbf{r},t), \quad S_0(\mathbf{r},t). \tag{1.29}$$

We shall seek a second solution in the form $\mathbf{v} = \mathbf{v}_0 + \sigma\mathbf{v}'$, $\rho = \rho_0 + \sigma\rho'$, $p = p_0 + \sigma p'$, and $S = S_0 + \sigma S'$, where the primes denote new unknown functions of $\mathbf{r}$ and t, either increments to the fundamental solution or perturbations, and σ is some small parameter.

If this solution is substituted in the system (1.7), (1.8), (1.14) we obtain, canceling out σ and recognizing that $\mathbf{v}_0$, ρ_0, p_0, and S_0 constitute the solution of the system,

$$\frac{\partial \rho'}{\partial t} + \operatorname{div}(\rho_0\mathbf{v}' + \mathbf{v}_0\rho' + \sigma\mathbf{v}'\rho') = 0,$$

$$\frac{\partial}{\partial t}(\rho_0\mathbf{v}' + \mathbf{v}_0\rho' + \sigma\mathbf{v}'\rho') + \operatorname{div}(\rho_0\mathbf{v}_0\mathbf{v}' + \rho'\mathbf{v}_0\mathbf{v}_0 + \rho_0\mathbf{v}'\mathbf{v}_0$$
$$+ \sigma\rho'\mathbf{v}_0\mathbf{v}' + \sigma\rho'\mathbf{v}'\mathbf{v}_0 + \sigma\rho_0\mathbf{v}'\mathbf{v}' + \sigma^2\mathbf{v}'\mathbf{v}'\rho') + \operatorname{grad} p' = 0, \tag{1.30}$$

$$\frac{\partial S'}{\partial t} + (\mathbf{v}_0\nabla)S' + (\mathbf{v}'\nabla)S_0 + \sigma(\mathbf{v}'\nabla)S' = 0,$$

$$p' = \frac{1}{\sigma}[p(\rho_0 + \sigma\rho', S + \sigma S') - p_0].$$

It follows from this that the new functions sought, $\mathbf{v}'$, ρ', etc., depend not only on $\mathbf{r}$ and t, but also on the parameter σ. Let the functions $\mathbf{v}'$, ρ', S', and p' as solutions of the exact equations (1.30), and also their derivatives in these equations, have a finite limit as $\sigma \to 0$. Then, going to the limit in Eqs. (1.30) we obtain the linearized system of the initial equations

$$\frac{\partial \rho'}{\partial t} + \operatorname{div}(\rho_0\mathbf{v}' + \mathbf{v}_0\rho') = 0,$$

$$\frac{\partial}{\partial t}(\rho_0\mathbf{v}' + \mathbf{v}_0\rho') + \operatorname{div}(\mathbf{v}'\rho_0\mathbf{v}_0 + \rho'\mathbf{v}_0\mathbf{v}_0 + \mathbf{v}_0\rho_0\mathbf{v}') + \operatorname{grad} p' = 0, \tag{1.31}$$

$$\frac{\partial S'}{\partial t} + (\mathbf{v}_0\nabla)S' + (\mathbf{v}'\nabla)S_0 = 0, \qquad p' = a_0^2\rho' + \left(\frac{\partial p}{\partial S}\right)_\rho S',$$

where $a_0^2 = (dp/d\rho)_s$ is the adiabatic speed of sound.

The procedure for obtaining equations of type (1.31) is called linearization about a known solution $\mathbf{v}_0, \rho_0, p_0$. The system (1.31) is the linearized set of gasdynamics equations. We will use the linearization method extensively.

If the fundamental solution is of the form

$$\mathbf{v}_0 = 0, \quad \rho_0 = \text{const}, \quad S_0 = \text{const}, \quad p_0 = \text{const},$$

we obtain from Eqs. (1.31) the known acoustics equations

$$\frac{\partial \rho'}{\partial t} + \rho_0 \operatorname{div} \mathbf{v}' = 0, \quad \frac{\partial}{\partial t}(\rho_0 \mathbf{v}') + \nabla p' = 0, \quad \frac{\partial S'}{\partial t} = 0, \quad p' = a_0^2 \rho'. \tag{1.32}$$

Assume $S' = \text{const}$. From Eqs. (1.32) we obtain the classical wave equation for the pressure

$$\frac{\partial^2 p'}{\partial t^2} - a_0^2 \Delta p' = 0. \tag{1.33}$$

It is customary to introduce in the theory of sound propagation the perturbation-energy density[5]

$$E' = \frac{\rho_0 v'^2}{2} - \frac{a_0^2}{2\rho_0} \rho'^2. \tag{1.34}$$

This energy corresponds to the difference between the total energy per unit volume of a moving liquid and the initial energy of the gas at rest to which the energy corresponding to the change of the amount of matter is added, namely,

$$E' = E - \rho_0 \varepsilon_0 - \left(\varepsilon_0 + \frac{p_0}{\rho_0}\right)\rho' + o(\sigma^2 \rho'^2).$$

From the system (1.27) we can obtain for the change of E' the equation

$$\frac{\partial E'}{\partial t} + \operatorname{div}(p'\mathbf{v}') = 0, \tag{1.35}$$

which is the energy-conservation law for the perturbation.

For one-dimensional motions, the set of acoustic equations takes the form

$$\frac{\partial \rho'}{\partial t} = -\rho_0 \frac{1}{r^{\nu-1}} \frac{\partial}{\partial r}(r^{\nu-1} v'), \quad p' = a_0^2 \rho', \quad \frac{\partial v'}{\partial t} = -\frac{1}{\rho_0} \frac{\partial p'}{\partial r}.$$

The wave equation for these cases is

$$\frac{\partial^2 p'}{\partial t^2} = a_0^2 \frac{1}{r^{\nu-1}} \frac{\partial}{\partial r}\left(r^{\nu-1} \frac{\partial p'}{\partial r}\right) \tag{1.36}$$

and has for the planar case the solution

$$p' = A_1 r + A_2 t + f_1\left(t - \frac{r}{a_0}\right) + f_2\left(t + \frac{r}{a_0}\right). \tag{1.37}$$

In the case of spherical symmetry we get

$$p' = \frac{f_1(t - r/a_0) + f_2(t + r/a_0)}{r}. \tag{1.38}$$

For a cylindrical wave the solution can be written in integral form

$$p' = \int_r^\infty \frac{f_1(t - R/a_0)}{\sqrt{R^2 - r^2}}\,dR + \int_r^\infty \frac{f_2(t + R/a_0)}{\sqrt{R^2 - r^2}}\,dR. \tag{1.39}$$

In the solutions (1.37)–(1.39), f_1 and f_2 are arbitrary functions of their arguments, $R^2 = r^2 + z^2$ (z is the coordinate along the axis of the cylindrical system), and A_1 and A_2 are arbitrary constants.

We consider now the case of simplified equations, when it is assumed merely that the density ρ is constant (incompressible fluid). Equation (1.8) remains unchanged, but Eq. (1.7) takes the form

$$\operatorname{div} \mathbf{v} = 0. \tag{1.40}$$

Now the energy equation with heat conduction taken into account[5] becomes

$$\frac{\partial}{\partial t}(\rho\varepsilon) + \operatorname{div}(\rho\varepsilon\mathbf{v}) - \operatorname{div}(\varkappa\nabla T) + \frac{\partial}{\partial t}\left(\frac{\rho v^2}{2}\right) + \operatorname{div}\left[\mathbf{v}\left(p + \frac{\rho v^2}{2}\right)\right] = 0.$$

With allowance for the continuity and momentum equations, we get

$$\frac{\partial}{\partial t}(\rho\varepsilon) + \mathbf{v}\nabla(\rho\varepsilon) - \operatorname{div}(\varkappa\nabla T) = 0. \tag{1.41}$$

If $\varkappa = \text{const}$, $\rho = \text{const}$, $\mathbf{v} = 0$ and $\varepsilon = c_v T + \text{const}$, Eq. (1.41) leads to the classical heat-conduction equation

$$\frac{\partial T}{\partial t} = \chi \Delta T, \tag{1.42}$$

where $\chi = \varkappa / c_v \rho$ and $\varkappa$ is the thermal conductivity. Now let $\mathbf{v} = 0$ and $\varkappa = \varkappa(\varepsilon)$. From Eq. (1.41) we get

$$\frac{\partial \varepsilon}{\partial t} = \operatorname{div}\left[\frac{\varkappa}{\rho}\left(\frac{dT}{d\varepsilon}\right)\nabla\varepsilon\right]. \tag{1.43}$$

Recognizing that $c_v = d\varepsilon/dT$, we can rewrite Eq. (1.43) as

$$\frac{\partial \varepsilon}{\partial t} = \operatorname{div}(\chi\nabla\varepsilon).$$

Equation (1.43) is an example of a nonlinear heat-conduction equation. For one-dimensional heat propagation, this equation takes the form

$$\frac{\partial \varepsilon}{\partial t} = \frac{1}{r^{\nu-1}}\frac{\partial}{\partial r}\left(r^{\nu-1}\chi\frac{\partial \varepsilon}{\partial r}\right). \tag{1.44}$$

In addition to the cited gasdynamic equations and some of their simplified forms, it is often necessary to use in the study of gas motion more complicated equations and take into account chemical reactions accompanied by heat release or absorption in the gas. We shall not write here the more complicated equations of gas motion for these cases, but shall indicate them in sections dealing with these topics.

2. Strong discontinuity conditions

2.1. Derivation of equations for strong discontinuities

We shall consider hereafter more general, discontinuous gas motion. If the gasdynamic functions are not continuously smooth, but are only piecewise smooth or piecewise continuous, they can no longer satisfy everywhere the system of gasdynamic differential equations. The gasdynamic equations in integral form, however, are valid for these equations. In this sense, piecewise continuous solutions are sometimes called discontinuous or generalized solutions, or else generalized motions. We consider discontinuous solutions for the motion of a gas with viscosity disregarded. A detailed analysis of various types of discontinuity is given in Ref. 6.

If the region in which the generalized motion is defined contains a hypersurface $s(\mathbf{r}, t)$ on which the quantities $\mathbf{v}$, ρ, p, and ε have first-order discontinuities and outside of which the motion is continuous, we have motion with strong discontinuity. The surface s with t fixed is called a strong-discontinuity surface. If, however, the gasdynamic functions are continuous, and only some of their derivatives with respect to coordinate and time are discontinuous, the motion has a weak discontinuity, with a corresponding weak-discontinuity surface.

We introduce now the concept of the velocity of a discontinuity surface in the space $\mathscr{E}_3$. We choose on the surface $s(t)$ at an instant t a certain point M and assume that a definite normal to s exists at the point M. We find next a point N where a normal drawn to s through the point M intersects the surface $s(t + \Delta t)$. Let $H(\Delta t)$ be the length of the vector $\mathbf{MN}$; the length is taken with a plus sign if the directions of $\mathbf{MN}$ and of the unit vector $\mathbf{n}$ normal to the surface s at the point M coincide, and with a minus sign if their directions differ.

The velocity of the surface s at the point M in $\mathscr{E}_3$ space is referred to as the vector $\mathbf{D}$ normal to s and defined as the limit

$$\mathbf{D} = \mathbf{n} \lim_{\Delta t \to 0} \frac{H(\Delta t)}{\Delta t}.$$

If the equation of the surface $s(t)$ is given in the form $s(\mathbf{r}, t) = 0$ and this surface is smooth, $\mathbf{D}$ can be calculated from the equation

$$\mathbf{D} = -\frac{\partial s}{\partial t}\mathbf{n} \Big/ |\mathrm{grad}\, s|. \tag{1.45}$$

Under coordinate transformations with changes to differently moving coordinate frames, the velocity vector $\mathbf{D}$ depends on the choice of the coordinate frame.

The hydrodynamic conservation laws impose special relations between $\mathbf{v}$, p, ρ, and ε. These can be easily derived by applying the conservation laws directly to the elements of the moving or immobile medium through which a shock wave propagates, or by using the integral gasdynamics equations (1.3)–(1.5). We present a brief derivation of the conditions on surfaces of a strong discontinuity, using Eqs. (1.3)–(1.5). As already noted, these equations can be applied to arbitrary volumes Ω, and particularly if an individual volume Ω^* contains a surface Σ_0 on which the hydrodynamic quantities become discontinuous. We introduce in addition to the moving Lagrangian volume Ω^* a moving volume Ω that coincides at the considered instant t with the volume Ω^* but moves with velocity D on the boundary (D is the normal velocity of the surface Σ bounding the volume Ω). The velocity D is in general not equal to the normal velocity v_n of the gas particles.

From mathematical analysis (see, e.g., Refs. 7 and 8) we know an equation for differentiating an integral with respect to a parameter

$$\frac{d}{dt}\int_\Omega \Phi\, d\Omega = \int_\Omega \frac{\partial \Phi}{\partial t} + \oint_\Sigma \Phi v_n\, d\Sigma, \tag{1.46}$$

where $\Phi(\mathbf{r}, t)$ is a bounded function integrable with respect to the coordinates and differentiable with respect to time, v_n is the normal component of the velocity of the Σ-surface particles, and $\Omega = \Omega(t)$ is a variable volume. If we write now Eq. (1.46) for the above moving volumes Ω^* and Ω, we get, after subtracting equal parts,

$$\frac{d}{dt}\int_\Omega \Phi\, d\Omega - \frac{d}{dt}\int_{\Omega^*} \Phi\, d\Omega$$

$$= \int_\Omega \frac{\partial \Phi}{\partial t}\, d\Omega - \int_{\Omega^*} \frac{\partial \Phi}{\partial t}\, d\Omega + \oint_\Sigma \Phi(D - v_n)\, d\Sigma. \tag{1.47}$$

For the instant t, in accord with the condition under which the volume $\Omega^*(t)$ was chosen, we have $\Omega(t) = \Omega^*(t)$; hence

$$\int_\Omega \frac{\partial \Phi}{\partial t}\, d\Omega = \int_{\Omega^*} \frac{\partial \Phi}{\partial t}\, d\Omega.$$

Equation (1.47) is therefore transformed into

$$\frac{d}{dt}\int_\Omega \Phi\, d\Omega = \frac{d}{dt}\int_{\Omega^*} \Phi\, d\Omega + \oint_\Sigma \Phi(D - v_n)\, d\Sigma. \tag{1.48}$$

Obviously, Eq. (1.48) is valid also for the vector function.

From Eqs. (1.3)–(1.5) we obtain with the aid of Eq. (1.48) the integral form of the gasdynamics equations for an arbitrarily moving volume $\Omega(t)$

$$\frac{d}{dt}\int_{\Omega}\rho\,d\Omega=\oint_{\Sigma}\rho(D-v_n)\,d\Sigma, \tag{1.49}$$

$$\frac{d}{dt}\int_{\Omega}\rho\mathbf{v}\,d\Omega=\int_{\Omega}\rho\mathbf{F}\,d\Omega-\oint_{\Sigma}p\mathbf{n}\,d\Sigma+\oint_{\Sigma}\rho\mathbf{v}(D-v_n)\,d\Sigma, \tag{1.50}$$

$$\frac{d}{dt}\int_{\Omega}\rho\left(\frac{v^2}{2}+\varepsilon\right)d\Omega=-\oint_{\Sigma}p\mathbf{v}\mathbf{n}\,d\Sigma+\int_{\Omega}\rho(\mathbf{v}\mathbf{F})\,d\Omega$$
$$+\oint_{\Sigma}\rho\left(\frac{v^2}{2}+\varepsilon\right)(D-v_n)\,d\Sigma-\oint_{\Sigma}\mathbf{q}\mathbf{n}\,d\Sigma. \tag{1.51}$$

In Eq. (1.51), $\mathbf{q}$ is the heat-flux vector. For one-dimensional motions we obtain

$$\frac{d}{dt}\sigma_\nu\int_{r_1}^{r_2}\rho r^{\nu-1}\,dr=\left[\rho r^{\nu-1}\sigma_\nu\left(\frac{dr}{dt}-v\right)\right]\Bigg|_{r_1}^{r_2},$$
$$\sigma_\nu=2(\nu-1)\pi+(\nu-2)(\nu-3),$$

$$\frac{d}{dt}\sigma_\nu\int_{r_1}^{r_2}\rho v r^{\nu-1}\,dr$$
$$=\sigma_\nu\int_{r_1}^{r_2}\rho F r^{\nu-1}\,dr+\sigma_\nu r^{\nu-1}\left[-p+\rho v\left(\frac{dr}{dt}-v\right)\right]\Bigg|_{r_1}^{r_2}, \tag{1.52}$$

$$\frac{d}{dt}\sigma_\nu\int_{r_1}^{r_2}\rho\left(\frac{v^2}{2}+\varepsilon\right)r^{\nu-1}\,dr=\sigma_\nu\int_{r_1}^{r_2}\rho(vF)r^{\nu-1}\,dr$$
$$+\sigma_\nu\left[-q-pv+\rho\left(\frac{v^2}{2}+\varepsilon\right)\left(\frac{dr}{dt}-v\right)r^{\nu-1}\right]\Bigg|_{r_1}^{r_2}.$$

Assume now that some isolated strong-discontinuity surface exists and that some arbitrary part Σ_0 of this surface divides the volume $\Omega(t)$ into two volumes, $\Omega_1(t)$ and $\Omega_2(t)$, and the surface $\Sigma(t)$ into two surfaces, $\Sigma_1(t)$ and $\Sigma_2(t)$, respectively. Let the volume Ω_1 be bounded by the surface $s_1=\Sigma_{01}+\Sigma_1$ and the volume Ω_2 by the surface $s_2=\Sigma_{02}+\Sigma_2$, where Σ_{01} and Σ_{02} are the sides of the surface Σ_0 that are external to the volumes Ω_1 and Ω_2. We use the subscripts 1 and 2 to designate the limiting values of the gasdynamic functions as a space point tends to the surface Σ_0 from the volumes Ω_1 and Ω_2, respectively.

Subtracting from Eq. (1.4) analogous equations written for the volumes Ω_1 and Ω_2, we obtain then

$$\int_{\Sigma}\rho(D-v_n)\,d\Sigma-\int_{s_1}\rho(D-v_n)\,d\Sigma-\int_{s_2}\rho(D-v_n)\,d\Sigma=0.$$

From this, recognizing that the normals on the Σ_{01} and Σ_{02} sides are oppositely directed, we have

$$\int_{\Sigma_0} [\rho_2(\mathbf{D} - \mathbf{v}_2) - \rho_1(\mathbf{D} - \mathbf{v}_1)]\mathbf{n}\, d\Sigma = 0,$$

where $\mathbf{n}$ is the outward normal unit vector to the surface Σ_0.

Since the parts of the discontinuity surface Σ_0 have been arbitrarily chosen, we obtain a conservation law for the mass flow through the discontinuity surface

$$\rho_2(D - v_{n2}) = \rho_1(D - v_{n1}). \tag{1.53}$$

We assume next, to generalize, that an energy Q per unit mass is fed or drawn from the direction of the volume Ω_2 in the vicinity of the discontinuity surface Σ_0. The energy Q can be regarded as the energy released instantaneously by combustion of a gas mixture, or absorbed upon evaporation of liquid droplets in a gas stream. We can then assume for the total energy E per unit volume on the Σ_{02} side of the Σ_0 surface

$$E_2 = \rho_2\left(\frac{v_2^2}{2} + \varepsilon_2 \pm Q\right),$$

where the plus and minus signs correspond respectively to energy flowing into and out of the discontinuity surface, and ε_2 is the internal energy of the medium in the state 2.

If the arguments by which the mass-conservation law was obtained are now applied to the momentum and energy equations (1.50) and (1.51), we obtain the remaining conditions on the strong-discontinuity surface:

$$\begin{gathered} \rho_2\mathbf{v}_2(D - v_{n2}) - \rho_1\mathbf{v}_1(D - v_{n1}) = (p_2 - p_1)\mathbf{n}, \\ \rho_2(D - v_{n2})\left(\frac{v^2}{2} + \varepsilon_2 \mp Q\right) - q_{n2} - p_2v_2 \\ = \rho_1(D - v_{n1})\left(\frac{v_1^2}{2} + \varepsilon_1\right) - q_{n1} - p_1v_1. \end{gathered} \tag{1.54}$$

2.2. Forms of discontinuity surfaces and their properties at $Q = 0$

With square brackets used to designate differences or discontinuities, for example $[\rho] = \rho_2 - \rho_1$, relations (1.53) and (1.54) take the form

$$\begin{gathered} [\rho(D - v_n)] = 0, \quad [\rho\mathbf{v}(D - v_n) - p\mathbf{n}] = 0, \\ \left[\rho(D - v_n)\left(\frac{v^2}{2} + \varepsilon\right) - q_n - pv\right] = 0. \end{gathered} \tag{1.55}$$

$D - v_{n1}$ and $D - v_{n2}$ can be regarded as the velocities of the discontinuity surface relative to particles on opposite sides of the discontinuity.

If $D - v_{n1} = 0$ and $D - v_{n2} = 0$, the particles of the medium do not cross from one side of the discontinuity to the other, and $v_{n1} = v_{n2}$. In this case the tangential components of the velocity can differ on the opposite sides of the discontinuity and an arbitrary density discontinuity $\rho_1 \neq \rho_2$ is possible. This is called a tangential or contact discontinuity. For a tangential discontinuity we have

$$[p] = 0, \quad [q_n] = 0, \quad [\rho] \neq 0, \quad [v_n] = 0. \tag{1.56}$$

If $[v_n] \neq 0$, the liquid particles go over from one side of the surface Σ_0 to the other and their characteristics, states, and motion change jumpwise.

Assume now that the particles go over from side 1 to side 2 of the discontinuity surface, so that $[v_n] > 0$, and we postulate the following:

Definition. Strong-discontinuity surfaces for which $[v_n] > 0$, $[p] > 0$, and $[\rho] > 0$ are called shock waves.

If, on the other hand, the particles go over from side 1 to side 2 of the surface Σ_0, but $[v_n] < 0$, $[p] < 0$, and $[\rho] < 0$, such discontinuities are called rarefaction shock waves.

When a strong discontinuity moves in an immobile medium ($v_{n1} = 0$), then, for a shock wave, the gas behind the jump Σ_0 is incident on the immobile medium and packing is produced. If, however, $v_{n2} < 0$, the fluid behind the discontinuity moves counter to the discontinuity-propagation direction in the immobile medium.

Let now $[\rho] = 0$ and $D - v_{n1} \neq 0$. It follows then from Eqs. (1.55) that

$$[p] = 0, \quad [v_n] = 0, \quad D - v_{n1} = \frac{[q_n]}{\rho_1 [\varepsilon]}. \tag{1.57}$$

Discontinuities for which $[\rho] = 0$, $D - v_{n1} \neq 0$, and the condition (1.57) is met are called phase-change fronts or phase fronts. Phase fronts take place when $q_n \neq 0$. If $q_n = 0$, it follows from Eqs. (1.57) that $[\varepsilon] = 0$ and none of the quantities undergo a jump on going through Σ_0, i.e., there is no strong discontinuity.

We shall study flows with strong discontinuities, of the type of ordinary shock waves, and flows in which energy is released on the discontinuity surface ($Q \neq 0$).

Let us consider now in greater detail the case of discontinuity surfaces that do not release or absorb heat by conduction ($Q = 0$, $q = 0$). In accordance with the law of entropy increase ($[S] > 0$), only shock waves are possible in the gas. We shall formulate this property below in greater detail.

That side of the Σ_0 surface on which the gas impinges will be called the forward side or the side ahead of the shock-wave front, while the opposite side will be called the back side or the side behind the front, with the shock-wave front taken to mean the discontinuity surface Σ_0.

We assume hereafter that the normal is directed ahead of the shock-wave front. The gas parameters ahead of the shock wave will be labeled by an index 1 or ∞ (sometimes also 0), while the gas parameters directly behind the shock-wave front will be labeled 2 or n (sometimes 1).

The index will be chosen to facilitate the investigation of the particular problem. Eliminating from Eqs. (1.55) the velocities v, $D - v_{n1}$, and $D - v_{n2}$ we arrive at the relation

$$\varepsilon_2 - \varepsilon_1 = \frac{1}{2}(p_2 + p_1)\left(\frac{1}{\rho_1} - \frac{1}{\rho_2}\right). \tag{1.58}$$

This relation connects the thermodynamic parameters ρ_1 and p_1 ahead of the shock wave with p_2 and ρ_2 behind, and is called the Hugoniot adiabat or the shock adiabat. For an ideal gas with constant heat capacities we obtain from Eq. (1.58) an equation for the shock adiabat

$$\frac{p_2}{p_1} = \frac{(\gamma+1)V_1 - (\gamma-1)V_2}{(\gamma+1)V_2 - (\gamma-1)V_1}, \quad V = \frac{1}{\rho}. \tag{1.59}$$

Note the main properties of shock waves. We designate the shock-wave intensity by $[p]$ (or by $[\rho]$) and determine, as usual, the adiabatic speed of sound from the relation $a^2(dp/d\rho)_s$.

Property 1. In media satisfying the inequality $d^2V/dp^2 > 0$, the entropy increases monotonically with the pressure p. It follows hence from the second law of thermodynamics that a shock wave is a compression jump for which $p_2 > p_1$ and $\rho_2 > \rho_1$. This is one of the basic conclusions of strong-discontinuity theory.

Property 2. The entropy jump in a shock wave is a small quantity of third order relative to the discontinuity intensity, i.e.,

$$\lim_{\rho_2 \to \rho_1} \frac{S_2 - S_1}{(\rho_2 - \rho_1)^3} = B > 0.$$

Property 3. The absolute value of the normal component of the gas velocity relative to a shock wave exceeds the speed of sound ahead of the front and is less than the speed of sound behind the front, i.e., $|v_{n1} - D| > a_1$ and $|v_{n2} - D| < a_2$. This conclusion is called the Zemplen theorem and is closely related to the condition of entropy growth in a shock wave.

Property 4. The tangential component of the velocity **v** undergoes no discontinuity on going through a shock wave.

Definition. A shock wave is called strong if $[p]/p_1 \gg 1$ and weak if $[p]/p_1 < 1$.

Property 5. A weak shock wave satisfies the following equations:

$$\lim_{\rho_2 \to \rho_1} \frac{p_2 - p_1}{\rho_2 - \rho_1} = a_1^2 = a_2^2.$$

Property 5 shows that when the discontinuity intensity tends to zero the relative gas velocity along the normal to the shock-wave surfaces tends to equal the sound velocity.

We are not considering here discontinuity-surface properties for $Q \neq 0$. They will be discussed for specific problems in Chap. 6.

2.3. Shock waves in one-dimensional flow

In one-dimensional gas flow only plane, cylindrical, or spherical shock waves are possible, in accord with the one-dimensional flow types considered above. The conditions on the shock waves can be obtained either from the integral equations (1.52) for one-dimensional motion, or from the general conditions (1.55) on a shock wave.

Assume no influx of heat ($\mathbf{q} = 0$). In the case of one-dimensional motions, the velocity vector is always normal to the discontinuity surface, i.e., $v = v_n$. The conditions on the shock waves, for one-dimensional motion without heat supply, can be written in the form

$$\begin{gathered} \rho_1(D - v_1) = \rho_2(D - v_2), \\ \rho_1 v_1(D - v_1) - p_1 = \rho_2 v_2(D - v_2) - p_2, \\ \rho_1(D - v_1)\left(\frac{v_1^2}{2} + \varepsilon_1\right) - p_1 v_1 = \rho_2(D - v_2)\left(\frac{v_2^2}{2} + \varepsilon_2\right) - p_2 v_2, \end{gathered} \tag{1.60}$$

where, as before, the subscripts 1 and 2 label quantities on the two sides of the discontinuity.

If it is assumed that the medium ahead of the shock wave is at rest, i.e., $v_1 = 0$, the conditions (1.60) simplify to

$$\begin{gathered} \rho_1 D = \rho_2(D - v_2), \quad p_2 = \rho_2 v_2(D - v_2) + p_1, \\ \rho_1 D \varepsilon_1 = \rho_2(D - v_2)(v_2^2/2 + \varepsilon_2) - p_2 v_2. \end{gathered} \tag{1.61}$$

For a perfect gas, introducing the speed of sound $a_1 = (\gamma p_1/\rho_1)^{1/2}$ in the gas at rest and replacing ε in accordance with Eq. (1.10), we transform Eqs. (1.61) into

$$\begin{gathered} v_2 = \frac{2}{\gamma + 1}(1 - q)D, \qquad \rho_2 = \frac{\gamma + 1}{\gamma - 1}\rho_1\left[1 + \frac{2q}{\gamma - 1}\right]^{-1}, \\ p_2 = \frac{2}{\gamma + 1}\rho_1 D^2\left[1 - \frac{\gamma - 1}{2\gamma}q\right]. \end{gathered} \tag{1.62}$$

where $q = a_1^2/D^2$; v_2, ρ_2, and p_2 are the velocity, density, and pressure directly behind a shock wave propagating with velocity D through a quiescent gas.

Replacing D by q in Eqs. (1.62), we get

$$\begin{gathered} v_2 = \frac{2}{\gamma + 1}\frac{1 - q}{\sqrt{q}}a_1, \quad \rho_2 = \frac{\gamma + 1}{\gamma - 1 + 2q}\rho_1, \\ p_2 = \frac{2\gamma - (\gamma - 1)q}{(\gamma + 1)q}p_1. \end{gathered} \tag{1.63}$$

From the last relation of Eqs. (1.63) we get

$$q = \frac{2\gamma}{(\gamma+1)p_2/p_1 + \gamma - 1}.$$

It follows hence that q, and hence also a_1/D, is small for large p_2/p_1, i.e., in very intense shock waves.

It is seen from Eq. (1.62) that if q is small the quantities in the square brackets differ little from unity. Thus, for $\gamma = 1.4$ these quantities differ from unity by less than 0.05 already if $q < 0.01$. If we put $q = 0$ in Eqs. (1.62) (this is equivalent to $p_1 = 0$), the errors of v_2, ρ_2, and p_2 are less than 5%. At $q = 0$, the conditions (1.62) on the shock wave take the simple form

$$v_2 = \frac{2}{\gamma+1}D, \quad \rho_2 = \frac{\gamma+1}{\gamma-1}\rho_1, \quad p_2 = \frac{2}{\gamma+1}\rho_1 D^2. \tag{1.64}$$

We shall frequently use below shock-wave conditions in the form of Eqs. (1.63) and (1.64). The conditions (1.60) and (1.61) were written under the assumption that no heat is transferred to the flow by conduction. If heat conduction is taken into account, it is necessary to add to the right- and left-hand sides of the last equation of the system (1.60), which expresses the energy-conservation law, the respective terms $\varkappa_2(\partial_T/\partial_r)|_2$ and $\varkappa_1(\partial_T/\partial_r)_1$ representing the heat flow into the shock wave ($\varkappa$ is the thermal conductivity).

Note that conditions of the type (1.64) hold also for non-one-dimensional motions in a shock wave propagating in a gas at rest, but it must then be assumed that $v_2 = v_{n2}$. This follows from the fact that the tangential component of the velocity is continuous, and since it is zero ahead of the wave it must be zero also behind the shock wave, i.e., $\mathbf{v}_2 = \mathbf{n}v_{n2}$.

3. Magnetohydrodynamic equations

3.1. System of differential equations of magnetohydrodynamics

The flow of an electrically conducting liquid or gas in an electromagnetic field is influenced both by electromagnetic forces and by energy exchange between the moving fluid and the electromagnetic field. On the other hand, the motion of the fluid can influence substantially the parameters of the electromagnetic field.

To describe the motions of an electrically conducting gas, with allowance for the interaction with the electromagnetic field, it is possible, just as in ordinary gasdynamics, to use Euler equations and Lagrange equations. The only difference is in the number of the equations sought, and it is necessary to add to the gasdynamic equations the system of Maxwell equations for the magnetic and electric fields in the moving conductors.

The magnetohydrodynamic equations consist of the set of Maxwell equations and the fluid-motion equations (the continuity, momentum, and energy equations), with account taken of the interaction between the electromagnetic field and the moving medium.

To describe the motion of electrically conducting gases, we shall use the Gaussian system of electromagnetic units. We write the Maxwell equations for the conductors in the form

$$\begin{aligned} \operatorname{rot} \mathbf{E} &= -\frac{1}{c}\frac{\partial \mathbf{B}}{\partial t}, \qquad \operatorname{div} \mathbf{B} = 0, \\ \operatorname{rot} \mathbf{H} &= \frac{4\pi}{c}\mathbf{j} + \frac{1}{c}\frac{\partial \mathbf{D}_u}{\partial t}, \qquad \operatorname{div} \mathbf{D}_u = 4\pi\rho_e, \end{aligned} \tag{1.65}$$

where $\mathbf{E}$ is the electric-field intensity vector, $\mathbf{H}$ the magnetic field intensity vector, $\mathbf{j}$ the current density, $\mathbf{B}$ the magnetic-induction vector, $\mathbf{D}_u$ the electric induction vector, ρ_e the macroscopic electric-charge density, and c the speed of light in a vacuum.

Let us clarify some of the concepts introduced. The electric current induced in a conductor, especially in an ionized gas (plasma), constitutes motion of charged particles. We denote by $\mathbf{v}_k$ the microscopic velocities of the electrons and ions, and by e_k their charges.

The current density $\mathbf{j}$ is introduced as the mean value, over a small volume Ω of the medium, of the sum of the products of the charges by the electron and ion velocities in the volume Ω: $\mathbf{j} = (\sum e_k \mathbf{v}_k)/\Omega$.

The vector $\mathbf{j}$ is usually written as the sum:

$$\mathbf{j} = \mathbf{j}^* + \rho_e \mathbf{v},$$

where $\mathbf{j}^*$ is the usual current produced in either moving or stationary conductors by an electromagnetic field and is called the conduction current, while the term $\rho_e \mathbf{v}$ is the current due to macroscopic charge transport. Since $\rho_e \mathbf{v} \ll \mathbf{j}^*$ for most conductors, we shall as a rule neglect the convective part of the current and assume $\mathbf{j} = \mathbf{j}^*$.

We define Ohm's law as the vector relation between the current density $\mathbf{j}$ and the characteristics of the magnetic field and the moving medium.

The simplest formulation of Ohm's law for mobile conductors is

$$\mathbf{j}^* = \sigma\left(\mathbf{E} + \frac{1}{c}(\mathbf{v} \times \mathbf{H})\right). \tag{1.66}$$

The scalar coefficient σ is called the conductivity of the medium and is assumed to be a function of its thermodynamic parameters.

The connections between the induction vectors $\mathbf{D}_u$ and $\mathbf{B}$ on the one hand and the field vectors $\mathbf{E}$ and $\mathbf{H}$ on the other are governed by the polarization and magnetization properties of the medium. For motions of media such as dielectrics and ferromagnets, these connections can be rather complicated and nonlinear. For isotropic media these connections take in many cases the simple form $\mathbf{B} = \mu\mathbf{H}$ and $\mathbf{D}_u = \varepsilon_e\mathbf{E}$, where μ is the magnetic permeability and ε_e is the dielectric constant. In magnetohydrodynamics the polarization and magnetization properties are usually disregarded and μ and ε_e are assumed constant[9,10] ($\mu = \varepsilon_e = 1$ in our system of units). Note that for electromagnetic waves propagating in a vacuum $\mathbf{B} = \mathbf{H}$ and $\mathbf{D}_u = \mathbf{E}$, i.e., $\mu = 1$ and $\mu_e = 1$. The second group of magnetohydrodynamic equations is in essence a generalization of the foregoing gasdynamic equations and, neglecting viscosity and non-electromagnetic mass forces, consists of

$$\frac{\partial\rho}{\partial t} + \operatorname{div}(\rho\mathbf{v}) = 0 \tag{1.67}$$

(the continuity equation),

$$\frac{\partial(\rho\mathbf{v})}{\partial t} + \operatorname{div}\hat{\mathbf{P}} = \mathbf{F}_L \tag{1.67a}$$

(the momentum equation), and

$$\frac{\partial}{\partial t}\left(\rho\varepsilon + \rho\frac{v^2}{2}\right) + \operatorname{div}\left[\rho\mathbf{v}\left(\varepsilon + \frac{v^2}{2}\right) + p\mathbf{v} + \mathbf{q}\right] = \mathbf{jE} \tag{1.67b}$$

(the energy equation).

In the system (1.67)–(1.67b), $\mathbf{F}_L$ is the Lorentz force or the ponderomotive mass action of the electromagnetic field on the medium,

$$\mathbf{F}_L = \rho_e\mathbf{E} + \frac{1}{c}(\mathbf{j}\times\mathbf{H}),$$

$\hat{\mathbf{P}}$ is a tensor with components $P_{ik} = -\delta_{ik}p + \rho v_i v_k$, $\mathbf{q}$ is the heat-flux vector, and the product $\mathbf{j}\cdot\mathbf{E}$ in Eq. (1.67b) is indicative of the energy change produced in the liquid by the presence of the electromagnetic field.

The system (1.67) can be recast in divergence form. Taking account of the Maxwell equation, these equations can be reduced to the form[10]

$$\begin{gathered}
\frac{\partial\rho}{\partial t} + \operatorname{div}(\rho\mathbf{v}) = 0,\\
\frac{\partial}{\partial t}\left(\rho\mathbf{v} + \frac{1}{4\pi c}(\mathbf{E}\times\mathbf{H})\right) + \operatorname{div}(\rho v_i v_k + p\delta_{ik} - \tau_{ik}) = 0,\\
\frac{\partial}{\partial t}\left[\rho\left(\varepsilon + \frac{v^2}{2} + \frac{H^2+E^2}{8\pi}\right)\right]\\
+ \operatorname{div}\left[\frac{c}{4\pi}(\mathbf{E}\times\mathbf{H}) + \mathbf{q} + q\mathbf{v} + \rho\mathbf{v}\left(\varepsilon + \frac{v^2}{2}\right)\right] = 0,\\
\tau_{ik} = \frac{1}{4\pi}(H_iH_k + E_iE_k) - \frac{1}{8\pi}(H^2+E^2)\delta_{ik}.
\end{gathered} \tag{1.68}$$

Equations (1.68) jointly with Eqs. (1.65) and (1.66) constitute the complete set of equations of motion of a compressible electrically conducting liquid.

We shall not dwell here on the validity limits of these equations for a description of the motion of ionized gases. We mention only that these equations of motion of an electrically conducting medium are very frequently used to consider nonrelativistic phenomena in space physics, such as the motion of ionized gases in the sun's atmosphere.

We consider now a simplified system of equations, which is frequently called also the system of magnetohydrodynamic (MHD) equations.

We neglect the displacement currents in the Maxwell equations, i.e., the term $\partial \mathbf{E}/\partial t$, and the convection current $\rho_e \mathbf{v}$. This is justifiable for noticeable conduction currents $\mathbf{j}^*$ and in the absence of very-high-frequency oscillations.[10] Estimates[9,10] show that in conducting media the ratio of electric energy per unit volume to the corresponding magnetic energy is of order v^2/c^2, which is quite low in view of the relation $v/c \ll 1$ for most phenomena.

The magnetohydrodynamic equations take under the foregoing conditions the form

$$\frac{\partial \rho}{\partial t} + \operatorname{div}(\rho \mathbf{v}) = 0, \tag{1.69}$$

$$\frac{\partial(\rho v_i)}{\partial t} = -\frac{\partial}{\partial x_k}\left[\rho v_i v_k + p\delta_{ik} - \frac{1}{4\pi}\left(H_i H_k - \frac{1}{2} H^2 \delta_{ik}\right)\right], \tag{1.70}$$

$$\frac{\partial}{\partial t}\left(\frac{\rho v^2}{2} + \rho\varepsilon + \frac{H^2}{8\pi}\right)$$
$$= -\operatorname{div}\left\{\rho \mathbf{v}\left(\frac{v^2}{2} + \varepsilon + \frac{p}{\rho}\right) + \frac{1}{4\pi}[\mathbf{H} \times (\mathbf{v} \times \mathbf{H})] + \mathbf{q}\right\}, \tag{1.71}$$

$$\frac{\partial \mathbf{H}}{\partial t} = \operatorname{rot}[\mathbf{v} \times \mathbf{H} - \nu_m \operatorname{rot} \mathbf{H}], \tag{1.72}$$

$$\operatorname{div} \mathbf{H} = 0, \tag{1.73}$$

where $v_m = c^2/(4\pi\sigma)$ is the magnetic viscosity.

Equation (1.72) above is a consequence of Maxwell's equation and Ohm's law, and is called (as in the Maxwell equations) the induction equation. A special role is assumed in the system (1.69)–(1.73) by Eq. (1.73). Indeed, taking the divergence of the induction equation, we find that in MHD

$$\frac{\partial}{\partial t} \operatorname{div} \mathbf{H} = 0.$$

To satisfy Eq. (1.73) in nonstationary problems it suffices therefore to stipulate that this equation be satisfied by the initial conditions for $\mathbf{H}$. Thus, the equation $\operatorname{div} \mathbf{H} = 0$ is satisfied by virtue of the induction equation and the initial conditions. In the solution of problems, however, it is convenient to use this equation in place of one of the projections of the induction equation.

We refer to a gas as ideally (or infinitely) conducting if the condition $\nu_m = 0$ is met. From the definition of magnetic viscosity, it follows that it is met when $\sigma \to \infty$. The MHD equations for an ideally conducting gas can be easily obtained from the system (1.69)–(1.73) with $\nu_m = 0$.

Note that for an ideally conducting gas Ohm's law takes the simple form

$$\mathbf{E} = -\frac{1}{c}(\mathbf{v} \times \mathbf{H}). \tag{1.74}$$

Let L be the characteristic linear dimension of the region in which the conducting gas flows and let v_* be the characteristic gas velocity. By analogy with the usual Reynolds number, we introduce the dimensionless number

$$R_m = v_* L / \nu_m \tag{1.75}$$

called the magnetic Reynolds number. If $R_m \ll 1$, we have motion with small magnetic Reynolds numbers, whereas $R_m \gg 1$ means motion with large Reynolds numbers. If $R_m \ll 1$, the effect of the gas motion on the electromagnetic field can be neglected.[12] For $R_m \gg 1$ the MHD equations for an ideally conducting gas can be used. In fact, by changing to dimensionless quantities in the induction equation and in Ohm's law and letting R_m tend to infinity we obtain, assuming all other quantities to be finite, the corresponding equations of an ideally conducting gas. For numbers R_m that are finite but large enough, the model of an infinitely conducting gas is used.

The complete set of equations (1.69)–(1.73) must be used for theoretical calculations if the magnetic Reynolds numbers are low, of order unity (as is very frequently the case in real processes).

We call the quantity $H^2/(8\pi)$ the magnetic pressure. The magnetic field in a moving gas with $H^2/(8\pi) \ll p$ [or $H^2/(8\pi) \ll \rho v^2$] is referred to as weak. If the R_m are not small, the Lorentz force and the electromagnetic energy can be neglected for weak fields, but account must be taken of the effect of the gas pressure on the parameters of the electromagnetic field. The foregoing approximate description of the motions will be called the weak-field approximation.

We now dwell briefly on the symmetry properties of gas flow in a magnetic field. The MHD equations do not admit at $H \neq 0$ spherically symmetric one-dimensional motions in which the medium interacts with the field. However, one-dimensional motions in planar and cylindrical geometry are perfectly feasible and will be considered by us later. Naturally, planar motions and motions with axial symmetry are encountered quite frequently in magnetohydrodynamics.

To conclude this section, we point out some important properties of the equations of motion of an ideally conducting liquid. As already mentioned, the induction equation takes in this case the form

$$\frac{\partial \mathbf{H}}{\partial t} = \operatorname{rot}\,(\mathbf{v} \times \mathbf{H}). \tag{1.76}$$

This equation is identical with the equation for the velocity curl in the hydrodynamics of an inviscid liquid; as is well known, it means that the vortex lines move together with the liquid.

It follows thus from Eq. (1.76) that the magnetic field varies as if the magnetic force line were rigidly fixed in the medium. It can be shown (see, e.g., Ref. 9) that Eq. (1.76) indicates that the magnetic flux linked with a closed contour is constant if each point of the contour moves with a local velocity $\mathbf{v}$. Motion along the force lines does not affect the field, but in transverse motion the force lines are displaced together with the substance. This property of the motion of an ideally conducting liquid is called magnetic flux freezing. Following Refs. 9 and 13, we present now some general results based on the freezing property. Using Eqs. (1.69) and (1.73), we can transform Eq. (1.76), which determines the field variation, into

$$\frac{d}{dt}\left(\frac{\mathbf{H}}{\rho}\right) = \left(\frac{\mathbf{H}}{\rho}\,\mathrm{grad}\right)\mathbf{v}, \tag{1.77}$$

where d/dt is the material derivative.

Let the position vector $\mathbf{r}_0$ of some particle become equal to $\mathbf{r}$ after some time, while $\mathbf{H}_0$ and ρ_0 assume values $\mathbf{H}$ and ρ.

Equation (1.77) leads to the following relation, called the freezing integral, between $\mathbf{r}_0$, $\mathbf{H}_0$, and ρ_0 on the one hand and $\mathbf{r}$, $\mathbf{H}$, and ρ on the other[9,13]:

$$\frac{\mathbf{H}}{\rho} = \left(\frac{\mathbf{H}_0}{\rho_0}\,\mathrm{grad}\right)\mathbf{r}, \tag{1.78}$$

or, in coordinate form

$$\frac{H_i}{\rho} = \frac{H_{0j}}{\rho_0}\frac{\partial x^i}{\partial x_0^j} = \frac{\partial x^i}{\partial x_0^j}\frac{H_{0j}}{\rho\,|\partial x^i/\partial x_0^j|}, \tag{1.79}$$

where $|\partial x^i/\partial x_0^j|$ is the determinant of the matrix $\|\partial x^i/\partial x_0^j\|$.

Let us consider one particularly important case of Eq. (1.79). Let the motion be planar and the vector $\mathbf{H}$ perpendicular to the plane of motion. It follows then from Eq. (1.77) or Eq. (1.78) that

$$\frac{\mathbf{H}}{\rho} = \frac{\mathbf{H}_0}{\rho_0}, \tag{1.80}$$

for in this case the length of the magnetic-force line element remains unchanged.

In the study of planar motions of an ideally conducting gas in a magnetic field perpendicular to the plane of motion, Eq. (1.80) makes it possible to eliminate $\mathbf{H}$ from the MHD equations. In addition, if the pressure and the internal energy are replaced by the functions $p^* = p + H^2/(8\pi)$ and $\varepsilon^* = \varepsilon$

$+H^2/(8\pi)$ the MHD equations take the same form as the equations of ordinary gasdynamics. It follows hence that the solutions of the MHD problems can be obtained in this case by recalculating the corresponding problems of ordinary gasdynamics. It is necessary only to introduce changes in the boundary and initial conditions of the problems.

We note there also the following important property of a perfect gas with $\gamma = 2$, viz., all the thermodynamic equations are unchanged by substitution of p^* for p.

3.2. MHD-approximation equations of motion of a low-density plasma

Equations similar to the hydrodynamic ones can be written also for the motion of a plasma of very low density in a magnetic field. The Boltzmann–Vlasov kinetic equation and the Maxwell equations[14,15] lead to the following hydrodynamic-approximation equations:

$$\frac{\partial \rho}{\partial t} + \operatorname{div}(\rho \mathbf{v}) = 0, \tag{1.81}$$

$$\frac{\partial(\rho v_i)}{\partial t} + \operatorname{div}\hat{\mathbf{T}} = 0, \tag{1.82}$$

$$\frac{\partial}{\partial t}\left(\varepsilon + \frac{\rho v^2}{2}\right) + \operatorname{div}\mathbf{Q} = 0, \tag{1.83}$$

$$\frac{\partial}{\partial t}(p_\perp/B) + \operatorname{div}(p_\perp \mathbf{v}/B) = 0, \tag{1.84}$$

$$\frac{\partial \mathbf{B}}{\partial t} + \operatorname{rot}(\mathbf{B}\times\mathbf{v}) = 0, \qquad \operatorname{div}\mathbf{B} = 0. \tag{1.85}$$

Here $\hat{\mathbf{T}}$ is the momentum-flux tensor with components

$$T_{ik} = \rho v_i v_k + p_\perp \delta_{ik} + \frac{p_\parallel - p_\perp}{B^2} B_i B_k - \frac{1}{4\pi}\left(B_i B_k - \frac{1}{2}B^2\delta_{ik}\right),$$

$\mathbf{Q}$ is the energy-flux vector

$$\mathbf{Q} = \rho\mathbf{v}\left(\frac{v^2}{2} + \varepsilon + \frac{p_\perp}{\rho}\right) + \frac{p_\parallel - p_\perp}{B^2}\mathbf{B}(\mathbf{vB}) + \frac{1}{4\pi}\left(\mathbf{B}\times(\mathbf{v}\times\mathbf{B})\right),$$

$\epsilon = (1/\rho)(p_\perp + p_\parallel)$ has the meaning of the plasma internal energy, while the quantities $p_\perp$ and $p_\parallel$ in the stress tensor, in the energy-flux vector, and in the expression for ϵ are called the transverse and longitudinal pressures. In the analysis of the motion of a low-density plasma we shall describe the magnetic field by the magnetic-induction vector $\mathbf{B}$, which is most frequently used in the relevant literature.

The system (1.81)–(1.85) describes roughly the behavior of a low-density plasma. It is frequently used, however, for an approximate investigation of the motion of a plasma of very low density in the presence of a magnetic field. The applicability of these equations to a physical problem is discussed in Ref. 16. Without dwelling on this topic, we note that the system (1.81)–(1.85) can be used only to describe motions of a low-density plasma in the presence of a magnetic field, and is not suitable for a description of the motion of a low-density nonconducting gas. We note furthermore that the energy equation (1.84) can be transformed with the aid of the other equations of the system (1.81)–(1.85) into

$$\frac{d}{dt}\left(\frac{p_{\parallel}B^2}{\rho^3}\right) = 0. \tag{1.86}$$

This equation can be used to replace the energy equation in the system (1.81)–(1.85).

The system (1.81)–(1.85) or its equivalent will hereafter be called the system of MHD equations for a low-density plasma.

The systems presented are basic for the investigation of many MHD and plasma-dynamics problems. The ionized gas is regarded in these equations as a single electrically conducting fluid, so that equations of this type are called equations in the single-fluid approximation.

It is possible to introduce a two-fluid or two-component model of the motion of fully ionized gases, i.e., to introduce an electron fluid for the electron flux and an ion fluid for the ion motion, and for forces that take into account their interaction with one another and with electromagnetic fields.

If the gas is not fully ionized, a three-fluid or three-component description is used for the motion of a gas consisting of electrons, ions (of one species), and neutral particles.

We shall not consider here the equations of two- or multicomponent MHD. For sufficiently dense gases these equations are derived in Refs. 17 and 18. For a strongly rarefied plasma they are given, for example, in Refs. 16 and 19.

Besides the equations in the form of conservation laws, plasma physics deals also with the motion of individual particles in electromagnetic fields, and the combined data on this motion are used to determine the currents and charges that enter in the Maxwell equations describing the variation of the electromagnetic field parameters. A special technique has been developed here for the averaging of the motions and for taking into account the electromagnetic field variation due to the particle motion.[20–25] We present now some data from the theory of the motion of a charged particle in an electromagnetic field.

Particle motion in an external electromagnetic field is described by the equations

$$\frac{d\mathbf{r}}{dt} = \mathbf{v}, \qquad \frac{d(m\mathbf{v})}{dt} = e\left(\mathbf{E} + \frac{1}{c}\,\mathbf{v}\times\mathbf{B}\right), \tag{1.87}$$

where $\mathbf{r}$ is the particle position vector, e is its charge, and m is its mass. At relativistic velocities, the mass m must be regarded as variable:

$$m = \frac{m_0}{\sqrt{1 + v^2/c^2}}, \tag{1.88}$$

where m_0 is the rest mass.

If $\mathbf{E} = 0$, the force $\mathbf{F} = (e/c)(\mathbf{v} \times \mathbf{B})$ acting on the particle has no component parallel to $\mathbf{B}$, so that the momentum $\mathbf{p}_e = m\mathbf{v}$ has a constant component parallel to $\mathbf{B}$. Further, the scalar product of Eqs. (1.87) and $\mathbf{p}_e$ yields $\mathbf{p}_e \, d\mathbf{p}_e/dt = 0$ and $|\mathbf{p}_e|^2 = \text{const}$. It follows from this and from Eq. (1.88) that m is constant and the length of the vector $\mathbf{v}$ is constant.

Let the magnetic field be constant and uniform. We resolve the total velocity $\mathbf{v}$ into components $\mathbf{v}_{\parallel}$ along the vector $\mathbf{B}$ and $\mathbf{v}_{\perp}$ perpendicular to $\mathbf{B}$. In a coordinate frame moving with velocity $\mathbf{v}_{\parallel}$ the particle trajectory is a circle. The particle motion perpendicular to the magnetic field is uniformly circular with an angular velocity

$$\omega = -\frac{e\mathbf{B}}{mc},$$

which is called the Larmor or cyclotron frequency. The trajectory and its radius are called the Larmor radius and circle, respectively. The Larmor radius $\mathbf{r}_L$ is given by

$$r_L = \left|\frac{v_{\perp}}{\omega}\right| = \left|\frac{mcv_{\perp}}{eB}\right|.$$

The angular-velocity vector ω is directed counter to $\mathbf{B}$ for a positive particle and along $\mathbf{B}$ for a negative one.

The magnetic moment of a particle is defined as the vector quantity

$$\mu = -\frac{mv_{\perp}^2}{2B^2}\mathbf{B}. \tag{1.89}$$

The magnetic moment μ has the meaning of the product of the average current produced by a rotating particle by the area of the Larmor circle: $\mu = \mathbf{I}s/c$.

From this we obtain Eq. (1.89) by putting

$$\mathbf{I} = \frac{e}{2\pi}\,\omega, \qquad s = \pi r_L^2.$$

In a constant and uniform magnetic field the particle moves along a cylindrical helix. The axis of the helix is parallel to the vector $\mathbf{B}$. The magnetic moment μ of the particle remains constant in the course of this motion. The particle motion becomes more complicated if $\mathbf{E} \neq 0$ and $\mathbf{B} \neq 0$, with $\mathbf{E}$ and $\mathbf{B}$ variable.

We shall not dwell in detail on particle motion in alternating electromagnetic fields. We note, however, that the values of $m\mu$ at relativistic velocities and of μ for nonrelativistic ones change very little in motion through weak electric fields and in response to slow variations of $\mathbf{B}$ in space and time. In other words, these quantities are approximate integrals of the particle's motion, or adiabatic invariants.

Returning to the MHD equations of a low-density plasma, let us clarify the meaning of Eqs. (1.81)–(1.86). We note, first, that these equations were derived from the Boltzmann–Vlasov equations under the assumption that the characteristic Larmor radius of the particle motion is much smaller than the characteristic inhomogeneity scale of the plasma, and the corresponding Larmor frequencies are higher than the characteristic temporal frequencies of the processes considered.

Next, the quantity $p_\perp/\rho B$ has the meaning of the particles' average magnetic moment per unit mass. The quantity $\mathbf{v}$ in the equations has the meaning of the average velocity, per unit volume, of an ensemble of noninteracting (not subject to close interactions or collisions) particles. Therefore, if we refer arbitrarily to a volume element moving with velocity $\mathbf{v}$ as an "individual" (or "liquid") volume, Eq. (1.84) means that the average magnetic moment $p_\perp/\rho B$ is preserved in an individual volume. When the plasma is compressed in the magnetic field direction, the values of $\mathbf{B}$ and $p_\perp$ do not change, while $p_\parallel$ and ρ are connected by the adiabatic relation with exponent $\gamma = 3$. Equation (1.86) thus plays the role of the adiabaticity condition in an individual volume for a medium moving along magnetic force lines. We note also that if we put formally $p_\parallel = 0$, the system (1.81)–(1.85) is transformed into a system of ideal MHD equations for a perfect gas with $\gamma = 2$, while $p_\perp$ plays the role of ordinary pressure.

3.3. Linearized system of MHD equations

The above procedure of linearizing hydrodynamic equations can be used to linearize a system of MHD equations about states with $\mathbf{v}_0 = 0$, $\rho = \text{const}$, and $\mathbf{H}_0 = \text{const}$ (uniform field). Let the vector $\mathbf{H}_0$ be directed along the z axis and let the conductivity be infinite. We then obtain by transformation from the set of equations (1.69)

$$\frac{\partial^4 \rho}{\partial t^4} = L\Delta\rho, \tag{1.90}$$

$$\frac{\partial^2}{\partial t^2} L_H \mathbf{H} = \frac{\mathbf{H}_0}{\rho_0}\frac{\partial^4 \rho}{\partial t^4} - \frac{H_0}{\rho_0}\nabla\left(L\frac{\partial \rho}{\partial z}\right), \tag{1.91}$$

$$\frac{\partial \mathbf{v}}{\partial t} = \frac{a_0^2\,\nabla\rho}{\rho_0} - \frac{H_0}{4\pi\rho_0}\left(\nabla H_z - \frac{\partial}{\partial z}\mathbf{H}\right), \tag{1.92}$$

where the differential operators L and L_H are given by

$$L = (a_0^2 + a_A^2)\frac{\partial^2}{\partial t^2} - a_0^2 a_A^2\frac{\partial^2}{\partial z^2}, \qquad L_H = \frac{\partial}{\partial t^2} - a_A^2\frac{\partial^2}{\partial z^2},$$

a_0^2 is the usual speed of sound, and $a_A = H_0/\sqrt{4\pi\rho_0}$ is the Alfvén velocity. Note that if $a_A = 0$ we have in place of Eq. (1.90) $(\partial^2/\partial t^2 - a_0^2\Delta) = 0$, i.e., the usual wave equation.

If, however, $a_0 = 0$, we also obtain from Eq. (1.90) a wave equation but with a_0^2 replaced by a_A^2.

In the particular case of an incompressible liquid (ρ = const) we obtain from Eq. (1.91)

$$L_H \mathbf{H} = 0. \tag{1.93}$$

Equation (1.93) describes the propagation of MHD or Alfvén waves. It shows that these waves are connected with propagation of disturbances along the force lines of the field frozen in an incompressible medium.

Analysis[9,10] of the equations for the propagation of small perturbations along the z axis shows that there are three propagation velocities, a_A, a_+, and a_-, and the last two satisfy the equation

$$a_{+,-}^2 = \frac{1}{2}\left(a_0^2 + \frac{H_0^2}{4\pi\rho_0} \pm \sqrt{\left(a_0^2 + \frac{H_0^2}{4\pi\rho_0}\right) - \frac{a_0^2 H_{z0}^2}{\pi\rho_0}}\right),$$

where a_+ and a_- are respectively the velocities of the fast and slow magnetosonic waves.

4. Conditions on MHD shock waves

4.1. MHD discontinuities

The conditions imposed on MHD shock waves can be obtained by standard procedures from the integral conservation laws written for either a moving liquid or an electromagnetic field. We shall not dwell here in detail on the derivation of these conditions for ordinary MHD, since it is similar to that described in Sec. 2 for gasdynamic shock waves. A derivation of these conditions is given, e.g., in Refs. 6, 9, 10, and 21.

The conditions for the electromagnetic field parameters are[6,9,26]

$$\mathbf{n} \times [\mathbf{H}] - \frac{1}{c} D[\mathbf{D}_u] = \mathbf{i}, \tag{1.94}$$

$$\mathbf{n} \times [\mathbf{E}] - D[\mathbf{B}] = 0, \tag{1.95}$$

$$\mathbf{n} \times [\mathbf{B}] = 0, \tag{1.96}$$

where D denotes the shock-wave velocity, $\mathbf{n}$ is the unit normal, and $\mathbf{i}$ is the surface current. Neglecting the displacement currents and putting $\mathbf{B} = \mathbf{H}$ and $\mathbf{D}_u = \mathbf{E}$, we have in accordance with the Maxwell equations and Ohm's law (1.66)

$$\mathbf{E} = \frac{1}{c}\,(\nu_m \operatorname{rot} \mathbf{H} - \mathbf{v} \times \mathbf{H}). \tag{1.97}$$

Assuming that no concentrated current flows over the shock-wave surface, Eq. (1.94) together with Eq. (1.96) leads to continuity of the magnetic field, for in addition to Eq. (1.96) the difference between the tangential components of the field **H** is also zero.[9] In the case of finite conductivity, Eqs. (1.95) and (1.97) thus yield

$$\mathbf{n} \times [\nu_m \operatorname{rot} \mathbf{H} - \mathbf{v} \times \mathbf{H}] = 0. \tag{1.98}$$

Since the magnetic field is continuous in the case of finite conductivity, the hydrodynamic conditions on the discontinuity surface reduce to conditions (1.55) above. The additional condition (1.98), however, must be taken into account for motion of a gas with finite electric conductivity. The conditions on the discontinuities are obtained for infinite conductivity simply from the MHD equation in integral form, for an arbitrary moving volume $\Omega(t)$. These equations are

$$\begin{gathered}
\frac{d}{dt}\int_\Omega \rho\, d\Omega + \oint_\Sigma \rho(v_n - D)\, d\Sigma = 0, \\
\frac{d}{dt}\int_\Omega \mathbf{G}\, d\Omega + \oint_\Sigma (\hat{\mathbf{T}}\mathbf{n})(v_n - D)\, d\Sigma = 0, \\
\frac{d}{dt}\int_\Omega W\, d\Omega + \int_\Sigma S_n(v_n - D)\, d\Sigma = 0.
\end{gathered} \tag{1.99}$$

Equations (1.99) follow from the analogous relations obtained in Ref. 9 for a constant volume, and for Eq. (1.46) for the differentiation of integral. In these equations,

$$\mathbf{G} = \rho\mathbf{v} + \frac{1}{4\pi c}\,(\mathbf{E} \times \mathbf{H}),$$

$\hat{\mathbf{T}}$ is the total momentum flux-density tensor with components

$$T_{ik} = \rho v_i v_k + p\,\delta_{ik} - \frac{1}{4\pi}\,(H_i H_k + E_i E_k) - \frac{1}{8\pi}\,(H^2 + E^2)\,\delta_{ik},$$

$W = \rho[\varepsilon + v^2/2 + (H^2 + E^2)/8\pi]$ is the total-energy density,

$$S_n = \mathbf{Sn} = \rho\left(\varepsilon + \frac{v^2}{2}\right)v_n + q_n - (\hat{\mathbf{T}}\mathbf{v})\mathbf{n} + \frac{c}{4\pi}\,(\mathbf{E} \times \mathbf{H})\mathbf{n},$$

D is the velocity of the boundary Σ, and **S** is the total energy-flux density.

The conditions on the discontinuities for ideally conducting media are easily obtained from Eqs. (1.99).

Recognizing that $\mathbf{E} = -(1/c)(\mathbf{v} \times \mathbf{H})$ and taking the electrodynamic relations (1.94)–(1.97) into account, the complete system of conditions on the discontinuity surface takes at $\mathbf{q} = 0$ the form[9,11]

$$[H_n] = 0, \tag{1.100}$$

$$[(v_n - D)\mathbf{H} - H_n\mathbf{v}] = 0, \tag{1.101}$$

$$[(v_n - D)\rho] = 0, \tag{1.102}$$

$$\left[(v_n - D)\rho\mathbf{v} + \left(p + \frac{H^2}{8\pi}\right)\mathbf{n} - \frac{H_n}{4\pi}\mathbf{H}\right] = 0, \tag{1.103}$$

$$\left[(v_n - D)\left(\frac{\rho v^2}{2} + \rho\varepsilon + \frac{H^2}{8\pi}\right) + v_n\left(p_n + \frac{H^2}{8\pi}\right) - \frac{H_n}{4\pi}(\mathbf{Hv})\right] = 0. \tag{1.104}$$

The first relation here is a consequence of the absence of magnetic charges, and the second follows from the field equations (1.95) and from the equation for $\mathbf{E}$. Relation (1.101) expresses in fact the flux-freezing law. The remaining relations are respectively the consequences of the mass, momentum, and energy conservation laws. Relations (1.100)–(1.104) are significantly more complicated than the analogous relations for ordinary hydrodynamics, and yield a greater variety of possible discontinuities. A detailed classification of the MHD discontinuities is given in Refs. 9, 10, 27, and 28.

If there is no flux of matter through the discontinuity surface, i.e., $m = \rho_1(v_{n2} - D) = \rho_2(v_{n2} - D) = 0$, the result is either a tangential discontinuity ($H_n = 0$) or a contact discontinuity ($H_n \neq 0$, $[p] = 0$, $[v_n] = 0$, $[H] = 0$). Rotational discontinuities and shock waves take place if $m \neq 0$. A rotational discontinuity is one for which $[v_n] = [\rho] = 0$. For this discontinuity we have from Eqs. (1.100)–(1.104)

$$m = \frac{H_n\sqrt{\rho}}{\sqrt{4\pi}}, \qquad v_n - D = \frac{H_n}{\sqrt{4\pi\rho}}.$$

All the thermodynamic parameters are continuous on a rotational discontinuity. Moreover, we have $[H^2] = 0$, i.e., the modulus of the vector $\mathbf{H}$ is continuous. The vector $\mathbf{H}$ changes direction on going through the discontinuity, with $[\mathbf{H}] = km(\frac{1}{2})(\mathbf{H}_1 + \mathbf{H}_2) \times \mathbf{n}$, where k is some constant.

Discontinuity surfaces for which $m \neq 0$ and $[\rho] \neq 0$ are called shock waves. They come in the following types: fast shock wave, slow shock wave, switch-on shock wave, and switch-off shock wave. The latter two are limiting cases of the first two.

The fast shock wave is an analog of an ordinary gasdynamic wave, with the magnetic field strength increasing on passing through the discontinuity. In the slow wave the magnetic field decreases. Note the following main properties of fast and slow shock waves:

Property 1. Shock waves are compression jumps. Rarefaction jumps are impossible, since they correspond to transitions with decrease of entropy.

Property 2. The entropy change is a quantity of third order in the shock-wave intensity, and the following equation holds:

$$T(S_2 - S_1) \approx \frac{1}{12}\left(\frac{\partial^2 V}{\partial p^2}\right)_S (p_2 - p_1)^3 - \frac{1}{16\pi}\left(\frac{\partial V}{\partial p}\right)_S (p_2 - p_1)(\mathbf{H}_{\tau 2} - \mathbf{H}_{\tau 1})^2.$$

Here T is the temperature, S the entropy, $V = 1/\rho$ is the specific volume, and $\mathbf{H}_\tau$ is the tangential component of the field $\mathbf{H}$.

Note that the shock-adiabat equation is defined as

$$\varepsilon_2 - \varepsilon_1 + \frac{p_2 + p_1}{2}(V_2 - V_1) + \frac{1}{16\pi}(V_2 - V_1)(\mathbf{H}_{\tau 2} - \mathbf{H}_{\tau 1})^2 = 0.$$

Property 3. The following inequalities hold for slow and fast waves, respectively:

$$a_{-,1} \le D - v_{n1} < a_{A1}, \qquad D - v_{n2} \le a_{-,2},$$
$$a_{+,1} \le D - v_{n1}, \qquad a_{A2} < D - v_{n2} \le a_{+,2}. \tag{1.105}$$

The conditions (1.105) follow from the condition that the shock wave is evolutionary. Equal signs in Eqs. (1.105) correspond to a shock wave of infinitesimal intensity.

We note one more important type of shock wave, for which $H_n = 0$, i.e., the magnetic field is parallel to the wave front. For such shock waves we find from conditions (1.101) and (1.102) that $[\mathbf{H}/\rho] = 0$.

The designation switch-on shock wave is customarily used for a shock wave for which $\mathbf{H}_{\tau 1} = 0$ but $\mathbf{H}_{\tau 2} \ne 0$. Passage of such a wave produces (switches on) a tangential magnetic field component previously absent ahead of the wave. A switch-on shock wave is essentially a fast shock wave whose intensity should be lower than a certain critical value and ahead of which the Alfvén wave is supersonic.

Note that in the switch-on wave there is produced, besides the tangential field component, also a tangential component of the gas velocity. A switch-on shock wave can be realized if the following inequality is met[21]:

$$p_1 < \frac{H_1^2}{4\pi\gamma}.$$

A feature of switch-off shock waves is that they have $\mathbf{H}_{\tau 2} = 0$ and $\mathbf{H}_{\tau 1} \ne 0$, i.e., the magnetic field tangential component present ahead of such a wave is made to vanish (is switched off) by the wave passage. A switch-off wave is slow and its intensity reaches a certain characteristic value, while the Alfvén wave behind the shock wave is supersonic.

The influence of finite conductivity on shock-wave propagation is treated in many papers (see, e.g., Ref. 31).

Also possible in MHD are the so-called shock waves with conductivity jumps. These shock waves arise in media for which the conductivity can be regarded as zero ahead of the shock wave front but nonzero or even infinite behind it.[32,33] Shock waves with conductivity jumps can take place in gases in which a shock wave passes and ionizes the gas. The conditions for shock waves of this type are quite diverse, depending on the relations between the dissipative coefficients (i.e., the viscosity, thermal conductivity, electric conductivity) after passage through the discontinuity. Note that if the decisive dissipative coefficient is magnetic viscosity, an additional condition that is imposed on fast ionizing shock waves is[32,33]

$$[\mathbf{H}] = 0.$$

We shall consider also this type of wave when solving blast-theory problems.

4.2. Conditions on shock waves for the hydrodynamic model of a low-density plasma

The study of the structure of shock waves in ordinary hydrodynamics, from the standpoint of the kinetic theory of gases or using the complete equations of motion of a viscous heat-conducting liquid, shows that the shock-wave width (i.e., the width of bands with steep velocity gradients and abrupt thermodynamic parameters) is of the order of several molecule mean free paths. A similar situation obtains in a dense plasma with relatively short molecule mean free path. In a plasma of quite low density the molecule mean free path is quite large (larger than the characteristic dimensions over which the average plasma parameters change), and from the molecular-kinetics standpoint an important role in the formation of bands with large enough gradients of the average plasma parameters is played by the collective plasma properties for which particle interaction is realized via electromagnetic fields and via instability mechanisms of a moving ensemble of particles.

Investigations of this subject (see Refs. 34 and 35) show that shock waves can exist having a width much smaller than the particle mean free path. For shock waves there are at present no universally accepted relations analogous to the classical Rankine–Hugoniot conditions.

In our analysis of rarefied plasma we use as the basis the macroscopic MHD equations (1.81)–(1.85) for a rarefied plasma. Since they are nonlinear, the feasibility of discontinuities follows here from the results of the general theory of partial differential equations,[36,37] and were already predicted for hydrodynamics by Riemann. Since little has been published on the conditions for shock waves of this type, we shall derive them briefly here. We start from the MHD equations for a rarefied plasma in integral form, such as Eqs. (1.99).

Note that the electrodynamic part of the conditions will take the form of Eqs. (1.100) and (1.101), inasmuch as here, too, we start from Maxwell's equations and from the condition of infinite conductivity. The hydrodynamic part of the equations in integral form consists of a mass-conservation equation, a momentum equation [in which the tensor $\hat{\mathbf{T}}$ must be replaced by the analogous tensor of Eq. (1.82)], and an energy equation (in which account must be taken of the dependence of the energy ε on $p_\perp$ and $p_\parallel$, and the energy-flux vector $\mathbf{Q}$ contained in Eq. (1.83) used in place of the vector $\mathbf{S}$). In addition, we add to the equations of type (1.99) a new equation,

$$\frac{d}{dt}\int_{\Omega}(p_{\perp}/B)\,d\Omega+\oint_{\Sigma}(p_{\perp}/B)(v_n-D)\,d\Sigma=0,$$

which corresponds to the law of conservation of the average adiabatic invariant $p_{\perp}/B$. Applying now to the foregoing integral conservation laws the standard procedure described in Sec. 2 and using for the electromagnetic field the conditions (1.100) and (1.101), we obtain

$$\begin{gathered}
[B_n]=0,\\
[(v_n-D)\mathbf{B}-B_n\mathbf{v}]=0,\\
[(v_n-D)\rho]=0,\\
\left[\rho\mathbf{v}(v_n-D)+\left(p_{\perp}\frac{B^2}{8\pi}\right)\mathbf{n}+\left(\frac{p_{\parallel}-p_{\perp}}{B^2}-\frac{1}{4\pi}\right)B_n\mathbf{B}\right]=0, \quad (1.106)\\
\left[(v_n-D)\left(\frac{\rho v^2}{2}\rho\varepsilon+\frac{B^2}{8\pi}\right)+v_n\left(p_{\perp}+\frac{B^2}{8\pi}\right)-\left(\frac{p_{\parallel}-p_{\perp}}{B^2}-\frac{1}{4\pi}\right)B_n(\mathbf{B}\mathbf{v})\right]=0,\\
\left[\frac{p_{\perp}}{B}(v_n-D)\right]=0.
\end{gathered}$$

The conditions (1.106) make it possible to analyze the changes of the fundamental quantities on going through the discontinuity. These conditions were derived under the important assumption that the average magnetic moment $p_{\perp}/B\rho$ is conserved in the shock wave. This condition was proposed by R. V. Polovin.[38] Favoring this assumption is the satisfaction of the evolutionality conditions in a compression wave,[38] and also the fact that the adiabatic invariant of a particle is conserved, with high accuracy, as the particle moves in electromagnetic fields and crosses a surface on which the field parameters have a discontinuity of the first kind (order).

We refer to Eq. (1.106) as the conditions on an MHD shock wave in a low-density plasma. Note certain low-density-plasma shock-wave properties that follow from these conditions. These conditions also exhibit discontinuities of the tangential, contact, and rotational type, analogous to those for the case of ideal MHD.

Both fast and slow shock waves are possible here, as are also switch-on waves. All the waves for which $B_{n1}\neq 0$ and $\mathbf{B}_{\tau 1}=0$ are switch-on waves ($\mathbf{B}_{\tau}\neq 0$ behind them).[40] Using for the entropy the equation[38] $S=(k/2)\times\ln(p_{\parallel}p_{\perp}^2/\rho^5)$, where k is the Boltzmann constant, the entropy in the shock wave will grow.

Gas motion in the presence of shock transition is most stable under the condition $p_{\parallel}-p_{\perp}>0$.[14,40]

The properties of magnetohydrodynamic shock waves in a rarefied plasma have been much less studied than the properties of analogous discontinuities in ordinary MHD.

5. Gasdynamics equations with account taken of heat transport by radiation

5.1. Radiation transport equation

Gas motion is strongly influenced at high temperatures by thermal motion and by heat transport by optical fluxes. It is known from physics that light propagation in media has a dual nature, exhibiting both wave and corpuscular properties. Of importance to us is only that the radiation field is described by parameters such as the wavelength λ or the cyclic frequency $\omega = 2\pi\nu$ ($\lambda = 2\pi c/\omega$), and also the radiation intensity I_ω and the optical-photon energy $\hbar$ ($\hbar = h/2\pi$, where h is Planck's constant).

We derive the radiation-transport equation by following Ref. 41. The optical energy ΔE_ω incident on an area $\Delta\sigma$ perpendicular to the beam, in the frequency interval $(\omega, \omega + \Delta\omega)$, in a solid angle Ω, and over a time Δt is assumed equal to

$$\Delta E_\omega = I_\omega\, \Delta\sigma\, \Delta\omega\, \Delta\Omega\, \Delta t. \tag{1.107}$$

The quantity I_ω in Eq. (1.107) is called the radiation intensity. Thus, I_ω is defined as the energy incident in a unit frequency interval and in a unit solid angle on a unit area perpendicular to the incident-radiation direction.

Assume now that the medium is capable of absorbing radiation, and has a temperature T and a density ρ. If the medium can absorb radiation, it will draw away from the energy ΔE_ω indicated above (Fig. 1), on a path Δs, a certain fraction proportional to Δs.

The energy absorbed is thus

$$k_\omega\, \Delta s\, \Delta E_\omega,$$

where the proportionality coefficient k_ω is called the absorption coefficient. This coefficient is usually assumed to depend on ω and on the thermodynamic parameters of the medium, say on T and ρ.

If the medium is also capable of emitting energy, the amount of energy received by a volume ΔV in a solid angle $\Delta\Omega$ in the interval from ω to $\omega + \Delta\omega$ in a time Δt will be proportional to $\Delta V\, \Delta\Omega \Delta\omega \Delta t$ and equal to $j_\omega \Delta V\, \Delta\Omega \Delta\omega \Delta t$. The quantity j_ω is called the emission coefficient and equals the energy radiation by a unit volume in a unit frequency interval over unit time into a unit solid angle.

We assume for the time being that the radiation field does not vary in time. We obtain now an equation describing the radiation transport. We take a unit cylinder with axis directed along the ray [see Fig. 1(b)] and let $\Delta\sigma$ be its cross section area and Δs its height; the elementary volume is then $\Delta V = \Delta\sigma\, \Delta s$.

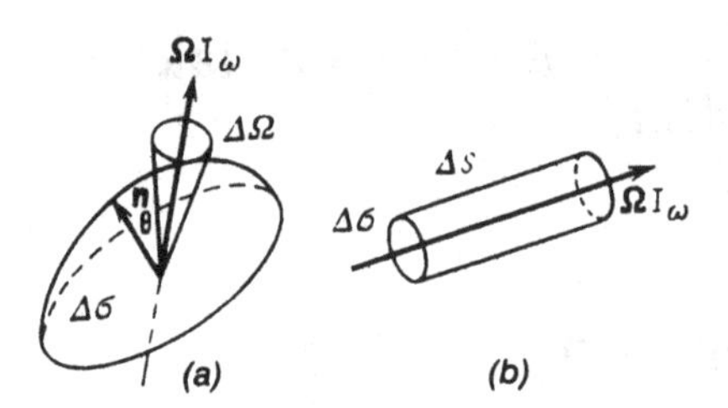

Figure 1. (a) Diagram of radiation propagation (n is a unit normal vector); (b) volume element along the beam.

The balance equation for the radiation entering and leaving the cylinder is

$$(I_\omega + \Delta I_\omega)\,\Delta\sigma\,\Delta\Omega\,\Delta\omega\,\Delta t = I_\omega\,\Delta\sigma\,\Delta\Omega\,\Delta\omega\,\Delta t - k_\omega\,\Delta s I_\omega\,\Delta\sigma\,\Delta\Omega\,\Delta\omega\,\Delta t + j_\omega\,\Delta\sigma\,\Delta s\,\Delta\Omega\,\Delta\omega\,\Delta t.$$

After canceling the common factors $\Delta\sigma\,\Delta\Omega\,\Delta\omega\,\Delta t$, gathering like terms, and taking the limit as $\Delta s \to 0$ we find (assuming that a limit exists) the equation

$$\frac{dI_\omega}{ds} = -k_\omega I_\omega + j_\omega. \tag{1.108}$$

We shall assume next, as a rule, the presence of local thermal equilibrium, meaning by the latter equilibrium in the substance as well as equilibrium between the substance and the radiation. This means formally that the temperature T will be used at a given point of space both to describe the medium and to determine the radiation parameters. In the case of equilibrium the system is subject to Kirchhoff's law

$$j_\omega = B_\omega(T)k_\omega,$$

where B_ω is the known Planck function:

$$B_\omega(T) = \frac{\hbar\omega^3}{2\pi^2c^2}\left(e^{\hbar\omega/kT} - 1\right)^{-1}, \tag{1.109}$$

which yields the equilibrium-radiation intensity, and c is the speed of light in the medium.

In the case of local thermodynamic equilibrium, Eq. (1.108) can be written in the form

$$\mathbf{\Omega}\nabla I_\omega + k_\omega I_\omega = k_\omega B_\omega, \tag{1.110}$$

where $\mathbf{\Omega}$ is a unit vector in the radiation-propagation direction [Fig. 1(a)].

Equation (1.110) is called the radiation transport equation. Recognizing that $\nabla(\mathbf{\Omega}I_\omega) = \mathbf{\Omega}\nabla I_\omega$, we can rewrite Eq. (1.110) in the form

$$\operatorname{div}(\mathbf{\Omega}I_\omega) + k_\omega I_\omega = k_\omega B_\omega. \tag{1.111}$$

Remark 1. It was assumed above that the radiation field and the radiation transport are stationary, and the time t entered as a parameter. This means assumption of an infinite rate of variation of I_ω, i.e., the entire field is instantaneously equalized. Actually, this is only an approximation, and the rate of perturbation of the optical field is finite and is equal to the speed of light c.

Assuming now that $ds = c\,dt$ and using the usual relations between the derivatives on going from the "Lagrange" to the "Euler" representation

$$\frac{dI_\omega}{dt} = \frac{\partial I_\omega}{dt} + c\boldsymbol{\Omega}\nabla I_\omega,$$

we obtain from Eq. (1.108) the nonstationary transport equation

$$\frac{1}{c}\frac{\partial I_\omega}{\partial t} + \boldsymbol{\Omega}\nabla I_\omega + k_\omega I_\omega = k_\omega B_\omega. \tag{1.112}$$

The term with $1/c$ in this equation is small in most gasdynamic problems. We shall therefore neglect it in subsequent gasdynamic studies.

Remark 2. The transport equation (1.110) takes into account only absorption and emission by the medium, but not scattering of the light. Scattering does play the principal role in some problems. Light is scattered by molecules, electrons, dust specks, or condensed-phase droplets. To take scattering into account it is necessary to include additional terms in the right-hand sides of Eqs. (1.110)–(1.112).

We present for completeness one possible form of a radiation transport equation with allowance for scattering

$$\frac{1}{c}\frac{\partial I_\omega}{\partial t} + \boldsymbol{\Omega}\nabla I_\omega + k_\omega I_\omega = k_\omega\left[(1-\sigma_\omega)B_\omega + \frac{\sigma_\omega}{4\pi}\int_{4\pi} I_\omega(\boldsymbol{\Omega}_*)g_\omega(\mu_0)\,d\boldsymbol{\Omega}_*\right],$$

$$k_\omega = k_{a\omega} + k_{s\omega}, \qquad \sigma_\omega = k_{s\omega}/k_\omega, \qquad \mu_0 = \boldsymbol{\Omega}\boldsymbol{\Omega}_*, \qquad \sum_{i=1}^{3}\Omega_i^2 = 1,$$

$$\sum_{i=1}^{3}\Omega_{*i}^2 = 1, \qquad g_\omega = g_\omega(\rho, T, \mu_0),$$

$$\frac{1}{4\pi}\int_{4\pi} g_\omega(\mu)\,d\boldsymbol{\Omega} = \frac{1}{4\pi}\int_{4\pi} g_\omega(\mu_0)\,d\boldsymbol{\Omega}_* = 1;$$

k_ω is here the effective absorption coefficient, or the attenuation coefficient for the medium, $k_{a\omega}$ is the absorption coefficient, $k_{s\omega}$ is the coefficient of absorption due to scattering, σ_ω is the albedo of the medium, and g_ω is the scattering indicatrix and is equal to the probability of scattering a photon of direction $\boldsymbol{\Omega}_*$ into a direction $\boldsymbol{\Omega}$.

The last equation of the above set is the normalization condition, which expresses the conservation of the number of photons in scattering.

5.2. Integral characteristics of radiation field

We denote by $\mathbf{q}_\omega$ the monochromatic-radiation flux vector summed over all directions (i.e., for a given frequency ω), and define it by the integral

$$\mathbf{q}_\omega = \int_{4\pi} I_\omega(\boldsymbol{\Omega})\boldsymbol{\Omega}\, d\Omega,$$

where the integration is over the entire solid angle.

We shall omit hereafter the argument $\boldsymbol{\Omega}$ of I_ω if no confusion results.

The quantity

$$\mathbf{q} = \frac{1}{2\pi}\int_0^\infty \int_{4\pi} \boldsymbol{\Omega} I_\omega\, d\Omega\, d\omega = \frac{1}{2\pi}\int_0^\infty \mathbf{q}_\omega\, d\omega \qquad (1.113)$$

will be called the total radiation-flux vector.

One can also introduce the radiation energy density

$$E_R = \frac{1}{2\pi c}\int_0^\infty \int_\Omega I_\omega\, d\Omega\, d\omega \qquad (1.114)$$

and the stress tensor due to the radiation

$$\hat{\mathbf{P}}_R = \frac{1}{2\pi c}\int_0^\infty \int_\Omega \boldsymbol{\Omega}\boldsymbol{\Omega} I_\omega\, d\Omega\, d\omega. \qquad (1.115)$$

Here $\hat{\mathbf{P}}_R$ is a symmetric tensor of second rank.

We show, as an example, that if l, m, and n are the direction cosines of the vector $\boldsymbol{\Omega}$ in a Cartesian frame x, y, z the components of the vector $\mathbf{q}$ and of the tensor $\hat{\mathbf{P}}_R$ take the form

$$q_x = \frac{1}{2\pi}\int_0^\infty \int_\Omega l I_\omega\, d\Omega\, d\omega \text{ etc.,}$$

$$P_{Rxx} = \frac{1}{2\pi c}\int_0^\infty \int_\Omega l^2 I_\omega\, d\Omega\, d\omega,$$

$$P_{Ryz} = R_{Rzy} = \frac{1}{2\pi c}\int_0^\infty \int_\Omega mn I_\omega\, d\Omega\, d\omega \text{ etc.}$$

We shall neglect hereafter, as a rule, the radiation-energy density and the stresses due to radiation, since they are small compared with the other corresponding properties of the medium.

5.3. Some simplified models of the transport equation

One of the important approximations in transport theory is the diffusion approximation, which obviates the need for taking into account the angular distribution of the intensity in the relevant equations.

We choose the transport equation in the form (1.111) and write for I_ω:

$$I_\omega = I^0_\omega + 3\Omega \mathbf{I}^1_\omega, \tag{1.116}$$

where I^0_ω and I^1_ω are new functions sought that do not depend on the angles. Substituting Eq. (1.116) in Eq. (1.111) and integrating the result over the angles, after multiplying it first by unity and then by $\mathbf{\Omega}$ or by a spherical harmonic containing the first Legendre polynomial P_1, we obtain for I^0_ω and $\mathbf{I}^1_\omega$ the system of differential equations

$$\begin{aligned} \operatorname{div} \mathbf{I}^1_\omega + k_\omega I^0_\omega &= k_\omega B_\omega, \\ \nabla I^0_\omega + 3k_\omega \mathbf{I}^1_\omega &= 0. \end{aligned} \tag{1.117}$$

Eliminating I^0_ω from Eqs. (1.117), we obtain the following equation for the vector $\mathbf{I}^1_\omega$ which is proportional in this case to the vector $\mathbf{q}_\omega$:

$$\nabla\left(\frac{1}{k_\omega} \operatorname{div} \mathbf{I}^1_\omega\right) - \nabla B_\omega - 3k_\omega \mathbf{I}^1_\omega = 0,$$

$$\mathbf{q}_\omega = \int_{4\pi} (I^0_\omega + 3\Omega \mathbf{I}^1_\omega)\,\Omega\, d\Omega = 4\pi \mathbf{I}^1_\omega.$$

This equation can thus be written in the form

$$\nabla\left(\frac{1}{k_\omega} \operatorname{div} \mathbf{q}_\omega\right) - 4\pi \nabla B_\omega - 3k_\omega \mathbf{q}_\omega = 0. \tag{1.118}$$

This is in fact the transport equation in the diffusion approximation relative to the flux vector $\mathbf{q}_\omega$.

Assume now that the absorption coefficient is independent of frequency, i.e., that there exists some average absorption coefficient $k = k(\rho, T)$. Then, integrating Eq. (1.118) over the entire spectrum, we obtain the known Milne–Eddington equation (see, e.g., Refs. 41–43) for the radiation flux of the gray-body gas in the diffusion approximation

$$\nabla\left(\frac{1}{k} \operatorname{div} \mathbf{q}\right) - 3k\mathbf{q} - 4\sigma \nabla T^4 = 0. \tag{1.119}$$

It was taken into account here that $\int_0^\infty \mathbf{q}_\omega\, d\omega = 2\pi \mathbf{q}$ and $\int_0^\infty B_\omega\, d\omega = 2\sigma T^4$, where $\sigma = \pi^2 k_B^4 / 60 c^2 \hbar^3$ is the Stefan–Boltzmann constant. Noting that $k(\rho, T)$ has the dimension of length, we propose now that $kL \gg 1$, where L is the characteristic length over which q, T, and R change substantially. Then, neglecting in Eq. (1.119) the first term which contains $1/k$, we obtain

$$\mathbf{q} = -\frac{4\sigma}{3k} \nabla T^4. \tag{1.120}$$

This is the expression for the flux, in the radiant heat-conduction approximation, for an optically thick medium. In the other limiting case $kL \ll 1$

it follows from Eq. (1.119) that $\nabla[(1/k) \operatorname{div} \mathbf{q} - 4\sigma T^4] = 0$ or, in the absence of sources, that

$$\operatorname{div} \mathbf{q} = 4\sigma k T^4. \tag{1.121}$$

This is the case of an optically thin medium.

5.4. Gasdynamics equations with allowance for radiation

In the general case the MHD equations (1.67)–(1.67b) are valid in the presence of magnetic fields also when account is taken of heat transport by the radiation, if it is now assumed that $P_{ik} = -\delta_{ik}p + \rho v_i v_k + P_{Rik}$, where P_{Rik} are the components of the radiation-induced stress tensor introduced by Eq. (1.115). In addition, the heat-flow vector of Eq. (1.67b) must now include the radiation flux, i.e., $\mathbf{q} = \mathbf{q}_T + \mathbf{q}_R$ ($\mathbf{q}_T$ is the vector of the heat flow due to thermal conductivity and $\mathbf{q}_R$ is the radiation-flux vector). We must also add here an expression for the variation energy, i.e., replace ε everywhere in this equation by $\varepsilon + E_R/\rho$, where E_R is defined by Eq. (1.114), and account must be taken of the work performed by the radiation forces. We shall not write out these equations for the general case.

In most problems, the radiation-tensor components are small compared with the pressure of the medium, and the internal energy of the medium is much higher than E_R. Therefore $\hat{\mathbf{P}}_R$ and E_R are as a rule disregarded in the gasdynamic equations. We ignore for simplicity the electromagnetic fields. Under these assumptions, the gasdynamic equations with allowance for the radiation heat transport become

$$\frac{\partial \rho}{\partial t} + \operatorname{div}(\rho \mathbf{v}) = 0,$$

$$\frac{\partial \rho v}{\partial t} + \operatorname{div}(\delta_{ik}p + \rho v_i v_k) = 0,$$

$$\frac{\partial}{\partial t}\left[\rho\left(\varepsilon + \frac{v^2}{2}\right)\right] + \operatorname{div}\left[\rho \mathbf{v}\left(\varepsilon + \frac{p}{\rho} + \frac{r^2}{2}\right) + \mathbf{q}\right] = 0, \tag{1.122}$$

$$\mathbf{q} = \mathbf{q}_T + \mathbf{q}_R, \qquad \mathbf{q}_T = -\varkappa \nabla T, \qquad \mathbf{q}_R = \frac{1}{2\pi}\int_0^\infty \int_{4\pi} I_\omega \boldsymbol{\Omega}\, d\Omega\, d\omega,$$

$$\boldsymbol{\Omega}\nabla I_\omega + k_\omega I_\omega = k_\omega B_\omega.$$

The transport equation is written in this system without allowance for scattering and for the nonstationary behavior of the transport process itself. The system (1.122) is frequently referred to as the radiative gasdynamic equations.

Let now the gas flow be such that the velocity v is low and can be disregarded. The heat is transported only by radiation. The energy equation is then replaced by

$$\frac{\partial(\rho\varepsilon)}{\partial t} + \operatorname{div} \mathbf{q}_R = 0. \tag{1.123}$$

The energy equations and the transport equations are in this particular case, assuming plane ($\nu = 1$) and spherical ($\nu = 3$) symmetry,

$$\frac{\partial(\rho\varepsilon)}{\partial t} + \frac{1}{r^{\nu-1}}\frac{\partial}{\partial r}(r^{\nu-1}q) = 0,$$

$$\mu\frac{\partial I_\omega}{\partial r} + \frac{\nu-1}{r}\left[\frac{5-\nu}{\nu+1}(1-\mu^2)\right]\frac{\partial I_\omega}{\partial \mu} = k_\omega(B_\omega - I_\omega),$$

$$q_\omega = \int_{4\pi} I_\omega \mu \, d\Omega = \int_0^{2\pi}\int_0^{\pi} I_\omega \cos\theta \sin\theta \, d\theta \, d\varphi,$$

$$q = \frac{1}{2\pi}\int q_\omega \, d\omega, \qquad \mu = \cos\theta,$$

where θ and φ are the polar and azimuthal angles, respectively. In the case of cylindrical geometry we have for the radiation-transport equation

$$\sin\theta\left[\mu_1\frac{\partial I_\omega}{\partial r} + \frac{(1-\mu_1^2)}{r}\frac{\partial I_\omega}{\partial \mu_1}\right] = k_\omega(B_\omega - I_\omega),$$

where $\mu_1 = \cos\varphi$. The radiation intensity in the stationary case thus depends here on three variables, i.e., $I_\omega = I_\omega(r, \theta, \varphi)$. This complicates the actual solution of many problems.

6. Elements of dimensionality theory and the concept of self-similar motions of a medium

To solve problems of point-blast theory we shall use extensively the conclusions and methods of similarity and dimensionality theory. We present below the necessary dimensionality-theory information and introduce the concept of self-similar motions.[44]

Physical quantities are measured in units that can be classified as basic and derived. Various physical quantities can be connected by different relations that express physical laws. If some of these quantities are taken to be basic and certain measurement units are set for them, the measurement units for all the remaining quantities are expressed in terms of these basic quantities, i.e., they are derivative measurement units. Expression of a derivative measurement unit in terms of basic ones is called dimensionality.

In the study of mechanical, thermal, and certain other physical phenomena it suffices to introduce only three basic measurement units—for the length, mass, and time. This system of basic units will be used for the most part in the study of blast phenomena.

The dimensionality of a quantity is expressed symbolically by a formula in which L stands for the unit length, M for the unit mass, and T for the unit time. The symbol for the dimensionality of some quantity a is $[a]$. The dimensionalities of all the physical quantities that we shall use hereafter can be expressed in the form $[a] = M^{n_1}L^{n_2}T^{n_3}$ with n_i constant. If $n_i = 0$ $(i = 1, 2, 3)$, then a is a dimensionless quantity.

A dimensional unit is said to have a dimensionality independent of all other quantities of a problem if the formula for its dimensionality cannot be represented by a power-law monomial of the dimensionalities of the remaining quantities. Thus, for example, the dimensionalities $[\rho] = ML^{-3}$ of the density and $[E] = ML^2T^{-2}$ of the energy are independent, while the dimensionalities of length L, velocity LT^{-1}, and acceleration LT^{-2} are dependent.

In solutions to problems of physics and mechanics we frequently encounter a case where some dimensional unit is a function of other dimensional units. Let a certain dimensional quantity a be a function of the dimensional quantities $a_1, a_2, \ldots, a_n$:

$$a = \varphi(a_1, a_2, \ldots, a_n). \tag{1.124}$$

We shall assume that Eq. (1.124) is a certain physical law, a fixed physical relation independent of the choice of the measurement-unit system.

Let only the first $k \le n$ of the dimensional quantities have independent dimensionalities. Since the number of basic units in the problems of interest to us is three, we have in our case $k \le 3$. Any physical relation of the form (1.124) can be regarded as a relation between dimensionless quantities. If it is known that this dimensionless quantity is a function of a number of dimensional quantities, this function can depend only on dimensionless combinations made up of the defining dimensional quantities.

According to the π-theorem of dimensionality theory, the functional connection between $(n + 1)$-dimensional quantities, which is independent of the system of measurement units, takes the form of a relation between $(n + 1 - k)$ dimensionless quantities that are dimensionless combinations of $(n + 1)$-dimensional quantities.

Naturally, the smaller the number of parameters that determine an investigated quantity, the more restricted the functional relation of type (1.124) and the easier it is to study. Thus, if the number of basic measurement units is equal to the number of defining parameters with independent dimensionalities, this dependence is determined by means of a constant. Indeed, if $n = k$ it is impossible to make up a dimensionless combination of the quantities $a_1, a_2, \ldots, a_n$. On the other hand, for the dimensionality of a we have the relation $[a] = [a_1]^{k_1}[a_2]^{k_2} \ldots [a_n]^{k_n}$, which is satisfied under the condition $a = c_k a_1 a_2 \ldots a_n$, where c_k is a dimensionless constant, and the exponents $k_1, k_2, \ldots, k_n$ can be easily obtained by comparing the dimensionalities of the right- and left-hand sides of the equation in question.

We begin the solution of some physical or mechanical problem by separating the main factors, the essential variables, and the constant parameters that define the quantities of interest to us. One of the first steps in an investigation of the functional dependences of the dimensionless quantities is tabulation of the main parameters that describe the phenomenon in question.

Dimensionality theory requires that the system of defining parameters include dimensional quantities in terms of which the dimensionalities of all the dependent quantities can be expressed. Thus, for example, it cannot be stated that the static state of a perfect gas is determined only by two dimensional quantities, the temperature T ($[T] = °C$) and density ρ, since the dimensionality of pressure cannot be expressed in terms of $[T]$ and $[\rho]$. We must add to these quantities one more dimensional constant, the gas constant R.

We proceed now to use the foregoing premises of dimensionality theory in problems involving the motion of continuous media. To describe motion of a compressible liquid in space we must determine the pressure p, the density ρ, and the velocity components v^i along the axes x^i as functions of the time, the coordinates, and the constant parameters of the conditions of the problem. In the general case of spatial motions, the system of the defining parameters can always be reduced to the form

$$x^i, t, a, b, c, \alpha_1, \alpha_2, \ldots, \alpha_n, \tag{1.125}$$

where a, b, and c are constants with independent dimensionalities, while $\alpha_1, \alpha_2, \ldots, \alpha_n$ are certain dimensionless constants.

Some quantity χ that defines some characteristic of the motion of the medium will be a function, independent of the measurement-units system, of the parameters (1.125). Let us apply the π-theorem to this function. We have $n = 7$ and $k = 3$. By making up a dimensionless combination of Eq. (1.125) with χ, we can introduce a dimensionless function $\tilde{\chi}$ that can depend in our case, according to the π-theorem, on four dimensionless variables. This follows also directly from the fact that it is possible to make up four independent dimensionless variable combinations of the parameters (1.125). We have thus in the general case

$$\tilde{\chi} = \psi(\tilde{x}^i, \tilde{t}, \alpha_1, \alpha_2, \ldots, \alpha_n),$$

where $\tilde{x}^i$ and $\tilde{t}$ are the dimensionless coordinates and the dimensionless time. The number of independent variables is not decreased and is equal to four.

Let now $c = 0$ and let the system of the defining parameters (1.125) have two constants a and b with independent dimensionalities

$$[a] = ML^kT^s, \qquad [b] = LT^{-\delta} \quad (\delta \neq 0). \tag{1.126}$$

Here $n = 6$, $k = 3$, and $n - k = 3$, i.e., the dimensionless function $\tilde{\chi}$ will depend on only three dimensionless combinations of variables. These dimensionless combinations take the form x^1/bt^δ, x^2/bt^δ, and x^3/bt^δ, and no other independent dimensionless combinations can be formed. Thus,

$$\tilde{\chi} = \psi\left(\frac{x^i}{bt^\delta}, \alpha_1, \alpha_2, \ldots, \alpha_n\right).$$

If $\delta = 0$ we have $[b] = L$, and then either a has the dimensionality $L^k T^s$ or there is one more dimensional constant with a dimensionality independent of $[a]$ and $[b]$. The second possibility leads to the general case. This follows from the fact that we must make up with the time t a dimensionless combination on which $\tilde{\chi}$ depends. If $[a] = L^k T^s$, only the two quantities a and b out of the set of parameters x^i, t, a, b, $\alpha_1, \ldots$, and α_n have independent dimensionalities and we have four dimensionless combinations on which $\tilde{\chi}$ will depend.

Definition. Unsteady motions of a continuous medium, in which combinations x^1/bt^δ, x^2/bt^δ, and x^3/bt^δ, with $[b] = LT^{-\delta}$, are called self-similar. In the particular case of one-dimensional self-similar motions the characteristics depend on one dimensionless variable combination r/bt^δ.

It follows from the above arguments that a sufficient condition for a problem to be self-similar is that the system of dimensional defining parameters, which is specified by the equations and supplementary conditions, particularly by the boundary or initial conditions, contain not more than two constants with independent dimensionalities other than length (and time), i.e., that the system of defining parameters have the form x^1, x^2, x^3, t, a, b, $\alpha_1, \ldots, \alpha_n$, where $\alpha_1, \alpha_2, \ldots, \alpha_n$ are dimensionless combinations, while $[a]$ and $[b]$ satisfy Eqs. (1.126) with k and s arbitrary constraints. It is possible to choose instead of a the constant $\tilde{a}$, with $[\tilde{a}] = ML^{\omega-3}$, where ω is an arbitrary number.

The system of defining parameters for one-dimensional self-similar motions must have the form r, t, a, b, α_1, $\alpha_2, \ldots, \alpha_n$. Note that in the initial formulation of the problem the number of dimensional defining constant parameters can be large, but only two of them, which we take to be a and b, should have independent dimensionalities. All the remaining dimensional constants can be made dimensionless with the aid of a and b and included among the parameters $\alpha_1, \ldots, \alpha_n$. Since one-dimensional self-similar motions depend on only one dimensionless variable $\lambda = r/bt^\delta$, the system of partial differential equations for one-dimensional self-similar motions is replaced, after reduction to dimensionless form, by an equivalent system of ordinary differential equations, thereby facilitating the solution.

Let us consider examples of self-similar one-dimensional motions of a perfect gas:

(1) *Propagation of a flame front or detonation front in the case of a zero-thickness combustion zone.* A homogeneous combustible mixture filling a space at constant pressure p_1 and constant density ρ_1 is ignited at the instant $t = 0$ at a point (spherical symmetry), along a line (cylindrical symmetry), or along a plane (planar symmetry). The flame or detonation front propagates through the mixture. Let their propagation conditions be such that the igniting energy can be neglected, the substance is assumed to burn up instantaneously on the flame or detonation front, and energy is released.

The defining parameters are r, t, ρ_1, p_1, γ, the heat Q released when a unit mass of gas burns up, and the known flame-front velocity U_1 if combustion takes place. The dimensionality of Q in mechanical units can be expressed in terms of the dimensionalities of ρ_1 and p_1. Thus, the number of constants with independent dimensionalities is two, i.e., the problem is self-similar. The solution of this problem is known.[44–46] Other examples of self-similar problems involving wave propagation in a combustible mixture of gases are considered in Chap. 6.

(2) *The inverse-pinch-effect problem.* Let a current that varies with time, $I = \sigma_1 t$ (σ_1 is constant), flow along a straight line in an infinitely conducting medium. The initial gas density ρ_1 and the initial pressure p_1 are constant. The initial velocity $v_1 = 0$. An initial field $H_1 = \text{const}$ is frozen in the medium and is parallel to the straight current-flow line.

In view of the freezing, the magnetic pressure of the field of the linear current

$$\frac{H_\varphi^2}{8\pi} = \frac{I^2}{2cr_0^2}$$

(r_0 is the distance from the symmetry axis to the considered point) will "push out" the gas from the region adjacent to the axis, and act as a piston having a radius $r_0(t)$ and expanding at a constant rate.[47] The ensuing compression has been named the inverse pinch effect. Since the dimensionality of the magnetic field H_1 is expressed in terms of the dimensionality of the pressure p_1 ($[H_1^2] = [p_1]$), the problem is self-similar. The solution of this problem was considered by us in Ref. 48.

Another important example from among the self-similar gasdynamic and MHD problems is that of a strong point blast, which we shall consider in detail later on.

7. Invariant and self-similar solutions of one-dimensional-motion equations

7.1. Connection between dimensionality analysis and transformation-group theory

From the mathematical viewpoint, dimensionality analysis is closely related to the deductions of continuous-transformation group theory. We present some known information from the theory of Lie groups.[49–51]

Definition. A group G is called a Lie group if G contains a subset G_r satisfying the following conditions:

1. G_r contains a unit element e of group G.
2. The products of the elements of G_r generate G.
3. The elements of G_r can be set in one-to-one correspondence with points α of an open r-dimensional sphere Q of Euclidean space, with the center O of the sphere corresponding to e. This condition introduces co-

ordinates in G_r. The element $g \in G$ corresponding to $\alpha \in Q$ will be designated g_α.

4. The product $g_\alpha \cdot g_\beta \in G_r$ is a certain element g_γ corresponding to the point $\gamma = \varphi(\alpha, \beta)$, and the notation $\gamma = \varphi(\alpha, \beta)$ stands for $\gamma^\lambda = \varphi^\lambda(\alpha^1, \ldots \alpha^r; \beta^i, \ldots \beta^r)\lambda = 1, 2, \ldots, r$.

5. In the region Q^r the function $\varphi(\alpha, \beta)$, defined in a certain vicinity of the points α and β, is triply continuously differentiable.

The subset G_r is called the group kernel or the local Lie group. The dimensionality r of the sphere Q is called the order of the local Lie group G_r. The associativity property for the Lie group takes the form $\varphi[\varphi(\alpha, \beta), \gamma] = \varphi[\alpha, \varphi(\beta, \gamma)]$. The study of the properties of a local Lie group reduces to a study of the properties, independent of the coordinate system, of the functions $\varphi^\lambda(\alpha, \beta)$ that specify the multiplication law.

We introduce now the concept of isomorphism: two groups, G and H, are called isomorphic if each element $A, B, C \ldots$ of the first group is in one-to-one correspondence with an element $a, b, c, \ldots$ of the second in such a way that $ab = c$ whenever $AB = C$.

Definition. A one-to-one mapping of a certain set $\mathscr{E}$ on itself is called a transformation of the set $\mathscr{E}$.

If T_1 and T_2 are two transformations of the set $\mathscr{E}$, their product $T_1 T_2 = T_3$ is determined by the relation

$$T_3(M) = T_1[T_2(M)] \qquad (1.127)$$

for arbitrary $M \in \mathscr{E}$. The role of unity is played by the identity transformation $e(M) = M, M \in \mathscr{E}$, with $eT = Te = T$. It can be shown[49,59] that the transformation product defined above is associative. The transformation T^{-1} inverse to T, by its definition, transforms any $T(M) \in \mathscr{E}$ element into an M element.

Thus, the aggregate G of the transformations of the set $\mathscr{E}$, which contains alongside any two transformations their product and alongside a transformation its inverse, is a group by virtue of the established multiplication law (1.127). Any such group is called a transformation group.

The following principle, by Weyl, holds: any group can be represented by a certain transformation group, i.e., there exists a transformation group that is isomorphic to the given group.

Let us consider the representation of an abstract Lie group by a transformation group. For example, the set G of all nondegenerate quadratic matrices of order n is a group with respect to ordinary matrix multiplication. The role of unity is played by the matrix $e = \|\delta^i_j\|$, where $\delta^i_i = 1, \delta^i_i = 1, \delta^j_i = 0, j \neq i$. Corresponding to the group of matrices $\|a^i_j\|$ are the transformations $x' = \|a^i_j\| x$, while multiplication of matrices corresponds to a sequence of transformations. Thus the matrix group is represented by linear transformations. Assume now that in the general case we have a local Lie group G_r and we have a set $\mathscr{E}(x^1, x^2, \ldots, x^n)$. We separate in this set a subset—a kernel. Assume that we have constructed transformations with the aid of group-G_r elements that "shift" the kernel of the set $\mathscr{E}$. This means that the equations for the transformations of the specified set will contain the coordinates of the group G_r, i.e., to each point α there will correspond a transformation

$$x'^i = f^i(x^1, \ldots, x^n, \alpha^i, \ldots, \alpha^r). \tag{1.128}$$

By the set $\mathscr{E}$ we shall mean hereafter an n-dimensional Euclidean space $\mathscr{E}_n$ of points x with coordinates $x_1, \ldots, x_n$. If f^i in Eq. (1.128) is triply continuously differentiable and the transformation is reversible, and if $f(x, 0) = x$ and the property $f[f(x, \alpha), \beta] = f[x, \varphi(\alpha, \beta)]$ takes place, then the set of transformations (1.128) forms a local Lie group of the point transformations $\mathscr{E}_n$. This group gives the representation of the group G_r. We denote the group of transformations by G_r^n.

The formula $T_\alpha T_\beta = T_{\varphi(\alpha,\beta)}$ for T_α, $T_\beta \in G_r^n$ corresponds then to the formula

$$f^i[f(x, \alpha), \beta) = f^i(x, \varphi(\alpha, \beta)].$$

Under the action of the transformation T_α, a certain function $F(x)$ defined on $\mathscr{E}_n$ is transformed into a new one: $F'(x) = T_\alpha F(x) = F(T_\alpha x)$.

Definition. The function $f(x) \not\equiv$ const is called an invariant of the group G_r if it is not altered by the action of any transformation $T_\alpha \in G_r^n$, i.e., $T_\alpha F(x) \equiv F(x)$ or $F(T_\alpha x) \equiv F(x)$.

We introduce the auxiliary functions

$$\xi_s^i(x) = \left.\frac{\partial f^i(x, \alpha)}{\partial \alpha^s}\right|_{\alpha=\alpha_0}, \quad \begin{array}{l} i = 1, 2, \ldots, n, \\ s = 1, 2, \ldots, r, \end{array} \tag{1.129}$$

and the matrix $A = \|\xi_j^i\|$ ($\alpha = \alpha_0$ corresponds to unity of the group).

Let the general rank of matrix A be R. The following holds:

Theorem.[51] *A group G_r^n has invariants if and only if $R < n$. If this inequality holds, there exist $t = n - R$ functionally independent invariants $I^k(x)$ ($k = 1, 2, \ldots, t$) of the group G_r^n such that any invariant of the group is a function of these invariants.*

This means that if $F(x)$ is an invariant, then

$$F(x) = F[I^1(x), I^2(x), \ldots, I^t(x)].$$

From this we can draw the following conclusion (the generalized π-theorem).[52]

Let the equation $F(x^i) = 0$ be invariant to the group G_r^n that transforms into itself a manifold of dimensionality l, which is an l-dimensional subspace of the space $\mathscr{E}_n$. The equation $F = 0$ is then equivalent to the equation

$$\Phi(I^1, I^2, \ldots, I^{n-l}) = 0 \tag{1.130}$$

of $n - l$ functions $(1 \le l \le r)$ $I^1, I^2, \ldots, I^{n-l}$, each of which is invariant to G_r. We remark that the dimensionality of l coincides with the rank of the matrix A.

Note that a generalization of these results to include tensor functions and applications of Lie groups to crystal theory is given in Ref. 53.

It is postulated in dimensionality theory that certain fundamental units $q_1, \ldots, q_r$ can be changed in arbitrary positive relations by transformations of the form

$$T_\alpha(q_i) = q_i' = \alpha_i q_i \qquad (\alpha_i > 0, \quad i = 1, \ldots, r), \qquad (1.131)$$

called scale changes. Also considered are homogeneous quantities $Q_1, \ldots, Q_n$ which are transformed by Eq. (1.131) as follows:

$$T_\alpha(Q_j) = Q_j' = \alpha_1^{b_{j1}} \ldots \alpha_r^{b_{jr}} Q_j. \qquad (1.132)$$

Thus, each of the Q_j has a dimensionality $q_1^{b_{j1}} \ldots q_r^{b_{jr}}$. Note that Eq. (1.131) is a particular case of Eq. (1.132) when $b_{ik} = \delta_k^i$ and $r = n$.

The transformations T_α are in one-to-one correspondence with vectors $\alpha = (\alpha_1, \ldots, \alpha_r)$ having positive components. In addition, if the following actions are introduced:

$$\alpha\beta = (\alpha_1\beta_1, \ldots, \alpha_r\beta_r), \qquad \alpha^{-1} = (\alpha_1^{-1}, \ldots, \alpha_r^{-1}), \qquad (1.133)$$

the following equations are obviously satisfied:

$$T_\alpha[T_\beta(Q_j)] = T_\beta[T_\alpha(Q_j)] = T_{\alpha\beta}(Q_j), \qquad T_{\alpha^{-1}}[T_\alpha(Q_j)] = Q_j.$$

It follows hence that the equalities (1.132) give a representation of a positive r-vector group defined in accordance with Eqs. (1.133), while the equalities (1.132) specify a group of linear transformations of the space of the vectors Q.

If we set up a matrix in accordance with Eq. (1.128), its rank will be equal to the rank of the matrix $\|b_{ik}\|$, i.e., to the number of quantities with independent dimensionalities. Using the beforementioned generalized π-theorem of group theory, we arrive at the π-theorem formulated in the preceding section for similarity and dimensionality theory. We call the group (1.132) a similarity group.

Of substantial help in the analysis of specific gasdynamics problems are applications of dimensionality theory, and this circumstance will be extensively used hereafter. However, the highly general character of group-theoretical methods makes it possible to extend the results and methods of dimensionality theory to more extensive classes of problems and to obtain new particular solutions of the gasdynamic equations.

7.2. Invariant solutions and their connection with self-similar solutions

Let a system of hydrodynamics equations be invariant with respect to group G_r^9 of a transformation in the space v^i, ρ, p, x^i, t, i.e., the system of hydrodynamic equations is a differential invariant of G_r^9. It is said in this case that the system of hydrodynamic equations admits a group G_r^9. Two trends can be distinguished in the theory of invariant solutions:

(1) The use of known groups (e.g., of the similarity group) to construct invariant solutions.[44,45]

(2) Finding the most extensive Lie group of transformations, which is allowed by the system of hydrodynamics equations, and the complete set of essentially different invariant solutions.[51]

We shall not consider these questions here in detail, but prove only that invariant solutions of the equations of one-dimensional hydrodynamics can be obtained from self-similar ones by elementary transformations and by limiting transitions.

It is shown in Ref. 51 that for arbitrary values of the ratio γ of the specific heats the system of equations (1.18)–(1.20) for one-dimensional motions with cylindrical and spherical symmetry has only the following invariant solutions:

$$v = u(r), \quad \rho = e^{\beta t}R(r), \quad p = e^{\beta t}P(r), \tag{1.134}$$

$$v = r + e^{t}u(\lambda), \quad \rho = e^{(\beta-2)t}R(\lambda), \quad p = e^{\beta t}P(\lambda), \quad \lambda = re^{-t}, \tag{1.135}$$

$$v = \frac{1}{t}u(r), \quad \rho = t^{2\beta+2}R(r), \quad p = t^{2\beta}P(r), \tag{1.136}$$

$$v = \frac{r}{t}u(\lambda), \quad \rho = r^{\beta+2\alpha-2}R(\lambda), \quad p = r^{\beta}P(\lambda), \quad \lambda = tr^{-\alpha}. \tag{1.137}$$

Invariant solutions of the type (1.135) and (1.137) were investigated by K. P. Stanyukovich.[45]

For the case of motions with plane waves, the following are added to these solutions:

$$v = u(t), \quad \rho = e^{\beta r}R(t), \quad p = e^{\beta r}P(t), \tag{1.138}$$

$$v = t + u(\lambda), \quad \rho = e^{\beta t}R(\lambda), \quad p = e^{\beta t}P(\lambda), \quad \lambda = r - \frac{t^2}{2}, \tag{1.139}$$

$$v = \frac{1}{t}u(\lambda), \quad \rho = t^{2\beta+2}R(\lambda), \quad p = t^{2\beta}P(\lambda), \quad \lambda = te^{-r}, \tag{1.140}$$

$$v = \frac{r}{t} + u(\lambda), \quad \rho = t^{\beta}R(\lambda), \quad p = t^{\beta}P(\lambda), \quad \lambda = te^{-r/t}, \tag{1.141}$$

$$v = \frac{r}{t} + u(t), \quad \rho = e^{\beta r/t}R(t), \quad p = e^{\beta r/t}P(t). \tag{1.142}$$

In Eqs. (1.134)–(1.142), α and β are constants.

Using dimensionality analysis methods, L. I. Sedov pointed out a class of self-similar solutions of the equations of one-dimensional gasdynamics, in the form

$$v = \frac{r}{t}V(\lambda), \quad \rho = \frac{a}{r^{k+3}t^{s}}R(\lambda), \quad p = \frac{a}{r^{k+1}t^{s+2}}P(\lambda), \lambda = \frac{r}{bt^{\delta}}. \tag{1.143}$$

Here a and b are dimensional constants while k, s and δ are certain numbers.

Let us establish the connection between the solutions (1.143) and the invariant solutions. We designate as limiting[44,54] those solutions which are obtained from self-similar ones by applying a time-shift (or coordinate-shift) transformation and by taking the limits of the parameters that enter in the self-similar solution (1.143). We have then the following[55]:

Theorem. All invariant solutions of the gasdynamics equations in spherical or cylindrical symmetry are either self-similar or limiting self-similar solutions.

Proof. The solutions (1.137) are in fact self-similar solutions, for by putting $\tilde{\lambda} = (\lambda/b)^{-1/\delta}$ in Eqs. (1.143) and transforming and redesignating the functions $R(\lambda)$, $P(\lambda)$, and $V(\lambda)$ and the factors preceding them, we arrive at a solution of the form (1.137).

Next, making the substitutions $t \to t + t_0$, $t_0 = \delta\tau$, $b = r_0(b\tau)^{-\delta}$, $V = \delta\tilde{V}$, $R = \delta^s\tilde{R}$, and $P = \delta^{s+2}P$, and taking the limit as $\delta \to \infty$, we obtain[44,54] solutions in the form

$$v = \frac{r}{t}\,\tilde{V}(\lambda), \qquad \rho = \frac{a}{r^{k+3}\tau^3}\,\tilde{R}(\lambda), \qquad p = \frac{a}{r^{k+1}\tau^{s+2}}\,\tilde{P}(\lambda), \qquad \lambda = \frac{r}{r_0}\,e^{-t/\tau}.$$

These solutions reduce readily to the form (1.135).

Putting $\delta = 0$, $s = -2\beta - 2$, and $a = b = 1$ in Eqs. (1.143) and redesignating the functions V, P, and R we arrive at solutions in the form (1.136). It remains to obtain solutions in the form (1.134). To this end we again replace in Eqs. (1.143) t by $t + t_0$ and put $t_0 = s\tau$, $a = a_0(\tau_s)^s$, $P = (\tau_s)^2\hat{P}$, $V = s\hat{V}$ and $R = \hat{R}$. We obtain then

$$v = \frac{r\hat{V}}{t/s + \tau}, \qquad \rho = \frac{a_0 r^{-(k+3)}}{(t/\tau s + 1)^s}\,\hat{R}(\lambda), \qquad p = \frac{a_0 r^{-(k+1)}}{(t/s + \tau)^2 (t/\tau s + 1)^s}\,\hat{P}(\lambda)$$

and, taking the limit as $s \to \infty$ and $\delta \to 0$, we get a solution in the form (1.134). This proves the theorem.

This theorem is valid also for solutions of equations of one-dimensional cylindrically symmetric motions in ideal MHD. The corresponding solutions were considered by the author in Ref. 55.

For planar symmetry, solutions of type (1.138) and (1.140) can be obtained as limits of self-similar ones by using a shift transformation with respect to the coordinate r.[44,54] To obtain the other invariant solutions it would be necessary to use also a Galileo group. We shall not consider in detail these limiting transitions and analyze the determination of all the invariant solutions for the case of planar symmetry.

7.3. Adiabatic-motion integrals and the energy integral

Let us consider one-dimensional motions of a gas with the dissipative processes disregarded. Self-similar motion is assumed.

Consider an integral conservation law of the type (1.48),

$$\frac{d}{dt}\int_\Omega \Phi\rho\, d\Omega = \oint_\Sigma \rho\Phi(D - v_n)\, d\Sigma,$$

so that

$$\frac{d}{dt}\int_{\Omega^*} \Phi\rho \, d\Omega = 0.$$

For one-dimensional motions we have

$$\frac{d}{dt}\sigma_\nu \int_{r'}^{r''} \Phi\rho r^{\nu-1}\, dr = \left[\Phi\rho r^{\nu-1}\sigma_\nu\left(\frac{dr}{dt} - v\right)\right]\Bigg|_{r'}^{r''}. \tag{1.144}$$

Let a and b be the significant dimensional constants of the problem, $[a] = ML^kT^s$, $[b] = LT^{-\delta}$.

We can write for the function Φ:

$$\Phi = a^\eta b^\xi G(\lambda), \qquad \lambda = r/bt^\delta.$$

Assume that the value of the integral

$$\mathcal{J} = \int_{r'}^{r''} \Phi\rho r^{\nu-1}\, dr, \qquad r' = bt^\delta\lambda', \qquad r'' = bt^\delta\lambda'' \tag{1.145}$$

is

$$\mathcal{J} = a^x b^y t^\mu K(\lambda', \lambda'').$$

We introduce for the functions v, p, and ρ the dimensionless variables V, P, and R in accordance with Eqs. (1.143).

If the function Φ is such that $\mu = 0$, relation (1.144) leads to the adiabatic integral[56]

$$G(\lambda)R(\lambda)[V-\delta]\lambda^{\nu-(k+1)} = \text{const.} \tag{1.146}$$

For adiabatic motion of a perfect gas we choose the function Φ in the form $(\rho/\rho^\gamma)^z = \exp(sz/c_V)$, where z is a certain number. By comparing dimensionalities, we obtain from Eqs. (1.145) a system of equations for z

$$(1-\gamma)z + 1 = x, \qquad (3\gamma - 1)z = kx + y + 3 - \nu, \qquad -2z = sx - \delta y.$$

The integral (1.146) takes the form

$$\left(\frac{P}{R^\gamma}\right)^z R(-\delta + V)\lambda^{\omega_1} = \varkappa_1,$$

$$\omega_1 = \nu - (k+1)[1 + z(1-\gamma)] + 1 - (k+1)[1 + (1-\gamma)z]. \tag{1.147}$$

This integral was obtained by M. L. Lidov.[57] We proceed now to the MHD equations with $\nu = 1$ and 2. Let

$$h_z = \frac{H_z^2}{8\pi} = ar^{-(k+1)}t^{-(s+2)}G_z,$$

$$h_\varphi = \frac{H_\varphi^2}{8\pi} = ar^{-(k+1)}t^{-(s+2)}G_\varphi, \qquad \mathbf{H} = \mathbf{H}(H_\varphi, H_z), \tag{1.148}$$

$$h_\varphi = 0 \qquad (\nu = 1).$$

In the case of infinite conductivity we obtain then from Eq. (1.146) the freezing integrals first obtained by us in Ref. 58,

$$G_\varphi^m(\lambda R)^{-2m} = \varkappa_2[R(V-\delta)]^{2-s-\delta(k+3)}\lambda^{-(k+5)m+(\nu-k+3)[2-s(k+3)]}, \qquad (1.149)$$

$$G_z R^{-2m} = \varkappa_3[R(V-\delta)]^{2-s-\delta(k+5)}\lambda^{-(2+\nu)s-2(k+3-\nu)},$$
$$m = s+\delta(k+3-\nu) \qquad (\nu = 1, 2). \qquad (1.50)$$

Consider now the MHD equations of a rarefied plasma for the case of one-dimensional cylindrical motions. Naturally, the integrals (1.140) and (1.150) are valid here. Added to them are two adiabatic-invariance integrals.

Let the system of dimensionless variables be given by Eqs. (1.143) and (1.148), with the equations for p replaced here by analogous ones for $p_\perp$ and $p_\parallel$. The adiabatic-invariance integrals are

$$P_\perp^m R^{-m} G^{-m/2} = \varkappa_4[R(V-\delta)]^{-1/2[s+2+\delta(k+s)]}\lambda^{-(s+k+1)}, \qquad (1.151)$$

$$(P_\parallel R^{-3}G)^m = \varkappa_5[R(V-\delta)]^{4-s-\delta(k+7)}\lambda^{-(6s+4k+4)}, \qquad G = G_\varphi + G_z. \qquad (1.152)$$

If $s+2-\delta(\nu-1-k)=0$, the self-similar systems have an energy integral.

This integral takes for the MHD equations the form[58]

$$\lambda^{\nu+2}\left[(P+G)V + (V-\delta)\left(\frac{1}{2}RV^2 + P\frac{1}{\gamma-1} + G\right)\right] = \varkappa_6. \qquad (1.153)$$

For a rarefied plasma, the integral (1.153) ($\nu = 2$) takes the form

$$\lambda^4[(P_\perp + G)V + (V-\delta)(\tfrac{1}{2}RV^2 + P_\perp + \tfrac{1}{2}P_\parallel + G)] = \varkappa_7. \qquad (1.154)$$

Here and elsewhere $\varkappa_i$ ($i = 1, 3, \ldots, 7$) are arbitrary constants.

The energy integral for the gasdynamics equations was first obtained by L. I. Sedov (see Ref. 44).

Assume $\delta = 2/(2+\nu-\omega)$, $s = 0$, and $k = \omega - 3$. Let us write down the adiabaticity and energy integrals for this case. For the adiabaticity integral we have

$$\left[\frac{2}{\gamma-1}\left(\frac{\gamma+1}{2} - \frac{1}{\lambda}\frac{v}{v_2}\right)\right]^{1-\omega\gamma/\nu}\left(\frac{p}{p_2}\right)^{1-\omega/\gamma}\left(\frac{r}{r_2}\right)^{\nu-\omega\gamma} = \left(\frac{\rho}{\rho_2}\right)^{\gamma-1},$$

$$p_2 = \frac{2\rho_1}{\gamma+1}D^2, \qquad v_2 = \frac{2D}{\gamma+1}, \qquad (1.155)$$

$$r_2 = bt^\delta, \qquad D = \frac{dr_2}{dt}, \qquad \rho_2 = \frac{\gamma+1}{\gamma-1}\rho_1.$$

We have chosen here an arbitrary constant $\varkappa_1$ so as to satisfy the condition (1.64) on a strong shock wave.

The energy integral (1.153) can be transformed into

$$\lambda^{\nu-1}\left[(\tilde{P} + \tilde{R}\tilde{V}^2)\left(\lambda - \frac{2}{\gamma+1}\tilde{V}\right) - \frac{2(\gamma-1)}{\gamma+1}\tilde{V}\tilde{P}\right] = \tilde{\varkappa}_6,$$

$$\tilde{p} = \frac{p}{p_2}, \quad \tilde{V} = \frac{v}{v_2}, \quad \tilde{R} = \frac{\rho}{\rho_1}\frac{\gamma-1}{\gamma+1}, \quad \lambda = \frac{r}{r_2}, \quad r_2 = bt^\delta, \quad D = \frac{dr_2}{dt}, \qquad (1.156)$$

where ρ_1 has the dimension of density. Note that p_2, v_2, and ρ_2 are functions corresponding to the conditions on a strong shock wave if r_2 is the shock-wave coordinate.

In the case $\tilde{\chi}_6 = 0$ the integral (1.156) satisfies the conditions on a strong shock wave and takes the form

$$\gamma \tilde{P}\left(\frac{\tilde{V}}{\lambda} - \frac{\gamma+1}{2\gamma}\right) = \tilde{R}\tilde{V}^2\left(\frac{\gamma+1}{2} - \frac{\tilde{V}}{\lambda}\right). \tag{1.157}$$

We note here also that the adiabatic integrals and the energy integral will exist also[44] for motions that lead in the limit to self-similar integrals, i.e., for other invariant solutions of the gasdynamics and MHD equations. They can be obtained, in particular, from Eqs. (1.146) and (1.153) by limiting transitions in analogy with the procedure used above in the investigation of the connection between self-similar and invariant solutions.

8. Formulation of point-blast problems

8.1. Some problems for linear equations of mathematical physics

(A) Wave equation for propagation of an instantaneous pointlike perturbation. A change of the pressure (or density) in a gas having low compressibility is described, in accord with Eqs. (1.32) and (1.33), by a wave equation that can be expressed for one-dimensional flows in the form (1.36):

$$\frac{1}{r^{\nu-1}}\frac{\partial}{\partial r}\left(r^{\nu-1}\frac{\partial p'}{\partial r}\right) = \frac{1}{a_0^2}\frac{\partial^2 p'}{\partial t^2}.$$

For simplicity, let the internal energy of the medium be proportional to the pressure, i.e., $\varepsilon = pf(\rho) + \text{const}$. Then, assuming that $\varepsilon' = \varepsilon - p_0 f(\rho_0)$, we get approximately

$$\varepsilon' = \left(f(\rho_0) + \left.\frac{df}{d\rho}\right|_{\rho=\rho_0}\frac{p_0}{a_0^2}\right)p',$$

i.e., we have for the perturbed internal energy ε' the equation

$$\frac{1}{r^{\nu-1}}\frac{\partial}{\partial r}\left(r^{\nu-1}\frac{\partial \varepsilon'}{\partial r}\right) = \frac{1}{a_0^2}\frac{\partial^2 \varepsilon'}{\partial t^2}. \tag{1.158}$$

Let a finite energy E_0 be released at the time $t = +0$ (instantaneously) in a medium with an initial state p_0, ρ_0, $\varepsilon_0(p_0, \rho_0)$ at a point, along a line, or over a plane (in the last two cases the energy E_0 is given per unit length or unit area, respectively). Assume that the energy-release point, line, or plane has respectively spherical ($\nu = 3$), cylindrical ($\nu = 2$), or planar ($\nu = 1$) symmetry. We assume that all the released at $r = 0$ energy E_0 went to increase the internal energy ε, and disregard the energy fraction used for the macroscopic motion of the medium, i.e., to increase the kinetic energy of the latter. Let the perturbation ε' produced in the medium satisfy Eq. (1.158). Assuming energy conservation, we have

$$E_0 = \int \rho_0 \varepsilon' \, d\Omega, \tag{1.159}$$

where $d\Omega$ is the volume element and the integration is over all of space.

This equation is valid for any $t \geq 0$. Our problem is thus to find the function under the initial conditions $\varepsilon' = 0$, $\partial\varepsilon'/\partial t = 0$ ($r > 0$, $t = +0$) and under the additional integral condition (1.159). Consider a three-dimensional case ($\nu = 3$, $d\Omega = dx^1 dx^2 dx^3$). In terms of generalized functions we can state that the initial data are specified for this problem in the form of a singular generalized function, namely, the Dirac δ function (for the definition and properties of the δ function see, e.g., Refs. 59 and 60)

$$\varepsilon' |_{t=+0} = \frac{E_0}{\rho_0} \delta(\mathbf{r}), \tag{1.160}$$

where $\delta(\mathbf{r})$ is the three-dimensional δ function.

A generalized solution of this problem can be easily expressed in terms of the generalized functions of the basic solution for a wave operator in the three-dimensional case. It takes the form[60]

$$\varepsilon' = \frac{\partial}{\partial t} \frac{E_0}{4\pi\rho_0} e(t) \frac{\delta(a_0^2 t^2 - r^2)}{a_0}, \tag{1.161}$$

where $e(t)$ is the Heaviside function ($e = 0$ for $t < 0$ and $e = 1$ for $t \geq 0$).

The solution (1.161) shows that the perturbation ε' will propagate as a spherical wave of radius $r = a_0 t$, moving with velocity a_0 in the state $\varepsilon' = 0$, and after the passage of this wave the unperturbed state is restored.

In the case of cylindrical symmetry ($\nu = 2$) the solution is

$$\varepsilon' = \frac{\partial}{\partial t} \frac{E_0}{\rho_0} \frac{e(a_0 t - r)}{\sqrt{a_0^2 t^2 - r^2}}. \tag{1.162}$$

It follows from this solution that the perturbation ε' due to the pointlike instantaneously acting source $(E_0/\rho)\delta'(x^1, x^2)\delta(t)$ will occupy at an instant $t > 0$ a cylindrical region of radius $r = a_0 t$. The ensuing perturbation, in contrast to the spatial case, does not vanish behind the wave front, but is preserved at a certain point of space after the passage of the wave. For the planar one-dimensional case ($\nu = 1$) the solution of the problem takes the form

$$\varepsilon' = \frac{E_0}{2\rho_0} \delta(a_0 t - |x^1|) \; (r = x^1). \tag{1.163}$$

It follows from this solution that by the instant of time the perturbation propagates from the source $(E_0/\rho_0)\delta'(x^1)\delta(t)$ in the form of a plane wave

$|x^1| \leq a_0 t$, the leading front of which $|x^1| = a_0 t$ moves with velocity a_0 in a direction perpendicular to the plane $x^1 = 0$. The leading front consists here of two planes, $x^1 = a_0 t$ and $x^1 = -at$, moving with velocity a_0 to the right and left from the initial plane, respectively.

It follows from Eq. (1.163) that the perturbation ε' will be located at the instant t at only two points $x = \pm a_0 t$, so that after passage of the wave front the unperturbed state is restored. Taking into account the equations relating the pressure and velocity, namely $\rho_0(\partial v/\partial t) + (\partial p/\partial r) = 0$, we can obtain the solution of the problem of instantaneous energy release also for the complete system of the acoustics equations, with allowance for conservation of the total acoustic energy expressed in terms of the squared energy E [Eqs. (1.34)].

(B) Problem of instantaneous heat source. As shown in Sec. 1, heat propagation in an immobile medium is described by the heat-conduction equation (1.44), which takes for the case χ = const the form

$$\frac{\partial \varepsilon}{\partial t} = \chi \frac{1}{r^{\nu-1}} \frac{\partial}{\partial r}\left(r^{\nu-1}\frac{\partial \varepsilon}{\partial r}\right). \tag{1.164}$$

Now let a certain amount E_0 of heat be instantaneously released at a time $t = 0$ from a point on a line or on a plane; just as in the preceding section, E_0 is taken to be the per-unit heat. If there are no other heat sources and $\varepsilon = 0$ at $t = +0$ and $r > 0$, we get from the energy-conservation law

$$E_0 = \rho_0 \int_0^{+\infty} \varepsilon \, d\Omega. \tag{1.165}$$

Thus we arrive again at the problem of an instantaneous pointlike δ-function perturbation

$$\varepsilon|_{t=+0} = \frac{E_0}{\rho_0}\,\delta(r). \tag{1.166}$$

The solution of this problem is well known and is given (for $t > 0$) by

$$\varepsilon = \frac{E_0}{\rho_0}\, e(t)\, \frac{1}{(2\sqrt{\chi\pi t})^{\nu}} \exp\left(-\frac{r^2}{4\chi t}\right). \tag{1.167}$$

It follows from the solution (1.167) and from the properties of the δ function[60] that $\varepsilon \to E_0\delta(r)/\rho_0$ as $t \to +0$, i.e., the initial condition (1.166) is met. Also satisfied is relation (1.165). Next, since $\varepsilon(r, t) > 0$ for all r and $t > 0$, the heat propagates in this model with infinite velocity. Note that the solution (1.167) is self-similar.

(C) Basic solution of the transport equation. Assume that there are no radiation sources in the medium (the thermal radiation of the medium is quite small).

The radiation-transport equation (1.112) takes then the form

$$\frac{1}{c}\frac{\partial I_\omega}{\partial t} + \Omega\nabla I_\omega + k_\omega I_\omega = 0, \tag{1.168}$$

$|\Omega| = 1$ and c is the speed of light in the medium. To find the perturbation produced by an instantaneous source located at the origin, we can construct a basic solution that is equivalent from the viewpoint of generalized functions,[60] i.e., a solution of Eq. (1.168) with a right-hand side of the form $\delta(\mathbf{r}, t)$.

A known[60] basic solution of the transport equation (1.168) for k_ω independent of Ω and $\mathbf{r}$ is

$$I_\omega(\Omega, t, \mathbf{r}) = ce(t)e^{-k_\omega ct}\delta(\mathbf{r} - c\Omega t). \tag{1.169}$$

That this is a basic solution can be verified by direct substitution in Eq. (1.168). The solution (1.169) can be found by a Fourier-transformation method in the class of generalized functions.

In the stationary case $\partial I_\omega/\partial t = 0$, the basic solution is transformed into

$$I_\omega(\Omega, \mathbf{r}) = \frac{e^{-k_\omega r}}{r^2}\,\delta\left(\Omega - \frac{\mathbf{r}}{r}\right).$$

8.2. Singular Cauchy problem for the Burgers–Cole–Hopf equation

The nonlinear analog of Eq. (1.164) in the planar case

$$\frac{\partial v}{\partial t} + v\frac{\partial v}{\partial r} = \chi\frac{\partial^2 v}{\partial r^2} \tag{1.170}$$

will be called the Burgers–Cole–Hopf (BCH) equation.

The physical meaning of the BCH equation is that it provides a model-based description of diffusion and viscosity. In particular, the description follows from the equation of motion of a Navier–Stokes system for a viscous liquid at negligibly small pressure gradients.

Assume that we have the initial condition

$$v|_{t=+0} = v_0\delta(r). \tag{1.171}$$

The problem considered is self-similar, for as in the preceding subsection it has only two constants with independent dimensionalities. We introduce the dimensionless variables

$$\lambda = r/\sqrt{\chi t}, \qquad \sqrt{t}v/\sqrt{\chi} = f(\lambda),$$

and obtain from Eq. (1.170)

$$\frac{d^2 f}{d\lambda^2} = \frac{df}{d\lambda}(f - \lambda) - f. \tag{1.172}$$

This equation has a first integral

$$\frac{df}{d\lambda} + \frac{1}{2}f^2 + \lambda f = \text{const} = 0.$$

With the aid of this integral Eq. (1.172) can be integrated completely. Taking into account the initial condition (1.171) or its equivalent $v_0 = \int_{-\infty}^{+\infty} v\, dr$ [where $v(r, 0) = 0, r \neq 0$], we can write the desired solution in the form

$$v = \sqrt{\frac{\chi}{t}}\, e(t) \frac{(e^{\alpha} - 1)e^{-\lambda^2/2}}{\sqrt{\pi} + (e^{\alpha} - 1) \int_{\lambda/2}^{\infty} e^{-\xi^2}\, d\xi}, \qquad \alpha = \frac{v_0}{2\chi}. \quad (1.173)$$

The properties of this solution were investigated in Refs. 61 and 62. It follows from the form of Eq. (1.173) that at small α the solution is close to a thermal wave. For $\alpha \gg 1$ analysis shows that the solution approaches a triangular profile, with a leading front in the form of a shock wave for the equation

$$\frac{\partial v}{\partial t} + v \frac{\partial v}{\partial r} = 0,$$

which propagates in accordance with the law $r = \sqrt{2v_0 t}$. Behind the wave front, v decreases linearly and vanishes at $r = 0$. With solution (1.173) as the example, the perturbation nonlinear-damping plot can be seen to have a single hump in the presence of diffusion. Note that a generalization of the BCH equation (1.170) to include the case of spherically symmetric processes is given in Ref. 63.

It would be possible also to investigate a similar problem for the Korteweg–de Vries equation, which can be written down by replacing the second derivative on the right by the third, i.e., by $\partial^3 v/\partial r^3$. In the linear case (neglecting the term $v\partial v/\partial r$) the solutions are expressed in terms of Airy functions.[61] No complete analysis has been made of the influence of nonlinear and dispersion effects on the solution.

We present now a definition of a point blast.[64]

Definition. A point blast is a process in which finite energy is instantaneously released from a point, or finite specific energy is released along a certain line or along a certain surface of three-dimensional space $\mathscr{E}_3$.

We shall refer to the problem of determining the disturbances produced by a point blast as the point-blast problem (PBP). The examples, that were given show that the PBP can be formulated for different equations and is a special Cauchy problem. We shall present below formulations of the PBP for the gasdynamic and MHD equations.

The problem of a point blast at a point ($\nu = 3$), along a straight line ($\nu = 2$), or along a plane ($\nu = 1$) will be called the fundamental problem. We shall not resort hereafter, as a rule, to the δ-function and step-function formalisms when writing down the initial and boundary conditions or the solution of the problems considered.

8.3. Point blast in an incompressible liquid

The simplest example of a medium for which the fundamental point-blast problem can be solved simply enough is an incompressible liquid. We consider spherical symmetry ($\nu = 3$) and assume that the initial pressure and the initial density are constant, with $\rho = \rho_1$. This problem was first formulated and solved by L. I. Sedov.[44] We describe below the method used in Ref. 4 to solve this problem.

Putting $\rho = \rho_1 = \text{const}$ and $\nu = 3$, we can rewrite the system (1.18)–(1.20) in the form

$$\frac{\partial v}{\partial t} + v\frac{\partial v}{\partial r} + \frac{1}{\rho}\frac{\partial p}{\partial r} = 0, \qquad \frac{\partial v}{\partial r} + 2\frac{v}{r} = 0, \qquad \rho = \rho_i. \tag{1.174}$$

This system is completely integrable, and its solution depends on two arbitrary functions of time, which we designate $\psi_1(t)$ and $\psi_2(t)$:

$$v = \frac{\psi_1(t)}{r^2}, \qquad p = \psi_2(t) - \frac{1}{2}\rho_1 v^2 + \frac{\rho_1}{r}\frac{d\psi_1}{dt}. \tag{1.175}$$

This solution can be used in the PBP for an incompressible liquid. The disturbances propagate in such a liquid with infinite velocity, so that the problem of a blast without a shock wave can be solved in this case. If the pressure in the liquid at the initial instant is p_1 and the liquid is at rest, a pressure equal to p_1 is preserved in the subsequent instants of time at $r = \infty$.

Since we have in the problem three defining dimensional parameters ρ_1, E_0, and p_1, the problem is self-similar if $p_1 = 0$. Let us consider this particular case: At $p_1 = 0$ all the blast energy E_0 goes to increase the kinetic energy of the liquid. Taking this into account, we have

$$E_0 = 4\pi \int_{r_*}^{\infty} \frac{\rho_1 v^2}{2} r^2\, dr,$$

where r_* is a certain radius that determines the perturbed-motion region. Substituting here v from Eqs. (1.175) we get

$$E_0 = 2\pi\rho_1\psi_1^2/r_*. \tag{1.176}$$

It follows from dimensionality considerations that

$$r_* = k\left(\frac{E_0}{\rho_1}\right)^{1/5} t^{2/5}. \tag{1.177}$$

We assume that r_* defines a sphere moving at the same velocity as the liquid particles. Then

$$\frac{dr_*}{dt} = v_* = \frac{\psi_1(t)}{r_*^2}. \tag{1.178}$$

From Eqs. (1.177) and (1.178) we get

$$\psi_1(t) = r_*^2\frac{dr_*}{dt} = \frac{2}{5}\frac{r_*^3}{t}.$$

Using Eq. (1.177), we can obtain

$$E_0 = \frac{8\pi\rho_1}{25}\frac{r_*^5}{t^2}.$$

Substituting here r_* from Eq. (1.177) we get the constant

$$k = \left(\frac{25}{8\pi}\right)^{1/5}.$$

For r_* we obtain finally

$$r_* = \left(\frac{25}{8\pi}\frac{E_0}{\rho_1}\right)^{1/5} t^{2/5} = \left(\frac{5}{2}\right)^{2/5}\left(\frac{E_0}{2\pi\rho_1}\right)^{1/5} t^{2/5}. \qquad (1.179)$$

Taking into account the boundary condition $p(\infty, t) = 0$ at infinity, we get from Eq. (1.175) $\psi_2 = 0$.

The pressure distribution is given by

$$\frac{p}{\rho_1} = \frac{2}{25}\frac{r_*^3}{t^2 r}\left[1 - \left(\frac{r_*}{r}\right)^3\right].$$

For $r = r_*$ we have $p = 0$. This means that an empty sphere is produced at the symmetry center—a cavity of radius r_* that expands with time in accordance with Eq. (1.179). The maximum pressure p_2 in the space is reached at $r = r_2 = 4^{1/3} r_*$. The time dependence of p_2 is determined from the equation

$$p_2 = \frac{3}{50}\left(\frac{5}{2}\right)^{4/5} 4^{-1/3}\rho_1\left(\frac{E_0}{2\pi\rho_1}\right)^{2/5} t^{-6/5} = 4^{-4/3}\frac{3E_0}{\pi} r_2^{-3}. \qquad (1.180)$$

We have thus obtained an exact solution of the problem of a strong blast in an incompressible liquid.

Having the general solution (1.175) of the system (1.174), we can obtain the solution of the non-self-similar PBP for $p_1 \neq 0$. The boundary and initial conditions of this problem are $p = p_1 = \text{const}$, $v = v_1 = 0$, and $r_* = 0$ for $t = 0$ and an energy release E_0 from the center of the blast:

$$\text{for } r = r_* \quad p = p_* = 0, \quad v = v_* = \frac{dr_*}{dt}, \qquad (1.181)$$

$$\text{for } r = \infty \quad p = p_\infty = p_1, \quad v = v_\infty = 0.$$

From Eqs. (1.175) and (1.178) we obtain, taking Eq. (1.181) into account,

$$\frac{p - p_1}{\rho_1} = -\frac{1}{2}\frac{r_*^4 (r'_*)^2}{r^4} + \frac{(r_*^2 r'_*)'}{r}, \qquad v = \frac{r_*^2 r'_*}{r^2}. \qquad (1.182)$$

The prime denotes differentiation with respect to time. For $r = r_*$ we get

$$\frac{p_1}{\rho_1} = \frac{1}{2}(r'_*)^2 - \frac{(r_*^2 r'_*)'}{r^*}. \qquad (1.183)$$

This equation gives the motion $r_*(t)$ of the boundary of the cavity produced by the blast. Integration yields

$$(r'_*)^2 = k_2 r_*^{-3} - \frac{2}{3}\frac{p_1}{\rho_1}. \qquad (1.184)$$

We determine the constant k_2 from the condition that the solution be self-similar for $p_1 = 0$. It turns out that

$$k_2 = \frac{E_0}{2\pi p_1}.$$

We define as the maximum cavity radius $r_{\max}$ that value of r_* at which $dr_*/dt = 0$. We get from Eq. (1.184)

$$r_{\max} = \left(\frac{3}{4\pi}\frac{E_0}{\rho_1}\right)^{1/3}.$$

In terms of the dimensionless variables

$$l = \frac{r}{r_{\max}}, \qquad \tau = \left(\frac{2}{3}\frac{p_1}{\rho_1}\right)^{1/2}\frac{t}{r_{\max}}, \qquad l_* = \frac{r_*}{r_{\max}},$$

Eq. (1.184) takes the form

$$\frac{dl_*}{d\tau} = \pm\frac{(1 - l_*^3)^{1/2}}{l_*^{3/2}}. \qquad (1.185)$$

The function $\tau(l_*)$ is then determined by the quadrature

$$\tau(l_*) = \pm\int_0^{l_*} \frac{l_*^{3/2}dl}{(1 - l_*^3)^{3/2}}. \qquad (1.186)$$

Equations (1.185) and (1.186) show that $l \leq 1$, and since τ increases, the cavity begins to move towards the center after reaching its maximum radius; the result is a distinctive oscillatory motion. Then $dl_*/d\tau$ goes through zero and becomes negative, i.e., the plus and minus signs in Eq. (1.186) correspond to motion away from and towards the center, respectively. After determining the cavity-boundary motion, the pressure and the velocity are obtained from the equations

$$\frac{p}{p_1} = 1 + \frac{1}{3}\left[2\frac{d}{d\tau}\left(l_*^2\frac{dl_*}{dt}\right) - l_*^4\left(\frac{dl_*}{d\tau}\right)^2 l^{-3}\right]l^{-1}, \qquad \frac{v}{v_*}\frac{l_*^2}{l^2}.$$

Finally, the period τ of the cavity oscillation is then

$$\tau = 2\tau(1) = 2\int_0^1 \frac{l_*^{3/2}dl_*}{(1 - l_*^3)^{1/2}}.$$

The change of variable $t_* = x^{1/3}$ transforms this integral into a Euler integral of form $B(p, q)$, which is expressed in terms of the Euler Γ function:

$$\tau = \frac{2}{3}\int_0^1 x^{-1/6}(1-x)^{-1/2}\,dx = \frac{2}{3}B\left(\frac{5}{6},\frac{1}{2}\right) = 2\sqrt{\pi}\,\frac{\Gamma(5/6)}{\Gamma(1/3)} \approx 1.5.$$

Changing to dimensional quantities, we obtain the following equations for the period of the oscillations:

$$T = 1.14\rho_1^{1/2}E_0^{1/3}p_1^{-5/6}.$$

8.4. Point blast in a gas

Let an energy E_0, per unit length or per unit area, be released at a point along a line or a plane from a linear or planar energy source, respectively. The initial state of the gas is strongly disturbed locally as a result. Let the undisturbed state of the gas be independent of time and let $\mathbf{v}_0 = \mathbf{v}(\mathbf{r})$, $p_0 = p_0(\mathbf{r})$, and $\rho_0 = \rho_0(\mathbf{r})$ at $t = 0$. Denoting by E the total energy per unit volume of gas, we obtain for $t \geq 0$ by virtue of the energy conservation law $E_0 = \int \Delta E\, d\Omega$, where $\Delta E = E - E^0$, $E^0 = (\rho_0 v_0^2/2) + \rho_0\varepsilon_0$ is the initial gas energy, and the integration is over all of space. Thus, just as in the cited examples, we arrive at the singular Cauchy problem

$$\mathbf{v}|_{t=0} = \mathbf{v}_0, \qquad p|_{t=0} = p_0, \qquad \rho|_{t=0} = \rho_0, \qquad \Delta E|_{t=t+0} = \delta(\mathbf{r})E_0.$$

Experiment and theory show that a blast produces in a gas a shock wave propagating in the gas from the blast point. From the experimental standpoint it is known that if the blast energy E_0 is high enough (say higher than the initial internal energy per unit gas volume for $\nu = 3$), and the volume in which the energy is released is significantly smaller than the volume in which E_0 becomes comparable with its initial energy, a shock wave propagates in the gas as a result of the abrupt increase of the pressure and temperature at the blast site.

It is known from theory, first, that even a Cauchy problem with smooth initial data can have discontinuous solutions,[36,37] all the more so if the initial data are specified in the form of a δ function for the energy source. Second, L. I. Sedov has shown[65] (see also Chap. 2) that the problem of a strong blast in a gas (i.e., in a gas with zero initial pressure) cannot have continuous solutions. The solution sought will thus in fact be a generalized solution in which the continuity regions will be separated by discontinuity surfaces.

Assume an equation $F(\mathbf{r}, t) = 0$ for a shock-wave front corresponding to the leading front of the perturbation in the gas. The conditions (1.53) and (1.54) should then be met on going through this front. The problem posed above can be formulated as a boundary-value problem with conditions (1.53) and (1.54) on a surface $F = 0$, and this surface contracts to a point or to a straight line, or else tends to the blast plane as $t \to 0$. In addition to these conditions we have the integral energy-conservation law, which we now write in the form

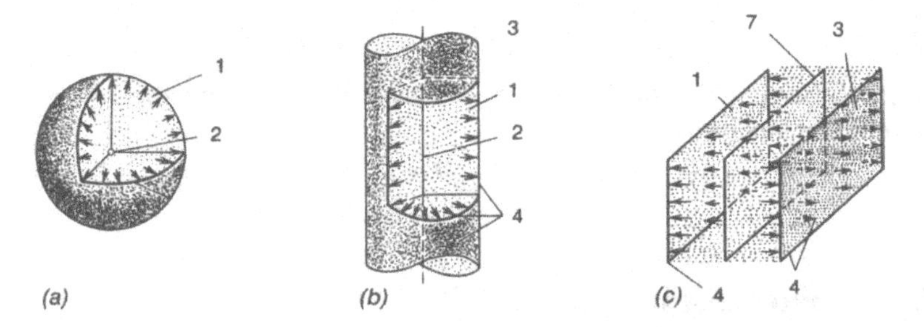

Figure 2. Motion of compressible medium following a blast: (a) spherical charge, (b) cylindrical charge, (c) planar charge. Key: (1) moving medium, (2) symmetry center (of blast), (3) medium at rest, (4) shock-wave front, (5) moving medium, (6) symmetry axis (of blast), (7) symmetry plane (of blast).

$$E_0 = \int_{\Omega(t)} \Delta E \, d\Omega, \tag{1.187}$$

where $\Omega(t)$ is the volume occupied by the perturbation as a result of the blast by the gas inside the shock-wave front.

Let us examine in greater detail the formulation of the fundamental problem for the case of a blast in a perfect gas at rest, when p_1 = const, $\rho_1 = \rho(r)$, and the gas motion behind the shock wave satisfies the adiabaticity condition. The motion is then one-dimensional and Eq. (1.187) takes the form

$$E_0 = \sigma_\nu \int_0^{r_2} \left(\frac{\rho v^2}{2} + \frac{p - p_1}{\gamma - 1} \right) r^{\nu - 1} \, dr, \tag{1.188}$$

where $\sigma_\nu = 2(\nu - 1)\pi + (\nu - 2)(\nu - 3)$, r_2 is the shock-wave radius, and the conditions (1.63) take place on the shock wave, i.e., at $r = r_2$.

At the initial instant of time $t = 0$ and for $r > 0$ we have $p = p_1$ = const, $\rho = \rho_1(r)$ $v_1 = 0$, and the additional conditions $r_2(0) = 0$ and that an energy E_0 is released at the symmetry center. To determine the functions sought we must integrate the system of partial differential equations (1.18)–(1.20) with the indicated initial conditions and with the conditions on the shock-wave front. If there is no mass source at the symmetry center (i.e., at $r = 0$), it follows from the conditions of the problem by virtue of the symmetry (see Refs. 4 and 44) that

$$v(0, t) = 0. \tag{1.189}$$

The problem is similarly formulated in Lagrange variables. The gas motion for a point blast is shown for one-dimensional cases in Fig. 2.

The PBP in a gas was first formulated and investigated by L. I. Sedov and by G. I. Taylor.[66,67] Among the first to investigate this problem were also K. P. Stanyukovich[68] and J. von Neumann.[69] The PBP in a gas can be naturally generalized to include motion of a gas with chemical reactions and for various types of nonadiabatic motions. If the medium in which the blast takes place is electrically conducting and furthermore an external electromagnetic field is applied, a point blast in such a medium will be accompanied by interaction of the magnetic field with the motion. The general formulation of the fundamental problem remains here similar to that for the case of blast in a gas without allowance for electromagnetic effects. The picture of the gas motion, however, becomes much more complicated. The main purpose of the analysis that follows is an investigation of the gasdynamic problems mentioned in the present section and certain applications of the results of these investigations to physical phenomena. We shall discuss also some generalizations of the fundamental problem to include asymmetric energy release by a blast.

Chapter 2
Strong blast in a gas

1. L. I. Sedov's exact solution of the self-similar problem of a strong blast in a perfect gas

A point blast produces in a gas a strong shock wave with a pressure p_2 behind its front much higher than the initial pressure p_1. If the initial gas pressure p_1 is neglected and the initial density is assumed to vary as

$$\rho_1 = Ar^{-\omega}, \tag{2.1}$$

where A is a constant with dimensionality $[A] = ML^{\omega-3}$ and ω an abstract constant, then the BPB in a perfect gas is self-similar.[1,2] In fact, among the defining parameters of the problem there are only two constants, E_0 ($[E_0] = ML^{\nu-1}T^{-2}$) and A, with independent dimensionalities.

Since it is self-similar, the system of gasdynamic equations that describe the disturbed motion of the gas behind a shock wave is transformed into a system of ordinary differential equations. Introducing the dimensionless variables

$$f = \frac{v}{D}, \qquad g = \frac{\rho}{\rho_1}, \qquad h = \frac{p}{\rho_1 D^2}, \qquad \lambda = \frac{r}{r_2}, \tag{2.2}$$

we obtain for the new functions sought, $f(\lambda)$, $g(\lambda)$, and $h(\lambda)$, from the gasdynamics equations (1.18)–(1.20) a system of three ordinary equations

$$\begin{gathered} g(f-\lambda)f' + h' + \frac{\omega-\nu}{2} fg = 0, \\ (f-\lambda)g' + \left(f' + \frac{\nu-1}{\lambda} f - \omega\right)g = 0, \\ (f-\lambda)h' + \gamma\left(f' + \frac{\nu-1}{\lambda} f\right)h - \nu h = 0. \end{gathered} \tag{2.3}$$

The boundary conditions on the front of the strong shock wave take the form

$$f(1) = \frac{2}{\gamma+1}, \qquad g(1) = \frac{\gamma+1}{\gamma-1}, \qquad h(1) = \frac{2}{\gamma+1}. \tag{2.4}$$

L. I. Sedov obtained[1,2] the energy integral (1.57) for the system (2.3), and this made possible an exact solution of the fundamental equation at zero backpressure. In Euler variables, this solution, with allowance for the conditions (2.4), can be written in the form[2–4]

$$\frac{r}{r_2} = \lambda(\mu) = \mu^{-\delta}\left[\frac{2\gamma}{\gamma-1}\left(\mu - \frac{\gamma+1}{2\gamma}\right)\right]^{\beta_2} \times \left\{\frac{2(\nu\gamma-\nu+2)}{3\nu-2-\gamma(\nu-2)-\omega(\gamma+1)}\left[\frac{\gamma+1}{\delta(\nu\gamma-\nu+2)} - \mu\right]\right\}^{-\beta_1}, \tag{2.5}$$

$$\frac{v}{v_2} = \frac{\gamma+1}{2} f = \lambda\mu, \tag{2.6}$$

$$\frac{\rho}{\rho_2} = \frac{\gamma-1}{\gamma+1} g(\mu) = \mu^{\delta\omega}\left[\frac{2\gamma}{\gamma-1}\left(\mu - \frac{\gamma+1}{2\gamma}\right)\right]^{\beta_3-\omega\beta_2} \times \left[\frac{2}{\gamma-1}\left(\frac{\gamma+1}{2} - \mu\right)\right]^{-\beta_5} \times \left\{\frac{2(\nu\gamma-\nu+2)}{3\nu-2-\gamma(\nu-2)-\omega(\gamma+1)}\left[\frac{\gamma+1}{\delta(\nu\gamma-\nu+2)} - \mu\right]\right\}^{\beta_4-\omega\beta_1}, \tag{2.7}$$

$$\frac{p}{p_2} = \frac{\gamma+1}{2} h = \mu^{\delta\nu}\left[\frac{2}{\gamma-1}\left(\frac{\gamma+1}{2} - \mu\right)\right]^{1-\beta_5} \times \left\{\frac{2(\nu\gamma-\nu+2)}{3\nu-2-\gamma(\nu-2)-\omega(\gamma+1)}\left[\frac{\gamma+1}{\delta(\nu\gamma-\nu+2)} - \mu\right]\right\}^{\beta_4-(\omega-2)\beta_1}, \tag{2.8}$$

$$\frac{T}{T_2} = \theta(\mu) = \mu^{\delta(\nu-\omega)}\left[\frac{2\gamma}{\gamma-1}\left(\mu - \frac{\gamma+1}{2\gamma}\right)\right]^{\omega\beta_2-\beta_3}\left[\frac{2}{\gamma-1}\left(\frac{\gamma+1}{2} - \mu\right)\right] \times \left\{\frac{2(\nu\gamma-\nu+2)}{3\nu-2-\gamma(\nu-2)-\omega(\gamma+1)}\left[\frac{(\nu+2-\omega)(\gamma+1)}{2(\nu\gamma-\nu+2)} - \mu\right]\right\}^{-2\beta_1}, \tag{2.9}$$

where

$$\beta_1 = \beta_2 + \frac{\gamma+1}{\nu\gamma-\nu+2} - \delta, \qquad \beta_2 = \frac{\gamma-1}{2\gamma-2+\nu-\omega\gamma},$$
$$\beta_3 = \frac{\nu-\omega}{2\gamma-2+\nu-\omega\gamma} = 1 + (\omega-2)\beta_2, \qquad \delta = \frac{2}{\nu+2-\omega}, \tag{2.10}$$
$$\beta_4 = \frac{2(\nu-\omega)\beta_1}{\delta(2\nu-\nu\gamma-\omega)}, \qquad \beta_5 = \frac{2\nu-\omega(\gamma+1)}{2\nu-\nu\gamma-\omega}.$$

The values of the density, pressure, and temperature directly behind the shock-wave front, i.e., at $\lambda = 1 - 0$, are obtained from the equations

$$v_2 = \frac{2\delta}{\gamma+1}\left(\frac{E_0}{\alpha A}\right)^{1/2} r_2^{\frac{\omega-\nu}{2}} = \frac{2\delta}{\gamma+1}\left(\frac{E_0}{\alpha A}\right)^{\delta/2} t^{\frac{\omega-\nu}{\nu+2-\omega}},$$

$$\rho_2 = \frac{\gamma+1}{\gamma-1} A r_2^{-\omega}, \qquad p_2 = \frac{2E_0}{\alpha}\delta^2\frac{r_2^{-\nu}}{\gamma+1} = \frac{2\delta^2 A}{\gamma+1}\left(\frac{E_0}{\alpha A}\right)^{\frac{2-\omega}{\nu+2-\omega}} t^{-\nu\delta}, \tag{2.11}$$

$$RT_2 = \frac{p_2}{\rho_2} = \frac{2(\gamma-1)}{(\gamma+1)^2}\delta^2\frac{E_0}{\alpha A} r_2^{\omega-\nu} = \frac{2(\gamma-1)}{(\gamma+1)^2} D^2, \qquad \delta = \frac{2}{\nu+2-\omega}.$$

For the shock-wave velocity and its coordinates we have the relations

$$D = \frac{dr_2}{dt} = \delta\frac{r_2}{t} = \delta\left(\frac{E_0}{\alpha A}\right)^{1/2} r^{\frac{\omega-\nu}{2}} = \delta\left(\frac{E_0}{\alpha A}\right)^{\delta/2} t^{\frac{\omega-\nu}{\nu+2-\omega}}, \tag{2.12}$$

$$r_2 = \left(\frac{E_0}{\alpha A}\right)^{\delta/2} t^{\delta}, \quad \delta = \frac{2}{\nu+2-\omega}. \tag{2.13}$$

The solution (2.1)–(2.5) describes the behavior of the gasdynamic functions behind the front of a shock wave propagating with variable velocity in a variable density immobile medium. From the structure of this solution it follows that it is expressed in terms of the parametric variable μ, with $\lambda(1) = 1$. The parameter α in the solution of the problem depends on γ, ω, and ν and is determined from the integral energy-conservation law

$$E_0 = \sigma_\nu \int_0^{r_2} \left(\frac{p}{\gamma-1} + \frac{\rho v^2}{2}\right) r^{\nu-1}\, dr. \tag{2.14}$$

Taking Eqs. (2.1)–(2.13) into account, we obtain from the integral energy-conservation law (2.14) an equation for $\alpha(\nu, \gamma, \omega)$:

$$\alpha(\nu,\gamma,\omega) = \frac{\delta^2\sigma_\nu}{\gamma-1}\int_0^1 \left(h + \frac{(\gamma-1)}{2} g f^2\right)\lambda^{\nu-1}\, d\lambda. \tag{2.15}$$

It follows from Eq. (2.15) that α is a functional of the solution.

If we take the Lagrangian coordinate to be the particle coordinate ξ, and recognize that $r_2 = \xi$ on the shock-wave front, we obtain from the conditions on a strong shock wave and from Eqs. (2.11) and (2.12) the following expression for the quantity p/ρ^γ, which is conserved in the particle:

$$\frac{p}{\rho^\gamma} = \frac{4\delta(\gamma-1)^\gamma E_0}{\alpha(\gamma+1)^{\gamma+1} A^\gamma \xi^{\nu-\omega\gamma}}. \tag{2.16}$$

The solution (2.5)–(2.10) and the relation (2.16) make it possible to find the connection between the Lagrangian coordinate ξ, the time t, and the parametric variable μ:

$$\xi = r_2\mu^{-\delta}\left[\frac{2\gamma}{\gamma-1}\left(\mu - \frac{\gamma+1}{2\gamma}\right)\right]^{\beta_6}\left[\frac{2}{\gamma-1}\left(\frac{\gamma+1}{2} - \mu\right)\right]^{\beta_8}$$
$$\times\left\{\frac{2(\nu\gamma-\nu+2)}{3\nu-2-\gamma(\nu-2)-\omega(\gamma+1)}\left[\frac{\gamma+1}{\delta(\nu\gamma-\nu+2)} - \mu\right]\right\}^{\beta_7}, \tag{2.17}$$

where

$$\beta_6 = \frac{\gamma}{2\gamma-2+\nu-\omega\gamma}, \quad \beta_7 = \frac{\nu\gamma-\nu+2}{2\nu-\nu\gamma-\omega}\beta_1,$$
$$\beta_8 = \frac{\gamma}{\nu\gamma-2\nu+\omega}. \tag{2.18}$$

Once the exact analytic solution is obtained, the problem arises of investigating its properties as functions of its parameters, and of determining the values of α.

We note here only certain known important properties of the solution[2,3] at a constant initial density ρ_1 = const and for $\gamma = 1.4$. For the case of constant density, the parameter μ ranges from 1 to $(\gamma+1)/2\gamma$ if the solution can be continued to the center, with $(\gamma+1)/2\gamma$ at the symmetry center.

An investigation of the distributions of the velocities, densities, pressures, and temperatures leads to the following conclusions:

(1) Since $v/v_2 = \mu\lambda$, and the parameter μ, with values on the segment $[(\gamma+1)/2\gamma, 1]$ at $\gamma = 1.4$ not differing greatly from unity, the velocity of the air particles at a fixed instant of time is approximately proportional to their distances r from the symmetry center.

(2) The density ρ decreases very rapidly in the direction from the shock wave to the symmetry center. Since $\rho_2 = (\gamma+1)/(\gamma-1)\rho_1$ on the shock-wave center, i.e., the density behind the front at $\gamma = 1.4$ is six times larger than the initial density, it follows that the bulk of the moving gas is concentrated in a rather narrow layer behind the shock-wave front.

(3) For $t > 0$ the pressure at the symmetry center is always finite and amounts to a certain fraction of its value on the shock wave. It is also typical that up to rather large distances (approximately $0.5r_2$) from the symmetry center, the pressure is constant along r: $p(r, t) \approx p(0, t)$, $r \leq 0.5r_2$. This region, in which the pressure is constant for fixed t, is sometimes called the "pressure plateau."

(4) The temperature tends to infinity as the symmetry center is approached. This is due to neglect of the thermal conductivity, which influences strongly the temperature distribution in the vicinity of the center.

The peculiarities of the spatial distributions of the velocity, pressure, and density, noted above for the case $\gamma = 1.4$, follow from the solution (2.2)–(2.10) and from the plots of the fundamental functions, shown in Figs. 3–6 (see Sec. 2).

2. Dependence of exact solution on the parameters ω, γ, and ν

2.1. Dependence of the solution on the parameters γ and ν for a constant initial density

The following important facts were established earlier[2] in investigations of the dependences of the solution (2.2)–(2.10) on γ and ν at $\omega = 0$. For spherical symmetry ($\nu = 3$) and $\gamma = 7$ the pressure and density vanish at the symmetry center. If $\gamma \geq 7$, there is produced near the center a spherical cavity, where $p = 0$ and $\rho = 0$, and as γ increases the solution tends to the solution, considered in the preceding chapter, for an incompressible liquid. If $\nu = 1$ or 2, the pressure near the center is not equal to zero for any finite γ. No complete answers, however, have been found for the detailed behavior of the basic functions as γ varies in the range $1 < \gamma \leq 7$ for all ν, and for the detailed behavior of the solution when $\gamma > 7$.

We have investigated in detail the dependences of the solution on γ for $1 < \gamma \leq 7$, and found the following: First, the solution (2.2)–(2.10) has a continuous dependence on γ. Analysis of this solution shows that the functions in it are continuous in γ everywhere except at the point $\gamma = 2$, at which the functions g and h have not been determined. If a finite limit for the solution exists as $\gamma \to 2$, we can additionally determine the functions g and h and set them equal at the point $\gamma = 2$ to their limiting values.

Further, the solution of the problem for $\gamma = 2$ can be determined directly from the differential equations and the boundary values. If the resultant solution agrees with the functions obtained by taking the limit, this means a continuous dependence of the solution at the point $\gamma = 2$. Since the solution is continuously dependent on γ at all other values $\gamma \in [1, 7]$, the solution will have a continuous dependence on γ.

Let us consider the limiting transition with respect to γ for the function $g(\mu)$. It follows from Eq. (2.7) that at $\omega = 0$

$$g(\mu) = \frac{\gamma+1}{\gamma-1}\left[\frac{2\gamma}{\gamma-1}\left(\mu - \frac{\gamma+1}{2\gamma}\right)\right]^{\beta_3}\left[\frac{2}{\gamma-1}\frac{\gamma+1}{2} - \mu\right]^{-\beta_5}$$
$$\times\left\{\frac{2(\nu\gamma-\nu+2)}{3\nu-2-\gamma(\nu-2)}\left[\frac{\gamma+1}{\delta(\nu\gamma-\nu+2)} - \mu\right]\right\}^{\beta_4}. \tag{2.19}$$

By simple transformations we can reduce this equation to the form

$$g(\mu) = \frac{\gamma+1}{\gamma-1}\left[\frac{2\gamma}{\gamma-1}\left(\mu - \frac{\gamma+1}{2\gamma}\right)\right]^{\beta_3}\left[\frac{2(\nu\gamma-\nu+2)}{3\nu-2-\gamma(\nu-2)}\right.$$
$$\left.\times\left(\frac{\gamma+1}{\delta(\nu\gamma-\nu+2)} - \mu\right)\right]^{\beta_4-\beta_5}$$
$$\times\left[\frac{(\gamma-1)(\nu\gamma-\nu+2)}{3\nu-2-\gamma(\nu-2)}\,\frac{[(\nu+2)(\gamma+1)/2(\nu\gamma-\nu+2)] - \mu}{(\gamma+1)/2-\mu}\right]^{\beta_5}.$$

The last factor can be transformed into

$$\left[1+\frac{1}{x}\right]^{-\frac{2\nu(\gamma+1)}{\nu+2}\frac{1-\mu}{[(\gamma+1)/2]-\mu}x},$$

where

$$x = \left(\frac{\gamma+1}{2} - \mu\right)(\nu+2)[(1-\mu)\nu(\gamma+1)(2-\gamma)]^{-1}.$$

Thus we seek the limit of the following expression:

$$\lim_{\gamma\to 2} g(\mu) = 3 \lim_{\gamma\to 2}\left\{\left[\frac{2\gamma}{\gamma-1}\left(\mu-\frac{\gamma+1}{2\gamma}\right)\right]^{\beta_3}\right\}$$

$$\times \lim_{\gamma\to 2}\left\{\frac{2(\nu\gamma-\nu+2)}{3\nu-2-\gamma(\nu-2)}\left(\frac{(\nu+2)(\gamma+1)}{2(\nu\gamma-\nu+2)}-\mu\right)^{\beta_4-\beta_5}\right\}$$

$$\times \lim_{\substack{x\to\infty\\ \gamma\to 2}}\left(1+\frac{1}{x}\right)^{-\frac{2\nu(\gamma+1)}{\nu+2}\frac{1-\mu}{[(\gamma+1)/2]-\mu}x}.$$

Taking the limit, we obtain

$$g(\mu) = 12\left(\mu-\frac{3}{4}\right)^{\nu/(\nu+2)}\left[2\left(\frac{3}{2}-\mu\right)\right]^{(\nu-2)/(\nu+2)}$$
$$\times \exp\left[-\frac{6\nu}{\nu+2}\frac{1-\mu}{3/2-\mu}\right]. \tag{2.20}$$

We can similarly obtain the limits of $h(\mu)$ and $\lambda(\mu)$, with the limiting transition for $\lambda(\mu)$ quite trivial. For $h(\mu)$ we get an expression similar to Eq. (2.20).

If, however, we substitute $\gamma = 2$ in the conditions (2.4) on the shock wave and integrate the system (2.3) with the conditions (2.4), we obtain the solution

$$\lambda = \mu^{-2/(\nu+2)}\left[2\left(\frac{3}{2}-\mu\right)\right]^{-2/(\nu+2)}\left[4\left(\mu-\frac{3}{4}\right)\right]^{1/(\nu+2)},$$

$$f = \frac{2}{3}\lambda\mu, \qquad g = g(\mu), \tag{2.21}$$

$$h = \frac{2}{3}\mu^{2\nu/(\nu+2)}\left[2\left(\frac{3}{2}-\mu\right)\right]^{2(\nu-2)/(\nu+2)}\exp\left[-\frac{6\nu}{\nu+2}\frac{1-\mu}{3/2-\mu}\right],$$

where the expression for $g(\mu)$ coincides with Eq. (2.20). From a comparison of the solution (2.21) with the solution obtained by taking the limit as $\gamma\to 2$ it follows that these two solutions are equal. Similarly, the limiting values of the functional $\alpha(\nu,\gamma)$ will equal the corresponding values obtained for the functions (2.21). It follows hence that the solution (2.5)–(2.13) is continuous in γ.

We have thus proved the following:

Theorem. The solution of the fundamental problem of a strong blast in a gas of constant initial density has a continuous dependence on the adiabatic exponent γ *in the range from 1 to 7.*

This result was first published in Ref. 5.

Note also, for the case $\gamma = 2$, an equation for the dimensionless pressure at the center of the blast, i.e., at $\mu = \mu_0 = \frac{3}{4}$:

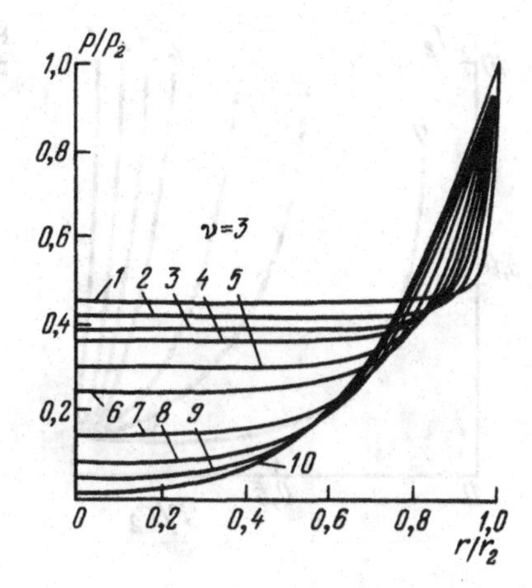

Figure 3. Distribution of the relative pressures for different γ: 1 – γ = 1.1; 2 – 1.2; 3 – 1.3; 4 – 1.4; 5 – $\frac{5}{3}$; 6 – 2; 7 – 3; 8 – 4; 9 – 5; 10 – 6.

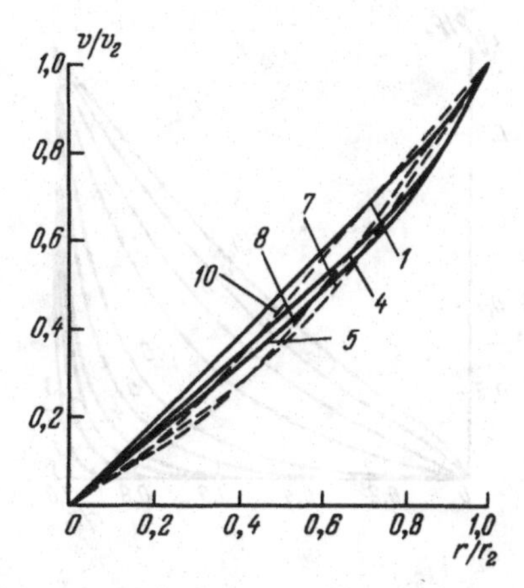

Figure 4. Distribution of relative velocities for different γ (see Fig. 3 for the labels).

$$h(\mu_0) = \tfrac{2}{3} 3^{4(\nu-1)/(\nu+2)} 4^{(2-3\nu)/(\nu+2)} e^{-2\nu/(\nu+2)}. \tag{2.22}$$

It follows from this equation that the pressure at the center is finite for all values of ν and differs from zero. Our analysis dispels all doubts concerning the behavior of a solution in the form (2.5)–(2.14) as $\gamma \to 2$, and yields for the problem at $\gamma = 2$ a complete solution that will be used in the investigation of MHD problems.

The influence of γ on the functions $f(\lambda)$, $g(\lambda)$, $h(\lambda)$, and $\alpha(\nu, \gamma)$ was investigated. Since the exact analytic solution of the strong-blast problem is given by rather complicated analytic equations, tables of self-similar functions were compiled.[6] These tables list the values of the functions sought for the following values of γ: 1.1, 1.2, 1.3, 1.4, $\frac{5}{3}$, 2, and 3. Self-similar functions were also calculated for γ = 4, 5, and 6.

Figures 3–6 show for p/p_2, v/v_2, and ρ/ρ_2 as functions of λ for $\nu = 3$ plots that describe the behavior of the solution for a wide range of γ. It is possible to trace here the very subtle variation of the curvature of the plot of ρ/ρ_2 for $4 < \gamma < 7$ as λ changes from zero to unity.

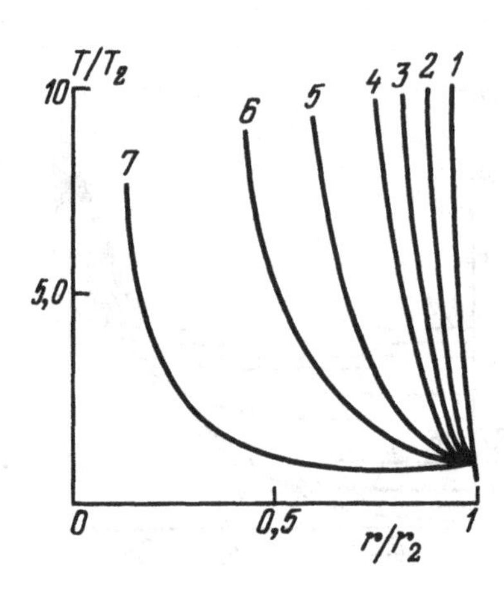

Figure 5. Distribution of relative temperatures for different γ (see Fig. 3 for the labels).

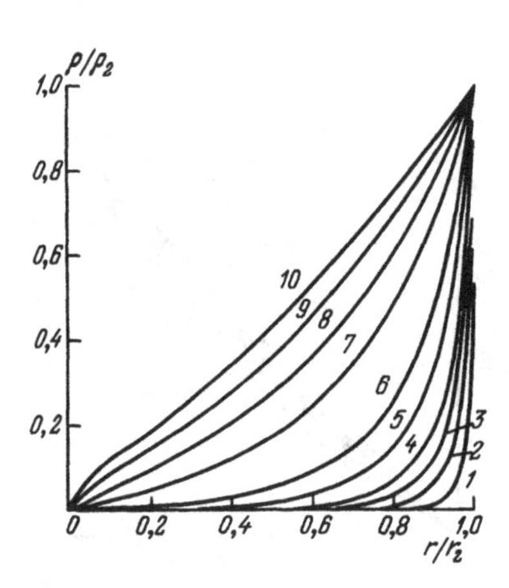

Figure 6. Distribution of the relative densities for different γ (see Fig. 3 for the labels).

As already mentioned, the solution contains the energy parameter $\alpha(\nu, \gamma)$ which is a functional of the solution (2.5)–(2.10) at $\omega = 0$. Knowledge of the numerical values of α is essential for applications of the theory to physical problems. The most accurate numerical values of α were obtained earlier[2,3] for $\gamma = 1.4$. We have calculated the parameter α with high accuracy for a wide range of γ and for $\nu = 1, 2,$ and 3.

To illustrate the range of the numerical values, we present the upper- and lower-bound estimates of this parameter. To this end, we write the equation for α in the form

$$\alpha(\nu, \gamma) = \frac{8\sigma_\nu}{(\nu + 2)^2(\gamma^2 - 1)} \int_0^1 \left[\frac{p}{p_2} + \frac{\rho}{\rho_2}\left(\frac{v}{v_2}\right)^2\right] \lambda^{\nu-1}\, d\lambda.$$

Since $p/p_2 \le 1$, $\rho/\rho_2 \le 1$, and $v/v_2 \le 1$, we obtain, replacing these quantities by unity, the following upper bound of the parameter:

$$\alpha < \alpha_1 = \frac{16\sigma_\nu}{\nu(\nu + 2)^2(\gamma^2 - 1)}.$$

If, however, we take p/p_2, ρ/ρ_2, and v/v_2 to be equal to their minimum values at $\lambda = 0$, we obtain the lower bound of the integral

$$\alpha > \alpha_2 = \frac{8\sigma_\nu h(0)}{(\nu+2)^2 2\nu(\gamma-1)}.$$

We have taken into account here that $v(0, t) = \rho(0, t) = 0$ for $\gamma \leq 7$. We have thus

$$\frac{4\sigma_\nu h(0)}{\nu(\gamma-1)(\nu+2)^2} < \alpha < \frac{16\sigma_\nu}{\nu(\gamma^2-1)(\nu+2)^2}.$$

To obtain accurate enough values of α, the integral in the expression for α is computed numerically.

If $\gamma < 7$, the simplest way to calculate approximately the integral in the equation for α is the following. The integrand is transformed by using the energy integral

$$\alpha = \frac{2\delta^2\sigma_\nu}{(\gamma-1)\gamma} \int_0^1 \frac{h\mu\lambda^{\nu-1}}{u(\mu)}\, d\lambda, \tag{2.23}$$

where

$$u(\mu) = \frac{2}{\gamma-1}\left(\frac{\gamma+1}{2} - \mu\right).$$

The integral in Eq. (2.23) is calculated by the trapezoidal method, with unequal subdivision of the segment of integration with respect to λ. The values of α calculated with highest accuracy (four correct significant figures) for $\gamma < 4$ and for various ν are tabulated in Ref. 6. Note that a direct calculation of the integral (2.23) with respect to μ is possible, but requires the use of special quadrature formulas, since the integrand will have a singularity (albeit integrable) at $\mu = (\gamma+1)/2\gamma$. It will follow from the sequel that the integral in Eq. (2.24) can be accurately calculated for $\gamma = 7$ and $\nu = 3$.

To calculate α for $\nu = 3$ and $\gamma > 7$ it is expedient to change to integration with respect to μ, for in these cases the solution does not reach the center and the integrand has no singularities. The equation for α reduces then to

$$\alpha = \frac{256}{25} \frac{\pi\gamma}{(\gamma-1)^2(\gamma-7)} \int_1^{(\gamma+1)/2} z(\mu)\, d\mu, \tag{2.24}$$

where

$$z(\mu) = \left\{\frac{2(3\gamma-1)}{\gamma-7}\left[\mu - \frac{5(\gamma+1)}{2(3\gamma-1)}\right]\right\}^a \left[\frac{2\gamma}{\gamma-1}\left(\mu - \frac{\gamma+1}{2\gamma}\right)\right]^b$$
$$\times \left[\frac{2}{\gamma-1}\left(\frac{\gamma+1}{2} - \mu\right)\right]^c \left(\mu^2 - \frac{\gamma+1}{\gamma}\mu + \frac{\gamma+1}{2\gamma}\right),$$

$$a = \beta_4 - 5\beta_1 - 1, \qquad b = 3\beta_2 - 1, \qquad c = \frac{2}{\gamma-2},$$

$$\beta_2 = \frac{\gamma-1}{2\gamma+1}, \qquad \beta_1 = \beta_2 + \frac{\gamma+1}{3\gamma-1} - \frac{2}{5}, \qquad \beta_4 = \frac{5}{2-\gamma}\beta_1.$$

The integral in Eq. (2.24) can be calculated by any standard method.

Some results of the calculation of α using Eqs. (2.23) and (2.24), for $\nu = 3$ and $\gamma > 3$, are the following:

γ	4	5	6	7	7.1
α	0.076625	0.050894	0.036676	0.027925	0.027248
γ	7.3	8	9	10	11
α	0.025956	0.022161	0.018122	0.015155	0.012899

Since values of γ other than those given in Ref. 6 and above may be encountered in applications, we have approximated the tabulated values by analytic formulas of the form

$$\alpha = k_1(\gamma - 1)^{k_2 + k_3 \lg(\gamma - 1)} \qquad (\gamma \leq 2),$$

$$\alpha = k_1 \gamma^{\varkappa} \qquad (\gamma > 2).$$

Calculations of k_1, k_2, k_3, and $\varkappa$ by the "mean values" and least-squares methods[7] yielded the following:

$\nu = 1$

$1.1 \leq \gamma \leq 3$, $\Delta_\alpha = 1.53\%$, $k_1 = 0.36011$, $k_2 = -1.2700$, $k_3 = -0.17912$,

$1.2 \leq \gamma \leq 2$, $\Delta_\alpha = 0.41\%$, $k_1 = 0.36594$, $k_2 = -1.2537$, $k_3 = -0.18471$;

$\nu = 2$

$1.1 \leq \gamma \leq 3$, $\Delta_\alpha = 1.56\%$, $k_1 = 0.34649$, $k_2 = -1.19796$, $k_3 = -0.14134$,

$1.2 \leq \gamma \leq 2$, $\Delta_\alpha = 0.35\%$, $k_1 = 0.35246$, $k_2 = -1.1768$, $k_3 = -0.13945$;

$\nu = 3$, $\gamma \leq 3$

$1.1 \leq \gamma \leq 3$, $\Delta_\alpha = 1.39\%$, $k_1 = 0.30774$, $k_2 = -1.1598$, $k_3 = -0.11917$,

$1.2 \leq \gamma \leq 2$, $\Delta_\alpha = 0.31\%$, $k_1 = 0.31246$, $k_2 = -1.1409$, $k_3 = -0.11735$;

$\nu = 3$, $\gamma > 2$

$2.5 \leq \gamma \leq 11.5$, $\Delta_\alpha = 0.50\%$, $k_1 = 1.238$,

$\varkappa = -2.1448 + 0.2325 \lg \gamma$.

Here Δ_α denotes the maximum relative error produced by the corresponding formula at the tabulated-data points for the indicated range of the adiabatic exponent γ. These formulas can be used for calculations in which α need not be known with high accuracy.

2.2. Analytic properties of the solution in the vicinity of the symmetry center for the case of constant initial density

To solve the linearized problem (which will be treated in Chap. 3) of a point blast with back pressure, and to study the behavior of the solution of the general non-self-similar problem (with allowance for back pressure) near a symmetry center, we need a solution of the self-similar problem in the form of series in powers of λ. This form can also be useful for the calculation of the functions f, g, and h for small values of λ.

In Sec. 1 of the present chapter it was shown that the dependence of λ on the parameter μ is given by [see Eqs. (2.5) and (2.10)]:

$$\lambda = \mu^{-\frac{2}{\nu+2}}\left[\frac{2(\nu\gamma-\nu+2)}{3\nu-2-\gamma(\nu-2)}\left(\frac{(\nu+2)(\gamma+1)}{2(\nu\gamma-\nu+2)}-\mu\right)\right]^{-\beta_1}$$
$$\times\left[\frac{2\gamma}{\gamma-1}\left(\mu-\frac{\gamma+1}{2\gamma}\right)\right]^{\beta_2}.$$

We shall consider the solution for $\gamma < 7$, i.e., the case when the solution reaches the symmetry center. At this center we have $\lambda = 0$ and $\mu = \mu_0 = (\gamma + 1)/2\gamma$. We transform the equation for λ by raising its left- and right-hand sides to the power $1/\beta_2$:

$$\lambda^{1/\beta_2} = \frac{2\gamma}{\gamma-1}\mu^{-2/(\nu+2)\beta_2}\left(\mu-\frac{\gamma+1}{2\gamma}\right)$$
$$\times\left[\frac{2(\nu\gamma-\nu+2)}{3\nu-2-\gamma(\nu-2)}\left(\frac{(\nu+2)(\gamma+1)}{2(\nu\gamma-\nu+2)}-\mu\right)\right]^{-\beta_1/\beta_2}.$$

Since the dependence of β_2 on γ and ν is given by $\beta_2 = (\gamma-1)/[2(\gamma-1)+\nu]$, we obtain

$$\frac{1}{\beta_2} = \frac{\nu}{\gamma-1} + 2 = s + 2, \qquad s = \frac{\nu}{\gamma-1}.$$

Introducing the notation $\lambda^{1/\beta_2} = \lambda^{s+2} = x$, we obtain for λ^{1/β_2} the equation

$$\Phi(x,\mu) = x - \frac{2\gamma}{\gamma+1}\mu^{-2/(\nu+2)\beta_2}(\mu-\mu_0)$$
$$\times\left[\frac{2(\nu\gamma-\nu+2)}{3\nu-2-\gamma(\nu-2)}\left(\frac{\nu+2}{2}\frac{\gamma+1}{\nu\gamma-\nu+2}-\mu\right)\right]^{-\beta_1/\beta_2} = 0. \qquad (2.25)$$

This equation defines μ as an implicit function of x, with $\Phi(0, \mu_0) = 0$ and $\dfrac{\partial\Phi}{\partial\mu}(0, \mu_0) \neq 0$.

It follows from its form that the function $\Phi(x, \mu)$ can be expanded in powers of x and $\mu - \mu_0$. It is known from the theory of implicit functions that in this case the function $\mu(x)$ can also be expanded into a convergent (in the vicinity of $x = 0$) series in powers of x. Using the theory of series, we can find the explicit dependences of the coefficients of the expansion of the function μ in powers of x on the values of γ and ν.

Equation (2.25) can be written in the form

$$\mu = \mu_0 + x\frac{\gamma-1}{2\gamma}\mu^{2/(\nu+2)\beta_2}\left[\frac{2(\nu\gamma-\nu+2)}{3\nu-2-\gamma(\nu-2)}\left(\frac{(\nu+2)(\gamma+1)}{2(\nu\gamma-\nu+2)}-\mu\right)\right]^{\beta_1/\beta_2}. \tag{2.25a}$$

It is proved in series theory[8] that if a function $y(x)$ is defined by an equation $y = a + x\psi(y)$, where the function $\psi(y)$ can be expanded in powers of $y - a$, the expansion of $y(x)$ in the vicinity of $x = 0$ takes the form

$$y = a + x\psi(a) + \frac{x^2}{2}\frac{d}{da}[\psi^2(a)] + \cdots + \frac{x^n}{n!}\frac{d^{n-1}}{da^{n-1}}[\psi^n(a)].$$

This series is a particular case of the Lagrange series.

Applying this formula to the implicit function $\mu(x)$ defined by Eq. (2.25), we obtain the expansion of μ in powers of x in the vicinity of $x = 0$:

$$\mu = \mu_0 + x\frac{\gamma-1}{2\gamma}\mu_0^{2/(\nu+2)\beta_2}\left[\frac{2(\nu\gamma-\nu+2)}{3\nu-2-\gamma(\nu-2)}\right.$$
$$\times\left.\left(\frac{(\nu+2)(\nu+1)}{2(\nu\gamma-\nu+2)}-\mu_0\right)\right]^{\beta_1/\beta_2} + \frac{x^2}{2}\frac{d}{d\mu_0}\left\{\left(\frac{\gamma-1}{2\gamma}\right)^2\mu_0^{4/(\nu+2)\beta_2}\right.$$
$$\times\left.\left[\frac{2(\nu\gamma-\nu+2)}{3\nu-2-\gamma(\nu-2)}\left(\frac{(\nu+2)(\gamma+1)}{2(\nu\gamma-\nu+2)}-\mu_0\right)\right]^{2\beta_1/\beta_2}\right\} + \cdots. \tag{2.25b}$$

Using the above rule we can, by differentiating, obtain any term of this series. Since $f = 2\mu\lambda/(\gamma+1)$ and $x = \lambda^{s+2}$, the equation obtained for μ makes it possible to expand the function $f(\lambda)$ in powers of λ. Using Eqs. (2.25b) and (2.5)–(2.9) we can find also expansions of $g(\lambda)$, $h(\lambda)$, and $\theta(\lambda)$ in powers of λ, in the form

$$\begin{aligned} f(\lambda) &= \frac{2}{\gamma+1}\lambda(\sigma_0 + \sigma_1\lambda^{s+2} + \sigma_2\lambda^{2(s+2)} + \cdots), \\ g(\lambda) &= \frac{\gamma+1}{\gamma-1}\lambda^s(\alpha_0 + \alpha_1\lambda^{s+2} + \alpha_2\lambda^{2(s+2)} + \cdots), \\ h(\lambda) &= \frac{2}{\gamma+1}(\delta_0 + \delta_1\lambda^{s+2} + \delta_2\lambda^{2(s+2)} + \cdots), \\ \theta(\lambda) &= \lambda^{-s}(\eta_0 + \eta_1\lambda^{s+2} + \eta_2\lambda^{2(s+2)} + \cdots). \end{aligned} \tag{2.26}$$

The dependences of the coefficients σ_0, α_0, δ_0, σ_1, δ_1, and α_1 on γ and ν are given in Ref. 3 and are not repeated here. Using the given equations we can determine the dependences of σ_i, α_i, δ_i, and η_i on γ and ν. The forms of the expansions (2.26) and the dependences of the coefficients on γ and ν are indicated in L. I. Sedov's book.[2] Our analysis can be regarded as a corroboration of the expansions (2.26) and as a new approach to the problem of determining the coefficients.

2.3. Dependence of the solution on ω, ν, and γ in the case of variable density

In the general case of variable density the solution contains the three parameters ω, ν, and γ. Let us examine the change of the properties of the solutions with change of these parameters. Let the condition that the initial mass be finite be satisfied in any finite volume containing the point $r = 0$. We consider the mass contained in the volume between the origin and the coordinate $r = \xi$:

$$M = \sigma_\nu \int_0^\xi \rho r^{\nu-1}\, dr = \begin{cases} \sigma_\nu A \left.\dfrac{r}{\nu-\omega}\right|_0^\xi & (\omega \neq \nu), \\ \sigma_\nu A \ln r|_0^\xi & (\omega = \nu), \end{cases}$$

where $\sigma_\nu = 2(\nu-1)\pi + (\nu-2)(\nu-3)$. Obviously, the requirement that the mass be finite inside this volume can be satisfied only if $\omega < \nu$. It follows from Eqs. (2.11) and (2.13) that v_2 and p_2 decrease with time if $\omega < \nu$ and the shock wave slows down. Since p/ρ^γ is proportional to $\exp(s/c_v)$, it follows from Eq. (2.16) that as the shock wave propagates the particle entropy behind the shock-wave front decreases if $\omega < \nu/\gamma$, increases if $\omega > \nu/\gamma$, and remains constant if $\omega = \nu/\gamma$ (Ref. 2). It is interesting to note that for ω in the range $\omega/\gamma < \omega < \nu$ the shock wave slows down on moving away from the blast location and the pressure behind its front falls, but the entropy increases nevertheless.[2,3] This effect is attributed to the strong decrease of the density.[2,3]

It follows from the solution (2.5)–(2.10) that the parametric variable μ can vary over the segment

$$\frac{\gamma+1}{2\gamma} \leq \mu \leq \frac{\gamma+1}{2}.$$

The value $\mu = \mu_2 = 1$ corresponding to a shock wave ($\lambda = 1$) lies in this segment, and the symmetry center corresponds to the value $\mu_0 = (\gamma + 1)/2\gamma$. If $\omega < \nu$, then for

$$\omega < \frac{3\nu - 2 + \gamma(2-\nu)}{\gamma+1}$$

the solution continues to the symmetry center and the variable μ takes on values in the range

$$\frac{\gamma+1}{2\gamma} \le \mu \le 1.$$

If $\nu = 3$ and $(7-\gamma)/(\gamma+1) < \omega < 6/(\gamma+1)$, a spherical cavity is produced around the blast center, expands with time, has on its boundary $\mu = \mu_* = (\gamma+1)/2$ and $\lambda = \lambda_*(\gamma, \omega)$, while the density, pressure, and temperature are zero. Asymptotic equations for finding the basic functions near the symmetry center and in the vicinity of the spherical cavity are given in Refs. 2 and 3.

E. V. Ryazanov and the author[5] have investigated the behavior of the solution in the vicinity of singular values of ω and γ, i.e., at those values of ω and γ for which either the coefficients of the right-hand sides of Eqs. (2.5)–(2.10) or the exponents β_j ($j = 1, 2, \ldots, 5$) become infinite.

It turns out that the following three singular values exist:

$$\omega_1 = \frac{3\nu - 2 + \gamma(2-\nu)}{\gamma+1}, \qquad \omega_2 = \frac{2(\gamma-1)+\nu}{\gamma}, \qquad \omega_3 = \nu(2-\gamma). \quad (2.27)$$

The solution of the problem for the first singular case, when $\omega = \omega_1$, was investigated by L. I. Sedov.[2] It has the simple form

$$f = \frac{2}{\gamma+1}\lambda, \qquad g(\lambda) = \frac{\gamma+1}{\gamma-1}\lambda^{\nu-2}, \qquad h = \frac{2}{\gamma+1}\lambda^{\nu}. \quad (2.28)$$

The form of the solution in the two other cases, i.e., for $\omega = \omega_2$ and $\omega = \omega_3$, and the continuous dependence of the solution on the parameters γ and ω in the vicinity of the $\omega_2(\gamma)$ and $\omega_3(\gamma)$ curves, are established in the same manner as for the case $\omega_3 = 0$ and $\gamma = 2$. We present here only the solution for $\omega = \omega_2$, namely,

$$\lambda = \mu^{-\frac{2\gamma}{\nu\gamma-\nu+2}}\left[\frac{2\gamma}{\gamma-1}\left(\mu - \frac{\gamma+1}{2\gamma}\right)\right]^{\frac{\nu-1}{\nu\gamma-\nu+2}} \exp\left[\frac{\gamma+1}{\nu\gamma-\nu+2}\,\frac{1-\mu}{\mu-(\gamma+1)/2\gamma}\right],$$

$$g(\mu) = \frac{\gamma+1}{\gamma-1}\,\mu^{\frac{2(\nu-2+2\gamma)}{\nu\gamma-\nu+2}}\left[\frac{2}{\gamma-1}\left(\frac{\gamma+1}{2} - \mu\right)\right]^{\frac{\nu-2(\gamma+1)}{\nu\gamma-\nu+2}}$$

$$\times\left[\frac{2\gamma}{\gamma-1}\left(\mu - \frac{\gamma+1}{2\gamma}\right)\right]^{\frac{4-\nu-2\nu}{\nu\gamma-\nu+2}}$$

$$\times \exp\left[-\frac{2(\gamma+1)}{\nu\gamma-\nu+2}\,\frac{1-\mu}{\mu-(\gamma+1)/2\gamma}\right], \quad (2.29)$$

$$h(\mu) = \mu^{\frac{2\nu\gamma}{\nu\gamma-\nu+2}}\left[\frac{2}{\gamma-1}\left(\frac{\gamma+1}{2} - \mu\right)\right]^{\frac{\gamma(\nu-2)}{\nu\gamma-\nu+2}}\left[\frac{2\gamma}{\gamma-1}\left(\mu - \frac{\gamma+1}{2\gamma}\right)\right]^{-\frac{\nu\gamma}{\nu\gamma-\nu+2}},$$

$$f = \frac{2\lambda}{\gamma+1}\,\mu.$$

In the third singular case, i.e., $\omega_3 = \nu(\gamma - 2)$, the solution is similar to the case $\gamma = 2$ considered above, and will not be given here. Note that for the singular case $\omega = \omega_1$ the energy parameter is

$$\alpha = \frac{2\sigma_\nu(\gamma+1)}{\nu(\gamma-1)(\nu\gamma-\nu+2)^2}. \tag{2.30}$$

For arbitrary γ and ω, the parameter α can be determined numerically. We note also that by limiting transitions in δ and by shift transformations with respect to time (or with respect to coordinate for $\nu = 1$) it is possible to obtain from the solution (2.5)–(2.10) solutions in the limit of self-similar motions. These limiting solutions were obtained by N. N. Kochina[9] by direct integration of the corresponding invariant gasdynamic equations.

3. Strong blast in a gas at zero temperature gradient

3.1. Point blast for the nonlinear thermal conductivity equation

The solution (2.15)–(2.10) of the strong-blast problem for adiabatic disturbed motions described by Eqs. (2.3) is characterized by large temperature gradients. For $\lambda = 0$ ($\omega = 0$) the temperature at the center of the blast is infinite and increases as the center is approached in proportion to $r^{-\nu/(\gamma-1)}$. This temperature distribution does not accord with a real distribution, since at substantial temperature gradients and high temperatures a major role is played by heat-conduction processes, primarily radiative conduction.[10] It is therefore of interest to consider other models of propagation of disturbances by point blasts, the simplest of which is the model of a nonlinearly heat-conducting body. Propagation of a heat wave in a medium whose thermal conductivity depends on temperature as a power law is considered in detail in Ref. 10. The formulation of the problem is similar to that considered in Sec. 7 of Chap. 1 for the usual heat-conduction equation. It is required to find the solution of Eq. (1.44) with initial condition (1.166). The solution of this problem for an arbitrary $\chi(T)$ dependence and nonzero initial energy u ($u = \rho_0\varepsilon$) has not been investigated in sufficient detail. Now let $\chi = \chi_0 u^n$ and $u|_{t=0} = 0$ $(r > 0)$, where $\varkappa_0$ is a constant with dimensionality $[\varkappa_0] = L^{n+1}T^{2n}M^{-n}$.

Equation (1.44) assumes the form

$$\frac{\partial u}{\partial t} = \frac{1}{r^{\nu-1}}\varkappa_0\frac{\partial}{\partial r}\left(r^{\nu-1}u^n\frac{\partial u}{\partial r}\right).$$

The problem has only two dimensional constants, $\varkappa_0$ and E_0, i.e., it is self-similar.

Note that we must take into account the integral energy-conservation law (1.165) and the condition at infinity $u|_{r=\infty} = 0$. The solution will depend on one dimensionless variable

$$\lambda = \frac{r}{\lambda_0(\varkappa_0 E_0^n t)^{1/(\nu n+2)}}. \qquad (2.31)$$

This solution was investigated in Ref. 10 for $\nu = 1$ and $\nu = 3$. Its form is

$$u = e(1-\lambda)u_0(1-\lambda^2)^{1/n}, \qquad (2.32)$$

where $e(1-\lambda)$ is the unit step function and $u_0 = u(0, t)$ is the value of the energy u at the center.

The leading front of the disturbance wave or the front of the thermal wave propagates as

$$r_2 = \lambda_0(\varkappa_0 E_0^n t)^{1/(\nu n+2)},$$

where λ_0 is a known constant determined from the integral energy-conservation law. In the case of spherical symmetry ($\nu = 3$) λ_0 and u_0 are given by

$$\lambda_0 = \left[\frac{3n+2}{2^{n-1}n\pi^n}\frac{\Gamma^n(5/2+1/n)}{\Gamma^n(1+1/n)\Gamma^n(3/2)}\right]^{1/(3n+2)},$$

$$u_0 = E_0\lambda_0^{3+2/n}\left[\frac{n}{2(3n+2)}\right]^{1/n} r_2^{-3},$$

where $\Gamma(x)$ is the Euler gamma function.

For $n > 1$ it follows from Eq. (2.32) that the energy u changes little in the region close to the center, and it decreases to zero only near $\lambda = 1$, i.e., $r = r_2$. If it is assumed that $u \sim T$, the temperature behaves similarly.

In the nonlinear-thermal-conductivity approximation we disregard the gas motion. For a strong blast in a gas this is true only at the very start of the process.[10]

If the coefficient $\varkappa_0$ is quite large, Eq. (1.44) can be approximately replaced by $\partial u/\partial r = 0$ or [since $u = u(T)$] by $\partial T/\partial r = 0$.

We arrive thus at the condition of the zero gradient of the temperature or of the internal energy. The solution is then given by

$$u = u(0, t) = u_0 \qquad (r < r_2). \qquad (2.33)$$

Thus, in media with a strong heat exchange between the particles, the zero-temperature-gradient condition can be used as an approximation.

Definition. Processes for which the condition $\partial T/\partial r = 0$ [$T = T(t)$] are met will be called homothermal. Note that if T is constant a homothermal process is isothermal.

3.2. Strong blast at zero temperature gradient

We consider now the problem of a strong point blast, for which we postulate the presence of intense heat exchange rather than adiabatic flow behind the shock-wave front. We assume therefore that there is no temperature gradient in the region where the gas moves, i.e., $\partial T/\partial r = 0$. This assumption corresponds to the initial stage of the evolution of a high-power (say, atomic) explosion, when the gas has in the flow region a high temperature, while radiation and heat conduction produce strong heat exchange between the gas particles. In an atomic explosion this will correspond approximately to the development stage when the shock-wave front has not yet become detached from the fireball. By virtue of the foregoing assumptions, the temperature in the flow region depends only on the time and does not depend on the distance to the blast center, i.e., $T = T(t)$, and the flow is homothermal.

We assume a constant initial density ρ_1 of the gas at rest. The formulation of the basic strong-blast problem for homothermal motions, and its solution in the case $\nu = 3$, were first published in our earlier paper.[11] A detailed analysis for all ν was published in Ref. 3. The case of spherical symmetry was considered independently by O. S. Ryzhov and T. I. Taganov.[12]

The system of partial differential equations describing the one-dimensional unsteady motion takes the form

$$\frac{\partial v}{\partial t} + v\frac{\partial v}{\partial r} + \frac{1}{\rho}\frac{\partial p}{\partial r} = 0, \qquad \frac{\partial T}{\partial r} = 0,$$
$$\frac{\partial \rho}{\partial t} + v\frac{\partial \rho}{\partial r} + \rho\left[\frac{\partial v}{\partial r} + (\nu - 1)\frac{v}{r}\right] = 0. \tag{2.34}$$

From the first equation of the system we exclude the pressure by using the equation of state of a perfect gas

$$p = R\rho T \tag{2.35}$$

thereby transforming the system (2.34) into

$$\frac{\partial v}{\partial t} + v\frac{\partial v}{\partial r} + \frac{RT}{\rho}\frac{\partial \rho}{\partial r} = 0, \qquad \frac{\partial T}{\partial r} = 0,$$
$$\frac{\partial \rho}{\partial r} + v\frac{\partial \rho}{\partial r} + \rho\left[\frac{\partial v}{\partial r} + (\nu - 1)\frac{v}{r}\right] = 0. \tag{2.36}$$

If the initial pressure p_1 is neglected, the system of the defining parameters of this problem consists of r, t, E_0, ρ_1, and γ, with the constant γ significant only for the calculation of the energy balance.

This gas motion is self-similar. All the dimensionless characteristics of the flow can be regarded as functions of the parameters

$$\lambda = \left(\frac{\rho_1}{E}\right)^{1/(\nu+2)} r t^{-2/(\nu+2)} \quad \text{and} \quad \gamma,$$

where E is a constant with the dimension of energy $[E] = ML^{\nu-1}T^{-2}$ and is connected with the blast energy E_0 by the relation $E_0 = \alpha E$, where α is a certain constant.

Since the coordinate r_2 is a function of the time t for a shock wave, we have $r_2 = \lambda_2(E/\rho_1)^{1/(\nu+2)}t^{2/(\nu+2)}$. We determine α from the condition that $\lambda_2 = 1$ on the shock wave. We have then $\lambda = r/r_2$. For the shock-wave propagation velocity we obtain

$$D = \frac{2}{\nu+2}\frac{r_2}{t} = \frac{2}{\nu+2}\left(\frac{E}{\rho_1}\right)r_2^{-\nu/2} = \frac{2}{\nu+2}\left(\frac{E}{\rho_1}\right)^{1/(\nu+2)} t^{-\nu/(\nu+2)}.$$

We introduce in the place of the variable λ a new independent variable Λ defined as

$$\Lambda = \frac{1}{\theta_2^{1/2}}\lambda, \tag{2.37}$$

where θ_2 is a certain constant. On a shock wave we have $\Lambda_2 = 1/\theta_2^{1/2}$, since $\lambda_2 = 1$.

We can write for the velocity, density, and temperature the equations

$$v = \theta_2^{1/2}Df(\lambda), \qquad \rho = \rho_1 g(\lambda), \qquad T = \frac{\theta_2}{R}D^2, \tag{2.38}$$

where $f(\lambda)$ and $g(\lambda)$ are the dimensionless velocity and density.

Using the connection (2.38) between the dimensional and dimensionless variables, as well as the equations $\partial/\partial r = (1/\sqrt{\theta_2})r_2(\partial/\partial\Lambda)$ and $\partial/\partial t = -(D\Lambda/r_2)(\partial/\partial\Lambda)$, we can transform the system (2.36) into an equivalent system of ordinary equations:

$$\frac{g'}{g} = (\Lambda - f)f' + \frac{\nu}{2}f, \qquad f' = (\Lambda - f)\frac{g'}{g} + (\nu-2)\frac{f}{\Lambda}. \tag{2.39}$$

The prime denotes differentiation with respect to Λ. Introducing the dimensionless pressure h given by $p = \rho_1(r_2^2/t^2)h(\Lambda)$, we obtain from Eqs. (2.35) and (2.38)

$$h(\Lambda) = \left(\frac{2}{\nu+2}\right)^2\theta_2 g(\Lambda). \tag{2.40}$$

This equation yields the pressure $h(\Lambda)$ if the function $g(\Lambda)$ and the constant θ_2 are known.

Note that by eliminating g'/g we can obtain from Eqs. (2.39) an equation for $f(\Lambda)$:

$$f' = \frac{(\nu/2)\Lambda f(\Lambda - f) - (\nu-1)f}{\Lambda[1-(\Lambda - f)^2]}. \tag{2.41}$$

From the first equation of Eqs. (2.39) we get

$$\frac{g'}{g} = -\left(\frac{f^2}{2}\right)' + (\Lambda f)' + \frac{\nu-2}{2}f.$$

Integrating this equation from a certain Λ to Λ_2, we get

$$\ln\frac{g}{g_2}=\frac{1}{2}(f_2^2-f^2)+\Lambda f-\Lambda_2 f_2+\frac{\nu-2}{2}\int_{\Lambda_2}^{\Lambda} f\,d\Lambda, \qquad (2.42)$$

where g_2 and f_2 are the values of g and f on the shock wave.

For $\nu = 2$, relation (2.42) yields the first integral of the system (2.39):

$$\ln\frac{g}{g_2}=\frac{1}{2}(f_2^2-f^2)+\Lambda f-\Lambda_2 f_2. \qquad (2.43)$$

It follows from all the foregoing that the solution of the strong-blast problem at a zero temperature gradient reduces in fact to finding the function $f(\Lambda)$ from the differential equation (2.41). Once $f(\Lambda)$ is found, the $g(\Lambda)$ dependence is determined from Eq. (2.42) or Eq. (2.43). The functions f and g must satisfy in this case certain boundary conditions. Let us derive them.

From the conservation of the momentum and mass on passing through the shock-wave front we have

$$\rho_1 D^2=\rho_2(v_2-D)^2+R\rho_2 T_2, \qquad -\rho_1 D=\rho_2(v_2-D). \qquad (2.44)$$

By changing the variables in Eqs. (2.44) to dimensionless variables using Eqs. (2.37) and (2.48), we obtain the boundary conditions for $f(\Lambda)$ and $g(\Lambda)$ at $\Lambda=\Lambda_2$:

$$f_2(\Lambda_2)=\tfrac{1}{2}\left(\Lambda_2+\sqrt{\Lambda_2^2-4}\right), \qquad (2.45)$$

$$g_2(\Lambda_2)=\frac{\Lambda_2}{\Lambda_2-f_2}. \qquad (2.46)$$

In addition, from the condition that the velocity vanish at the symmetry center we have one more boundary condition for $f(\Lambda)$:

$$f(0)=0. \qquad (2.47)$$

From the boundary conditions on the shock wave it follows that the solution of Eq. (2.41) which satisfies the condition (2.45) depends on $\Lambda_2 = 1/\theta_2^{1/2}$ as a parameter. We choose Λ_2 to have the integral curve $f(\Lambda)$ satisfy the boundary condition (2.47). In addition, it follows from Eq. (2.45) that the condition $\Lambda_2 \geq 2$ must be met.

It should be noted that in this problem the dimensionless functions f and g are independent of γ, a fact that follows from the system (2.39) and the boundary conditions (2.45)–(2.47). The influence of γ comes into play only in the calculation of the energy balance.

Let us express the constant E in the equations for the characteristics of the motion in terms of the blast energy E_0 (equal in our formulation to the total energy of the disturbed gas). For the total energy we have the equation

$$E_0=\sigma_\nu\int_0^{r_2}\frac{\rho v^2}{2}r^{\nu-1}\,dr+\sigma_\nu\int_0^{r_2}\frac{R\rho T}{\gamma-1}r^{\nu-1}\,dr.$$

The first and second terms correspond here to the kinetic and thermal energies of the gas, respectively. Changing to dimensionless variables we get

$$\frac{E_0}{E} = \alpha = \left(\frac{2}{\nu+2}\right)^2 \frac{\sigma_\nu}{\Lambda_2^{\nu+2}} \left[\int_0^{\Lambda_2} \frac{gf^2}{2} \Lambda^{\nu-2}\, d\Lambda + \frac{1}{\gamma-1} \int_0^{\Lambda_2} g\Lambda^{\nu-1}\, d\Lambda\right]. \quad (2.48)$$

To calculate the second terms in the square brackets we can use relations that follow from the mass conservation of the gas set in motion,

$$\sigma_\nu \int_0^{r_2} \rho r^{\nu-1}\, dr = \frac{\sigma_\nu}{\nu} r_2^\nu. \quad (2.49)$$

In dimensionless variables, Eq. (2.49) takes the form

$$\frac{\sigma_\nu}{\Lambda_2^\nu} \int_0^{\Lambda_2} g\Lambda^{\nu-1}\, d\Lambda = \frac{\sigma_\nu}{\nu}. \quad (2.50)$$

Taking Eq. (2.50) into account, we can express Eq. (2.48) in the form

$$\frac{E_0}{E} = \left(\frac{2}{\nu+2}\right)^2 \frac{\sigma_\nu}{\Lambda^2} \left[\frac{e_\nu}{\Lambda_2^\nu} + \frac{1}{\nu(\gamma-1)}\right], \quad (2.51)$$

where

$$e_\nu = \int_0^{\Lambda_2} \frac{gf^2}{2} \Lambda^{\nu-1}\, d\Lambda. \quad (2.52)$$

Equations (2.51) and (2.52) express the constant E in terms of the blast energy E_0 and γ.

We proceed now to specific particular cases of this problem. We shall investigate the solutions of Eq. (2.41), which we rewrite as

$$f' = \frac{(\nu/2)(\Lambda - f)\Lambda - (\nu-1)\,f}{1-(\Lambda - f)^2} \frac{f}{\Lambda}. \quad (2.53)$$

To investigate the behavior of Eq. (2.53), let us examine the field of the integral curves of this equation at different values of ν. This enables us to choose the set of integral curves, the only one that satisfies the boundary conditions (2.45) and (2.46).

a) Case of spherical symmetry. For $\nu = 3$, Eq. (2.53) has in the region of the (Λ, f) plane the following singular points:

$$O(0,0), \qquad A(1,0), \qquad B = (\tfrac{4}{3}, \tfrac{1}{3}).$$

The singular point O is a saddle. Entering it are two integral curves, the straight lines $\Lambda = 0$ and $f = 0$. The singular point A is a node, into which the integral curve $f = 0$ enters. At the singular point B we have a saddle into which enter two integral curves with slopes $N_1 = 1.14$ and $N_2 = -0.385$. The boundary condition on the shock wave

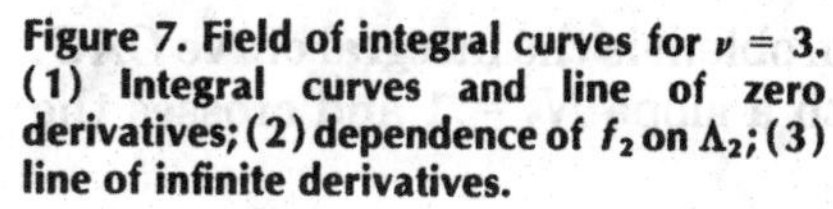

Figure 7. Field of integral curves for $\nu = 3$. (1) Integral curves and line of zero derivatives; (2) dependence of f_2 on Λ_2; (3) line of infinite derivatives.

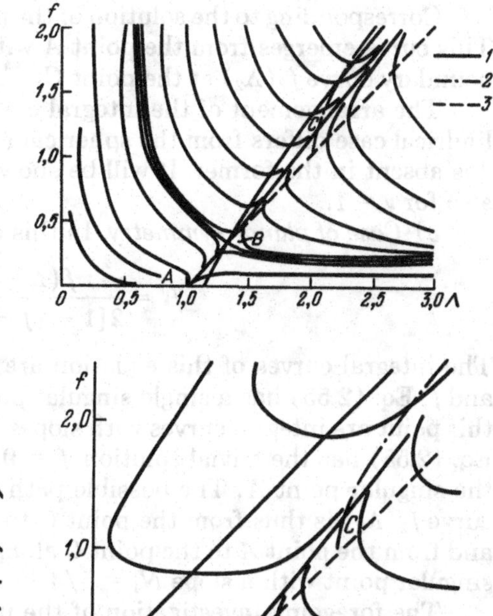

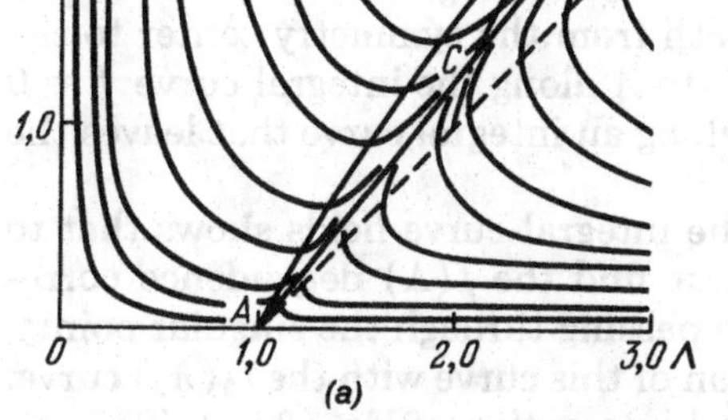

Figure 8. Fields of integral curves: (a) $\nu = 2$; (b) $\nu = 1$ (see Fig. 7 for the notation).

$$f_2(\Lambda_2) = \tfrac{1}{2}(\Lambda_2 + \sqrt{\Lambda_2^2 - 4})$$

yields in the (Λ, f) plane a curve that should cross the integral curve sought.

The complete picture of the field of the integral curves is shown in Fig. 7. It follows from an examination of this field that the only curve that can satisfy all the boundary conditions is $OABC$, which begins at the point $O(0, 0)$, passes through the singular points A and B, and crosses the curve $f_2(\Lambda_2)$ at the point C.

b) Case of cylindrical symmetry. For $\nu = 2$ we have from Eq. (2.53)

$$f' = \frac{f}{\Lambda}\frac{\Lambda(\Lambda - f) - 1}{1 - (\Lambda - f)^2}. \tag{2.54}$$

For finite values of Λ this equation has two singular points, $O(0, 0)$ and $A(1, 0)$. The singular point O is a saddle, into which the straight lines $\Lambda = 0$ and $f = 0$ enter. The point A is a singularity of rational character.[13] The critical directions, i.e., the directions along which the curves enter the singular point, are $N_1 = 0$ and $N_2 = 1$. A single integral curve enters the singular point along each of these critical directions. The field of the integral curves of Eq. (2.54) is shown in Fig. 8(a).

Corresponding to the solution of the problem is the integral curve OAC. This curve emerges from the point A with a slope $N_2 = 1$ and crosses the boundary curve $f_2(\Lambda_2)$ at the point C.

The arrangement of the integral curves in the (f, Λ) plane in the cylindrical case differs from the spherical in that the singular point B of Fig. 7 is absent in the former. It will be shown below that the same holds true also for $\nu = 1$.

c) Case of planar symmetry. In this case we have

$$f' = \frac{f(\Lambda - f)}{2[1 - (f - \Lambda)^2]}. \qquad (2.55)$$

The integral curves of this equation are shown in Fig. 8(b). For finite Λ and f, Eq. (2.55) has a single singular point, the saddle $A(1, 0)$. Entering this point are integral curves with slopes $N_1 = 5/4$ and $N_2 = 0$. In addition, Eq. (2.55) has the trivial solution $f = 0$. The integral curve $f = 0$ enters the singular point A. The possible path from the symmetry center to the curve $f_2(\Lambda_2)$ is thus from the point O to A along the integral curve $f = 0$ and from the point A to the point C along an integral curve that leaves the singular point with a slope $N_1 = 5/4$.

The foregoing investigation of the integral-curve fields shows that to solve the problem completely we must find the $f(\Lambda)$ dependence corresponding to the sought integral curve passing through the singular points, and determine the point of intersection of this curve with the $f_2(\Lambda_2)$ curve. This problem was solved by numerical integration of Eq. (2.53). The calculations yielded the following values of the constants Λ_2 and θ_2:

$$\Lambda_2 = 2.024, \quad \theta_2 = 0.244 \quad (\nu = 3),$$

$$\Lambda_2 = 2.040, \quad \theta_2 = 0.240 \quad (\nu = 2),$$

$$\Lambda_2 = 2.076, \quad \theta_2 = 0.232 \quad (\nu = 1).$$

The $g(\Lambda)$ dependence was likewise obtained in accordance with Eq. (2.42).

The solutions of the problem for all three symmetries are illustrated by the plots of $p/p_2(\lambda)$, $\rho/\rho_2(\lambda)$, and $v/v_2(\lambda)$ in Fig. 9. The solution is characterized, first, by the fact that a region of quiescent gas exists near the center up to the value $r_* = \theta_2^{1/2} r_2$. This region expands with time, and corresponds in the (f, Λ) plane to the integral curve $f = 0$. For $r > r_*$ the gas velocity is directed away from the center and increases almost linearly with r. In the central quiescent region the gas density is constant, while the pressure and temperature vary with time as $t^{-2\nu/(2+\nu)}$. For values $r > \theta_2^{1/2} r_2$ the density and pressure increase with r and reach maximum values on the shock-wave front. As the shock wave is approached, the gradients of the density and of the pressure increase. The parameter e_ν ranges from 1 to 2.5, with $e_\nu = 1.088$ obtained at $\nu = 3$. Assuming that the density varies in accordance with Eq. (2.1), the problem remains self-similar. The behavior of the solution of this problem for various ω was investigated by E. V.

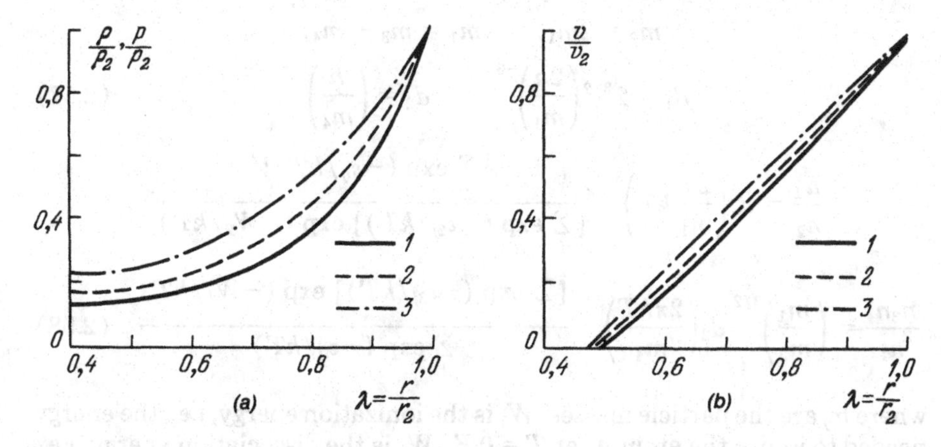

Figure 9. (a) Dependence of the dimensionless pressure and of the density on λ: (b) dependence of the dimensionless velocity on λ. (1) $\nu = 3$; (2) $\nu = 2$; (3) $\nu = 1$.

Ryazanov.[3] Besides the PBP, other problems were also investigated for the homothermal region (see, e.g., Ref. 14). An application of the homothermal model to electric discharges in gases is described in Refs. 15 and 16.

4. Allowance for high-temperature effects in the strong-blast problem

4.1. Thermodynamic properties of gases at high temperatures

Propagation of strong shock waves in a gas produces high pressures and temperatures. At high temperatures the gas motion is influenced by molecule dissociation and atom ionization, as well as by radiation-induced effects.

We assume that dissociation and ionization are equilibrium processes, and disregard radiation for the time being. We consider by way of example the case of single ionization of a diatomic gas, i.e., when it can be assumed that $\gamma = 1.4$ ahead of the shock-wave front. For the thermal equation of state of the gas-particle mixture we can write

$$p = \sum kn_j T, \tag{2.56}$$

where $j = 1, 2, 3, 4$; $n_1, \ldots, n_4$ are the numerical densities of the neutral atoms, molecules, positive ions, and electrons, respectively, with $n_4 = n_3$ by virtue of the quasineutrality; k is the Boltzmann constant.

An equilibrium chemical process $A_2 = n_1A_1 + n_2A_2 + n_3A_3 + n_4A_4$ is subject to the law of effective masses[17]

$$m_2 = 2m_1, \qquad m_1 = m_3 + m_4,$$

$$d_1 = 2^{3/2}\left(\frac{2h}{m_1}\right)^{-3}, \qquad d_2 = \left(\frac{h}{m_4}\right)^{-3}, \tag{2.57}$$

$$\frac{n_1^2}{n_2} = d_1\left(\frac{2\pi}{m_1}kT\right)^{3/2} \cdot \frac{[\sum_j \exp(-\varepsilon_{ij}/kT)]^2}{[\sum_j \exp(-\varepsilon_{2j}/kT)]\exp(-W_2/kT)},$$

$$\frac{n_3 n_4}{n_1} = \left(\frac{m_1}{m_3}\right)^{3/2} d_2\left(\frac{2\pi kT}{m_4}\right)^{3/2} \frac{[\sum_j \exp(-\varepsilon_{3j}/kT)]\exp(-W_1/kT)}{\sum_j \exp(-\varepsilon_{ij}/kT)}, \tag{2.58}$$

where m_j are the particle masses, W_1 is the ionization energy, i.e., the energy needed to ionize the atom A_1 at $T = 0$ K, W_2 is the dissociation energy, i.e., the energy needed to dissociate the molecule at $T = 0$ K, h is Planck's constant, and ε_{ij} is the energy of the particle A_i in various states. For the gas mass density we have

$$\rho = \sum_j m_j n_j, \tag{2.59}$$

with $m_4/m_3 \ll 1$ and $m_2 \approx m_3$.

With p and ρ known, Eqs. (2.56)–(2.59) can yield the unknowns n_1, n_2, n_4, and T. The thermodynamic properties of the gas enter in its equations of motion via the connection between the internal energy, density, and pressure. For the internal energy ϵ we can write down the relation[10,17,18]

$$\rho\varepsilon = \frac{3}{2}nkT + \sum_i n_i(u_i + w_i), \tag{2.60}$$

where

$$u_i = kT^2\frac{\partial}{\partial T}\ln z_i, \qquad z_i = \sum_s \exp\frac{\varepsilon_{is}}{kT}, \qquad i = 1, 2, 3; \qquad z_4 = 1,$$

$$w_1 = \frac{1}{2}W_2, \qquad w_2 = w_3 = 0, \qquad w_4 = W_1. \tag{2.61}$$

A simple example of a gas for which the solution of the blast problem with allowance for the effects at high temperature is meaningful is hydrogen. Equations (2.56)–(2.61) yield for hydrogen the thermodynamic functions all the way to total ionization. No account is taken here, to be sure, of the Coulomb interaction and of the appearance of negative ions,[19,20] but these effects can be neglected in a number of cases.

For a more complicated gas mixture and for multiple ionization of atoms, the equations for the thermodynamic functions become greatly complicated.[10] The thermodynamic functions of air have now been calculated in a wide range of temperatures and pressures by N. M. Kuznetsov.[21] The following procedures can be invoked to obtain rough estimates and to take approximate account of the influence of ionization and dissociation on the variation of the temperature and density.

1. The gas is assumed to be perfect with a certain "effective" adiabatic exponent γ. Estimates made by various authors show that during the strong-shock-wave stage the range of the effective adiabatic exponents of gases is $1.2 \le \gamma \le 1.3$. The thermodynamic functions of the gas reduce in this case to those of a perfect gas with constant specific heat. Temperatures can be roughly estimated by assuming that the changes of the thermodynamic properties of the gas at high temperature have little influence on the pressure and density distributions in the gas stream. Then, solving the problem for the effective γ and assuming p and ρ to be known, equations such as Eqs. (2.56)–(2.61) or tables of thermodynamic functions can be used to determine T and the concentrations of the gas-mixture components. Such estimates of the temperatures for singly ionized air were obtained in Ref. 22.

2. To determine the gas properties more accurately, the dependence of the internal energy ε on p and ρ, or on ρ and T, can be approximated by analytic equations and the differential equations solved by approximate methods. This device is extensively used in various gasdynamic problems. This approach was used for a strong blast in air in Refs. 23 and 24.

Examples of approximations of the internal energy by simple analytic relations are given also in Ref. 10. An approximate formula proposed here for this purpose is

$$\varepsilon = \varepsilon_0 T^\alpha \rho^\beta, \qquad \varepsilon_0 = \text{const.} \tag{2.62}$$

Thus, if the air density varies in the range $10\rho_0 - 10^{-3}\rho_0$ (ρ_0 is the normal density), and the temperature ranges from 10^4 K to 2.5×10^5 K, Eqs. (2.2) can be written in the form

$$\varepsilon = \varepsilon_0 \left(\frac{T}{10^4}\right)^{1.5} \left(\frac{\rho_0}{\rho}\right)^{0.12}, \tag{2.63}$$

where ε_0 is the internal energy at $T = 10^4$ K and $\rho = \rho_0$.

We have approximated the relations on a strong shock wave and the internal energy of air. The basic data on the thermodynamic functions were taken from the tables of Ref. 21. The tabulated data were approximated for the relations on a shock wave by two methods. In the first we approximated the dependence of the enthalpy on the shock-wave velocity D, and in the second we approximated the dependence of the pressure p_2 behind the shock wave on the velocity v_2 behind the shock wave.

Let us consider first the first approximation method. As noted in Ref. 10, the relation between the enthalpy i and D^2 depends little on the thermodynamic properties of air at high temperatures. In fact, it follows from the conditions on the discontinuity that

$$i_2 = \frac{D^2}{2}\left[1 - \left(\frac{\rho_1}{\rho_2}\right)^2\right],$$

and in the case of strong compression the ratio i_2/D^2 changes little. We introduce the dimensionless variables

$$\mathcal{J} = \frac{\rho_1 i_2}{p_1}, \qquad \bar{D} = \frac{D}{\sqrt{p_1/\rho_1}}$$

and approximate roughly the $\mathcal{I}(D)$ dependence by the equation

$$\mathcal{I} = \frac{k_j}{2}\bar{D}^2 + j\mathcal{I}_0, \qquad j = 0, 1. \tag{2.64}$$

We shall use Eqs. (2.64) both in the case of the cruder approximation $j = 0$ and for the more accurate one with $j = 1$. Actual calculations were performed only for the initial parameters of a standard atmosphere at sea level: $p_1 = 1.01325 \times 10^6$ dyn/cm^2 and $\rho_1 = 1.224 \times 10^{-3}$ g/cm^3.

For $j = 0$, approximation of the tables yields $k_0 = 0.992$, i.e.,

$$\mathcal{I} = 0.496\bar{D}^2. \tag{2.65}$$

Comparison with the tabulated data has shown that at temperatures from 1300 K to 2×10^6 K behind the shock wave the error in the determination of i from Eq. (2.65) does not exceed 3%. Naturally, for temperatures lower than 1000 K the error increases greatly. More accurate results are obtained using the approximation with Eq. (2.64), where $j = 1$. In this case we obtain $k_1 = 2 \times 0.49446$ and $\mathcal{I}_0 = 0.32$, i.e., Eq. (2.64) takes the form

$$\mathcal{I} = 0.49446\bar{D}^2 + 0.32. \tag{2.66}$$

This equation is more accurate than Eq. (2.65). For T in the range from 1×10^3 to 3×10^6 K the error of Eq. (2.66) does not exceed 3%.

Thus, these equations, together with the dynamic conditions on the shock wave, can replace completely the tables of Ref. 21 for p_2 and ρ_2. It is necessary here to use Eqs. (2.64) in place of the energy-conservation law on going through discontinuities. On the other hand, the temperatures T_2 must be found from the tables, using p_2 and i_2 or p_2 and ρ_2 known from the approximate equations.

The second method for approximating the data involves finding an approximate relation between $P_2 = p_2/p_1$ and $v_2/a_1 = V_2$. For normal initial density and pressure at sea level we have

$$\gamma_{\text{eff}} = -0.3875(T_2 \cdot 10^{-4}) + 1.4155$$

(T_2 is the temperature in K),

$$\theta_2 = a + \sqrt{\frac{T_2}{T_1} + a^2}, \qquad a = \frac{1}{2}\left(\frac{\gamma_{\text{eff}} + 1}{\gamma_{\text{eff}} - 1} - 6\frac{T_1}{T_2}\right),$$
$$\frac{p_2}{p_1} = \theta_2\frac{T_2}{T_1}\left(1 \le \frac{p_2}{p_1} \le 88.47\right), \tag{2.67}$$

$$P_2 = (1.65V_2^2 + 1) - [0.0243(\ln V_2)^{6.791}]\ (88.47 < P_2 \le 77\,410), \tag{2.68}$$

$$P_2 = 1.65V_2^2 + 1\ (77\,410 < P_2 \le 567\,400), \tag{2.69}$$

$$P_2 = 1.65V_2^2 + 1 + 222.56V_2 - 137\,750(567\,400 < P_2 \le 936\,600). \tag{2.70}$$

Equations (2.67)–(2.70) correspond to the range of T_2 from 288.16 to 3×10^6 K, and their deviation from the tabulated data does not exceed 3%.

No simple high-accuracy approximating equations could be found for the dependences of the thermodynamic functions on the density and pressure or on the density and temperature. We note here only one result.

We seek an approximation of the internal energy ε in the form

$$\varepsilon = \frac{p}{\rho}\varphi\left(\rho, \frac{p}{\rho}\right). \tag{2.71}$$

The function φ must be taken from the tables. If φ depends only on ρ, we obtain in this case only a very crude approximation of $\varepsilon(p, \rho)$. A more accurate approximation can be obtained, for example, using equations in the form ($T > 5000$ K)

$$\varphi = b \operatorname{cth}\left(\frac{g + \delta_0}{a}\right), \qquad g = \frac{\rho}{\rho_0}, \tag{2.72}$$

where b, a_0, and δ_0 depend on the ratio p/ρ and vary little when p/ρ changes, ρ_0 is the characteristic density ($\rho_0 \approx \rho_1$), and the pressure is referred to p_1. Thus, a fair approximation is obtained from the relations

$$b = \frac{1}{2.013}\left(\frac{p}{g}\right)^{-0.06687}, \quad a = 2.8996\left(\frac{p}{g}\right)^{0.0342}, \quad \delta_0 = 5.606\left(\frac{p}{g}\right)^{-0.0641}. \tag{2.73}$$

Naturally, Eqs. (2.72) and (2.73) can be further refined and generalized.

We present one more approximation of the caloric equation of state. By analogy with a perfect gas, we introduce the quantity

$$\gamma = 1 + p/\rho\varepsilon = 1 + 1/\varphi(p, \rho). \tag{2.74}$$

We introduce next a new parameter μ defined by the relation

$$\gamma = \gamma(p, \rho) = (\mu + 1)/(\mu - 1). \tag{2.75}$$

An approximation, by H. Brode,[23,24] of American tabulated data yields for μ the approximate formula

$$\mu = 1 + \frac{Ay + 3}{By + 1} + \sum_{i=1}^{3} \frac{C_i(1 - y)y}{D_i y^2 + 1}, \tag{2.76}$$

$A = 25.895$; $B = 4.7790$; $C_1 = 861$; $D_1 = 3000$;

$C_2 = 2356$; $D_2 = 90\,000$; $C_3 = 41\,000$; $D_3 = 12\,000\,000$.

The notation in Eq. (2.76) is

$$y = p_0\rho/p\rho_0, \qquad p_0 = 1.013\,25 \text{ bar}, \qquad \rho_0 = 1.225 \times 10^{-3}\,\text{g/cm}^3.$$

Note that Eqs. (2.74)–(2.76) yield $\gamma = \frac{5}{3}$ for $y = 0$ and $\gamma \approx 1.4$ for $y = 1$.

It is frequently preferable to replace the above approximate formulas by tables that yield, by linear or quadratic interpolation, thermodynamic-function values adequate for applications.

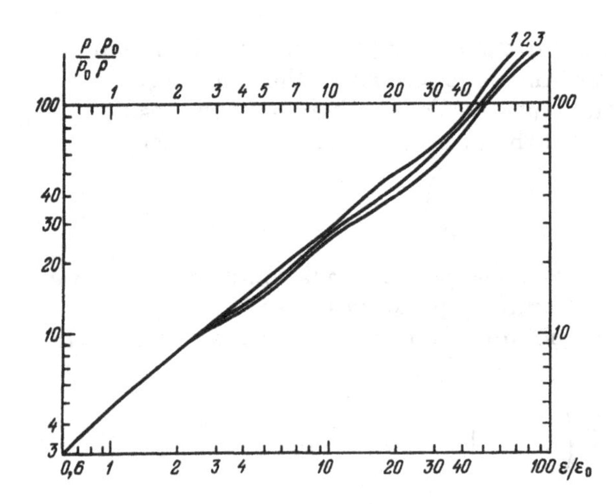

Figure 10. Plot of $(p/p_0)(\rho_0/\rho)$ vs $\varepsilon/\varepsilon_0$ for three values of g (air): (1) $g = 10$; (2) $g = 1$; (3) $g = 0.1$.

By way of illustration, we consider the case of air in the temperature range from 273 to 30 000 K.[25] A log–log plot of $p\rho_0/p_0\rho$ against $\varepsilon/\varepsilon_0$ is shown for air in Fig. 10. Here $\rho_0 = 1.2921 \times 10^{-3}$ g/cm^3, $p_0 = 1$ bar, and $\varepsilon_0 = 1$ kJ/g. It is seen from the figure that $p/p_0 g$ depends little on g in a wide range of the internal energy. Moreover, a slight change of the variables makes the log–log plots straight lines. This indicates that it is expedient to write the caloric equation of state in logarithmic form, a frequent practice in numerical solutions.

At high temperatures it is sometimes important to take into account the kinetics of the chemical reactions between various gas components also in the dissociation and ionization processes, i.e., to go beyond the equilibrium approximations. The influence of disequilibrium and of the rates of chemical reactions on the composition of air subject to a strong blast is described in Ref. 10. We shall not dwell here on these questions. Related topics will be discussed in Chap. 6, where fast chemical-combustion reactions are studied.

4.2. Strong blast in a gas with allowance for dissociation and ionization

To study the influence of various physical processes in gases on the evolution of a point blast, it is of definite interest to formulate and investigate the PBP using more complicated models of gaseous media. We discuss in the present section models in which account is taken of equilibrium dissociation and of ionization, neglecting the influence of radiation and heat conduction. The general formulation of the problem remains unchanged, but it must be borne in mind that in this case it is necessary to solve, jointly with an initial set of equations such as Eqs. (1.18)–(1.20), also equations of type (2.56)–(2.61) to determine the gas composition and the internal energy. Naturally, the approximate equations of state indicated above can be used. It is accordingly necessary to subject the shock waves to boundary conditions more general than for the usual perfect gas.

The simplest case of a problem of this type is that of a strong blast in atomic hydrogen, or in more complex but singly ionized atoms. In fact, the dependences of the internal energy on the temperature and pressure are given in this case by an expression of the type (2.60)

$$\varepsilon = \frac{3}{2}(1+\alpha)R_0T + \alpha\frac{W_1}{m_0}, \tag{2.77}$$

$$\alpha = \left[\frac{h^3}{(2\pi m_e)^{3/2}}\frac{p}{(kT)^{5/2}}\frac{g_0}{2g_1}\exp\left(\frac{W_1}{kT}\right)+1\right]^{-1/2}, \tag{2.78}$$

where α is the degree of ionization, m_e is the electron mass, m_0 is the atom mass, W_1 is the ionization energy, R_0 is the gas constant per unit mass, and g_0 and g_1 are the statistical weights of the ground level of the neutral atom and ion.

The thermal equation of state is

$$p = \rho(1+\alpha)R_0T. \tag{2.79}$$

The set of parameters that define the problem acquires two additional dimensional constants, W_1/m_0 and a pre-exponential factor in Eq. (2.78). It follows hence that the problem is no longer self-similar.* In L. I. Sedov's book[2] are considered the classes of gas caloric equations of state for which the problem of a strong blast in a medium of constant density remains self-similar. To preserve self-similarity, the dependence of the internal energy on p and ρ should be of the form

$$\varepsilon = \frac{p}{\rho_0}\varphi\left(\frac{\rho}{\rho_0}\right), \tag{2.80}$$

where ρ_0 is a certain constant with dimensionality of pressure. If the initial density is variable, the function (2.80) reduces in the self-similar case simply to that for a perfect gas

$$\varepsilon = \beta\frac{p}{\rho}, \tag{2.81}$$

where β is an abstract constant.

Thus, the solution of the strong-blast problem for an equation of state in the form (2.77) does not reduce to integration of ordinary differential equations, and calls for the use of numerical methods.

If, however, this problem is considered in the approximation with an effective adiabatic exponent γ, it can have the solution (2.5)–(2.13) considered above. As for the tabulated thermodynamic functions of air, it follows

* Approximate solutions of the PBP for a so-called "ideally dissociating" gas, whose thermodynamic functions are similar to Eqs. (2.77)–(2.79), were obtained by F. Higashino.[26]

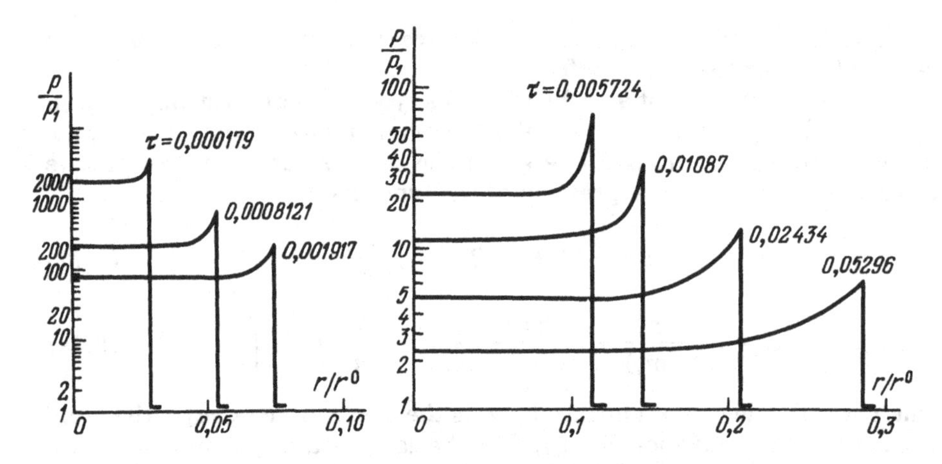

Figure 11. Distribution of the pressures behind a shock wave in air for a number of values of τ.

from the discussion above that only a crude approximation of these tables can make the strong-blast problem self-similar.

Non-self-similar strong-blast problems have so far not been sufficiently well investigated. In fact, each gas mixture requires here a separate solution in view of the differences between the equations of state.

A strong spherical blast in air was investigated by H. Brode in the following approximate formulation.

The chosen caloric equation of state was an analytic approximation similar to Eqs. (2.75) and (2.76) (for more details see Ref. 24). The initial conditions used were the solution for an ideal gas with $\gamma = 1.25$ at an instant of time corresponding to a pressure of 4500 atm on the shock-wave front. The problem was solved numerically using an artificial-viscosity shock-capturing calculation method. Note that the calculations were carried out prior to the later stage, when the back pressure is important, and that for pressures lower than $p_2 = 1.2p_1$ the problem was solved as for an ideal gas with $\gamma = 1.4$.

Without dwelling here on the calculation details (see Appendix I), we present only some results for a spherical blast (cylindrical blasts are considered in Refs. 27 and 28, while the planar case has not been investigated in detail).

Figure 11 shows the spatial distributions [as functions of $R = r/r^0$, where $r^0 = (E_0/p_1)^{1/3}$] of the relative pressures p/p_1 for various dimensionless times $\tau = t/t^0$, where $t^0 = r^0(\rho_1/p_1)^{1/2}$ (the numbers at the curves correspond to $= 0.000\,179$, $0.000\,8121$, $0.001\,917$, $0.005\,724$, $0.010\,87$, $0.024\,34$, and $0.052\,96$). The plots are given for $p_2/p_1 > 6$, where the dissociation and ionization effects are most pronounced.

A table given in Ref. 24 for wave-front parameters is not reproduced here in view of the inaccuracy of the assumed equation of state.

Let us dwell briefly on the main conclusions that follow from the numerical solutions.

The dependence of p_2/p_1 on $R_2 = r_2/r^0$ is similar to that for a perfect gas. The (p_2/p_1) (R_2) curve for the case of air during the strong blast stage is in fair agreement with the corresponding curve for an ideal gas with $\gamma = 1.4$, if the shock-wave radius is multiplied by 0.843. Such a decrease of the dimensionless radius corresponds to a decrease of the effective energy to 60% of the total energy E_0. Thus, for maximum excess pressures, the solution for "real" air has during the initial stage an efficiency equal to about 60% of that obtained from the solution for an ideal gas. The cause is the energy loss to dissociation and ionization. The central region is heated to high temperatures, and the recombination processes proceed nonuniformly in the flow region and slowly feed the energy back to the motion zone. In addition, irreversible entropy-induced losses proceed here differently than for an ideal gas with $\gamma = 1.4$. In the case of strong wave damping, for equal distances, the difference between the values of p_2/p_1 decreases, but not to zero. In the case of a real blast, an important role is played at high temperatures, in addition to dissociation and ionization, by non-equilibrium chemical reactions and by radiative heat transfer. The study of the point-blast problem for an equilibrium model of a medium, in which only reversible dissociation and ionization take place, is mainly of theoretical interest.

4.3. Effect of thermal radiation on gas motion

At temperatures on the order of several thousand degrees and higher, the gas motion is strongly influenced by thermal radiation. The gasdynamic equations with allowance for radiation were given in Sec. 5 of Chap. 1. The formulation of the problem remains actually the same as for the purely gasdynamic variant. To be sure, certain additional boundary conditions are imposed on the radiation flux. They reduce to vanishing of the radiation flux at the symmetry center and at infinity.

Let us consider a number of models used to describe the flow in a radiating and an absorbing gas.

a) Gasdynamic equilibrium model. This model involves a mixture of gas particles with a photon "gas," and transport processes are neglected.[29] It is assumed that the thermodynamic equations for a gas in thermodynamic equilibrium with the radiation it contains are

$$\varepsilon^* = \varepsilon + \frac{1}{\gamma_1 - 1}\frac{p_R}{\rho}, \qquad S^* = S + \frac{4a}{3}\frac{T}{\rho^{\gamma_1 - 1}},$$

$$p = p(\rho, T), \qquad p_R = \frac{a}{3}T^{\gamma_1/(\gamma_1 - 1)}, \tag{2.82}$$

$$p^* = p + p_R, \qquad a = \frac{4\sigma}{c}, \qquad \gamma_1 = \frac{4}{3}.$$

Let the gas have the equation of state of an ideal gas. Relation (2.82) is greatly simplified if the adiabatic exponents of the gas and of the radiation are equal, $\gamma = \gamma_1$. This occurs if $\gamma = \frac{4}{3}$ and if the temperature dependence of the radiation pressure is given by

$$p_R = (\gamma - 1)\bar{a}T^{\gamma/(\gamma-1)}, \qquad \bar{a} = \text{const.}$$

In these cases

$$\begin{aligned} \varepsilon^* &= \frac{1}{\gamma - 1}\frac{p + p_R}{\rho} = \frac{p^*}{\rho(\gamma - 1)}, \\ p^* &= R\rho T + (\gamma - 1)\bar{a}T^{\gamma/(\gamma-1)}, \\ S &= c_v \ln \frac{T}{\rho^{\gamma-1}} + \frac{4a}{3}\frac{T}{\rho^{\gamma-1}}. \end{aligned} \tag{2.83}$$

Since

$$\frac{p}{R\rho^\gamma} = \frac{T}{\rho^{\gamma-1}}\left[1 + (\gamma - 1)\frac{\bar{a}}{R}\left(\frac{T}{\rho^{\gamma-1}}\right)^{1/(\gamma-1)}\right],$$

the expression in Eqs. (2.83) for the entropy S yields $S = F(p/\rho^\gamma)$. Thus we see that the solution of the strong-blast problem with the functions p^*, ρ, and v remains the same as if there were no allowance for radiation. All that changes is the temperature distribution, since the temperature must be determined from the second relation of Eqs. (2.83). We note here that the problem of determining p^*, ρ, and v remains self-similar, even though the temperature distribution is no longer self-similar in view of the new dimensional constant a.

Since the effective value of γ for air at high temperatures is close to $\gamma_1 = \frac{4}{3}$, it can be assumed that even if $\gamma \neq \gamma_1 = 4.3$ the radiation will have little effect on the distributions of p^*, ρ, and v. These functions can be chosen the same as for an ordinary gas. In the calculation of the temperature, however, one must use the equation

$$p^* = R\rho T + \frac{a}{3}T^4.$$

Since the temperatures in the vicinity of the symmetry center are very high, the second term in this equation can play an important role.

We shall not dwell in detail on this solution,[3,29] but note only that for this model the temperatures at the center of the blast turn out to be finite at $t > 0$ and their time variation is given by

$$T(0, t) = \left[\frac{3}{a}\frac{8\rho_1}{(\nu + 2)^2(\gamma + 1)}\frac{p}{p_2}(0)\right]^{\frac{1}{4}}\left(\frac{E_0}{\alpha\rho_1}\right)^{\frac{1}{2(\nu+2)}} t^{-\frac{\nu}{2(\nu+2)}}.$$

In the vicinity of the blast center, the temperature depends little on the coordinate.

b) Radiant heat-conduction model. In the radiant heat-conduction approximation the radiation energy flux is proportional, under local equilibrium conditions, to the temperature gradient, i.e., the radiation transport has the character of heat conduction, with the thermal conductivity dependent on the temperature and on the density.

In the radiant heat conduction model, the gasdynamic equations retain their usual form if the pressure p is taken to mean p^*, a term $\varepsilon_R = 4\sigma T^4/c$ is added to the internal energy $\rho\varepsilon$, and the heat-flux vector is defined as

$$\mathbf{q} = -[\varkappa_T + \varkappa_R]\nabla T,$$

where $\varkappa_T$ is the thermal conductivity of the gas mixture and $\varkappa_R$ is the previously introduced (see Sec. 5, Chap. 1) radiant thermal conductivity given by

$$\varkappa_R = \frac{16}{3}\frac{\sigma}{k_R}T^3.$$

Here k_R is the Rosseland mean absorption coefficient of the gas:

$$k_R = \int_0^\infty k_\omega^{-1}\frac{dB_\omega(T)}{dT}\,d\omega \bigg/ \int_0^\infty \frac{dB_\omega(T)}{dT}\,d\omega.$$

The quantity $l_R = 1/\varkappa_R$ is indicative of the mean free path of the radiation. The values of l_R for air are given in Refs. 21 and 30–32 for a wide range of temperatures and in Ref. 33 for the temperature range $5000\ \mathrm{K} \le T \le 20\,000\ \mathrm{K}$.

Since $\varkappa_T \ll \varkappa_R$ at high temperature, the usual thermal conductivity of the gas can be disregarded (note that the electronic thermal conductivity can play an important role at very high temperatures, but this will be immaterial below).

Assume now that the pressure of the equilibrium radiation (together with its energy) is disregarded. It is possible then to study self-similar formulations of the problems.

Let the initial density be constant, and let the dependence of the thermal conductivity on the density and temperature be given by

$$\varkappa = \varkappa_1 \rho^{a_1} T^{a_2}. \tag{2.84}$$

In this case, just as in the case $a_1 = 0$ considered by G. M. Bam-Zelikovich,[34] the problem is self-similar for all a_1 if $a_2 = (\nu - 2)/2\nu$ and the gas is perfect.

We have obtained[3,35,36] the self-similarity conditions for the case of a blast in a medium of variable density $\rho_1 = Ar^{-\omega}$, when $a_1 = 0$. Dimensionality analysis of the constants of the problem shows here that the strong-blast problem is self-similar if the relation between the parameters ν, a_2, and ω is

$$\omega + \nu - 2 - 2a_2(\nu - \omega) = 0. \tag{2.85}$$

A self-similar strong-blast problem was investigated in Refs. 35–37 for the case $a_1 = 0$, $\nu = 3$, and constant density.

Let us dwell briefly on the results of the investigation of this problem. The present author considered this problem earlier[35,36] in an approximate formulation, without allowance for the thermal wave propagating through the cold gas ahead of the shock wave. The following conclusions were drawn. The gas is set in motion by the shock wave away from the center and is contained in a sphere whose radius decreases with time as

$$r_2 = \left(\frac{E_0}{\alpha\rho_1}\right)^{1/5} t^{2/5}. \tag{2.86}$$

The temperature at the center is finite for $t > 0$ and decreases monotonically with increasing r, having a maximum at the center. The particle velocities near the center are low. At $r < 0.5r_2$ the gas particles are nearly at rest. The pressure distribution remains qualitatively the same as in the case of the adiabatic and homothermal model of the flow. The following asymptotic equations hold near the center of the blast:

$$v = \frac{1}{\varkappa_1} b_1(t) r^3 + o(r^3), \qquad \rho = \rho(0, t) + \frac{1}{\varkappa_1} b_2(t) r^2 + o(r^2),$$
$$T = T_0(0, t) + \frac{b_3}{\varkappa_1} r^2 + o(r^2). \tag{2.87}$$

If the thermal conductivity is low, it will affect the temperature substantially only in some small vicinity of the center. Analysis of the solution of the problem has shown that as $\varkappa_1$ increases the region of weak variation of the temperature expands, i.e., the flow becomes close to homothermal.

V. E. Neuvazhaev[37] carried out detailed calculations of the problem with account taken of the thermal wave and the gas flow induced by it ahead of the shock wave. Note that the present author had proposed[35] for small $\varkappa_1$ an approximate method of solving the problem, based on "joining" the solution of the heat-conduction problem near $r = 0$ to the solution of the adiabatic problem. A similar approach was subsequently developed in Refs. 38 and 39.

The case of a strong cylindrical blast, with constant $\varkappa$, was investigated by I. O. Bezhaev.[40] The density changes in the shock wave, obtained by studying the homothermal model, are substantially lower than in the adiabatic theory, with the pressure changing by somewhat less than a factor of two. The same takes place also for large but finite $\varkappa$. The case of variable density was investigated by V. P. Shidlovskii.[41]

In the homothermal point-blast model described above it was assumed in essence that $\varkappa = \infty$ behind the shock-wave front and that $\varkappa = 0$ ahead of the shock wave.

We consider now a rough scheme for taking into account the thermal wave ahead of the shock wave for homothermal motion inside the wave. Assume that the blast is spherical, the density is constant, and self-similarity obtains. The problem has then an energy integral[2,3]

$$r^2\left[\left(\frac{2}{5}\frac{r}{t} - v\right)\left(\frac{\rho v^2}{2} + \rho\frac{\tilde{T}}{\gamma - 1}\right) + \rho\tilde{T}v + \frac{\varkappa}{R}\frac{\partial\tilde{T}}{\partial r}\right] = C. \tag{2.88}$$

Here C is an arbitrary constant and $\tilde{T} = RT$. Recognizing that $v = T = \varkappa(T) = 0$ for sufficiently large r, we set the constant C equal to zero. Changing to the dimensionless variables $f = v/D$, $g = \rho/\rho_1$, $\theta = \tilde{T}/D^2$, and $\lambda = r/r_2$ we get

$$B\theta^{1/6}\frac{d\theta}{d\lambda} + g(\lambda - f)((\gamma - 1)f^2 + \theta) = (\gamma - 1)gf\theta, \qquad (2.89)$$

where B is a dimensionless constant proportional to the coefficient $\varkappa$ (Ref. 3).

Assuming the temperatures to be equal on both sides of the discontinuity (as is the case when heat conduction is taken into account), we have for Eq. (2.89) the boundary conditions

$$\theta(1) = \theta_2. \qquad (2.90)$$

We neglect now the velocity and density changes due to the thermal wave, i.e., we put $g = \text{const} = g_0$ and $f = 0$ for $\lambda > 1$. The integration of Eq. (2.89) is then elementary. With allowance for the boundary condition (2.90) we get

$$\theta = \left[\theta_2^{1/6} + \frac{g_0}{12B}(1 - \lambda^2)\right]^6. \qquad (2.91)$$

Let θ vanish at $\lambda = \lambda_T$; we obtain then for the coordinate λ_T of the leading front of the thermal wave

$$\lambda_T = \sqrt{1 + \frac{12B}{g_0}\theta_2^{1/6}}, \qquad (2.92)$$

where the values of θ_2 are given by Eq. (2.37) and $g_0 = 1$; λ_T is large for large $B > 1$. As $B \to 0$ the boundary of the thermal wave approaches the shock-wave front. If the effective value of $\varkappa_1$ is small, we have $B < 1$ and λ_T is close to unity. The temperature then drops abruptly to zero in a narrow region ahead of the wave, while at $\lambda = \lambda_T$ it vanishes together with its five derivatives with respect to λ. Neglecting the velocities v and the change of the density ρ at $\lambda > 1$ is justified here.

For the dimensional temperature T we have the following distribution in space: $T = T_2(t)$ for $r \le r_2$: for $r > r_2$ the temperature decreases with r and vanishes at $r = \lambda_T r_2$. In addition, since $\theta \neq 0$ in the region $1 \le \lambda \le \lambda_T$, some fraction of the energy E_0 will remain here. From an integral energy conservation law such as Eq. (1.184) it follows then that the constant $\alpha = E_0/E$ introduced above is increased by $\Delta\alpha$ above the value $\alpha = \alpha_{ht}$ calculated from Eq. (2.48). Denoting by α^* the new value of E_0/E, we have $\alpha^* = \alpha_{ht} + \Delta\alpha$. For $\Delta\alpha$ we can write in this approximation

$$\Delta\alpha = \frac{8}{25}\frac{\sigma_\nu}{(\gamma^2 - 1)g_2\theta_2}\int_1^{\lambda_T}\theta\lambda^2\,d\lambda,$$

where g_2 and θ_2 are known from the solution of the homothermal problem (see Sec. 3).

c) Gray-body-gas model in the diffusion approximation. Self-similar solution.[42]

As before, we confine ourselves to one-dimensional flows of the emitting gas. We choose the basic equations in the form (see Sec. 5, Chap. 1)

$$\frac{\partial v}{\partial t} + v\frac{\partial v}{\partial r} = -\frac{1}{\rho}\frac{\partial p}{\partial r}, \qquad \frac{\partial \rho}{\partial t} + \frac{1}{r^{\nu-1}}\frac{\partial}{\partial r}(r^{\nu-1}\rho v) = 0,$$

$$\frac{\partial}{\partial t}\left[r^{\nu-1}\rho\left(\varepsilon + \frac{v^2}{2}\right)\right] + \frac{\partial}{\partial r}\left\{r^{\nu-1}\left[\rho v\left(\varepsilon + \frac{v^2}{2} + \frac{p}{\rho}\right) + q\right]\right\} = 0, \tag{2.93}$$

$$\frac{1}{k}\frac{\partial}{\partial r}\left[\frac{1}{kr^{\nu-1}}\frac{\partial}{\partial r}(r^{\nu-1}q)\right] - 3q - \frac{16\sigma T^3}{k}\frac{\partial T}{\partial r} = 0. \tag{2.94}$$

Here $k = k(\varepsilon, \rho)$ is the average absorption coefficient and q is the radiation flux. The connection between the temperature and the internal energy is specified as $T = T_0\theta(\varepsilon, \rho)$, where T_0 is a certain constant with the dimension of temperature and θ is a differentiable function. Replacing T in Eq. (2.94) by this relation we get

$$\frac{1}{k}\frac{\partial}{\partial r}\left[\frac{1}{k}r^{\nu-1}\frac{\partial}{\partial r}(r^{\nu-1}q)\right] - 3q - 3\varkappa T_0\left(\theta'_\varepsilon\frac{\partial \varepsilon}{\partial r} + \theta'_\rho\frac{\partial \rho}{\partial r}\right) = 0, \tag{2.95}$$

where $\varkappa$ is the radiant thermal conductivity coefficient [$\varkappa = (16/3) \times (\sigma T_0^3\theta^3/k)$]. It follows from Eq. (2.95) that this approximation is more general than the radiant-thermal-conductivity approximation.

We make next two important assumptions: (1) we approximate the absorption coefficient by the power-law dependence

$$k = k_0\rho^{\beta_1}\varepsilon^{\beta_2}, \tag{2.96}$$

where k_0, β_1, and β_2 are certain constants; (2) we put $\theta'_\rho = 0$ and assume for the coefficient $\varkappa_1 = 3\varkappa T_0\theta'_\varepsilon$ the power-law approximation

$$\varkappa_1 = \varkappa_0\rho^{\delta_1}\varepsilon^{\delta_2}. \tag{2.97}$$

The radiation transport is thus described in this model by relations (2.95)–(2.97). We assume next for simplicity that the initial density of the medium is constant (the results are easily generalized to include the case $\rho_1 = Ar^{-\omega}$), and the connection between p, ρ, and ε is given by $p = (\gamma - 1)\rho\varepsilon$, where γ is the effective adiabatic exponent.

The system of defining parameters for a strong point blast in a gas is

$$r,\ t,\ E_0,\ \rho_0,\ k_0,\ \varkappa_0,\ \gamma,\ \beta_i,\ \delta_i \qquad (i = 1, 2). \tag{2.98}$$

It follows from simple dimensionality considerations that a solution for this equation can be sought in a self-similar form provided that

$$\beta_2 = \frac{1}{\nu}, \qquad \delta_2 = \frac{(\nu - 2)}{2\nu}.$$

Now we consider briefly a method of solving the self-similar problem and some results of this solution. We introduce new unknown functions defined as

$$v = Df(\lambda), \qquad p = \rho_1 D^2 h(\lambda), \qquad \rho = \rho_1 g(\lambda),$$

$$\varepsilon = D^2\psi(\lambda), \qquad q = \rho_1 D^3 \Sigma(\lambda),$$

$$D = \delta \frac{r_2}{t}, \qquad r_2 = \lambda_2 \left(\frac{E_0}{\alpha\rho_1}\right)^{\delta/2} t^{\delta}, \tag{2.99}$$

$$\delta = \frac{2}{\nu + 2}, \qquad \alpha = \text{const.}$$

From the hydrodynamic equations of a radiating gas, with allowance for Eqs. (2.99), follows the set of ordinary equations

$$h' + g(f - \lambda)f' - \frac{\nu}{2} gf = 0,$$

$$gf' + (f - \lambda)g' + \frac{\nu - 1}{2} fg = 0, \tag{2.100}$$

$$g[(f - \lambda)\psi' - \nu\psi] + hf' + \frac{\nu - 1}{\lambda} fh + k^* g^{\beta_1}\psi^{\beta_2} F = 0,$$

$$\frac{1}{k^*} F' g^{-\beta_1}\psi^{-\beta_2} - 3\Sigma - \varkappa^* g^{\delta_1}\psi^{\delta_2}\psi' = 0,$$

$$k^* g^{\beta_1}\psi^{\beta_2} F = \Sigma' + \frac{\nu - 1}{\lambda}\Sigma, \tag{2.101}$$

$$h = (\gamma - 1) g\psi, \tag{2.102}$$

where

$$k^* = k\delta^{2\beta_2}(\rho_0)^{\beta_1 - \beta_2}(\rho^*)^{-\beta_1}(\varepsilon^*)^{-\beta_2}(E_0/\alpha)^{\beta_2},$$

$$\varkappa^* = \varkappa\delta^{\delta_2 - 1}(\rho_0)^{\delta_1 - \delta_2 - 1/2}(\rho^*)^{-\delta_1}(\varepsilon^*)^{-\delta_2}(E_0/\alpha)^{\delta_2 - 1/2},$$

while ρ^* and ε^* are the characteristic density and the characteristic internal energy.

When a shock wave is produced, the parameters of the flux on both sides of the discontinuity surface are related by the conservation laws

$$[g(\lambda^* - f)] = 0, \qquad [h] = g_1\Delta_1[f],$$

$$g_1\Delta_1\left(\frac{f_1^2}{2} + \psi_1\right) - h_1 f_1 - \Sigma_1 = g_1\Delta_1\left(\frac{f_2^2}{2} + \psi_2\right) - h_2 f_2 - \Sigma_2, \tag{2.103}$$

$$h_i = (\gamma_i - 1) g_i \psi_i,$$

where $i = 1$ and 2, $\Delta_1 = \lambda^* - f_1$, the subscripts 1 and 2 refer to parameters ahead of and behind the shock wave, and λ^* is the shock-wave coordinate.

It is necessary to supplement the conditions (2.103), depending on the character of the problem, by the following conditions: either $[\Sigma] = 0$ or else $[\theta] = 0$ (Refs. 10 and 32). The first condition means continuity of the radiation flux, and the second continuity of the temperature (isothermal jump). We assume below the condition

$$[\Sigma] = 0. \tag{2.104}$$

To solve the above problem, the system (2.100)–(2.102) must be integrated under the following conditions: $\psi(\lambda)$ is bounded at the center, $f(0) = 0$ (there is no mass or energy source), $\Sigma(0) = 0$ (the radiation flux is symmetric), and $\Sigma(\infty) = 0$ (there is no source or sink of radiant energy at infinity); in the initial state of the gas, $f_1 = \psi_1 = h_1 = 0$.

Note that the conditions $\Sigma(0) = 0$ and $\Sigma(\infty) = 0$ are essential and do not follow from the usual formulation of the point-blast problem. The parameter α can be determined from the total-energy-conservation law by the integral condition

$$\alpha(\nu, \gamma) = \frac{8\sigma_\nu}{(\nu+2)^2(\gamma^2-1)} \int_0^{r_2} (h + gf^2)\lambda^{\nu-1}\, d\lambda. \tag{2.105}$$

$r_2(t) \lesssim \infty$ is the forward boundary of the perturbation front.

At $k^* = \infty$ we have the case considered above, a strong blast in a heat-conducting gas. We know from the study of this problem that the leading front of the perturbation can be either a shock wave or a thermal (absorption) wave. At a constant thermal conductivity ($\nu = 2$) the perturbation goes to infinity. We shall consider hereafter only the case $\nu = 3$.

This problem contains an energy integral that takes, with allowance for the boundary condition, the form

$$g(f-\lambda)(\psi + f^2/2) + hf + \Sigma = 0. \tag{2.106}$$

The problem can be solved numerically.[42] The shock wave is determined by finding the coordinate λ^* at which the conditions (2.103) are satisfied with the specified accuracy at the discontinuity.

The solutions for certain values of the problem parameters are plotted in Fig. 12 for $\nu = 3$, $\gamma = 1.4$, $\beta_1 = 0$, $\delta_1 = 1$, $k^* = 0.0666$, $\varkappa_* = 1990.9$, and two values of g_0 and ψ_0. Besides the functions $h(\lambda)$, $\psi(\lambda)$, $\Sigma(\lambda)$, and $g(\lambda)$ the figure shows plots of the relative quantity $T/T_0 = (\psi/\psi_0)^{2/3}(g/g_0)^{0.08}$, where $T_0 = T(0, t)$ is the temperature at the symmetry center.

The dependence of ε/D^2 on r/r_2 is not monotonic and has a maximum behind the shock-wave front. Preceding the shock wave is a heated thermal-wave zone whose width depends on the parameters of the problem. On the leading front (at $\lambda > 2$ in the examples considered) of the heat wave we have $\varepsilon = 0$.

The radiation flux $q/\rho_1 D^3$ has a maximum ahead of the shock-wave front and tends to zero as $r \to \infty$. Thus, the flux at infinity is zero in accordance with the formulation of the problem.

The distribution of the pressures, density, and velocity are qualitatively close to those for the homothermal and heat-conduction problems.

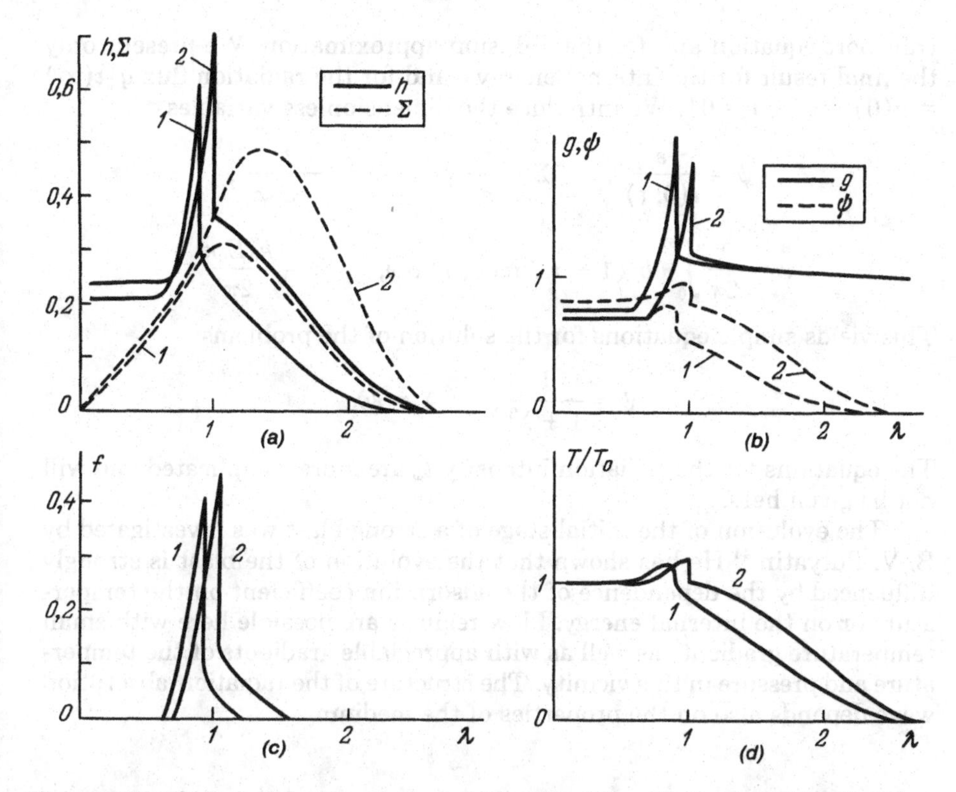

Figure 12. Distributions of the dimensionless pressures h and the radiation flux Σ (Fig. a), of the densities g and the internal energy ψ (Fig. b), of the velocities f (Fig. c), and of the relative values T/T_0 (Fig. d) for $\nu = 3$. Curves 1 are for $g_0 = 0.76$ and $\psi_0 = 0.69$; curves 2—for $g_0 = 0.71$ and $\psi_0 = 0.83$.

The self-similar solution above describes the gas flow produced by concentrated blasts, when the heat wave overtakes the shock wave. For certain parameters it is possible here to obtain also solutions in which the shock wave precedes the heat wave (approximate methods of obtaining such solutions were considered in Refs. 35 and 39).

The problem of local energy release in a self-similar facility for gray-body gas was investigated in Ref. 43 where, however, the motion of the medium was not taken into account at all.

The problem above can be generalized to include a solution that depends on the radiation frequency ω. Indeed, assuming $k = k_0(\omega)\rho^{\beta_1}\varepsilon^{\beta_2}$ and $4\pi(B_\omega)'_\varepsilon = k_\omega B_0(\omega)\rho^{\delta_1\delta}\varepsilon^{\delta 2}$, the problem remains self-similar. It is necessary here, however, to solve the transport equation (1.118) with the frequency taken into account.

In the case $v = 0$ (the motion of the medium is disregarded) and for $k_\omega = k_0\varkappa(\eta)\varepsilon$, $B_\omega = B_0 b(\eta)\varepsilon$, and $\eta = \omega/\omega_0$ (ω_0 is a certain frequency), the problem of instantaneous energy release along a plane has an exact analytic solution obtained by B. V. Putyatin[44,45] both for the complete stationary transport equation and for the diffusion approximation. We present only the final result for the internal energy ε and for the radiation flux $q[\varepsilon(\infty) = q(0) = q(\infty) = 0]$. We introduce the dimensionless variables

$$\psi = \frac{\varepsilon}{\varepsilon(0, t)}, \qquad \Sigma = \frac{\pi t}{E_0} q, \qquad \lambda = \frac{r}{4\pi B_0 \varphi t},$$

$$\varphi = \frac{1}{2\pi} \int_0^\infty b\tau(1 - \tau \operatorname{Arcctg} \tau)\, d\omega, \qquad \tau = \frac{k_0 E_0 \varkappa}{2\pi}.$$

This yields simple equations for the solution of the problems:

$$\psi = \frac{1}{1 + \lambda^2}, \qquad \Sigma = \psi\lambda.$$

The equations for the radiation intensity I_ω are more complicated and will not be given here.

The evolution of the initial stage of a strong blast was investigated by B. V. Putyatin.[44] He has shown that the evolution of the blast is strongly influenced by the dependence of the adsorption coefficient on the temperature or on the internal energy. Flow regimes are possible here with small temperature gradients as well as with appreciable gradients of the temperature and pressure in this vicinity. The structure of the radiation-absorption wave depends also on the properties of the medium.

4.4. Numerical solutions of the problem of a strong blast in an emitting gas

In the absence of self-similarity, the problem of a strong blast in an emitting gas can be solved numerically by a finite-difference method.

The initial data can be specified by using the following models: (1) the self-similar formulations described above; (2) expansion of a uniformly compressed volume of immobile gas; (3) expansion of a non-uniformly compressed gas volume with specified distributions of the velocity, pressures, and densities behind the gas; (4) source of instantaneous or prolonged radiation that is either monochromatic or has a continuous spectrum.

We shall discuss here in detail the expansion of a uniformly compressed volume. Since the initial stage does not play an important role in blast development over scales larger than two or three characteristic initial linear dimensions, we shall not dwell in detail on non-uniform initial data.

It would be most consistent in the considered theory to specify the initial data in terms of solutions of strong-point-blast problems, but there are as yet no detailed calculations for such formulations. Note that the first results in which account is taken of radiation in the gray-body-gas approximation were published by H. L. Brode.[30]

We consider now the expansion of small homogeneous air spheres heated to 30 000 K. Expansion of the high-pressure plasma formation produces a blast wave. An important contribution to the energy transport in the flow region is made by the radiation.

To imitate laser-blast processes (more in Sec. 5 below) we assume the blast energy to be low (from 0.1 to 300 J) and the initial sphere radii to be $r_0 < 1$ cm. Now we formulate the mathematical problem.

The radiation gasdynamics equations in Lagrangian variables for spherical ($j = 2$), cylindrical ($j = 1$), and planar symmetries take the form

$$\frac{\partial r}{\partial t} = u, \qquad \frac{\partial u}{\partial t} = -r^j \frac{\partial p}{\partial m}, \tag{2.107}$$

$$\frac{\partial \varepsilon}{\partial t} + p\frac{\partial V}{\partial t} + V\int_{4\pi}\int_0^\infty k_\omega(B_\omega - I_\omega)\, d\omega\, d\Omega = 0, \tag{2.108}$$

$$\begin{gathered} \Omega\nabla I_\omega = k_\omega(B_\omega - I_\omega), \qquad k_\omega = k(T, V, \omega), \\ p = p(T, V), \qquad \varepsilon = \varepsilon(T, V). \end{gathered} \tag{2.109}$$

Here m is the Lagrange mass, r the Euler coordinate, $V = 1/\rho$ the specific volume, and the remaining symbols are obvious from the equations.

The initial conditions at $t = 0$ are

$$p = p^* \quad \text{for} \quad m \le m_0, \qquad p = p_0 \quad \text{for} \quad m > m_0,$$

$$\rho = \rho_0, \qquad u = 0 \qquad \text{for all } m \qquad (p^* \gg p_0).$$

The boundary-value conditions are

$$u(0, t) = 0, \qquad p(\infty, t) = p_0, \qquad \rho(\infty, t) = \rho_0, \qquad u(\infty, t) = 0,$$

$$I_\omega(0, t, \mathbf{\Omega}) = I_\omega(0, t, -\mathbf{\Omega}), \qquad I_\omega(\infty, t, \mathbf{\Omega}) = 0 \qquad \text{for} \qquad (\mathbf{r}\mathbf{\Omega}) < 0.$$

In view of the complexity of the system (2.107)–(2.109), this problem can be solved only numerically, by finite-difference or finite-element methods.

The qualitative picture of the flow is the following. The blast wave attenuates with time and, starting at certain instants, the parameters of its front become equal to those of a point blast. The gas emits its energy into the surrounding medium, with a substantial fraction of the radiation going to the ultraviolet, so that the temperatures of the central part remain higher than 10 000 K for an appreciable time, while the characteristic linear dimensions of the entire process remain less than 1 m. The examples that follow were calculated by the finite-difference method described in Refs. 30, 45, and 47. The gasdynamics calculations are described in Appendix I. The calculations were made in the diffusion approximation, both with allowance for the radiation selectivity (multigroup method) and for a gray-body gas.

We present a few typical relations from the calculation results* for the spherical case $j = 2$. The calculations were made for the following conditions:

* All these computations were made by B. V. Putyatin, who compiled the algorithms and programs in FORTRAN-IV.

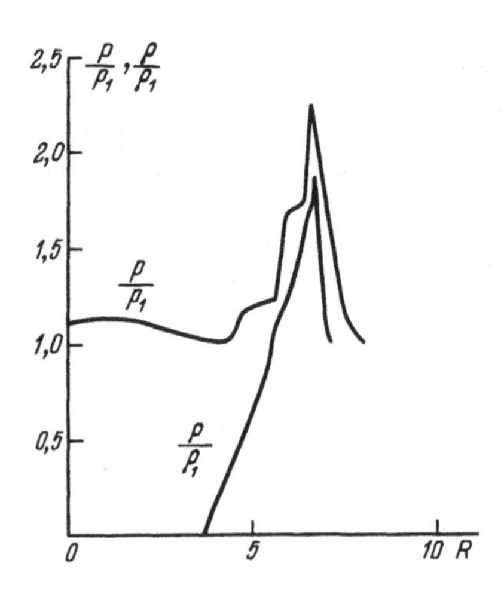

Figure 13. Pressure and density distributions for $t = 42\ \mu s$, $p_1 = 1$ atm, $\rho_0 = 0.00129\ g/cm^3$, and $r_0 = 0.07$ cm (selectively emitting gas).

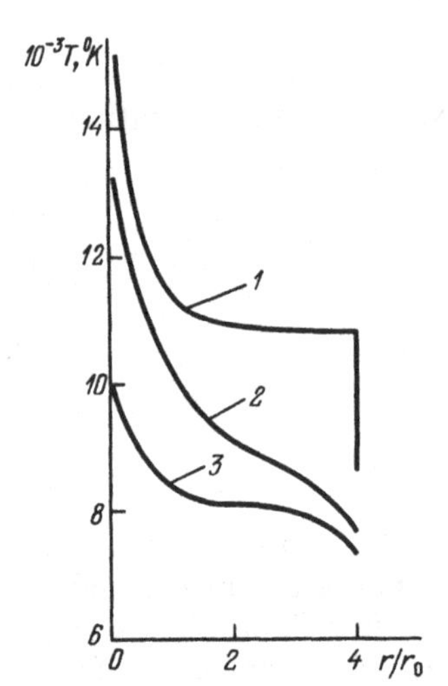

Figure 14. Gas-temperature distribution at $t = 42\ \mu s$ in the vicinity of the symmetry center after an instantaneous spherical detonation of air at $p_1 = 1$ atm, $\rho = 0.00129\ g/cm^3$, $r_0 = 0.07$ cm, and $T_0 = 20\ 000$ K: 1—without emission, 2—selectively emitting gas, 3—gas emitting gray-body radiation.

an air sphere of radius $r_0 = 0.07$ cm was "detonated instantaneously" and acquired an approximate initial temperature $T_0 = 20\ 000$ K at a total energy $E_0 = 180$ J. The blast takes place in a quiescent atmosphere under standard initial conditions with $p_1 = 1$ atm. Figure 13 shows the pressure and density distributions for the instant of time $t = 42\ \mu s$. These distributions correspond qualitatively to a homothermal blast. Figure 14 shows the temperature distribution in the vicinity of the blast center for the same instant of time, with and without taking account of the radiation. The temperature profile is seen to be substantially altered by the radiation. Figure 15 shows plots of the total radiation flux from the high-temperature region. The figure shows calculated data both for the gray gas and with allowance for the radiation selectivity (diffusion approximation). These plots show that the one-group approximation yields qualitatively correct results, but overesti-

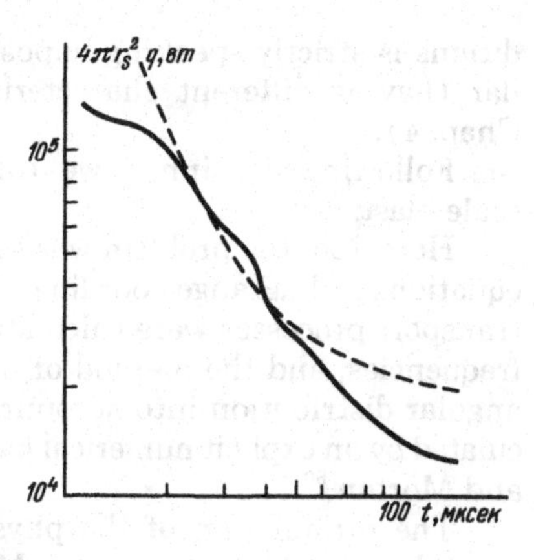

Figure 15. Power lost by radiation from an air plasma (spherical case); r_s is the radius of blast wave. Dashed curve—gray-body gas.

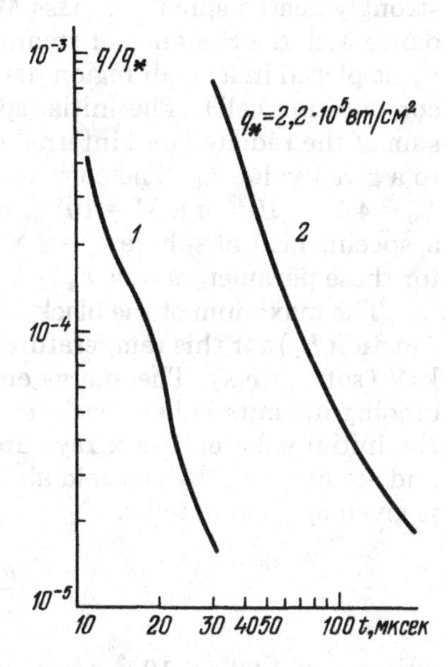

Figure 16. Radiation flux q from an expanding air plasma in the visible and ultraviolet bands. The initial conditions are the same as for Figs. 13–15; 1 – $0.4 < \lambda < 0.78$ μm, 2 – $0.14 < \lambda < 0.4$ μm.

mates somewhat the energy radiated from the plasma. This conclusion is in full agreement with the results of Latko and Viskanta[48] for the propagation of a planar blast.

Figure 16 shows a plot of the radiation flux from a disturbed region in the visible band ($0.4 < \lambda < 0.78$ μm) and in the ultraviolet ($0.14 < \lambda < 0.4$ μm). It is seen from these data that an appreciable fraction of the energy goes off from the heated-gas region to the ultraviolet band. This radiation, however, is fenced in by the cold air and subsequently an appreciable fraction returns to the moving zone behind the propagating shock wave.

The plot of the shock-wave motion at $r > 2r_0$ is close to that for a point blast, with $r_2 \sim t^{2/5}$ during the strong stage of wave propagation. Note that extrapolation of the present results to high energies and other initial conditions is strictly speaking impossible, since the phenomena are not similar (having different characteristic optical thicknesses; for details see Chap. 4).

Following J. Zinn,[49] we turn now to the calculation of a large-scale blast.

Here, too, the problem was solved by using the radiative-gasdynamics equations in Lagrange coordinates [see Eqs. (2.107)–(2.109) above]. The transport processes were calculated in a multigroup approximation in the frequencies, and the method of discrete coordinates was used to take the angular distribution into account. The hydrodynamic equations were calculated by an explicit numerical method described in the book of Richtmeyer and Morton.[50]

The formulation of the physical model problem is the following: a strongly heated sphere of mass M, emitting into the ambient air in accordance with the Stefan–Boltzmann law at a surface intensity σT^4 for a time τ_*, is placed in a small region near a symmetry center (of size equal to one computation cell). The initial sphere temperature is chosen to make the sum of the radiated and internal energies of this sphere (of mass M) equal to a given value E_0. The computation was made for the initial parameters $E_0 = 4.18 \times 10^{22}$ erg, $M = 10^6$ g, sphere area $4\pi r_0^2 = A = 10^5$ cm^2, $\tau_* = 10^{-7}$ s, specific heat of sphere $c_v = 2 \times 10^8$ erg/(g deg). The initial temperature for these parameters was $T_* = 1.8 \times 10^7$ K (1.6 keV).

The maximum of the black-body radiation curve (relative to the Planck function B_ω) for this temperature corresponds to a photon energy $\hbar\omega = 4.5$ keV (soft x rays). The energy emitted in the time T_* and during the succeeding instants is held back in the vicinity of the symmetry center, near the initial sphere. The x rays are subsequently reradiated in the hot gas and are absorbed by the cold air. The mean free path of the x-ray photons is given approximately by[30]

$$\lambda_\omega = \frac{\rho}{4\rho_0} (\hbar\omega)^{-2.78} \text{ cm};$$

where $\rho_0 = 1.293 \times 10^{-3}$ g/cm^3, and $\hbar\omega$ is the photon energy in keV ($\hbar\omega > 0.55$ keV). The ambient cold air is heated to high temperatures (of order 10^6 K). The result is an almost isothermal sphere of fully ionized gas with an abrupt temperature boundary, but no substantial gas flow sets in. The fully ionized gas does not block the x-ray photons.

Outside the isothermal sphere there is a region where the temperature drops rapidly (exponentially). This region is due in part to the hard x-ray photons present in the Planck distribution. A fast supersonic absorption wave propagates thus through the gas. Examples of the variation of the optical coefficients of the air are shown in Fig. 17 in the temperature range from 10^7 to 10^3 K. It follows from the plots that in the range from 300 to 20 000 K the absorption coefficients increase very rapidly with temperature, and for the characteristic linear dimensions of the problem the air is in fact opaque above 8000 K.

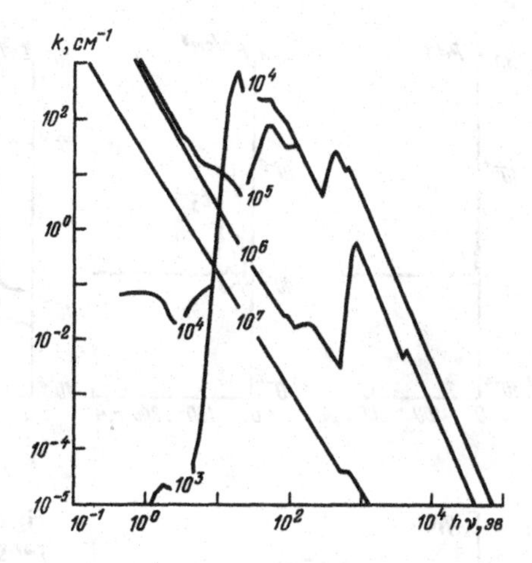

Figure 17. Absorption coefficient k of air for different temperatures versus the photon energy $\hbar\omega$ (Ref. 49): $\hbar\nu = \hbar\omega$, $h = 6.62491 \times 10^{-27}$ erg/s, ρ is the frequency, and $= 1.293 \times 10^{-3}$ g/cm³. The temperatures (in K) are marked on the curves.

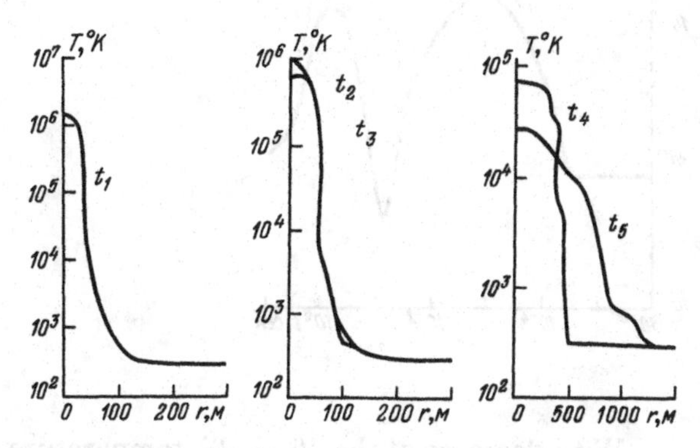

Figure 18. Temperature distribution behind the shock-wave front[49] ($E_0 = 4.2 \times 10^{22}$ erg): $t_1 = 10^{-6}$ s, $t_2 = 5 \times 10^{-5}$ s, $t_3 = 3 \times 10^{-4}$ s, $t_4 = 9 \times 10^{-2}$, $t_5 = 0.8$ s.

The air region heated to 8000 K extends much farther than the isothermal-sphere boundary.

Plots of the temperature and density distributions in space for the instants $t_1 = 10^{-6}$ s (1), $t_2 = 5 \times 10^{-5}$ s and $t_3 = 4 \times 10^{-4}$ s (2), and $t_4 = 9 \times 10^{-2}$ s and $t_5 = 0.8$ s (3) are shown in Figs. 18 and 19. It can be assumed that at the instant t_1 the radius of the visible fireball corresponds to the coordinate where the temperature is 8000 K. At $t = 10^{-4}$ the isothermal sphere grows via radiation transport. The internal parts emit photons of approximate energy 10 eV or higher (vacuum ultraviolet). For the temperature interval 10^4–10^6 K the coefficients of air absorption for these frequencies decrease as the temperature is lowered. The photons leaving the isothermal sphere have a large mean free path until they reach the cold air layers. The emission from the isothermal sphere and absorption at the boundary with the cold air cause the heated region to grow rapidly. A small temperature gradient is then maintained inside the hot sphere as the average value of this temperature decreases. An abrupt increase of the radiation in the visible band is observed in that location in the hot region where the temperature reaches 8000 K.

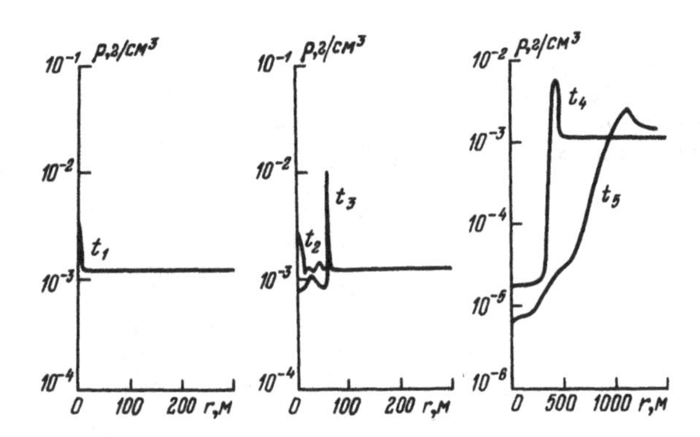

Figure 19. Density distribution behind the shock-wave front (the energy and the times are the same as in Fig. 18).

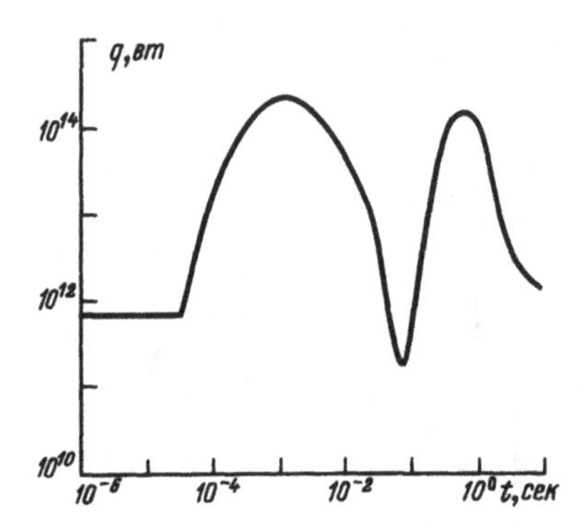

Figure 20. Radiation flux in the visible frequency range.[49]

With decrease of the "fireball" temperature, conditions are reached ($T \approx 500\,000$ K) at which the rate of radiative expansion is less than the speed of sound, and the mechanism, noted above, of disturbance propagation with the aid of a supersonic absorption wave no longer operates. A strong shock wave is produced and the hydrodynamic phase of the blast development begins. By the instant $t = 3 \times 10^{-4}$ s the shock wave is already completely formed. The process evolves then qualitatively in the same way as described for the self-similar case for a gray-body gas, with the shock wave gradually overtaking the leading front of the thermal absorption wave. Note that on going from the supersonic to the subsonic regime, the absorption wave goes through a detonation state in which the Jouguet condition $D = a_2 + v_2$ is met for the shock-wave velocity.

Of considerable interest is the time dependence of the total energy yield $q = q_v$ in the visible band (Fig. 20). During the initial stage the heat flux q_v changes little and is due to radiation by the high-temperature fireball. After the onset of the strong shock wave, the air is heated not only by reradiation from the central regions, but also by shock compression in this wave. The shock wave attenuates with time, and the gas heating in it weakens. On the other hand, the area of the wave front increases continuously. The energy output in the optical band increases as a result of these processes approximately to a time $t = 10^{-3}$ s, and then begins to decrease. The reason for this decrease is that the compressed air near the shock wave becomes optically opaque and "blocks" the radiation from the inner heated regions. When the shock-layer temperature decreases with time, the radiation from it also decreases and, notwithstanding the decreased opacity of this layer and the increased energy output from the fireball, the total heat flux decreases and reaches a minimum at $t = 9 \times 10^{-2}$ s. Starting with this minimum, the radiation flux from the inner region exceeds the radiation flux from the shock layer. The shock layer next becomes transparent to visible radiation, and the radiation power increases strongly. Next, as the fireball expands and cools, the radiated-energy flux goes through a maximum by the instant 0.8 s, and begins to decrease. A radiative cooling wave passes through the fireball, and the ball cools to below 8000 K and becomes gradually transparent. At instants $t > 10$ s the radiation power becomes low ($q < 10^{12}$ W). However, in view of the long action time, a noticeable fraction of the total blast energy is emitted during this stage.

These are the characteristic features of a large-scale blast in air from a pointlike energy source. For other gases the picture varies with their thermodynamic properties and in connection with substantial changes of the absorption coefficients.

Note also that during the initial stage the temperature distribution and the shock-wave formation can be influenced by the electronic thermal conductivity. Zinn,[49] whose procedure we followed, did not take the electronic thermal conductivity into account.

The expansion of large spheres (100–200 m in radius) and cylinders ($j = 1$) was investigated in Ref. 51. It was found that if the sphere temperature exceeds 10^4 K, but is lower than 3×10^4 K, approximately 10% of the total energy of the initially compressed sphere is radiated from the heated zone. It was found that solution of the problem in the single-group approximation (gray body), and in the 10-group approximation with allowance for the spectral properties, overestimates by 1.5–3 times the energy emission obtained in the gray-body theory. The flow patterns in the spherical and cylindrical cases are quite similar.

Expansion of a flat layer ($j = 0$) of heated gas was investigated by I. V. Nemchinov and co-workers.[52]

Note also V. V. Svettsov's result.[53] He has shown by a numerical method that in expansion of a sphere with energy 10^{11} erg (10 kJ) there are observed two regimes of the initial motion, depending on the density of the surrounding air. At characteristic low optical thicknesses ($\rho = 0.01\rho_0$, $r_0 l_p^{-1} \approx 0.001$, $T_0 \approx 25 \times 10^4$ K), the heating layer ahead of the shock wave is large and the compression in the shock wave is small. If the optical thicknesses are large ($\rho = 3\rho_0$, $r_0 l_R^{-1} \approx 10^2$, $T_0 \approx 25 \times 10^4$ K), the character of the radiation transport is close to radiant heat conduction, and the solution of the problem recalls a homothermal point blast with large compression in the shock wave (the density increases almost tenfold).

This result is in qualitative agreement with the analytic investigation[44] discussed above.

The results above show that much progress was made during the last decade in research into blasts with allowance for radiation transport. Many unsolved problems remain, however.

5. Some applications of strong-blast theory

5.1. Initial stage of gas motion in a strong blast

A major stimulus in the development of strong-blast theory was its application to atomic (nuclear) blasts. We shall not dwell here in detail on this question, since it is expounded in Refs. 2, 3, 10, 30, and 54–56.

We note only the following:

(1) Experimental data on the time dependence of the shock-wave blast is well approximated by an equation of the type

$$\frac{5}{2}\lg r_2 - \lg t = C_1, \tag{2.110}$$

where C_1 is a constant.

According to the data[54] obtained in 1945 for the atomic bomb in New Mexico, a constant $C_1 = 11.915$ should be assumed.

The theoretical equation $r_2 = (E_0/\alpha\rho_1)^{1/5}t^{2/3}$ agrees with Eq. (2.110) if it is assumed that

$$0.5 \lg \frac{E_0}{\alpha\rho_1} = 11.915.$$

If ρ_1 and α are known, this equation can be used to estimate the energy fraction E_0 used to move the gas.

The numerical value of α (meaning also the theoretical estimate of E_0) depends on the flow model assumed.

Thus, for an adiabatic model of a perfect gas with constant γ we have for $\nu = 3$ (see Sec. 2)

$$\alpha_{ad} = 0.31246(\gamma - 1)^{-(1.1409+0.11735\lg(\gamma-1))} \qquad (1.2 \le \gamma \le 2).$$

For the homothermal model of flow at $\nu = 3$ we have $\alpha_{ht} = 0.0643 + 0.163/(\gamma - 1)$. The inequality $\alpha_{ad} > \alpha_{ht}$ holds for equal γ. It follows hence that the energies E_0 consumed in gas motion (obtained from α_{ad}) will be overestimated if heat conduction is not taken into account. Since the dissociation and ionization energies have been taken into account by using a rather approximate γ, exact allowance for these processes can also change the value of E_0 (Refs. 24 and 57).

(2) The temperature distribution in the region between the center and the leading front of the perturbation becomes closer to the real one if account is taken of the heat transport by the radiation. At the first instants after

the blast (when the temperature is higher than 3×10^5 K) the velocity of the heat wave exceeds that of the shock wave[10,30,49] and the temperature distributions are close to those given by Eq. (2.32). Next, the solutions of the homothermal problem can be approximately used up to the instant of detachment of the shock wave from the fireball.[55] For later instants of time, at a density close to the normal one near the earth's surface, the temperature can be estimated by using the gray-body gas model considered above (see also Ref. 30).

With further decrease of the temperatures in the strong-wave stage, the flow can be adequately described by the adiabatic theory with allowance for dissociation and ionization. The temperatures near the center can be calculated in this case by the method proposed in Ref. 39 for splicing the asymptotic expansions.

The most correct and consistent theoretical solution leads to allowance for the radiation within the framework of radiative gasdynamics in which the thermal emission and absorption are frequency selective, in accordance with Ref. 49. For a qualitative analysis of the initial stage of the blast development it is useful to use the modern theory of propagation of absorption waves (of ionization fronts).

(3) The total energy radiated from the hot zone is not monotonic in time. This is clearly seen from Zinn's theoretical solution considered above. H. L. Brode[30] cites the following empirical rules for the calculation of the thermal pulse:

time of first maximum = $W^{1/3}\eta$ (ms),

time of first minimum = $60W^{0.4}\eta$ (ms),

time of second maximum = $0.90(W_\eta)^{0.42}$ (ms).

power in second maximum – $1.2.10^{15}W^{0.6}\eta^{-0.42}W$,

ratio of thermal momentum to total energy = $0.4 + 0.06W^{1/2}/(1 + 0.5W^{1/2})$,

ratio of thermal momentum in the visible frequency band to the total energy

$$=0.27 + 0.06\eta\frac{0.12W^{1/2}}{1 + W^{1/2}} + \frac{114\eta}{8200\eta^2 + 1},$$

where W is the TNT-equivalent energy in MT (1 MT $\approx 4.2 \times 10^{22}$ erg), $\eta = \rho/\rho_0$, and ρ_0 is the characteristic initial density.

As already noted, heat-conduction effects influence strongly not only the temperature distribution but also the velocity, pressure, and density distributions.

5.2. Electric explosion of conductors and discharges in a gas

The conclusions of the strong-blast theory can be used also to consider gas disturbances produced by electric discharges along conductors such as thin metallic wires and plates.[27,58–61] In view of the relatively low temperatures, it is frequently possible here to use an adiabatic model with single ionization.

The experimental results lead to the following conclusions. For thin straight conductors (wire thickness less than 1 mm) the explosion energy is of the order of 3×10^8 erg/cm, and the dependence of the coordinate of the primary shock wave (as a rule, secondary shock waves are produced in the stream) is close to the self-similar $r_2 = (E_0/\alpha\rho_1)^{1/2}t^{1/2}$ if $r_2 > 10r_0$, where r_0 is the radius of the conductor and t is the time after the end of the discharge. This is confirmed also by Rose's calculations[27] of the theoretical model of an exploding wire in air with allowance for dissociation and ionization. In particular, it is shown here that the ratio α_a of the energy parameters for air with allowance for equilibrium ionization and dissociation for the case of a perfect gas with $\gamma = 1.4$ is equal to 2.18, i.e., $\alpha_a = 2.18\alpha(\gamma)$. This corresponds to an effective exponent γ_{eff} for air close to 1.2.

The secondary shock waves produced by exploding wires are due to expansion of the metal vapor. This phenomenon cannot be described by the usual theory of a point blast without a mass supply. Similar conclusions can be drawn also for the case of explosion of thin metallic plates, if this process is compared with a planar strong point blast.

Besides the explosion of wires and plates, high temperatures are obtained by producing in a gas discharges between planar or cylindrical electrodes. Consider a spark discharge in air between two electrodes. The general qualitative picture is here the following. Ionization by the electric field produces in an air-filled discharge gap a thin current-conducting channel. In this channel the air is Joule-heated to temperatures on the order of several times ten thousand degrees and is ionized. The rapid temperature rise increases the pressure drastically and a strong cylindrical shock wave propagates through the gas. Spark discharge phenomena were investigated by S. L. Mandel'shtam and his associates[62] and also by Bobrov and Shats.[15,16] Without dwelling on this phenomenon in detail, we point out only that measurements of the density distribution behind the shock-wave front have shown that the gas motion is close to that due to a cylindrical strong blast, and the average density in the region of the spark channel is approximately 10^{-3} of the initial air density. Similar phenomena are produced in the atmosphere by lightning.

Processes close to planar explosion are observed in electric discharges along planar current-carrying channels[63] or in high-current discharges in T-shaped tubes.[64,65] The discharge is produced here in the short part of the T-shaped tube. The abrupt rise of the temperature in the part of the T-shaped tube perpendicular to the discharge tube initiates propagation of a planar strong shock wave.

Strong planar blast theory was not used to interpret Cash's experimental results[64] on discharges in air. It is indicated there, however, that the relation between the time of arrival of the shock wave and its position is well approximated by $t = ax^{8/5}$, where x is the distance from the discharge and a is a constant.

For a planar strong explosion we have

$$t = (\alpha\rho_1/E_0)^{1/2} r_2^{3/2}. \tag{2.111}$$

Since the exponent $\frac{8}{5}$ differs from $\frac{3}{2}$ by only $\frac{1}{10}$, it is clear that reduction of the data by using Eq. (2.111) would also agree well with experiment. Kolb's experiments[65] on discharges in hydrogen and deuterium have shown that in a certain range of distances from the point of discharge there is satisfactory agreement between the strong-blast theory and measurements of shock-wave velocities along a tube. Using Kolb's data one can determine the specific energy E_0 consumed by gas motion behind the wave front.

Kolb's paper[65] contains in fact a table according to which $D = 5.6 \times 10^6$ cm/s at $r_2 = 9$ cm. Since all the energy goes to one side of the blast point, we have $2E_0 = \alpha\rho_1(3D/2)^2 r_2$. Assuming for γ the value 1.3 and an initial density $\rho_1 = 1.8 \times 10^{-7}$ g/cm^3 (the initial pressure was 0.7 Torr), we get $E_0 \sim 10^8$ erg/cm^2. Note also that the data in that table indicate that the product $D^2 r_2$ varies little in the distance range from 3 to 11 cm along the tube (the total tube length is 12 cm).

The foregoing analysis allows us to conclude that the strong-blast theory can be applied also to this group of problems of practical importance.

5.3. Local heating of a gas by a laser beam

Recently, strong point-blast theory has found an important new application for processes connected with gas heating by a focused pulsed laser beam. The laser beam can produce in the focusing region quite high gas temperatures (about a million degrees) in a volume on the order of 10^{-3} cm^3. This produces a strong shock wave that ionizes the gas and propagates at an approximate velocity 10^8 cm/s. The time needed to supply the energy is quite short and can be of the order of 10^{-7} s. (We have in mind here not the time in which the laser supplies the energy, on the order of 5×10^{-9} s, but the time of shock-wave formation.) The laser produces in a pulse an energy 0.1–100 J and higher. Two principal methods are used to obtain high temperatures in gases: breakdown (or "spark") induced in the gas by the laser, or focusing on a target placed in a sufficiently rarefied gas. In the first method the energy release in the gas is as a rule not isotropic in the vicinity of the focal spot (see, e.g., Refs. 66 and 67). It is possible to produce[68,69] even a long spark, about 1 m and more, with a variable specific energy input to the gas, i.e., we have here energy release along a straight line, with $E_0 = E(z)$, where z is the coordinate along this line. If edge effects are disregarded, a cylindrical explosion will be produced along a straight line at times longer than the shock-wave formation time, when $E_0 = E(z)$. We shall also deal with problems of this type in Chap. 5.

When a target is used in the focal region, a strong shock wave is produced as a rule.[69,70] Let us estimate the energy going into the shock wave and into gasdynamic motion during the detonation stage. It follows from

Ref. 71 that at an instant of time $t = 1.5 \times 10^{-6}$ s after the breakdown in air (at atmospheric pressure) the shock-wave coordinate measured in the experiment along the laser-beam axis is approximately 1 cm. Using Eqs. (2.13) we have

$$E_0 = \alpha \rho_1 r_2^5 / t^2.$$

Choosing approximately $\alpha = 1.5$, which corresponds to $\gamma = 1.25$ and $\rho_1 = 1.25 \times 10^{-3}$ g/cm^3, we find from this equation that $E_0 \approx 7$ J. This estimate agrees well with the experimental pulse energy, 10 J. Recognizing that the detonation energy did not spread over the total 4π solid angle, but in a smaller part of space (this follows from the pictures of the waveform), we obtain a somewhat smaller value of E_0, i.e., approximately 70% goes to shock-wave formation. The remaining energy is used to produce the breakdown and goes off in the form of optical radiation and other losses.

Motion-picture photographs of the evolution of a laser blast in air are given in Ref. 72. It follows from these results that at an energy $E_0 \approx 0.5$ J the "detachment" of the shock wave from the brightly glowing plasma zone lasts 20 μs. After 40 μs the shock-wave diameter reaches 4 cm (the initial breakdown disturbance is approximately 1 cm long and the time to supply the energy is 5×10^{-9} s).

The foregoing data point to new applications of point-blast theory and to the need for formulating new problems in this theory. We mention only one such problem. Consider a spherical volume of radius r_0 filled with a quiescent plasma having parameters ρ_1, p_1, and T_1, while outside the volume the gas parameters are ρ_0, p_0, and T_0 ($\rho_0 < \rho_1$).

An energy $E_0 4\pi r_0^2$ is released at the instant $t = 0$ in a narrow spherical layer of thickness $\Delta r = r_0 - r_1 \ll r_0$. It is required to determine the resultant motion. This problem can be encountered in spherical focusing of a laser beam on a plasma target, if the laser-light absorption length is small and the energy-release time is considerably shorter than $r_0(\rho_1/\gamma p_1)^{1/2}$. As $\Delta r \to 0$, we have the problem of a surface blast on the interface of two media, with specific energy E_0 (see in this connection also Sec. 5 of Chap. 3). Initially, a contact surface and shock waves traveling to and from the center are produced, and zones with quite high densities and temperatures can evolve in the course of the plasma motion.

Chapter 3
Linearized non-self-similar one-dimensional problems

1. Linearization of gasdynamic equations

In the preceding chapter we considered the exact solution of the self-similar problem of a strong point blast in a gas at rest. It was assumed that the back pressure p_1 can be neglected. This assumption is valid only during the initial stage of the blast, i.e., for short instants of time. In the course of propagation of the shock wave, the influence of the initial pressure becomes substantial. The initial conditions of the problem acquire therefore an additional dimensional parameter p_1, by virtue of which the problem ceases to be self-similar and all the dimensional characteristics of the flow no longer depend on one but rather on two variables. A non-self-similar problem can be solved by numerically integrating the system of gasdynamics equations. These topics will be considered in Chap. 4. If $(p_2 - p_1)/p_1 < 1$, however, the problem can be solved by a simpler method[1-4] based on linearizing the initial equations about a self-similar solution. More generally speaking, we assume[5,6] that the initial pressure p_1 is variable, and that

$$\rho_1 = Ar^{-\omega}, \tag{3.1}$$

$$p_1 = Cr^{-\omega_*}. \tag{3.2}$$

We dispense also with the condition that the initial gas velocity v_1 be zero, and put[5,6]

$$v_1 = Br^{\omega-\nu+1}. \tag{3.3}$$

The case where account is taken of the heat released on the shock-wave front will be dealt with later (in Chap. 6).

The question of realizing the initial states (3.1)–(3.3) will not be discussed here. We note only that the density ρ_1 and the velocity v_1 satisfy a stationary continuity condition. The distributions (3.1) and (3.2) can be used to approximate (for $\omega = \omega_*$) the pressures and densities in isothermal stellar atmospheres. If the state (3.1)–(3.3) is not strictly in equilibrium, we shall neglect the nonstationary flow that can appear ahead of the blast wave. Naturally, for $B = 0$ and $\omega_* = 0$ the initial state can be considered to be in equilibrium. Since the dimensionalities of the constants C and B for arbitrary ω and ω_* are not expressed in terms of the dimensionalities of E_0 and A, the point-blast problem in a gas with an initial state (3.1)–(3.3) becomes non-self-similar if at least one of the constants, C or B, differs from zero. The general formulation of the problem of linearizing the gasdynamic equations was considered in Sec. 1 of Chap. 1. Let us dwell on this question in greater detail for the case where the principal motion corresponds to the exact solution (2.5)–(2.12).

We use a zero subscript to label functions corresponding to this solution, which we seek in the form

$$\begin{aligned} v &= v_0(r,t) + \varepsilon v_{(1)}(r,t), \\ p &= p_0(r,t) + \varepsilon p_{(1)}(r,t), \\ \rho &= \rho_0(r,t) + \varepsilon \rho_{(1)}(r,t), \end{aligned} \tag{3.4}$$

where ε is a small parameter that can be taken to be the constant C or B; we use the subscript 1 to label the new unknown functions (additions to the main solution, or perturbations). By the standard linearization procedure indicated in Sec. 1 of Chap. 1, we arrive at a system of linear partial differential equations with coefficients that depend on v_0, p_0, and ρ_0. We assume that in the linearized equations obtained from a system in the form (1.31) the functions $v_{(1)}$, $p_{(1)}$, and $\rho_{(1)}$ and their derivatives approach finite limits as $\varepsilon \to 0$. If this assumption is correct, the linearization is justified. We shall therefore strive to obtain for the solutions of the linearized system conditions under which the conditions that $v_{(1)}$, $p_{(1)}$, and $\rho_{(1)}$ be finite are met. In certain problems these conditions are not met, and as a rule locally (e.g., in the vicinities of certain singular points). In these cases the linearization becomes strictly speaking meaningless, but in some problems, far from certain singular points, the solution of the linearized equations can correctly describe the considered motion qualitatively and quantitatively. From this standpoint solutions that are unbounded at individual points may turn out to be useful. We use hereafter a formalism in which the equations are made nondimensional and small dimensionless parameters proportional to C or B are introduced.

We choose the initial system of gasdynamics equations for adiabatically perturbed motions of an ideal gas in the form

$$\begin{aligned} \frac{\partial v}{\partial t} + v\frac{\partial v}{\partial r} + \frac{1}{\rho}\frac{\partial p}{\partial r} &= 0, \\ \frac{\partial \rho}{\partial t} + v\frac{\partial \rho}{\partial r} + \rho\left(\frac{\partial v}{\partial r} + (\nu-1)\frac{v}{r}\right) &= 0, \\ \frac{\partial p}{\partial t} + v\frac{\partial p}{\partial r} + \gamma p\left(\frac{\partial v}{\partial r} + (\nu-1)\frac{v}{r}\right) &= 0. \end{aligned} \tag{3.5}$$

We denote by $r_2(t)$ the leading front of the gas disturbance which we shall regard as an ordinary shock wave. We have then as before, besides the initial conditions (3.1)–(3.3), the condition $r_2(0) = 0$ and the condition that an energy E_0 be released at $r = 0$. In accordance with Eq. (1.60), the conditions on the shock wave take at $r = r_2$ the form

$$\rho_1(D - v_1) = \rho_2(D - v_2),$$
$$\rho_1 v_1(D - v_1) - p_1 = \rho_2 v_2(D - v_2) - p_2, \tag{3.6}$$
$$\rho_1(D - v_1)\left(\frac{v_1^2}{2} + \frac{p_1}{\rho_1}\frac{1}{\gamma - 1}\right) - p_1 v_1 = \rho_2(D - v_2)\left(\frac{v_2^2}{2} + \frac{p_2}{\rho_2(\gamma - 1)}\right) - p_2 v_2.$$

In addition to Eqs. (3.6) we can write down the conditions at the symmetry center of the flow. Thus, if there is no active source at the center, we have

$$v(0, t) = 0. \tag{3.7}$$

We consider separately two cases: (a) $B \neq 0$, $C = 0$ and (b) $C \neq 0$, $B = 0$. We introduce the system of dimensionless variables

$$f = v/D, \qquad g = \rho/\rho_1, \qquad h = p/\rho_1 D^2. \tag{3.8}$$

We have used similar variables in Chap. 2. Since there is no self-similarity, the functions (3.8) will now depend on two dimensionless variables.

We introduce two new dimensionless variables:

$$y = v_1/D, \qquad q = \gamma p_1/\rho_1 D^2. \tag{3.9}$$

For the case $\omega \geq \nu - 1$ the parameter y is small for finite B and for small t. By virtue of the property w (Sec. 2, Chap. 1) we have for shock waves $D - v_1 \geq a_1$ and $D \geq a_1 + v_1$, i.e., we always have $D > v_1$ or $y < 1$. The parameter q is the ratio of the square of the sound velocity in the unperturbed medium to the square of the shock wave. This parameter is also small in the initial stage of the blast development.

The velocity and the pressure p_1 of a propagating blast wave can be treated separately. For problem (a) we introduce in place of r and t the new independent variables

$$\lambda = r/r_2, \qquad y, \tag{3.10}$$

and assume that f, h, and g are functions of λ and y. When only back pressure is taken into account [problem (b)], we introduce the independent variables

$$\lambda = r/r_2, \qquad q, \tag{3.11}$$

where f, h, and g are functions of λ and q. To write down directly the dimensionless equations for the variable systems (3.8), (3.10) and (3.8), (3.11), we introduce the notation $z_1 = y$ and $z_2 = q$.

For simplicity we assume for the time being that $\omega = \omega_*$. The gasdynamics-equations system (3.5), with allowance for the relations for the derivatives $\partial/\partial r = (1/r_2)(\partial/\partial\lambda)$ and

$$\partial/\partial t = (D/r_2)[r_2(dz_i/dr_2)(\partial/\partial z_i) - \lambda(\partial/\partial\lambda)],$$

takes in the new variables the form

$$(f_i - \lambda)\frac{\partial f_i}{\partial \lambda} + \eta_i\left(-\frac{f_i}{iz_i} + \frac{\partial f_i}{\partial z_i}\right) + \theta_i f_i + \frac{1}{g_i}\frac{\partial h_i}{\partial \lambda} = 0, \tag{3.12}$$

$$(f_i - \lambda)\frac{\partial g_i}{\partial \lambda} - \omega g_i + \eta_i\frac{\partial g_i}{\partial z_i} + g_i\left(\frac{\partial f_i}{\partial \lambda} + (\nu - 1)\frac{f_i}{\lambda}\right) = 0, \tag{3.13}$$

$$(f_i - \lambda)\frac{\partial h_i}{\partial \lambda} + \varkappa_i h_i + \eta_i\left(\frac{\partial h_i}{\partial z_i} - \frac{2h_i}{iz_i}\right) + \gamma h_i\left(\frac{\partial f_i}{\partial \lambda} + (\nu - 1)\frac{f_i}{\lambda}\right) = 0, \tag{3.14}$$

where

$$\begin{gathered}\eta_i = r_2\frac{dz_i}{dr_2}, \qquad \theta_1 = \omega - \nu + 1, \qquad \theta_2 = 0,\\ \varkappa_1 = \omega - 2\nu + 2, \qquad \varkappa_2 = -\omega, \qquad i = 1, 2.\end{gathered} \tag{3.15}$$

To investigate the influence of the velocity v_1 or of the pressure p_1 we seek a solution of the systems (3.12)–(3.15) in the form

$$\begin{aligned} f_i(\lambda, z_i) &= f_0(\lambda) + z_i f_{i1}(\lambda) + o(z_i),\\ g_i(\lambda, z_i) &= g_0(\lambda) + z_i g_{i1}(\lambda) + o(z_i),\\ h_i(\lambda, z_i) &= h_0(\lambda) + z_i h_{i1}(\lambda) + o(z_i),\end{aligned} \tag{3.16}$$

$$\frac{dz_i}{d\ln r_2} = \frac{b_i z_i}{1 + A_{i1}z_i + o(z_i)}. \tag{3.17}$$

Here f_0, g_0, and h_0 are known self-similar functions, $b_1 = (\nu + \omega - 4)/2$, $b_2 = \nu - \omega$, $o(z_i)$ denotes quantities of higher order of smallness compared with z_i as $z_i \to 0$, and A_{i1} are constants to be determined. Substituting the solution (3.16), (3.17) in the system (3.12)–(3.14) and neglecting all quantities of order higher than z_i, we obtain the following system of linear ordinary differential equations for the functions f_{i1}, g_{i1}, and h_{i1}:

$$g_0[mf'_{i1} + (f'_0 + b_{i1})f_{i1} + s_i A_{i1} f_0] + h'_{i1} + \left(m_0 f'_0 - b_2\frac{f_0}{2}\right)g_{i1} = 0, \tag{3.18}$$

$$mg'_{i1} + g_0\left(f'_{i1} + \frac{\nu - 1}{\lambda}f_{i1}\right) + g'_0 f_{i1} + \left(k_i + f'_0 + \frac{(\nu - 1)f_0}{\lambda}\right)g_{i1} = 0, \tag{3.19}$$

$$\begin{aligned} &mh'_{i1} + \gamma h_0\left(f'_{i1} + (\nu - 1)\frac{f_{i1}}{\lambda}\right) + h'_0 f_{i1}\\ &\qquad + \left(m_{i1} + \gamma f'_0 + (\nu - 1)\gamma\frac{f_0}{\lambda}\right)h_{i1} + m_{i0}A_{i1}h_0 = 0,\end{aligned} \tag{3.20}$$

where

$$m = f_0 - \lambda, \qquad b_{11} = \omega - \nu + 1, \qquad b_{21} = \frac{b_2}{2}, \qquad s_i = \frac{b_i}{i}, \qquad k_i = b_i - \omega,$$

$$m_{11} = b_1 - \nu, \qquad m_{21} = -\omega, \qquad m_{i0} = \frac{2b_i}{i}, \qquad f'_0 = \frac{df_0}{d\lambda}, \qquad f'_{i1} = \frac{df_{i1}}{d\lambda} \quad \text{etc.}$$

When considering the linearized solutions we must linearize also the boundary conditions of the problem.

We write the conditions on the shock wave (at $\lambda = 1 - 0$) in dimensionless form

$$(1 - y) = g_2(1 - f_2), \qquad y(1 - y) - \frac{q}{\gamma} = g_2 f_2(1 - f_2) - h_2,$$
$$(3.21)$$
$$(1 - y)\left(\frac{y^2}{2} + \frac{q}{\gamma(\gamma - 1)}\right) - \frac{yq}{\gamma} = (1 - f_2)\left(g_2\frac{f_2^2}{2} + \frac{h_2}{\gamma - 1}\right) - f_2 h_2.$$

Linearizing with respect to the parameter y (at $q = 0$) or with respect to q (at $y = 0$) we obtain the conditions for the functions f_{i1}, q_{i1}, and h_{i1}:

$$f_{11}(1) = \frac{\gamma - 1}{\gamma + 1}, \qquad g_{11}(1) = 0, \qquad h_{11}(1) = -\frac{4}{\gamma + 1}, \qquad (3.22)$$

$$f_{21}(1) = -\frac{2}{\gamma + 1}, \qquad g_{21} = -\frac{2(\gamma + 1)}{(\gamma - 1)^2}, \qquad h_{21} = -\frac{\gamma - 1}{\gamma(\gamma + 1)}. \qquad (3.23)$$

The symmetry condition takes the form

$$f_{i1}(0) = 0. \qquad (3.24)$$

The question of finding the increments to the self-similar solution reduces thus to a solution of the systems (3.18)–(3.20) with boundary conditions (3.22) and (3.24) for $i = 1$ or (3.23) and (3.24) for $i = 2$.

In the general case of arbitrary γ, ω, and ν the system (3.18)–(3.20) has no exact analytic solution, and numerical methods must be used to investigate the linearized problem. Since this system contains the unknown constant A_{i1}, the problem is not a Cauchy problem and the condition (3.24) or its equivalents is necessary to determine the constant A_{i1}. It can thus be treated as a two-point boundary-value problem for the system (3.18)–(3.20). To determine the constant we can use in place of Eqs. (3.24) the conservation laws

$$\frac{E_0}{\sigma_\nu} = \int_0^{r_2}\left(\frac{\rho v^2}{2} + \frac{p}{\gamma - 1}\right) r^{\nu - 1}\, dr - \int_0^{r_2}\left(\frac{\rho_1 v_1^2}{2} + \frac{p_1}{\gamma - 1}\right) r^{\nu - 1}\, dr, \qquad (3.25)$$

which is the integral energy-conservation law, and

$$\int_0^{r_2} (\rho - \rho_1) r^{\nu-1}\, dr = \int_0^t \rho_1 v_1 r_2^{\nu-1}\, dt, \tag{3.26}$$

which is the integral mass-conservation law.

By writing these laws in dimensionless form and linearizing them we obtain certain integral relations containing the functions f_{i1}, g_{i1}, and h_{i1} and the constant A_{i1}. If the constants A_{i1} are determined by using condition (3.24), the integral relations for the energy and mass can be used to monitor the computation accuracy or to obtain approximate solutions.

We examine now in greater detail the numerical methods of solving the linearized problems, using as an example the blast problem with account taken of back pressure at constant density, i.e., for the case $\omega = 0$ and $v_1 = 0$ (see Sec. 5). At variable density we are interested mainly in the case of spherical symmetry ($\nu = 3$). The dimensionless variables (3.8) are sometimes replaced by the functions

$$\bar{f} = v/v_2, \qquad \bar{g} = \rho/\rho_2, \qquad \bar{h} = p/p_2, \tag{3.27}$$

where v_2, ρ_2, and p_2 are the values of the respective functions ($v_1 = 0$) on the shock wave. Let us show the connection between f_0, g_0, h_0 and $\bar{f}_0$, $\bar{g}_0$, $\bar{h}_0$, and also between f_{21}, g_{21}, h_{21} and $\bar{f}_1$, $\bar{g}_1$, $\bar{h}_1$, i.e., with only the back pressure taken into account. Straightforward calculations yield

$$\bar{f}_0 = \frac{\gamma+1}{2} f_0, \qquad \bar{f}_1 = \frac{\gamma+1}{2}(f_0 + 2f_{21}), \qquad \bar{g}_0 = g_0 \frac{\gamma-1}{\gamma+1},$$

$$\bar{g}_1 = \frac{2g_0}{\gamma+1} + \frac{\gamma-1}{\gamma+1} g_{21}, \qquad \bar{h}_0 = \frac{\gamma+1}{2} h_0, \tag{3.28}$$

$$\bar{h}_1 = \frac{\gamma+1}{2}\left(\frac{\gamma-1}{2\gamma} h_0 + h_{21}\right).$$

For linear problems we choose the following system of dimensionless variables: λ, y, f, g, and h if $v_1 \neq 0$ and $p_1 = 0$, and λ, q, f, g, and h if $v_1 = 0$, $p_1 = Cr^{-\omega}$ and $\rho_1 = Ar^{-\omega}$.

The corresponding systems of linear equations can be obtained from Eqs. (3.18)–(3.20), with allowance for Eq. (3.28), or else directly from the gasdynamic equations. Note certain general properties of the system (3.18)–(3.20).

(1) This system (and its equivalent at $\omega \neq \omega_*$) has an adiabaticity integral whose existence was proved by M. L. Lidov.[7]

(2) For the singular self-similar solution (2.28) with

$$\omega = \omega_1 = \frac{3\nu - 2 + \gamma(2-\nu)}{\gamma+1}, \qquad f_0 = \frac{2}{\gamma+1}\lambda,$$

$$g_0 = \frac{\gamma+1}{\gamma-1}\lambda^{\nu-1}, \qquad h_0 = \frac{2}{\gamma+1}\lambda^{\nu} \tag{3.29}$$

the system (3.18)–(3.20) and its equivalents have exact analytic solutions.[4,8]

(3) To find the connection between the shock-wave coordinate and the time, we must use the expression for η_i and the equation $dr_2/dt = D$.

We proceed now to a more detailed consideration of specific cases.

2. Allowance for back pressure in the case of an isothermal atmosphere of variable density

We choose the dimensionless variables λ, q, f, g, and h, and put $\nu = 3$. We seek a solution in the form (3.16), (3.18) for $i = 2$ [problem (b) of Sec. 1]. We follow Refs. 5 and 6 in the investigation of this problem.

The system of linearized equations (3.18)–(3.20) with $i = 2$ has an adiabaticity integral, i.e., an integral of Eqs. (3.19) and (3.20) which can be written, with allowance for the boundary conditions (3.23) on the shock wave, in the form

$$H_{(2)} + kG_{(2)} - (\gamma + k)F_{(2)} = -A_{21} + \tilde{C}\exp(\omega - \nu)\varphi, \quad \varphi = \int_1^\lambda \frac{d\lambda}{m},$$

$$k = -\left[\frac{\gamma}{2} + \frac{\nu}{2}\frac{\gamma - 1}{\nu - \omega}\right],$$

$$\tilde{C} = A_{21}H_{(2)}(1) - kG_{(2)}(1) - (\gamma + k)F_{(2)}(1), \tag{3.30}$$

$$(\lambda - f_0)F_{(2)} = f_{21}(\lambda), \qquad g_0 G_{(2)} = g_{21}(\lambda), \qquad h_0 H_{(2)} = h_{21}(\lambda).$$

The function φ can be explicitly expressed in terms of g_0, h_0, and λ.

The boundary-value problem for the system (3.18)–(3.20) with $i = 2$ can be solved numerically. For $\omega = \omega_1$ there is an analytic solution that satisfies the condition $f_{21}(0) = 0$, namely,

$$f_{21} = \frac{\gamma - 1}{\gamma + 1}\lambda[F_0 + C_2\lambda^{-n_2[(\gamma+1)/(\gamma-1)]}],$$

$$g_{21} = \frac{\gamma + 1}{\gamma - 1}\lambda\left[C_1\lambda^4 + \frac{n_2 - 3 + \omega_1}{n_2 + 3 - \omega_1}C_2\lambda^{-n_2[(\gamma+1)/(\gamma-1)]} + G_0\right], \tag{3.31}$$

$$h_{21} = \frac{2}{\gamma + 1}\lambda^3\left[\frac{3}{7}C_1\lambda^4 + \frac{\gamma n_2 - 3(\gamma - 1)}{n_2 + 3 - \omega_1}C_2\lambda^{-n_2[(\gamma+1)/(\gamma-1)]} + H_0\right],$$

where F_0, G_0, and H_0 are known functions of ω_1, A_{21}, and γ (Ref. 6), while n_2 is the smallest root of the quadratic equation

$$n_2^2 - \frac{\gamma - 1}{(\gamma + 1)^2}(17\gamma - 3)n - \left(\frac{\gamma - 1}{\gamma + 1}\right)^2\left(\frac{10 - 14\gamma}{\gamma + 1}\right) = 0. \tag{3.32}$$

The constants C_1, C_2, and A_{21} are obtained from the boundary conditions (3.23). For solution (3.31) we calculated the functions f_{21}, g_{21}, and h_{21} for different γ and for $\omega = \omega_1$. The constants A_{21}, C_1, C_2, F_0, H_0, and n_1 are

Table 1. Values of the constants in Eqs. (3.31).

γ	A_{21}	C_1	C_2	F_0	H_0	n_2
1.2	7.8750	−19.863	−0.73533	−9.2647	7.412	0.037 482
1.4	4.5411	−9.7302	−1.2158	−3.7842	2.271	0.083 213
5/3	3.2711	−5.7455	−1.0374	−1.9626	0.654	0.139 81

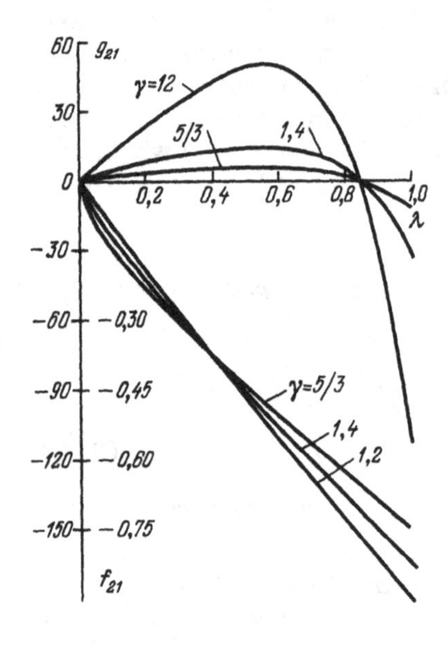

Figure 21. Plots of the functions $g_{21}(\lambda)$ and $f_{21}(\lambda)$ for various γ.

listed in Table 1. Figures 21 and 22 show plots of f_{21}, g_{21}, and h_{21} for various γ, while Fig. 23 shows plots of f, g, and h for $q = 0.05$ at $\gamma = 1.2$.

Introducing the dimensionless radius $R = r_2/r^0$ $[r^0 = (E_0/c)^{1/(3-\omega)}]$, and the dimensionless time $\tau_0 = t/t_0$ [where $t_0 = r^0(A/C)^{1/2}$], we can obtain the following relations for the motion $R(\tau_0)$ of the shock wave:

$$R^{3-\omega} = \frac{\delta^2}{\alpha\gamma} q e^{A_{21}q}, \qquad \delta = \frac{2}{5-\omega}, \qquad z = \frac{1}{3-\omega} - \frac{1}{2}, \tag{3.33}$$

$$\tau_0 = \gamma^{-1/2}\left(\frac{\delta^2}{\alpha\gamma}\right)^{1/(3-\omega)} \frac{q^{z+1}}{(z+1)(3-\omega)} \times \left[1 + \frac{z+1}{z+2}\left(1 + \frac{1}{3-\omega}\right)A_{21}q\right].$$

Note that for a self-similar blast we have

$$R = \alpha^{-1/(5-\omega)}\tau_0^{\delta}, \qquad R^{3-\omega} = \frac{\delta^2}{\alpha\gamma} q.$$

The $R(\tau_0)$ dependence given by Eqs. (3.33) for $\gamma = 1.2$ and for $\omega = \omega_0$ is shown in Fig. 24. The differential properties of the solution become worse

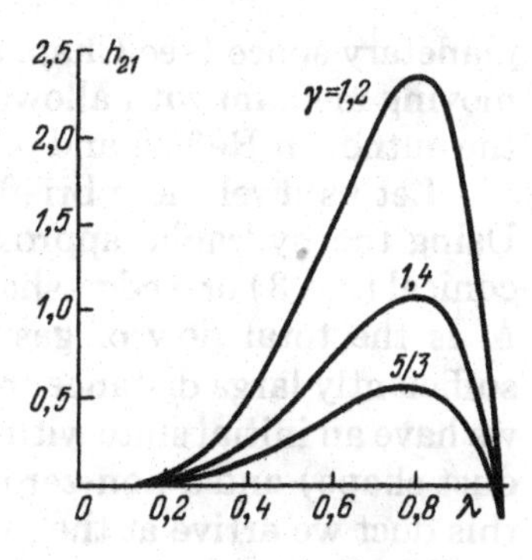

Figure 22. Plots of the functions $h_{21}(\lambda)$ for various γ.

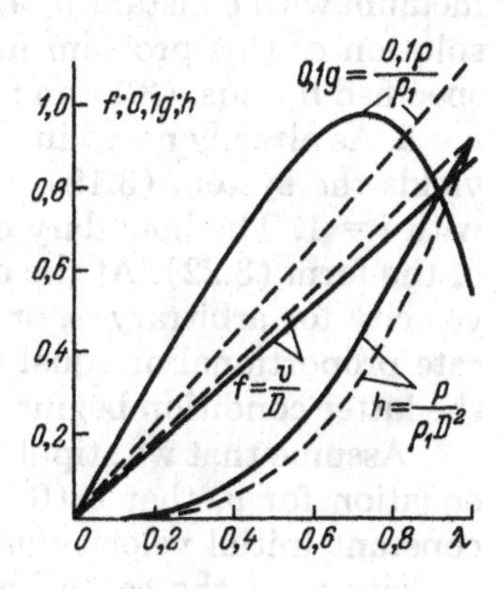

Figure 23. Distributions of the dimensionless pressures, densities, and velocities ($q = 0.05$, $\gamma = 1$ and 2). The dashed curves correspond to the self-similar solution.

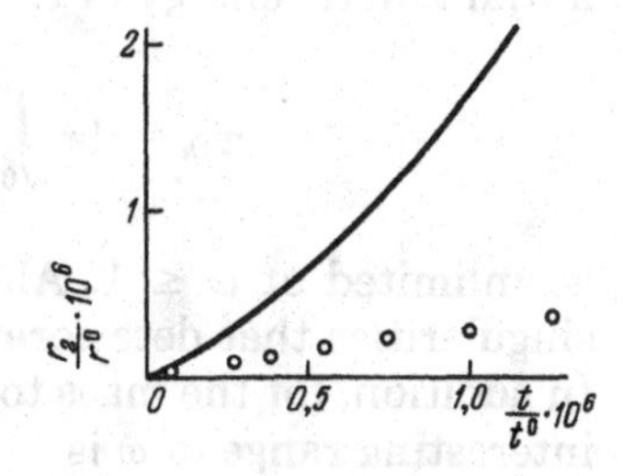

Figure 24. Dimensionless shock-wave radius versus the dimensionless time (the circles correspond to the self-similar solution).

in the vicinity of $\lambda = 0$ than for $\rho_1 = \text{const}$. The functions $f_{21}(\lambda)$ and $g_{21}(\lambda)$, being bounded on the segment $[0, 1]$, have infinite derivatives at the point $\lambda = 0$.

3. Strong blast in a moving gas

In a number of cases it may be of interest to investigate a blast in a moving ideal gas at high pressure p, escaping from a vessel to an ambient medium of low pressure p_1. This problem is encountered also when account is taken of the initial gas motion in solar-flare disturbance propagations in interplanetary space (see Chap. 8). One-dimensional point-blast problems in a moving medium with allowance for the initial velocity were formulated by the author in Refs. 5 and 6.

Let us dwell here briefly on the case of a gas escaping from a vessel. Using the hydraulic approximation,[9] we get for a gas flowing through a conical ($\nu = 3$) or wedge-shaped ($\nu = 2$) duct a density $\rho_1 = M/v_1 r^{\nu-1}$, where M is the total flow of gas mass. When p_1 is small, the velocity v_1 at a sufficiently large distance from the escape point is close to a maximum and we have an initial state with variable density ρ_1 ($\omega = \nu - 1$ for the considered duct shape) and a non-zero initial velocity. For a blast in a narrow end of this duct we arrive at the problem, formulated above, of a variable-density medium with constant v_1 and p_1 and with $\omega = \nu - 1$. We consider now the solution of this problem in a linearized formulation, when v_1 and ρ_1 are specified by Eqs. (3.1) and (3.3), the initial pressure is zero ($p_1 = 0$), and $\nu = 3$. As already noted in Sec. 1, linearization of the system (3.12)–(3.15) yields the system (3.18)–(3.20) of linear ordinary differential equations with $i = 1$. The boundary conditions for this system on a shock wave are of the form (3.22). At the center of the blast one can either stipulate zero velocity for arbitrary ω, or postulate that the source is reestablished at a rate proportional or equal to v_1. Of course, the motion may be such that the latter condition begins to be met only for sufficiently large t.

Assume that we stipulate the condition $v(0, t) = 0$. It follows from the equation for v_1 that $v_1(0) = 0$ if $\omega > 2$. For $\omega = 2$ we have the case of constant initial velocity v_1, while the case $\omega < 2$ corresponds to infinite velocity v_1 at the center, and is not considered here. Note also that the initial kinetic energy of the gas inside a certain volume of radius r,

$$E_{1k} = 4\pi \int_0^r \frac{\rho_1 v_1^2}{2} r^2\, dr = 2\pi A B^2 \int_0^r r^{\omega-2}\, dr,$$

is unlimited at $\omega \leq 1$. All this indicates that the solution acquires new singularities that deteriorate its differential properties with decrease of ω. In addition, for the mass to be finite we must require $\omega < 3$. Thus, the most interesting range of ω is

$$2 \leq \omega < 3. \tag{3.34}$$

Assume that the condition (3.34) is met. The system of linearized equations (3.18)–(3.20), as in the preceding case, has an adiabaticity integral (i.e., an integral of the last two equations of this system). For an integral control over the calculations we can use the mass-conservation law (3.26). Carrying out the linearization, we obtain from Eq. (3.26)

$$\int_0^1 g_{11} \lambda^2\, d\lambda = -\frac{2}{5-\omega}. \tag{3.35}$$

In the particular case $\omega = \omega_1$ the system (3.18)–(3.20) with $i = 1$ has an exact solution[6] in the form

Table 2. Values of the constants in Eq. (3.36).

γ	A_{11}	C_1	C_2	n_2
1.2	0.583 41	0.638 25	−1.6254	−0.80980
4/3	0.680 72	0.396 47	1.46413	−0.90401
1.4	0.693 35	0.352 82	1.6933	−0.94495
5/3	0.562 86	0.505 91	1.4221	−1.0779

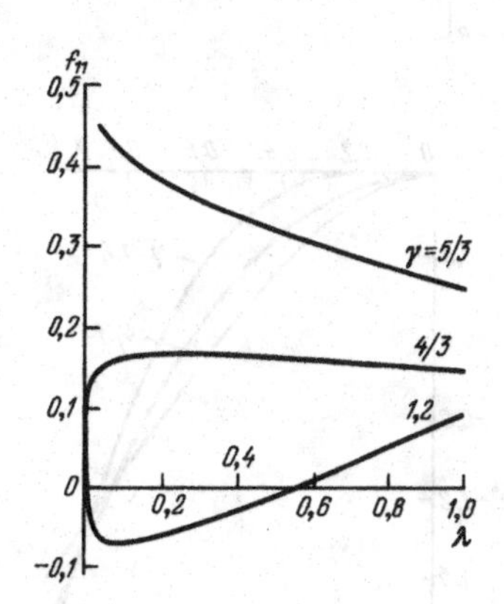

Figure 25. Plots of the functions $f_{11}(\lambda)$ for various γ.

$$f_{11} = \frac{\gamma - 1}{\gamma + 1}\lambda(C_2\lambda^{n_2} + F_0),\ n_1 = \frac{3-\gamma}{\gamma - 1},$$

$$g_{11} = \frac{\gamma+1}{\gamma-1}\lambda\left[C_1\lambda^{n_1} + \frac{n_2+4}{n_2-n_1}C_2\lambda^{n_2} + G_0\right],$$

$$h_{11} = \frac{2\lambda^3}{\gamma+1}\left[\frac{3}{n+3}C_1\lambda^{n_1} + \frac{\gamma^{n_2} + 3(\gamma-1)}{n_2-n_1}C_2\lambda^{n_2} + H_0\right], \tag{3.36}$$

$$n_2 = -\frac{a}{2} + \sqrt{\frac{a^2}{4} - b},\quad a = \frac{13\gamma+1}{\gamma+1},\quad b = \frac{20\gamma-12}{\gamma-1}.$$

For the constants F_0, G_0, and H_0 we have

$$G_0 = \frac{2(\omega_1 - 3)}{\omega_1 - 1}F_0,\qquad H_0 = -(3\gamma-3)\frac{2}{\omega_1-1}F_0 - 2A_{11},$$

$$F_0 = 0.5\left(12 - \frac{\gamma+1}{\gamma-1}(\omega_1-1)\right)\left[\gamma - 1 - \frac{6(\omega_1 - 6 + 3\gamma)}{\omega_1 - 1}\right]^{-1}A_{11}.$$

The constants A_{11}, C_1, and C_2 are determined from the boundary conditions (3.22). The calculation results are given in Table 2.

Plots of the functions f_{11}, g_{11}, and h_{11} for various γ are shown in Figs. 25–27. Figure 28 shows the functions f, g, and h for $y = 0.1$ and $\gamma = 1.2$.

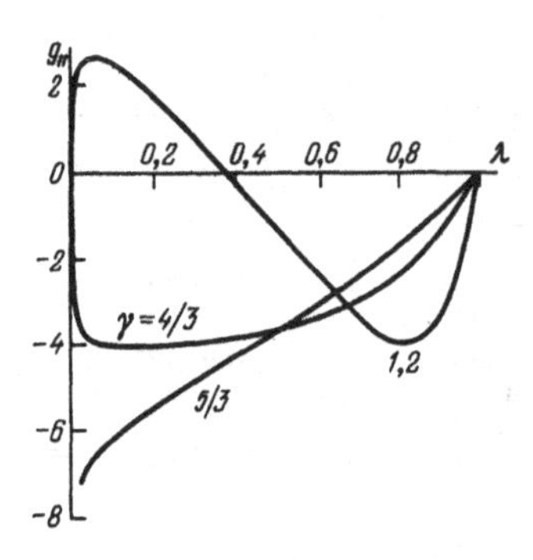

Figure 26. Plots of the functions $q_{11}(\lambda)$ for various γ.

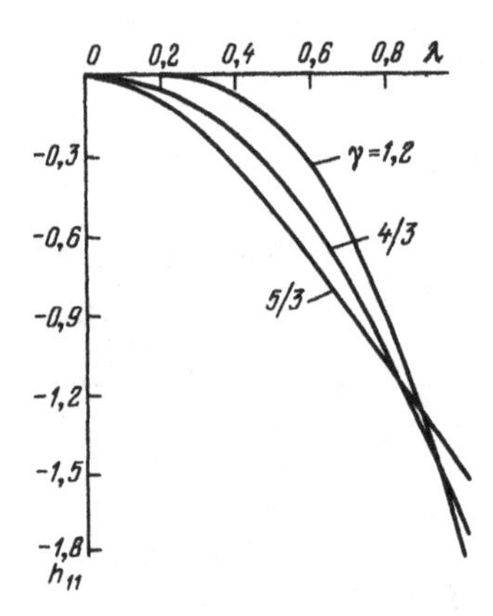

Figure 27. Plots of the functions $h_{11}(\lambda)$ for various γ.

The calculations show that for $\gamma > 1.2$ there is near $\lambda = 0$ an extensive region in which the solution becomes meaningless, since the density or pressure passes through zero and becomes negative. This is explained by the singular character of the initial solution f_0, g_0, and h_0 at $\omega = \omega_1$, by the behavior of the solution, and by the shortcomings of the linearization method. It is possible here to consider also a solution variant in which a spherical cavity with zero pressure is introduced in the vicinity of the point $\lambda = 0$. One can hope, however, that such inherently integral characteristics of the motion as the constant A_{11} will provide accurate values for the correction to the law of motion of the shock wave and will not depend strongly on the local properties of the solution near the symmetry center.

We introduce a dimensionless radius $x_2 = r_2/r^*$ and a dimensionless time $\tau = t/t^*$, where $r^* = (E_0/AB^2)^{1/(\omega-1)}$ and $t^* = (r^*)^{3-\omega}/B$. For the relation between x_2 and τ we get then

$$x_2 = (\delta\alpha^{-1/2})^{2/(\omega-1)} y^{2/(\omega-1)} \exp\frac{2A_{11}}{\omega-1} y,$$
$$\tau = \delta(\delta\alpha^{1/2})^{2(3-\omega)/(\omega-1)} y^{(5-\omega)/(\omega-1)}\left[1 + \frac{(5-\omega)^2}{4(\omega-1)} A_{11} y\right]. \quad (3.37)$$

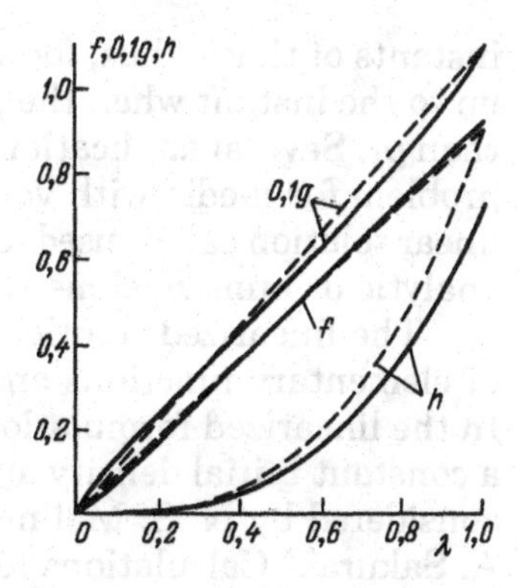

Figure 28. Distribution of the dimensionless pressures, densities, and velocities ($y = 0.1$, $\gamma = 1.2$). The dashed curves correspond to the self-similar solution.

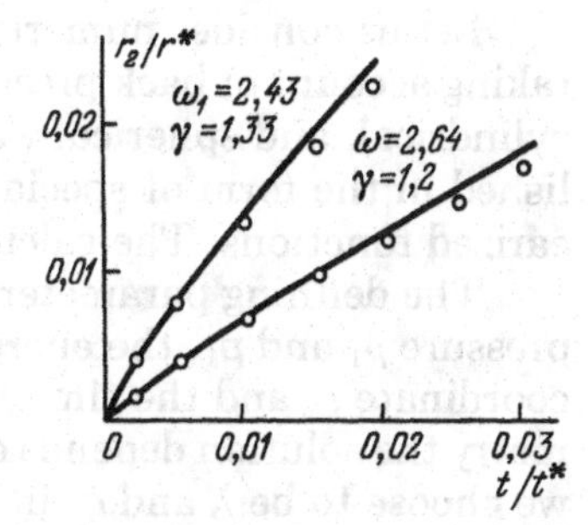

Figure 29. Dependence of the radius r_2/r^* of the shock wave on the time for $\gamma = 1.33$ and $\gamma = 1.2$. (The circles correspond to the points of the self-similar solution.)

In the self-similar case we have

$$x_2 = \alpha^{-1/(5-\omega)}\tau^{\delta}, \qquad y = \frac{\sqrt{\alpha}}{\delta} x_2^{(\omega-1)/2}.$$

Plots of Eqs. (3.37) for the cases $\omega = \omega_1$, $\gamma = 1.2$, and $\gamma = \frac{4}{3}$ are shown in Fig. 29.

4. Allowance for back pressure in the initial stage of a planar, cylindrical, or spherical blast in a gas of constant initial density

4.1. Basic equations and some of their properties

It has already been noted that the solution of the self-similar problem of a strong blast in a gas at rest, without allowance for back pressure, describes the behavior of the blast at the initial instants of time, when the pressure p_2 behind the shock-wave front is much higher than the initial gas pressure p_1. Solution of the linearized problem gives the functions sought for later instants of time. Thus, for a blast in air ($\gamma = 1.4$) this solution can be used up to the instant when the pressure p_2 is approximately eight times higher than p_1. Several applications of point-blast theory require a solution of the problem for media with various adiabatic exponents γ. The results of the linear solution can be used to calculate the complete problem by approximate analytic or numerical methods (see Chap. 4).

The linearized solutions for arbitrary γ cannot be expressed in terms of elementary functions and can be obtained only by numerical methods. In the linearized formulation, the problem of a blast with back pressure, at a constant initial density and with an adiabatic exponent $\gamma = 1.4$, was first considered by N. S. Mel'nikova[1] (see also Ref. 2), and independently by A. Sakurai.[3] Calculations for cylindrical and planar blasts at $\gamma = 1.4$ were made in Ref. 10.

Let us consider numerical methods of solving the linear blast problem taking account of back pressure, for a large range of γ in the case of planar, cylindrical, and spherical waves. The results of the calculations were published in the form of special tables[11] containing both self-similar and linearized functions. The calculation method is described in Ref. 12.

The defining parameters of the problem are the initial gas density and pressure ρ_1 and p_1, the energy E_0, the adiabatic exponent γ, the geometric coordinate r, and the time t. These parameters show that at a given symmetry the solution depends on γ and on two dimensionless variables, which we choose to be λ and q, in the ranges $0 \le \lambda \le 1$ and $0 \le q \le 1$.

We choose as the sought functions the dimensionless velocity f, the density g, and the pressure h:

$$f = v/D, \qquad g = \rho/\rho_1, \qquad h = qp/\gamma p_1. \tag{3.38}$$

Additional unknowns are $R_2 = r_2/r_0$ and $\tau = t/t^0$, where r^0 and t^0 are, respectively, the dynamic length and time defined as

$$r^0 = (E_0/p_1)^{1/\nu}, \qquad t^0 = r^0(\rho_1/p_1)^{1/2}. \tag{3.39}$$

The system of differential equations that describe one-dimensional unsteady motions of an ideal gas takes in the chosen dimensionless variables the form (3.12)–(3.14) with $i = 2$ and $\omega = 0$. This system yields the functions $f(\lambda, q)$, $g(\lambda, q)$, and $h(\lambda, q)$, subject to the following boundary conditions: On the shock wave (for $\lambda = 1$)

$$f(1, q) = \frac{2}{\gamma + 1}(1 - q), \qquad g(1, q) = \frac{\gamma + 1}{\gamma - 1 + 2q},$$
$$h(1, q) = \frac{2\gamma - (\gamma - 1)q}{\gamma(\gamma + 1)}. \tag{3.40}$$

At the symmetry center (for $\lambda = 0$)

$$f(0, q) = 0. \tag{3.41}$$

The initial conditions (for $q = 0$) are

$$f(\lambda, 0) = f_0(\lambda), \qquad g(\lambda, 0) = g_0(\lambda), \qquad h(\lambda, 0) = h_0(\lambda),$$

where $f_0(\lambda)$, $g_0(\lambda)$, and $h_0(\lambda)$ are functions corresponding to the self-similar solutions.

The fourth boundary condition (3.41) helps to find the shock-wave radius $R_2(q)$. The latter can also be obtained from the integral energy-conservation law (3.25). The $\tau(q)$ dependence is next calculated by quadrature using the equation

$$\frac{d\tau}{dq} = \left(\frac{q}{\gamma}\right)^{1/2} \frac{dR_2}{dq}.$$

For the initial stage of a blast with back pressure (i.e., for small q), the solution can be sought in the form (3.16), (3.17) with $i = 2$ and $\omega = 0$:

$$f = f_0 + qf_1 + o(q), \qquad g = g_0 + qg_1 + o(q), \qquad h = h_0 + qh_1 + o(q),$$

$$\frac{R_2}{q}\frac{dq}{dR_2} = \frac{\nu}{1 + A_1 q + o(q)}, \tag{3.42}$$

where A_1 is a constant to be determined.

If we substitute the functions (3.42) in the system (3.12)–(3.14), neglect terms of order q^2 and higher in the boundary conditions, and then equate to zero terms of like powers in q, we obtain systems of ordinary differential equations and boundary conditions for the self-similar and linearized problems.

Terms of order q^2 were taken into account in Ref. 13 (see also Refs. 14–16). The results call for the following comments. The solutions (3.42) can be expressed in the form $f = f_0 + qf_1 + q^2 f_2 + \ldots$ (and similarly for g and h), i.e., they can be regarded as series in powers of q. The region of convergence of expressions (3.42) as series in q has not yet been determined. At any rate, one can apparently expect such a series to diverge at $q \geq (\gamma - 1)/2$, since the expansion for the function $g(1, q)$ from Eq. (3.40) converges only if $q < (\gamma - 1)/2$. Thus, if $\gamma < 2$ these representations of the functions can be used only if q is small, but then inclusion of terms of order q^2 can hardly improve significantly the linearized solution. The function $1/g$, to be sure, can also be used to analyze the solutions with the aid of series in q. At $\lambda = 1$ this function is linear in q and the objections above are not as serious.

We introduce in place of λ a new independent variable μ such that

$$\lambda = \mu^{-\delta} u_1^{\beta_2} u_2^{-\beta_1}, \tag{3.43}$$

where $\lambda = 0$ corresponds to $\mu = \mu_0 = (\gamma + 1)/2\gamma$ and $\lambda = 1$ corresponds to $\mu = 1$.

As indicated in Chap. 2, for $\gamma \neq 2$ the solution of the self-similar problem for a strong blast can be expressed with the aid of the parametric variable μ in the form:

$$f_0 = \frac{2}{\gamma + 1}\mu\lambda, \quad g_0 = \frac{\gamma + 1}{\gamma - 1} u_1^{\beta_3} u_2^{\beta_4} u_3^{-\beta_5},$$
$$h_0 = \frac{2}{\gamma + 1} \mu^{\nu\delta} u_2^{\beta_4 - 2\beta_1} u_3^{1-\beta_5}. \tag{3.44}$$

The case $\gamma = 2$ is special, and we have here for g_0 and h_0

$$g_0 = 3\varkappa u_1^{\delta\nu/2} u_3^{\delta(\nu-2)/2}, \qquad h_0 = \tfrac{2}{3}\varkappa \mu^{\delta\nu} u_3^{\delta(\nu-2)}. \tag{3.45}$$

We use in Eqs. (3.43)–(3.45) the notation

$$u_1 = \frac{2\gamma}{\gamma - 1}\left(\mu - \frac{\gamma+1}{2\gamma}\right), \qquad u_2 = \frac{\nu\gamma - \nu + 2}{3\nu - 2 - \gamma(\nu - 2)}\left[\frac{(\gamma+1)(\nu+2)}{\nu\gamma - \nu + 2} - 2\mu\right],$$

$$u_3 = \frac{2}{\gamma - 1}\left(\frac{\gamma+1}{2} - \mu\right), \qquad \varkappa = \exp\left(6\delta\,\frac{\mu - 1}{3 - 2\mu}\right), \qquad \delta = \frac{2}{\nu + 2},$$

$$\beta_1 = \beta_2 + \frac{\gamma + 1}{\nu\gamma - \nu + 2} - \delta, \qquad \beta_2 = \frac{\gamma - 1}{2\gamma - 2 + \nu}, \qquad \beta_3 = 1 - 2\beta_2,$$

$$\beta_4 = \frac{\nu + 2}{2 - \gamma}\,\beta_1, \qquad \beta_5 = \frac{2}{2 - \gamma}.$$

On the other hand, $R(q)$ and $\tau(q)$ are obtained in the self-similar problem from the expressions $R_{20}^{\nu} = \delta^2 q/\gamma\alpha$ and $\tau_{(0)} = (q/\gamma)^{1/2}\,\delta R_{20}$, where the constant α is determined from Eq. (2.25).

The system of linear differential equations for the functions $f_1(\lambda)$, $g_1(\lambda)$, and $h_1(\lambda)$ and for the constant A_1 takes in the linearized problem the form (3.18)–(3.20) with $i = 2$, $\omega = 0$, and $A_{21} = A_1$ (see Sec. 1).

The boundary conditions for this system are

$$f_1(1) = -\frac{2}{\gamma + 1}, \qquad h_1(1) = -\frac{\gamma - 1}{\gamma(\gamma + 1)},$$
$$g(1) = -\frac{2(\gamma + 1)}{(\gamma - 1)^2}, \qquad f_1(0) = 0. \tag{3.46}$$

It can be assumed that the last of these conditions can be used to find the constant A_1.

For the functions $R_2(q)$ and $\tau(q)$ in the linearized problem we obtain

$$R_2^{\nu} = R_{20}^{\nu}\exp(A_1 q), \qquad \tau = \tau_{(0)}\left[1 + \frac{(\nu\delta + 2)A_1}{2\nu\delta\,(\nu\delta + 1)}\,q\right]. \tag{3.47}$$

These relations yield, in parametric form, the law of shock-wave motion, i.e., the function $R_2(\tau)$. Using Eqs. (3.47) and the relations (3.46) on the shock wave, we can determine the dependences of all the characteristics on the shock-wave front on the radius or on the time. For example, for the pressure ratio p_2/p_1 as a function of R_2 we have

$$\frac{p_2}{p_1} = \frac{\gamma - 1}{\gamma + 1} + \frac{2\gamma}{\gamma + 1}\left(A_1 + \frac{\delta^2}{\gamma\alpha}\,R_2^{-\nu}\right).$$

4.2. Numerical solution of the problem

The solution of the linearized problem reduces to numerical integration of the system (3.18)–(3.20) with $i = 2$ and $\omega = 0$ under the conditions (3.46). We consider only cases with γ in the range $1 < \gamma < 7$. Note that this problem has an exact analytic solution for $\gamma = 7$ ($\nu = 3$) (see Sec. 1).

To recast the linearized system in a form suitable for numerical calculations, we introduce new unknown functions $F(\lambda)$, $G(\lambda)$, and $H(\lambda)$, connected with $f_1(\lambda)$, $g_1(\lambda)$, and $h_1(\lambda)$ by the relations

$$f_1 = -mF, \qquad g_1 = g_0 G, \qquad h_1 = g_0 H, \qquad m = f_0 - \lambda.$$

After transformations, the system (3.18)–(3.20) takes the form

$$\begin{gathered} mF' - \frac{h_0}{g_0 m} H' + \left(2f_0' + \frac{\nu - 2}{2}\right) F - \left(f_0' - \frac{\nu}{2}\frac{f_0}{m}\right)(G - H) - \frac{\nu}{2}\frac{f_0}{m} A_1 = 0_1, \\ m(F' - G') - \nu(F + G) = 0, \qquad (3.48) \\ m(\gamma F' - H') - \nu[(\gamma - 1)F + H + A_1] = 0. \end{gathered}$$

On the other hand, the conditions (3.47) become

$$\begin{gathered} F(1) = -\frac{2}{\gamma - 1}, \qquad G(1) = -\frac{2}{\gamma - 1}, \\ H(1) = -\frac{\gamma - 1}{2\gamma}, \qquad F(0) = F_0, \end{gathered} \qquad (3.49)$$

where the constant F_0 depends only on γ and ν [see Eqs. (3.52) below].

Just as in the preceding problems, the system (3.48) has an integral that can be represented, with allowance for Eq. (3.49), in the form

$$F - 2H + (2\gamma - 1)G - 2A_1 = -\left(2A_1 + \frac{3\gamma - 1}{\gamma}\frac{\gamma + 1}{\gamma - 1}\right)\mathcal{J}, \qquad (3.50)$$

where

$$\mathcal{J}(\mu) = \frac{2}{(\gamma + 1)h_0}\left(\frac{\gamma - 1}{\gamma + 1} g_0\right)^{\gamma}.$$

We change now in the system (3.48) to the new independent variable defined by Eq. (3.43). The resultant system can be reduced to

$$\begin{gathered} \frac{dF}{d\mu} = \left[(\gamma - 1)F + H - \frac{\gamma x \Omega}{\nu}\Phi_1 + \left(1 + \frac{\gamma}{\gamma + 1} x\mu\right)A_1\right]\frac{\nu \Phi_2}{\Omega(x\Omega - 1)\gamma}, \\ \frac{dG}{d\mu} = \frac{dF}{d\mu} + \frac{\nu \Phi_2}{\Omega}(F + G), \quad \frac{dH}{d\mu} = \gamma\frac{dF}{d\mu} + \frac{\nu \Phi_2}{\Omega}[(\gamma - 1)F + H + A_1]. \end{gathered} \qquad (3.51)$$

We use here the notation

$$x = \frac{\gamma+1}{\gamma-1}\left(\mu - \frac{\gamma+1}{2\gamma}\right)\frac{1}{\mu^2}, \qquad \Omega = 1 - \frac{2\mu}{\gamma+1},$$

$$\Phi_1 = -\left[\frac{4}{\gamma+1}\left(\mu + \frac{1}{\Phi_2}\right) + \frac{\nu-2}{2}\right]F$$

$$+\left[\frac{2}{\gamma+1}\left(\mu + \frac{1}{\Phi_2}\right) + \frac{\nu}{\gamma+1}\frac{\mu}{\Omega}\right](G-H),$$

$$\Phi_2 = \frac{\gamma+1}{\gamma-1}\frac{\beta_2}{x\mu^2} - \frac{\delta\beta_1}{\delta(\mu-1)-\beta_1+\beta_2} - \frac{\delta}{\mu}.$$

In the vicinity of the blast center we can write for the sought solution of the system (3.51) the asymptotic relations

$$F = F_1 + B(a_{01} + a_{11}\lambda^k) + O(\lambda^{2k}),$$

$$G = G_1 - \frac{2}{2\gamma-1}\left[\left(A_1 + \frac{2\gamma-1}{2\gamma}\frac{\gamma+1}{\gamma-1}\right)\mathcal{J}_0\right.$$

$$\left. + B\left(\frac{a_{01}-2}{2} + \frac{a_{11}-2b_{11}}{2}\lambda^k\right)\right] + O(\lambda^{2k}), \tag{3.52}$$

$$H = H_1 + B(1 + b_{11}\lambda^k) + O(\lambda^{2k}),$$

where F_1, G_1, and H_1 are independent of λ and are particular solutions of the system (3.51), B is an arbitrary constant, $k = 1/\beta_2$, and $\mathcal{J}_0$ stands for the principal term of the expansion of $\mathcal{J}(\lambda)$ in the vicinity of $\lambda = 0$. For $1 < \gamma < 7$ (with the case $\gamma = 2$ excluded) the constants in the asymptotic form (3.52) are

$$a_{01} = -\frac{1}{\gamma-1}, \quad a_{11} = \frac{(\gamma-1-n)b_{11}}{(\gamma-1)(\gamma+n)}, \quad b_{11} = \left(m_0 + \frac{l}{\gamma-1}\right)\frac{D_0(\gamma-1)}{\gamma^2 h(\mu_0)},$$

$$n = \gamma\nu\beta_2, \qquad m = B_2\frac{3-2\gamma}{2\gamma-1}\left(1 + \frac{\nu}{2}\frac{\gamma}{\gamma-1}\right),$$

$$l = \beta_2\left[\left(2 + \frac{\nu-2}{2}\gamma + \frac{1}{2\gamma-1}\left(1 + \frac{\nu}{2}\frac{\gamma}{\gamma-1}\right)\right)\right],$$

$$D_0 = \frac{\gamma+1}{\gamma-1}[u_2'(\mu_0)]^{\beta_1/\beta_2(2-\gamma)}[u_3(\mu_0)]^{-\beta_5}\mu_0^{\delta\beta_3/\beta_2},$$

$$F_1 = -G_1 = \left[2 + \frac{\nu-2}{2}\gamma - (\gamma-2)\left(1 + \frac{\nu}{2}\frac{\gamma}{\gamma-1}\right)\right]^{-1}A_1,$$

$$H_1 = -A_1 - (\gamma-1)F_1.$$

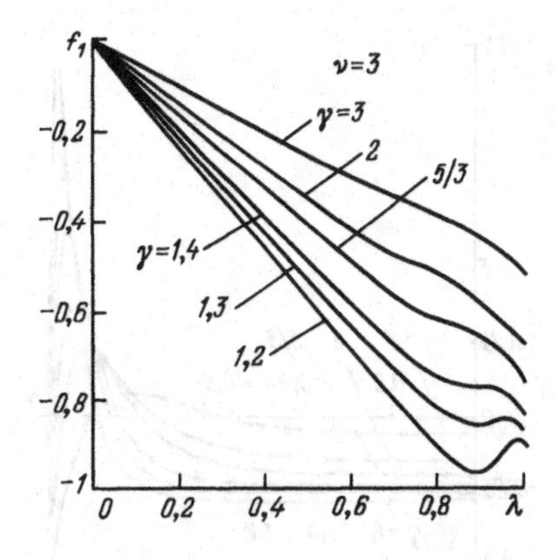

Figure 30. Plots of $f_1(\lambda)$ for different γ in the case of spherical symmetry.

If $\gamma = 2$, only the equation for D_0 is changed, i.e.,

$$D_0 = 3\varkappa(\mu_0)[u_2(\mu)]^{\frac{\nu}{\nu+2}\frac{\beta_1}{\beta_2}}[u_3(\mu_0)]^{\frac{\nu-2}{\nu+2}}\mu_0^{\frac{\delta\nu}{(\nu+2)\beta_2}}.$$

For numerical integration of the system of differential equations (3.51) we used Eqs. (3.52) and a special trial-and-error method (Ref. 12).

4.3. Calculation results

K. V. Sharovatova wrote a computer program for a large volume of computations for $\gamma = 1, 2, 3$ and for $\gamma = 1.1, 1.2, 1.3, 1.4, \frac{5}{3}, 2$, and 3. From these computations we obtained, besides the functions f_1, g_1, and h_1 and the constants A_1, the self-similar function f_0, g_0, and h_0, and also

$$e_0 = \frac{\gamma}{2} f_0^2 g_0 + \frac{\gamma h_0}{\gamma - 1}, \qquad e_1 = \gamma f_0 g_0 f_1 + \frac{\gamma}{2} f_0^2 g_1 + \frac{\gamma h_1}{\gamma - 1},$$

$$\theta_0 = \frac{\gamma h_0}{g_0}, \qquad \theta_1 = \frac{\gamma h_1 - \theta_0 g_1}{g_0}, \qquad \sigma_0 = \frac{\gamma h_0}{g_0^\gamma}, \qquad \sigma_1 = \frac{\gamma(h_1 - \theta_0 g_1)}{g_0^\gamma}.$$

The last functions serve to compute the total energy, temperature, and entropy of the gas in the linearized approximation. In addition, from the calculated data one can determine the pressure, momentum, and other integral characteristics, as well as the connection between the Euler and Lagrange coordinates (with the aid of σ_0 and σ_1). All these functions were computed to five significant figures. Tables of all the functions and constants in the asymptotic equations for these functions near a blast center (in the vicinity of $\lambda = 0$) were printed in Ref. 11.

We present here for illustration only a few diagrams containing isolated results. Figure 30 shows a plot of $f_1(\lambda)$ for different values of γ in the spherical case, while Fig. 31 shows the corresponding plot for $\gamma h_1(\lambda)$. Plots of $\gamma h_1(\lambda)$ with $\gamma = \frac{5}{3}$ vs ν are given in Fig. 32. Figure 33 shows plots of the relative pressure p/p_1 at $q = 0.1$ for various γ and for $\nu = 3$. Table 3 lists

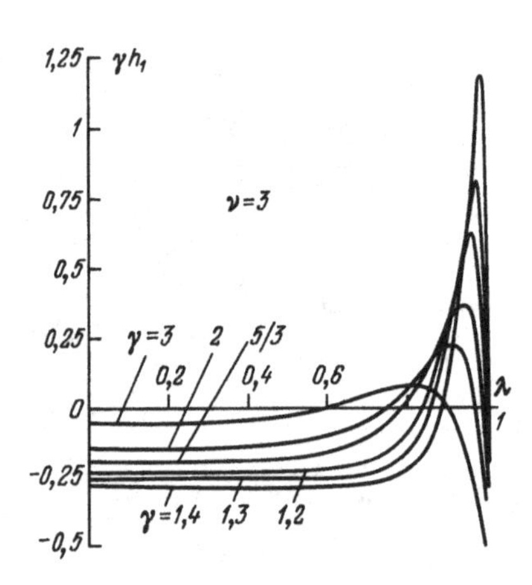

Figure 31. Plots of $\gamma h_1(\lambda)$ which describe the linear corrections to the pressure distributions at various γ.

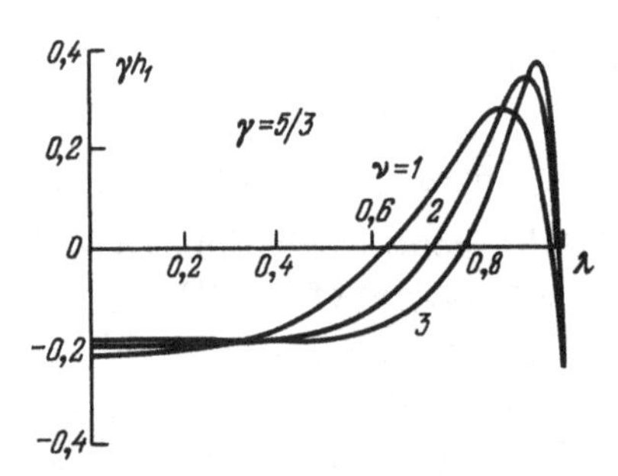

Figure 32. Comparison of the $h_1(\lambda)$ distributions for different symmetries.

the values of the constant A_1 obtained for a number of adiabatic exponents γ and for $\nu = 1$, 2, and 3.

The results presented here and in Ref. 11 point to a substantial dependence of the solution on γ. They also indicate that the linear increment functions $f_1(\lambda)$, $g_1(\lambda)$, and $h_1(\lambda)$ are bounded and linearization is meaningful everywhere if $0 \leq \lambda \leq 1$.

It follows from the foregoing that the problem considered above can be generalized[4,8] to include a variable initial density $\rho_1 = Ar^{-\omega}$, with $\omega < \nu$.

It is interesting to note that at $\omega = \omega_1$ and $\gamma \leq 3$ it follows from the exact solution of the linearized problem that the functions f_1, g_1, h_1 and their derivatives with respect to λ are bounded everywhere on the segment [0, 1]. Thus, it can be regarded as proved that for a medium with variable density the point-blast problem has at $\omega = \omega_1$ and $\gamma \leq 3$ bounded derivatives of v/c, p/p_2, and ρ/ρ_2 with respect to q.

For other $\omega < \nu$ ($\omega \neq \omega_1$) similar conclusions can be drawn only from a numerical analysis of the linearized systems.

4.4. Variant of a sweeping method of solving the linearized system

A drawback of the method considered in Ref. 12 for numerically solving the linearized system is that to choose the coefficient A_1 it is necessary to

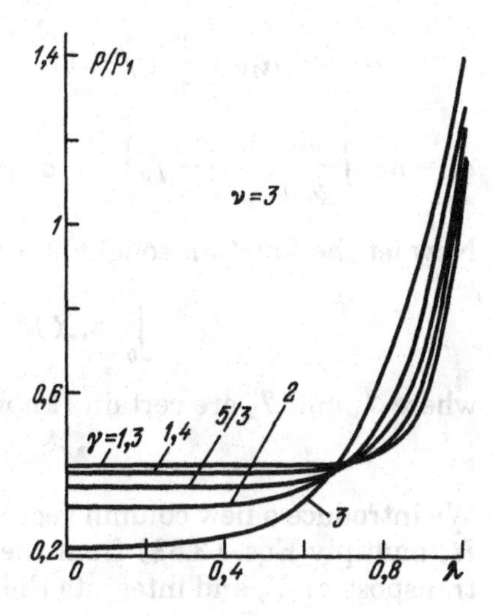

Figure 33. Distribution of the relative pressures in space (case $\nu = 3$, $q = 0.1$).

Table 3. Values of the constant A_1.

ν	γ						
	1.1	1.2	1.3	1.4	5/3	2	3
1	2.3257	2.2437	2.1862	2.1433	2.0683	2.0143	1.9407
2	2.0866	2.0424	2.0092	1.9836	1.9374	1.9043	1.8632
3	2.0010	1.9666	1.9396	1.9182	1.8785	1.8498	1.8141

solve the system (3.51) numerically several times. It is possible to get around this shortcoming by using a special method similar to the sweeping methods of Ref. 17. The present section is devoted to this method.

We express the system (3.48) in matrix form:

$$X' = Q_1 X + A_1 Q_2,$$

$$X = \begin{matrix} F \\ G \\ H \end{matrix}, \qquad Q_1 = \begin{matrix} \alpha_{11} & \alpha_{12} & \alpha_{13} \\ \alpha_{21} & \alpha_{22} & \alpha_{23} \\ \alpha_{31} & \alpha_{32} & \alpha_{33} \end{matrix}, \qquad Q_2 = \begin{matrix} a_1 \\ a_2 \\ a_3 \end{matrix},$$

$$\alpha_{11} = \frac{1}{\xi}\left[\left(2f'_0 + \frac{\nu-2}{2}\right)g_0 m + \frac{\nu h_0(\gamma-1)}{m}\right], \qquad \xi = g_0 m^2 - \gamma h_0, \quad (3.53)$$

$$\alpha_{12} = \frac{g_0 m}{\xi}\left(f'_0 + \frac{\nu}{2}\frac{f_0}{m}\right), \qquad \alpha_{13} = \frac{1}{\xi}\left[\frac{\nu h_0}{m} + g_0 m\left(f'_0 - \frac{\nu}{2}\frac{f_0}{\Omega}\right)\right],$$

$$\alpha_{21} = \alpha_{11} - \frac{\nu}{m}, \qquad \alpha_{22} = \alpha_{12} - \frac{\nu}{m}, \qquad \alpha_{23} = \alpha_{13},$$

$$\alpha_{31} = \gamma\alpha_{11} - \frac{\nu}{m}(\gamma - 1), \qquad \alpha_{32} = \gamma\alpha_{12}, \qquad \alpha_{33} = \gamma\alpha_{13} - \frac{\nu}{m},$$

$$a_1 = mg_0\left(\frac{\nu h_0}{g_0 m^2} - \frac{\nu}{2m} f_0\right), \quad a_2 = a_1, \quad a_3 = \gamma a_1 - \frac{\nu}{m} A_1, \quad m = f_0 - \lambda.$$

Now let the function sought satisfy the integral relation

$$\int_0^1 MX\lambda^{\nu-1}\, d\lambda = \mathcal{J}_1 + A_1\mathcal{J}_2, \tag{3.54}$$

where $\mathcal{J}_1$ and $\mathcal{J}_2$ are certain (known) numbers and M is the row matrix

$$M = \|M_1, M_2, M_3\|. \tag{3.55}$$

We introduce a new column vector Y with unknown components $\tilde{F}$, $\tilde{G}$, and $\tilde{H}$, multiply Eqs. (3.53) from the left by the row matrix Y^* which is the transpose of Y, and integrate the result with respect to λ:

$$\int_0^1 Y^*X'\, d\lambda = \int_0^1 Y^*Q_1X\, d\lambda + A_1\int_0^1 Y^*Q_2\, d\lambda.$$

Integrating by parts, we get

$$\int_0^1 (Y^{*\prime} + Y^*Q_1)X\, d\lambda = [Y^*X]_0^1 - A_1\int_0^1 Y^*Q_2\, d\lambda. \tag{3.56}$$

We choose Y to satisfy the equation

$$Y^{*\prime} + Y^*Q_1 = M\lambda^{\nu-1}. \tag{3.57}$$

Relation (3.56), with allowance for Eqs. (3.54) and (3.57), can be written in the form

$$\mathcal{J}_1 + A_1\mathcal{J}_2 = [Y^*X]\,|_0^1 - A_1\int_0^1 Y^*Q_2\, d\lambda. \tag{3.58}$$

We now choose Y to satisfy at $\lambda = 0$ the condition

$$[Y^*X]\,|_{\lambda=0} = 0$$

or

$$(\tilde{F}F + \tilde{G}G + \tilde{H}H)\,|_{\lambda=0} = 0. \tag{3.59}$$

If F, G, and H are finite for $\lambda = 0$, it can be assumed that

$$\tilde{F}(0) = \tilde{G}(0) = \tilde{H}(0) = 0 \tag{3.60}$$

and the condition (3.59) will be met.

Equation (3.57) does not contain the constant A_1 and can be integrated from $\lambda = 0$ to $\lambda = 1$ (for example, numerically). From the obtained values of Y we determine $[Y^*X]_{\lambda=1}$ and calculate the integrals contained in Eq. (3.58), after which we obtain A_1 from

$$A_1 = -\frac{\mathcal{J}_1 - [Y^*X]|_{\lambda=1}}{\mathcal{J}_2 + \int_0^1 Y^*Q_2\,d\lambda}.$$

Once A_1 is found, integration of the system (3.53) with the initial data for $\lambda = 1$ solves the problem.

By way of the simplest integral relation of type (3.54) we choose the linearized mass-conservation law. In the absence of a mass source at the center we have

$$\int_0^1 g\lambda^{\nu-1}\,d\lambda = 1. \tag{3.61}$$

From this we get for the linearized solution $\int_0^1 g_1\lambda^{\nu-1}\,d\lambda = 0$ or $\int_0^1 g_0 G\lambda^{\nu-1}\,d\lambda = 0$. For the elements of the matrix M and for the numbers $\mathcal{J}_1$ and $\mathcal{J}_2$ we have $M_1 = 0$, $M_2 = g_0$, $M_3 = 0$, and $J_1 = J_2 = 0$. For other integral equations of the type (3.54) we can use the linearized relation for the energy integral

$$\gamma \int_0^1 \left(-f_0 g_0 mF + \frac{h_0}{\gamma - 1} H + \frac{f_0^2 g_0}{2} G \right) \lambda^{\nu-1}\,d\lambda$$

$$= -A_1\mathcal{J}_0 + \frac{1}{\nu(\gamma-1)}, \tag{3.62}$$

$$\mathcal{J}_0 = \gamma \int_0^1 \left(\frac{g_0 f_0^2}{2} + \frac{h_0}{\gamma - 1} \right) \lambda^{\nu-1}\,d\lambda,$$

and the relation obtained by multiplying the adiabaticity integral (3.50) by $\lambda^{\nu-1}$ and integrating from 0 to 1. If Eq. (3.61) or Eqs. (3.62) are used, the adiabaticity integral (3.50) can be used to check on the computational accuracy.

Thus, knowing certain integral and local (near $\lambda = 0$) properties of the solution, we succeed in finding the constant A_1 by integrating the auxiliary equation in the direction from the center to the shock wave, and determine next all the sought functions by integrating the basic equations from the shock wave to the center. In this sense the method considered can be regarded as a special variant of the sweeping method used to solve boundary-value problems.[17] A method close to the one considered above was considered in Ref. 14 for a system of two equations such as Eqs. (3.53) and the particular case where the integral energy-conservation law is used.

Note that the method we have developed can be modified for the case where other independent variables (say, μ) are used. It can be employed to solve other linearized problems. These include the problem of a strong blast in a detonating medium (see Chap. 6), the problem of piston motion with allowance for back pressure (in which case the limits in the integral equation are not 0 and 1 but correspond to the coordinates of the piston and the shock wave), and the problem of the influence of the magnetic field on gas flow in the blast.

5. Initial stage of converging blast waves

We consider now the propagation of spherical or cylindrical converging waves produced by a point blast. Let a spherical or cylindrical volume with pressure p_1 and density ρ_1 be contained in a chamber of radius r_0 with walls assumed for simplicity to be nondeformable. A planar blast with total energy E_0 (for the sphere) or with energy E_0 per unit length (for the cylinder) is produced along the wall at the instant t_0. Clearly, at the first instant after the onset of the blast the gas flow will be close to that of a plane strong point blast. As it propagates towards the center, however, the plane wave is influenced by the converging flow geometry. After some damping, the blast wave becomes enhanced by the convergence of the flow, which changes over in the vicinity of the symmetry center $r = 0$ to the self-similar Guderley–Landau–Stanyukovich regime of a converging strong shock wave.[18–20]

We shall consider here only the initial propagation of the converging blast wave. The case of planar geometry coincides with that of a planar strong blast (up to the instant of wave collision) and will not be treated here.

We seek the solution of the hydrodynamics equations with the Rankine–Hugoniot boundary conditions on the wave front and with the zero-velocity condition at the wall, i.e., $v(r_0, t) = 0$, by the method of linearization about a planar self-similar solution.[21]

We introduce the dimensionless variables

$$\frac{v}{D} = f(\xi, x_s), \qquad \frac{p}{\rho_1 D^2} = h(\xi, x_s), \qquad \frac{\rho}{\rho_1} = g(\xi, x_s),$$

where

$$D = \frac{dr_s}{dt}, \qquad x_s = \frac{r_s - r_0}{r_0}, \qquad \xi = \frac{r - r_0}{r_s - r_0},$$

and r_s is the shock-wave coordinate. The gasdynamics equations (3.5) take in these new variables the form

$$(f = \xi)\frac{\partial t}{\partial \xi} + \theta f + \frac{1}{g}\frac{\partial h}{\partial \xi} = -x_s \frac{\partial f}{\partial x_s},$$

$$(f - \xi)\frac{\partial g}{\partial \xi} + g\frac{\partial f}{\partial \xi} + \frac{(\nu - 1)}{1 + x_s \xi} x_s f g = -x_s \frac{\partial g}{\partial x_s}, \tag{3.63}$$

$$(f - \xi)\frac{\partial h}{\partial \xi} + \gamma h \frac{\partial f}{\partial \xi} + 2\theta h + \gamma \frac{(\nu - 1)x_s}{1 + x_s \xi} h f = -x_s \frac{\partial h}{\partial x_s},$$

where $\theta = x_s \ddot{x}_s / \dot{x}_s^3$, $\dot{x}_s = dx_s/dt$, and $\ddot{x}_s = d^2 x_s/dt^2$.

It is required to find a solution of Eqs. (3.63) in the region bounded by the wall ($\xi = 0$) and the shock wave ($\xi = 1$), with the flow starting from $x_s = 0$ (the wall) and converging to the center at an instant of time corresponding to the coordinate $x_s = -1$.

The integral energy-conservation law takes in dimensionless variables the form

$$I(x_s) - a_{\nu*}^2[1 - (1 + x_s)^\nu]\left[\gamma(\gamma - 1)\nu x_s\left(\frac{dx_s}{d\tau}\right)^2\right]^{-1} = \frac{1}{\nu x_s}\, t_{\nu*}^{-2}\left(\frac{dx_s}{d\tau}\right)^{-2}, \tag{3.64}$$

where

$$a_{\nu*}^2 = a_1^2\frac{t_{\nu*}^2}{r_0^2}, \qquad t_{\nu*}^2 = \frac{\rho_1\sigma_\nu r_0^{\nu+2}}{\nu E_0}, \qquad \tau = \frac{t}{t_{\nu*}},$$

$$I(x_s) = \int_1^0\left(\frac{h}{\gamma - 1} + \frac{1}{2}gf^2\right)(1 + x_s\xi)^{\nu-1}\, d\xi.$$

The solution of the problem can be sought in the form

$$\begin{aligned} f(\xi, x_s) &= f_0(\xi) + f_1(\xi)x_s + f_2(\xi)x_s^2 + f_3(\xi)x_s^3 + \ldots,\\ h(\xi, x_s) &= h_0(\xi) + h_1(\xi)x_s + h_2(\xi)x_s^2 + h_3(\xi)x_s^3 + \ldots,\\ g(\xi, x_s) &= g_0(\xi) + g_1(\xi)x_s + g_2(\xi)x_s^2 + g_3(\xi)x_s^3 + \ldots; \end{aligned} \tag{3.65}$$

$$\begin{aligned} x_s\left(\frac{dx_s}{d\tau}\right)^2 &= \varphi_0 + \varphi_1 x_s + \varphi_2 x_s^2 + \varphi_3 x_s^3 + \ldots,\\ \theta(x_s) &= \theta_0 + \theta_1 x_s + \theta_2 x_s^2 + \theta_3 x_s^3 + \ldots. \end{aligned} \tag{3.66}$$

The quantities with subscript 1 give the solution of the linearized equation. We examine this solution in greater detail for the case where the term with a_1^2 in Eq. (3.64) (i.e., the initial gas energy or the back pressure) can be neglected.

For the functions f_0, f_1, h_0, h_1, g_0, and g_1 we have from the system (3.63) the equations

$$\begin{aligned} &(f_0 - \xi)f_0 + \frac{h_0'}{g_0} = -\theta_0 f_0, \qquad (f_0 - \xi)g_0' + g_0 f_0' = 0,\\ &(f_0 - \xi)h_0' + \gamma h_0 f_0' = 2\theta_0 h_0, \qquad \varphi_0 = \frac{1}{\nu}I_0,\\ &I_0 = \int_1^0\left(\frac{h_0}{\gamma - 1} + \frac{1}{2}g_0 f_0^2\right)d\xi, \qquad \theta_0 = \frac{1}{2},\\ &(f_0 - \xi)f_1' + \frac{h_1'}{g_0} = -\left\{f'(f_0' + 1 + \theta_0) + \theta_1 f_0 - g_1\frac{h_0'}{g_0^2}\right\}, \end{aligned} \tag{3.67}$$

$$(f_0 - \xi)g_1' + g_0 f_1' = \{g_1(f_0' + 1) + f_1 g_0' + (\nu - 1) f_0 g_0\},$$

$$(f_0 - \xi)h_1' + \gamma h_0 f_1' = -\{h_1(\gamma f_0' + 1 + 2\theta_0) + 2\theta_1 h_0 + f' h_0' + (\nu - 1)\gamma h_0 f_0\},$$

$$\varphi_1 = \varphi_0 \frac{I_1}{I_0}, \qquad \theta_1 = \frac{1}{2}\frac{\varphi_1}{\varphi_0},$$

$$I_1 = \int_1^0 \left[(\nu - 1)\xi \left\{ \frac{h_0}{(\gamma - 1)} + \frac{1}{2} g_0 f_0^2 \right\} + \frac{h_1}{\gamma - 1} + \frac{1}{2} f_0^2 g_1 + g_0 f_0 f_1 \right] d\xi. \tag{3.68}$$

The boundary conditions on the shock-wave front follow from the conditions (3.6) (at $v_1 = 0$); they are

$$f_0(1) = h_0(1) = \frac{2}{\gamma + 1}, \qquad g_0(1) = \frac{\gamma + 1}{\gamma - 1}, \tag{3.69}$$

$$f_1(1) = 0, \qquad h_1(1) = 0, \qquad g_1(1) = 0. \tag{3.70}$$

Note that Eqs. (3.69) represent the conditions for the self-similar functions, while the conditions (3.70) follow from the neglect of the back pressure.

To find the linear corrections f_1, h_1, and g_1 we must integrate numerically the system (3.67), starting from the point $\xi = 1$, where the zero boundary conditions are met. The constant θ_1 should be found either by using the boundary condition $f(0) = 0$ at the center of the blast or with the aid of the integral relation (3.68).

Calculations have shown[21] that θ_1 can be easily obtained by linear "*regula falsi*" interpolation.

To find the law of shock-wave motion we use the relations

$$\tau = \frac{t}{t_{\nu *}}, \qquad \frac{dx_s}{d\tau} = \dot{x}_s, \qquad \tau = \int_0^{x_s} \frac{1}{\dot{x}_s}\, dx_s,$$

$$\tau = -\int_0^{x_s} \sqrt{\{\varphi_0/x_s + \varphi_1 + \varphi_2 x_s + \varphi_3 x_s^2\}^{-1}}\, dx_s,$$

from which we have for the linear approximation

$$\tau = B_0 |x_s|^{3/2}(1 + B_1 x_s),$$

$$B_0 = 2\{3\sqrt{|\varphi_0|}\}, \qquad B_1 = -\frac{3}{10}\frac{\varphi_1}{\varphi_0}.$$

We present now some calculation results for $\gamma = 1.4$.
For $\theta(x_s)$ we have

$$\theta = -\left[0.5 + \frac{0.612\,84}{4 - \nu} x_s\right], \qquad \nu = 2, 3.$$

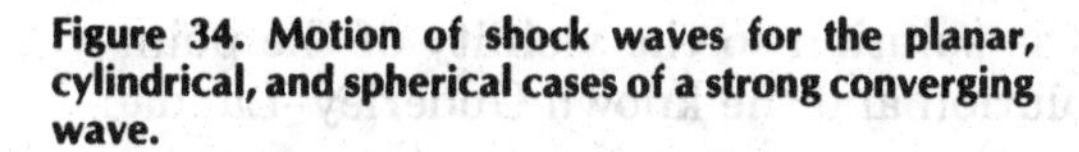

Figure 34. Motion of shock waves for the planar, cylindrical, and spherical cases of a strong converging wave.

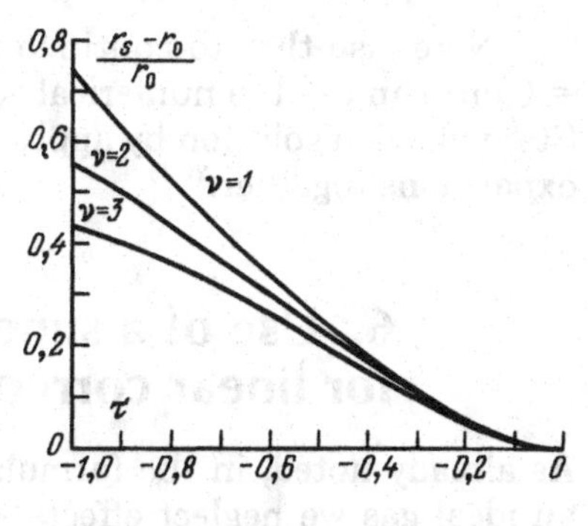

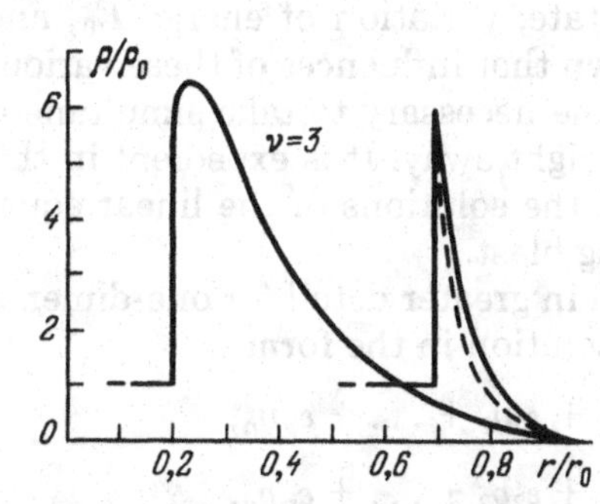

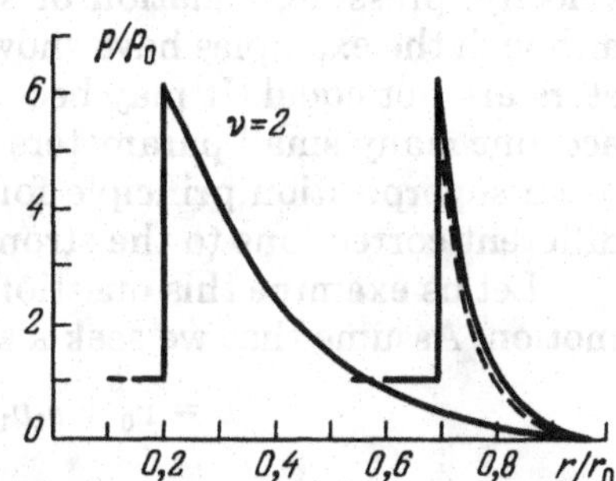

Figure 35. Density distribution behind strong cylindrical and spherical waves for two instants of time. The dashed curve corresponds to a strong planar blast.

The calculations yield $B_0 = 1.271\,31$, $B_1 = 0.367\,70$ for $\nu = 3$ and $B_0 = 1.038\,02$, $B_1 = 0.183\,85$ for $\nu = 2$.

Plots of the laws of motion (more accurately, of $r_s/r_0 - 1$ against $t/t_{\nu*}$) of a strong shock wave are shown in Fig. 34 for the planar, spherical, and cylindrical cases. We see from this figure that the flow geometry comes into play already at $r_s/r_0 - 1 = 0.3$. The density distribution profiles are shown in Fig. 35 in the spherical and cylindrical cases for two instants of time corresponding to $r_s/r_0 = 0.7$ and 0.3, respectively. The dashed curve illustrates for comparison the planar case. It is clearly seen here that the density behind the shock wave begins to increase, passes through a maximum, and decreases at large r/r_0. The pressure has approximately the same behavior. The velocity profile behaves qualitatively as in a strong blast (monotonic decrease behind the wave front). After the wave has passed through about half the distance to the center, however, this profile already deviates from linearity with respect to the coordinate. Note that in Ref. 21 there were obtained, besides the linear approximation, also the corresponding functions and constants for the expansions (3.65) and (3.66) up to third order inclusive, but it follows from the calculations there that at $|r_s/r_0 - 1| < 0.5$ the main contribution to the solution sought is given by the linear approximation. The density distributions plotted in Fig. 35 take into account corrections up to third order inclusive.

Note also that to continue the solution to the vicinity of the point $r = 0$ one can use the numerical solution and the known Guderley–Landau–Stanyukovich solution by applying the procedure of joining the asymptotic expansions together.[22,23]

6. Use of a superposition principle for linear corrections

As already noted, in the formulation of the main point-blast problem for an ideal gas we neglect effects such as the possible changes of the initial velocity, pressure, equation of state, variation of energy E_0, and others, although the examples have shown that influences of these various parameters are not equal. It may become necessary to take simultaneously into account many small parameters right away. It is expedient in this case to use a superposition principle for the solutions of the linear equations for different corrections to the strong blast.

Let us examine this question in greater detail for one-dimensional gas motion. Assume that we seek a solution in the form

$$
\begin{aligned}
v &= v_0 + \varepsilon_1 v_1 + \varepsilon_2 v_2 + \ldots + \varepsilon_\pi v_n, \\
\rho &= \rho_0 + \varepsilon_1 \rho_1 + \varepsilon_2 \rho_2 + \ldots + \varepsilon_n \rho_n, \\
p &= p_0 + \varepsilon_1 p_1 + \varepsilon_2 p_2 + \ldots + \varepsilon_n p_n,
\end{aligned} \tag{3.71}
$$

where $\varepsilon_1, \varepsilon_2, \ldots, \varepsilon_n$ are independent parameters. We substitute the solution (3.71) in the system (3.5) and take v_0, ρ_0, and p_0 to be the solution of this system.

As a result, we obtain, for example, from the momentum equation

$$
\frac{d_0}{dt} \Delta v + (\Delta v) \frac{\partial v_0}{\partial r} - \frac{\Delta \rho}{\rho_0(\rho_0 + \Delta \rho)} \frac{\partial p_0}{\partial r} + \frac{1}{\rho_0 + \Delta \rho} \frac{\partial \Delta p}{\partial r} + \Delta v \frac{\partial \Delta v}{\partial r} = 0,
$$

$$
\Delta p = \sum \varepsilon_i p_i, \qquad \Delta v = \sum \varepsilon_i v_i, \qquad \Delta \rho = \sum \varepsilon_i \rho_i, \qquad \frac{d_0}{dt} = \frac{\partial}{\partial t} + v_0 \frac{\partial}{\partial r}.
$$

Since the parameters ε_i are independent, we arrive, by setting all but a single ε_k equal to zero, at a system of equations of type (1.31) for the perturbations of p_k, v_k, and ρ_k. The same holds true also for the boundary and initial conditions. We obtain thus a linearized system and conclude perforce that the linear corrections must satisfy linear-equation systems corresponding to one perturbation with its own parameter ε_k and subject to their own boundary and initial conditions. On the other hand, if all the linear systems corresponding to individual parameters ε_k are satisfied by certain solutions v_k, ρ_k, and p_k, we satisfy, accurate to terms $\varepsilon_i \varepsilon_j$ $(i, j = 1, \ldots, n)$, also the initial system. Naturally, all the corrections must be bounded to justify the linearization. On the basis of the foregoing we can formulate a rule for simultaneously taking several corrections into account. We express the solution of the problem in the form

$$v = v_0(r, t) + \sum \varepsilon_i v_i(r, t) + o(l),$$
$$p = p_0(r, t) + \sum \varepsilon_i p_i(r, t) + o(l), \tag{3.72}$$
$$\rho = \rho_0(r, t) + \sum \varepsilon_i \rho_i(r, t) + o(l), \qquad l = \sqrt{\varepsilon_1^2 + \varepsilon_2^2 + \ldots + \varepsilon_n^2}.$$

We solve the particular linearized problems for the parameters ε_i. Substituting these particular solutions in Eqs. (3.72), we obtain by summation the complete solution that takes into account the corrections to the standard point-blast problem. As shown in the preceding paragraphs, it is technically simpler to linearize with respect to some variable time-dependent parameter $q(t)$. Since this means almost always that we seek a partial increment in the form $\varepsilon_k \varphi_k(r, t) = \varepsilon_k \varphi_{1k}(t) \varphi_{2k}(\lambda)$, this leads in fact to a correction-summation method based on the use of the constant parameters ε_k.

By way of example we consider allowance for the parameters B and C in the point-blast problem for variable back pressure p_1 and for an initial velocity v_1 (Secs. 1, 3, 4).

The independent correction can be taken into account here by means of the solution

$$f = f_0(\lambda) + y f_{11}(\lambda) + q f_{21}(\lambda) + o(y, q),$$
$$g = g_0(\lambda) + y g_{11}(\lambda) + q g_{21}(\lambda) + o(y, q),$$
$$h = h(\lambda) + y h_{11}(\lambda) + q h_{21}(\lambda) + o(y, q),$$
$$x = (\delta \alpha^{-1/2})^{2/(\omega-1)} y^{2/(\omega-1)} \exp\left(\frac{2A_{11}}{\omega - 1} y + \frac{A_{21}}{3 - \omega} q\right).$$

To calculate the solution for the same instants of time, given A, B, and C, we must take into account the connection between the parameters q and y:

$$q = \gamma(\delta \alpha^{-1/2})^{-2(\omega-2)} \alpha_3^{\omega-3} y^{-\frac{4(\omega-2)}{\omega-1}+2},$$

where $\alpha_3 = r^0/r^*$ is the ratio of the characteristic lengths.

Other examples of solution superposition will be considered in sections that follow.

We have considered above the one-dimensional linearized problem. The linearization method, however, can be used also for two- and three-dimensional flows. Thus, we can take into account by this method the density variation with height for a blast in a layered atmosphere, the influence of the wind in the case of an altitude-dependent velocity, or the influence of the momentum (see Chap. 5). Linearized solutions can be used to describe complicated flows by joining together the asymptotic expansions.[23,24]

Chapter 4
Spherical, cylindrical, and planar blasts with allowance for back pressure at a constant initial density

1. Asymptotic behavior of solution near a symmetry center

We solve in this chapter the nonlinear point-blast problem with allowance for back pressure. Before describing the results, we consider in the first two sections of this chapter the theoretical concepts most useful for the study of the PBP.

We consider first the behavior of the PBP solution near the blast center.

The behavior of the solutions of Eqs. (3.5), which are singular at the symmetry center, was first considered by L. I. Sedov.[1] We have shown in Chap. 2 that in a self-similar problem the functions sought are represented near the center by the series (2.26) which can be expressed in dimensional variables in the form

$$
\begin{aligned}
v_0 &= r[\varphi_{00}(t) + \varphi_{10}(t)r^{s+2} + \ldots],\\
\rho_0 &= r^s[\omega_{00}(t) + \omega_{10}(t)r^{s+2} + \ldots],\\
p_0 &= \psi_{-10} + r^{s+2}[\psi_{00}(t) + \psi_{10}(t)r^{s+2} + \ldots],
\end{aligned}
\tag{4.1}
$$

where $s = \gamma/(\gamma - 1)$.

Assuming that when back pressure is taken into account the expansion (4.1) of the function $\rho = (r, t)$ in powers of r, refined by linearization in accordance with Eqs. (3.52), remains unchanged, we represent the non-self-similar problem near a symmetry center by the following asymptotic expression for the gas density:

$$
\rho = r^s[\omega_0(t) + \omega_1(t)r^{s+2} + \omega_2(t)r^{2(s+2)} + \omega_1^* r^{s+\nu} + \ldots]. \tag{4.2}
$$

To obtain asymptotic forms of the other functions, we transform the system (3.5). The momentum equation can be reduced, with the aid of the continuity equation, to the form (1.19)

$$
\frac{\partial}{\partial t}(\rho v) + \frac{\partial}{\partial r}(\rho v^2) + (\nu - 1)\frac{\rho v^2}{r} + \frac{\partial p}{\partial r} = 0.
$$

We write the continuity equation and the last equation of Eqs. (3.5) in the form

$$\frac{\partial}{\partial t}(\rho r^{\nu-1}) + \frac{\partial}{\partial r}(\rho r^{\nu-1} v) = 0,$$

$$\frac{\partial}{\partial t}(p^{1/\gamma} r^{\nu-1}) + \frac{\partial}{\partial r}(p^{1/\gamma} r^{\nu-1} v) = 0.$$

Assuming that $p(0, t) \neq 0$, we obtain from the foregoing equations the following set of integro-differential relations equivalent to Eqs. (3.5):

$$p = p(0, t) - \rho v^2 - (\nu - 1)\int_0^r \frac{\rho v^2}{r}\,dr - \int_0^r \frac{\partial(\rho v)}{\partial t}\,dr, \tag{4.3}$$

$$v = -\frac{1}{r^{\nu-1}\rho}\int_0^r r^{\nu-1}\frac{\partial \rho}{\partial t}\,dr, \tag{4.4}$$

$$v = -\frac{1}{r^{\nu-1}p^{1/\gamma}}\int_0^r \frac{\partial p^{1/\gamma}}{\partial t} r^{\nu-1}\,dr. \tag{4.5}$$

It follows then that if $\rho(r, t)$ takes the asymptotic form (4.2), the expansions indicated in Eqs. (4.1) are valid for the functions $p(r, t)$ and $v(r, t)$. This was first noted in Refs. 1 and 2, and can be verified by substituting Eq. (4.2) in Eqs. (4.3) and (4.4) and integrating with respect to r.

Thus, with our assumption concerning $\rho(r, t)$ near the center, we find that the solution has the following asymptotic form:

$$\begin{gathered} v = r[\varphi_0(t) + \varphi_1(t) r^{s+2} + \ldots], \\ \rho = r^s[\omega_0(1) + \omega_1(t) r^{s+2} + \omega_1^* r^{s+\nu} + \ldots], \\ p = \psi_{-1}(t) + r^{s+2}[\psi_0(t) + \psi_1(t) r^{s+2} + \ldots]. \end{gathered} \tag{4.6}$$

Note that a solution in the form (4.6) was used in Refs. 3 and 4 to calculate gas flow in the vicinity of a blast center for $\nu = 3$ and $\gamma = 1.4$ [equations in the form (4.6) were obtained by another method by O. S. Golinskii].

The first term of the expansion of $v(r, t)$ in Eqs. (4.6) is linear in r. Recognizing that $p(0, t) \neq 0$ and $\partial p/\partial t)_{r=0} \neq 0$, this fact follows from Eq. (4.5) regardless of the assumptions made concerning the character of the expansion of $\rho(r, t)$.

Using Eqs. (4.6), we can obtain an asymptotic relation between the Euler and Lagrange coordinates near the center. According to Eq. (1.24) the momentum equation in Lagrangian form is

$$\frac{\partial p}{\partial \xi} = -\rho_1 \frac{\xi^{\nu-1}}{r^{\nu-1}}\frac{\partial v}{\partial t}. \tag{4.7}$$

From Eqs. (4.6) and (4.7) we have

$$(s + 2) r^{s+\nu-1}\psi_0 \frac{\partial r}{\partial \xi} = -\xi^{\nu-1}\rho_1\left(\varphi_0^2 + \frac{d\varphi_0}{dt}\right) + \ldots,$$

whence

$$r = \xi^{\nu/(s+\nu)}\zeta_1(t) + \zeta_2(t),$$

where $\zeta_1(t)$ and $\zeta_2(t)$ are certain functions of the time. Since $r(0, t) = 0$, it follows that $\zeta_2(t) = 0$. The connection between the Euler and Lagrange coordinates is thus given by

$$r = \xi^{1-1/\gamma}\zeta_1(t) + \dots. \tag{4.8}$$

Equations (4.6)–(4.8) lead to certain qualitative conclusions concerning the behavior of the solution. Thus, for example, it follows from these equations that the temperature $T = p/R\rho$ (R is the gas constant) is of order r^{-s}. If the function $\omega_0(t)$ is bounded during the entire time of motion, the temperature at the blast center is always infinite. An asymptotic equation can also be obtained for the density in the vicinity of the center at large t. Assuming that the entropy near the center, calculated from the self-similar solution, differs little from the true entropy, we have $p_0\rho_0^{-\gamma} = p\rho^{-\gamma}$. Using Eqs. (4.1) and (4.6), we can write

$$\rho = \omega_{00}(t)\left(\frac{\psi_{-1}(t)}{\psi_{-10}(t)}\right)^{1/\gamma} r^s.$$

After a long time, the pressure near the center becomes close to atmospheric and we can put therefore $\psi_{-1} = p_1$. For the density near the center at large t, we obtain the asymptotic equation[3,5]

$$\rho = \omega_{00}(t)r^s\left(\frac{p_1}{\psi_{-10}(t)}\right)^{1/\gamma}. \tag{4.9}$$

Since ψ_{00} and ψ_{-10} are known from the self-similar solution, we can, using Eq. (4.9), calculate the approximate values of the density near the center for large t.

One more remark concerning the dependence of the solution near the blast center on γ: It is clear from the definition of s that as γ ($\gamma > 1$) decreases the function $s(\gamma)$ increases. Thus, for $\nu = 3$ at $\gamma = 1.4$ we have $s = 7.5$, and at $\gamma = 1.2$ we get $s = 15$. Thus, for $\gamma = 1.2$ the density near the center is proportional to r^{15}, and the pressure is practically independent of r much farther from the center than for $\gamma = 1.4$. The order of the Euler coordinate with respect to ξ depends only on γ and is independent of ν.

2. Damping of shock waves at large distances from the blast point

In view of the dissipation of the energy on the shock-wave front (and the increase of its surface area in the case of spherical and cylindrical symmetry), the shock wave attenuates as it propagates from the blast center in a quiescent medium with initial pressure p_1. The laws governing the damping of a point-blast shock wave and the values of all the gas-flow parameters will be described in detail in the following sections of this chapter. The problem of shock-wave damping can be solved in the entire time range [from the start of the blast to almost its total damping and conversion into a low-amplitude (acoustic) wave] can be solved numerically by modern computer techniques. Most useful for a qualitative investigation of shock-wave behavior and for quantitative estimates of the shock-wave front parameters at large distances are asymptotic equations for shock-wave damping limits. The laws of degeneracy of planar shock waves were established in 1913 by Crussard (see, e.g., Ref. 6). The laws governing damping of spherical and cylindrical shock waves were first formulated by L. D. Landau.[7] These laws are based on the assumption that the gas motion behind the shock-wave front weakens and tends to become a traveling wave having a triangular profile and having a more accurately determined speed of sound. L. I. Sedov[2] proposed a method, based on the use of the conditions on the shock waves, of obtaining for these waves the asymptotic basic functions and their derivatives with respect to the coordinate r. We present here only a few results of these investigations.

In the case of plane waves ($\nu = 1$) and cylindrical waves ($\nu = 2$) the following asymptotic relations are valid to first order for the basic functions and for the shock-wave equation of motion $r_2(t)$:

$$v_2 = \frac{2a_1 C_\nu}{(\gamma + 1) r_2^{(\nu+1)/4}}, \qquad \rho_2 = \rho_1 \left[1 + \frac{2C_\nu}{(\gamma+1) r_2^{(\nu+1)/4}} \right], \tag{4.10}$$

$$p_2 = p_1 \left[1 + \frac{2\gamma C_\nu}{(\gamma+1) r_2^{(\nu+1)/4}} \right], \qquad r_2 = a_1 (t - t^*) + \nu C_\nu r_2^{1/2\nu}. \tag{4.11}$$

Here C_ν are arbitrary constants, t^* are certain constants with dimension of time, and a_1 is the speed of sound.

The following asymptotic laws were obtained for spherical shock waves:

$$v_2 = \frac{2C_3 a_1}{(\gamma+1) r_2 \sqrt{\ln (r_2/r^*)}}, \qquad \rho_2 = \rho_1 \left[1 + \frac{2C_3}{(\gamma+1) r_2 \sqrt{\ln (r_2/r^*)}} \right], \tag{4.12}$$

$$\begin{gathered} p_2 = p_1 \left[1 + \frac{2\gamma C_3}{(\gamma+1) r_2 \sqrt{\ln (r_2/r^*)}} \right], \\ r_2 = a_1 (t - t^*) - C_3 \sqrt{\ln (r_2/r^*)}. \end{gathered} \tag{4.13}$$

Assuming the solution to be smooth in a certain vicinity of the shock wave, Sedov's method was developed further in Refs. 8 and 9. It was shown there, in particular, that if the constants C and r^* are suitably defined the refined equations for the pressure p_2 on the shock-wave front are of the form

$$\frac{p_2}{p_1} = 1 + \frac{2\gamma}{\gamma+1} \left(\frac{C_{21}}{r_2^{3/4}} + \frac{C_{22}}{r_2^{5/4}} \right) + O(r_2^{-3/2}) \qquad (\nu = 2), \tag{4.14}$$

$$\frac{p_2}{p_1} = 1 + \frac{2\gamma}{\gamma+1}\left(\frac{C_{31}}{r_2\sqrt{\ln(r_2/r^*)}} + \frac{C_{32}}{r_2[\ln(r_2/r^*)]^2}\right)$$

$$+ O\left[r_2^{-1}\ln\left(\frac{r_2}{r^*}\right)^{-7/2}\right] \qquad (\nu = 3). \qquad (4.15)$$

Here $C_{\nu i}$ are new constants whose values depend on the initial form of the wave. The remainder $O[f(r)]$ stands for quantities that tend to zero as r increases, and the order of the tendency to zero is $f(r)$.

3. Numerical solution of the problem

3.1. Brief survey of methods and results

Various numerical and approximate methods are used to calculate the parameters of blast waves and nonstationary flows produced by a blast. We consider in this section in greater detail the calculation of a point blast in a perfect unbounded gas for a constant initial density $\rho_1 = \rho_\infty$ in the presence of back pressure, in the spherical, cylindrical, and planar cases. The gas is assumed to be inviscid, not heat conducting, and having a constant adiabatic exponent γ. The formulation of the basic point-blast problem (PBP) was presented in the preceding chapters. Its complete solution for various ν and γ is of great importance both for the theory itself and for its various applications.

To consider the time range in which the calculation is to be performed, we introduce the dimensionless time τ defined, as in Chap. 3, by the equations

$$\tau = \frac{t}{t^0}, \quad t^0 = r^0\left(\frac{\rho_\infty}{p_\infty}\right)^{1/2}, \quad r^0 = \left(\frac{E^0}{p_\infty}\right)^{1/\nu}, \qquad (4.16)$$

where E^0 is the blast energy, p_∞ is the initial pressure (in general, the subscript ∞ will be used here to label the parameters of the unperturbed gas, and we shall find it convenient to designate the gas parameters directly behind the shock-wave front by the subscript n).

Among the numerical methods employed one should note the finite-difference method and its various modifications as applied to the calculation of shock waves. Examples of a numerical solution of the PBP with back pressure by the finite-difference method were given in Refs. 3, 4, 10, and 11. It was considered in the works only the particular case of a spherical blast with $\gamma = 1.4$. The cited authors specified initial data from the self-similar solution for different values of τ and accordingly for different pressure ratios p_n/p_∞ at the transition through the front of a spherical shock wave.

D. E. Okhotsimskii *et al.*[3] used $\tau_{\text{init}} = 0.000\,39$ ($p_n/p_\infty = 1743$); Goldstine and von Neumann[4] used $\tau_{\text{init}} = 0.004\,25$ ($p_n/p_\infty = 100$), while Brode[10] used

$\tau_{\text{init}} = 0.000\,42$ ($p_n/p_\infty = 1601$). The calculation was carried out up to the following values of $\tau = \tau_k$ and corresponding p_n/p_∞: $\tau_k = 18.8$ ($p_n/p_\infty = 1.008$) (Ref. 3); $\tau_k = 9.526$ ($p_n/p_\infty = 1.017$) (Ref. 4), and $\tau_k = 2.905$ ($p_n/p_\infty = 1.06$) (Ref. 10).

Note also that a calculation for $\tau > 18.8$ is given in Ref. 11. These data show that in these references the calculations were carried out up to a stage close to degeneracy of the shock wave into an acoustic one. Note that the most complete and most accurate results were obtained in Refs. 3 and 11.

In addition to finite-difference methods, point blasts were calculated also by the method of characteristics[3,12] (general aspects of the method of characteristics are treated in Refs. 13–15).

Nonstationary gas flows can be calculated by approximate methods[1,2,16,17] based on introducing special interpolation equations for one or several of the unknown functions. The unknown parameters contained in these interpolation equations are determined from the integral conservation laws and from the gasdynamic equations.

The approximate calculation methods proposed by Sedov[1,2] and based on introducing interpolation equations and using integral mass- and energy-conservation laws were further developed for applications to point blasts in Refs. 18–22. Similar methods were used also in Refs. 23–25. They are simple to use, but are not accurate enough. The method developed by us to solve the PBP is close to these approximate methods, but is based on an integral-equation computation scheme proposed by A. A. Dorodnitsyn,[26] and widely used to calculate stationary flows by O. M. Belotserkovskii, P. I. Chushkin, and others (see Refs. 27 and 28 for reviews). Development of methods of solving blast-theory problems using an integral-equation scheme was initiated by the present author in 1960. Preliminary results of the first stage of the work are contained in a 1962 article[19] as well as in a paper[29] written jointly with V. P. Karlikov and E. V. Ryazanov. The second stage, dating from 1962–1967, was carried out together with P. I. Chushkin and featured the use of higher approximations in the integral-equations scheme. This method was extended there to include blasts with allowance for the influence of a magnetic field and for a blast in a detonating gas. These will be dealt with in the chapters that follow.

Preliminary calculations for the solution of the PBP by the above method were published in Ref. 30. A more complete exposition of the method and of the results is given in Refs. 31–33. We note also that Ref. 33 was followed by publication, by a group of authors,[34] of results of calculations, by the finite-difference method, of the planar, cylindrical, and spherical problem. A detailed numerical analysis of the cylindrical problem was made by N. A. Arkhangel'skii.[35]

3.2. Basic equations and conditions of the problem

We choose the equations of nonstationary one-dimensional gas flow in Euler variables in the form

$$\frac{1}{\nu}\frac{\partial}{\partial t}\left(\frac{\rho u}{r^{\nu-2}}\right)+\frac{\partial}{\partial \zeta}(\rho v^2+p)+\frac{\nu-1}{\nu\zeta}\rho v^2=0, \tag{4.17}$$

$$\frac{1}{\nu}\frac{\partial \rho}{\partial t}+\frac{\partial}{\partial \zeta}(\zeta\rho u)=0, \tag{4.18}$$

$$\frac{1}{\nu}\frac{\partial E}{\partial t}+\frac{\partial}{\partial \zeta}[\zeta u(E+p)]=0, \tag{4.19}$$

$$\frac{1}{\nu}\frac{\partial p^{1/\gamma}}{\partial t}+\frac{\partial}{\partial \zeta}(\zeta u p^{1/\gamma})=0, \tag{4.20}$$

where $\zeta = r^\nu$, $u = v/r$, ρ is the density, p the pressure, v the velocity, and E the total energy per unit volume of the gas, the latter given in our case by

$$E=\frac{\rho v^2}{2}+\frac{p}{\gamma-1}.$$

The first equation of the set (4.17)—(4.20) is the momentum equation, the second the continuity equation, the third the energy-conservation law, and the last the flow adiabaticity condition. Only three equations of this set are independent. One of the four equations (4.17)–(4.20) can be regarded as a consequence of the others.

The boundary conditions of the problem are the condition at the blast center

$$v(0,t)=0 \tag{4.21}$$

and the condition on the shock wave

$$p_n = p_\infty\frac{2\gamma-(\gamma-1)q}{(\gamma+1)q}, \qquad v_n=\frac{2a_\infty}{\gamma+1}\frac{1-q}{\sqrt{q}},$$
$$\rho_n=\rho_\infty\frac{\gamma+1}{\gamma-1+2q}, \qquad E_n=\frac{p_n}{\gamma-1}+\frac{1}{2}\rho_n v_n^2. \tag{4.22}$$

Here a is the speed of sound, $q = a_\infty^2/D^2$, D the shock-wave velocity, and n labels the parameters of the shock-wave front.

The following integral equations also hold:

$$\rho_\infty r_n^\nu = \nu\int_0^{r_n}\rho r^{\nu-1}\,dr, \tag{4.23}$$

$$E^0=\sigma_\nu\int E r^{\nu-1}\,dr-\frac{\sigma_\nu p_\infty r_n^\nu}{\nu(\gamma-1)}. \tag{4.24}$$

These relations can be obtained either from the differential equations and the boundary conditions (see Ref. 5) or by applying the integral mass- and energy-conservation laws to the mass of the gas flow.

To solve the problem it is convenient to introduce a system of dimensionless variables, a procedure suitable for all values of E^0, p_∞, and ρ_∞. In dimensionless variables, the functions sought and independent variables can be referred either to appropriate dimensional constants or to certain dimensional functions. We assume the new dimensionless independent variables and the dimensionless functions to be the following:

$$q,\ \xi = \frac{\zeta}{\zeta_n}, \quad \varphi = ur_n\frac{1}{D}, \quad g = \frac{\rho}{\rho_\infty}, \quad e = \frac{Eq}{p_\infty}, \quad \psi = \left(\frac{pq}{p_\infty}\right)^{1/\gamma}. \tag{4.25}$$

From the kinematic condition $D = dr_n/dt$ follows a connection between the dimensionless time $\tau = t/t^0$ and q:

$$\tau' = R_n\frac{\mu}{\nu}\sqrt{\frac{q}{\gamma}}, \left(\mu = \frac{\delta'}{\delta} + \frac{1}{q}\right). \tag{4.26}$$

Here and elsewhere a prime denotes a derivative with respect to q. In addition, we introduce the notation $R = r/r^0$ and $\delta = \sigma_\nu R_n^\nu/\nu q$.

In the new variables (4.25), the system (4.17)–(4.20) takes the form

$$\frac{\partial m}{\partial q} - \frac{m}{2q} + \frac{\mu}{\nu}\left[m + \nu\frac{\partial}{\partial\xi}\left(gf^2 + \frac{\psi^\gamma}{\gamma} - m\xi\right) + \frac{\nu-1}{\xi}gf^2\right] = 0, \tag{4.27}$$

$$\frac{\partial g}{\partial q} + \mu\left[\frac{\partial}{\partial\xi}(g\varphi\xi) - \frac{\partial}{\partial\xi}(g\xi) + g\right] = 0, \tag{4.28}$$

$$\frac{\partial e}{\partial q} - \frac{e}{q} + \mu\left[\frac{\partial}{\partial\xi}(e\varphi\xi) - \frac{\partial}{\partial\xi}(e\xi) + \frac{\partial}{\partial\xi}(\psi^\gamma\varphi\xi) + e\right] = 0, \tag{4.29}$$

$$\frac{\partial\psi}{\partial q} - \frac{\psi}{\gamma q} + \mu\left[\frac{\partial}{\partial\xi}(\psi\varphi\xi) - \frac{\partial}{\partial\xi}(\psi\xi) + \psi\right] = 0, \tag{4.30}$$

where

$$m = g\varphi\xi^{(2-\nu)/\nu}, \qquad f = \varphi\xi^{1/\nu}.$$

The connection between the functions g, φ, ψ, and e is given by

$$e = \frac{\gamma}{2}g\varphi^2\xi^{2/\nu} + \frac{\psi^\gamma}{\gamma-1}. \tag{4.31}$$

The boundary conditions (4.21) and (4.22) can be reduced to

$$f(0, q) = 0, \tag{4.32}$$

$$\varphi_n = \frac{2}{\gamma+1}(1-q), \qquad g_n = \frac{\gamma+1}{\gamma-1+2q},$$

$$\psi_n^\gamma = \frac{2\gamma - (\gamma-1)q}{\gamma+1}, \tag{4.33}$$

$$e_n = \frac{2\gamma}{\gamma+1}\left[\frac{(1-q)^2}{\gamma-1+2q} + \frac{1}{\gamma-1}\right] - \frac{q}{\gamma+1}.$$

Obviously, to solve the problem it suffices to obtain from the set (4.27)–(4.30) only three out of the four sought functions g, e, φ, and ψ defined by Eqs. (4.25). The fourth unknown function can be obtained from the algebraic relation (4.31).

To solve the considered problem it is necessary to integrate the set of gasdynamics equations (4.27)–(4.30) in the ranges $0 \le q \le 1$ and $0 \le \xi \le 1$, with boundary conditions (4.32) and (4.33). Note that the value $q = 0$ corresponds to the known solution of the strong-blast problem, while the solution of the linearized problem holds for small q.

When the problem is solved in Lagrange variables, the initial system of gasdynamics equations (see Sec. 1 of Chap. 1) can be taken in the form

$$g\frac{\partial R^{\nu}}{\partial \eta} = 1, \qquad \frac{\partial R}{\partial \tau} = \sqrt{\gamma}F, \qquad \frac{\partial F}{\partial \tau} = -\frac{\nu}{\sqrt{\gamma}}R^{\nu-1}\frac{\partial P}{\partial \eta},$$
$$\frac{\partial}{\partial \tau}[\chi(\theta)] = 0 \qquad (0 \le \tau < \infty, 0 \le \eta < \infty), \tag{4.34}$$

where $\eta = r_0^{\nu}/r^{0\nu}$, $F = v/a_{\infty}$, $P = p/p_{\infty}$, $\theta = g/P^{1/\gamma}$, $\chi(\theta)$ is the entropy function, and r_0 is the initial coordinate of the particle. In Lagrange variables we have, besides the conditions (4.21) and (4.22),

$$R(0, \tau) = 0, \qquad R(\eta_n, \tau) = \eta_n^{1/\nu}. \tag{4.35}$$

From the first equation of the system (4.34) we obtain the relation

$$\eta = R_n^{\nu}\left(1 - \int_{\xi}^{1} g\, d\xi\right), \tag{4.36}$$

between the Lagrangian coordinate η and the relative Euler coordinate ξ.

3.3. Calculation by the integral-equation method in Euler variables*

From the results of Chap. 2 and of Sec. 1 of Chap. 4 it follows that the sought functions g, e, ψ, and φ are described by the asymptotic relations

$$g = \frac{\gamma+1}{\gamma-1}(C\,\delta\psi_0^{\gamma}\,\xi)^{1/(\gamma-1)} + O(\xi)^{2d\nu-1/\nu}, \qquad e = e_0 + O(\xi^d),$$
$$\psi = \psi_0 + O(\xi^d), \qquad \varphi = \varphi_0 + O(\xi^d), \tag{4.37}$$

with

$$e_0 = \frac{\psi_0^{\gamma}}{\gamma-1}, \qquad \varphi = \frac{1}{\mu}\left(\frac{1}{\gamma q} - \psi_0^{-1}\psi_0'\right),$$

$$d = \frac{\nu + 2(\gamma-1)}{\nu(\gamma-1)}, \qquad C = \frac{\alpha\nu(\nu+2)^2(\gamma+1)^2}{8\gamma\sigma_{\nu}},$$

* For details of the calculation procedure and the peculiarities of the numerical realization of the method see Appendix II.

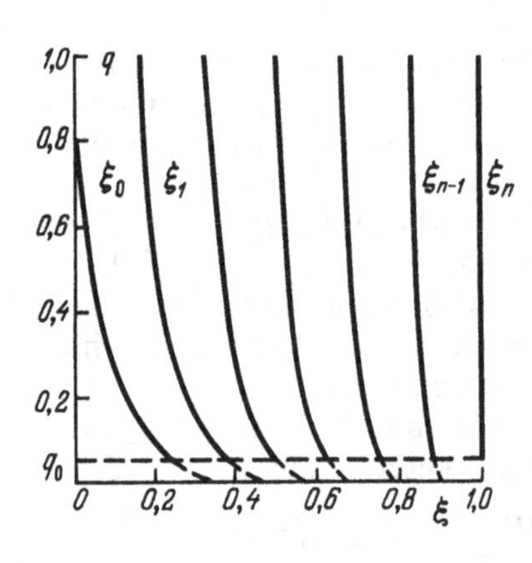

Figure 36. Subdivision of the integration region into strips.

where α is a constant determined from the integral energy-conservation law in the self-similar case (see Chap. 2).

We distinguish in the integration regions $0 \le \xi \le 1$ and $0 \le q \le 1$ a central interval bounded by the lines $\xi = 0$ and $\xi = \xi_0(q)$, in which the asymptotic equations (4.37) are valid. To choose the value of ξ_0 we fix the Lagrange coordinate of the particle $\eta = \eta_0 = (r_{00}/r^0)^\nu$ such that during the initial stage of the blast the corresponding asymptotic relation in Eqs. (4.37) yields ψ_0 at the specified accuracy. Then, using the system continuity equation (4.34) and Eqs. (4.37), we obtain for ξ_0

$$\xi_0 = \frac{1}{C\,\delta\psi_0}\left[\frac{\gamma\sigma_\nu C\eta_0}{\nu q(\gamma+1)}\right]^{(\nu-1)/\gamma}. \tag{4.38}$$

Note that other methods of choosing the central interval are also possible.

We determine the solution in the region between the boundary $\xi = \xi_0(q)$ of the central interval and the shock wave $\xi = \xi_n = 1$ by the integral-equation method. In the nth approximation we break up this region into n strips by drawing $n-1$ intermediate lines (Fig. 36):

$$\xi_i(q) = \frac{n-i}{n}\,\xi_0 + \frac{i}{n} \qquad (i = 1, 2, \ldots, n-1).$$

If each equation of the system (4.27)–(4.30) is integrated with respect to ξ from the value $\xi_l(q)$ ($l = 0, 1, \ldots, n-1$) to the shock wave $\xi_n = 1$, we obtain from this system, with allowance for the boundary conditions, $4n$ integral equations

$$\mathcal{J}'_{1l} + m_l\xi'_l - \frac{1}{2q}\,\mathcal{J}_{1l} + \mu\left[\frac{1}{\nu}\,\mathcal{J}_{1l} + m_l\xi_l - g_l f_l^2 + \frac{1}{\gamma}\,(q - \psi_l^\gamma)\right] + \frac{\nu-1}{\nu}\,\mathcal{J}_{0l} = 0, \tag{4.39}$$

$$\mathcal{J}'_{2l} + g_l\xi'_l - \mu[1 + g_l\xi_l(\varphi_l - 1) - \mathcal{J}_{2l}] = 0, \tag{4.40}$$

$$\mathcal{J}'_{3l} + e_l\xi'_l = \frac{1}{q}\mathcal{J}_{3l} - \mu\left[-\mathcal{J}_{3l} + e_l\xi_l(\varphi_l - 1) + \psi_l^{\gamma}\varphi_l\xi_l + \frac{q}{\gamma - 1}\right] = 0, \quad (4.41)$$

$$\mathcal{J}'_{4l} + \psi_l\xi'_l - \frac{1}{\gamma q}\mathcal{J}_{4l} - \mu\left[-\mathcal{J}_{4l} + \psi_l\xi_l(\varphi_l - 1) + \frac{\psi_n}{g_n}\right] = 0. \quad (4.42)$$

We use in these equations the notation

$$\mathcal{J}_{0l} = \int_{\xi_l}^{1} gf^2\frac{d\xi}{\xi}, \qquad \mathcal{J}_{1l} = \int_{\xi_l}^{1} m\,d\xi, \qquad \mathcal{J}_{2l} = \int_{\xi_l}^{1} g\,d\xi,$$

$$\mathcal{J}_{3l} = \int_{\xi_l}^{1} l\,d\xi, \qquad \mathcal{J}_{4l} = \int_{\xi_l}^{1} \psi\,d\xi.$$

Note that this system contains only $3n$ independent integral equations. The sought functions contained in the integral terms of these equations are approximated by interpolating Lagrange polynomials with interpolation nodes on the lines ξ_k ($k = 0, 1, \ldots, n$). Substituting the corresponding quadrature equations for the integrals into the integral-equation system (4.39)–(4.42) we arrive at an approximating system of $4n$ ordinary differential equations in q (only $3n$ of which are independent). Suitably choosing the approximating system of $3n$ equations, we obtain the calculation scheme of the integral-equation method.

The approximating system of ordinary differential equations is integrated by some numerical method with a computer. For this system there exists a Cauchy problem, with the initial data for a certain small value $q = q_0$ taken from the linearized solution given in Chap. 3 for the blast problem. The approximating system is integrated up to a value of q close to unity.

In this method, the solution of the problem is determined for different numbers n of the strips. Comparison of the results serves as an additional check on the method and is indicative of the accuracy of the computed solution.

The computations show that a satisfactory accuracy is obtained in practice when $n = 4$.

For a more accurate description of the behavior of the front parameters of a shock wave degenerating into a sound wave, it is useful to take into account the asymptotic shock-wave-damping laws (4.10)–(4.13) considered in the preceding section. Using these damping laws we obtain the following equations for $\delta(q)$ at q close to unity:

$$\frac{\delta'}{\delta} = \frac{2\sqrt{R_r}}{\alpha_{1*}} - \frac{1}{q} \qquad (\nu = 1),$$

$$\frac{\delta'}{\delta} = \frac{8}{3}\,\alpha_{2*}R_n^{3/4} - \frac{1}{q} \qquad (\nu = 2), \qquad (4.43)$$

$$\frac{\delta'}{\delta} = \frac{6(\ln R_n + \alpha_{3*}) - 3}{2(\ln R_n + \alpha_{3*})(1 - q)} - \frac{1}{q} \qquad (\nu = 3),$$

where α_{ν_*} are constants determined for sufficiently large q from the condition that the function $\delta(q)$ obtained from the approximating system and from the asymptotic relations (4.43) have a smooth transition.

Let us examine the method used to calculate the flow fields and the blast-wave parameters. As a result of a numerical solution of the differential equations we obtain the sought functions on the lines $\xi_i(q)$. The values of all the sought functions, for $\xi > \xi_0$ and for fixed τ or q, can be calculated next by using the assumed interpolation polynomials or by using, for large n, an interpolation of lower order.

The variation of the flow parameters at fixed points of space can be calculated by the following procedure. The value $R_n = R^*$ is fixed for some instant of time and the quantity $R = r/r^0$ is determined from it; that is to say, it is assumed that $R = R^*$. Recognizing that $\xi = (R/R_n)^\nu$, the quantities $\xi^* = (R^*/R_n)^\nu$ are calculated for a fixed $R = R^*$ at some other instant of time. These values of ξ^* are used to calculate $z = n(\xi^* - \xi_0)/(1 - \xi_0)$, and from the values of the sought functions on the strip boundaries we obtain the time dependences of the functions at fixed points of space.

Great interest attaches, in particular, to a determination of the dimensionless pressure $P = (p/p_\infty) = \psi^\gamma/q$. From the values $P(R^*, \tau) = P^*(\tau)$ one can calculate the total momenta of the excess pressures

$$J_p = \int_{\tau_*}^{\tau_k} (P^* - 1)\, d\tau,$$

and also the positive momenta

$$J_p^+ = \int_{\tau_*}^{\tau_k} (P^* - 1)\, d\tau, \qquad P^* - 1 > 0,$$

where τ_* is the time of arrival of the shock wave at the given point, and τ_k is the final calculated time.

The dimensionless momentum J_p is connected with the dimensional momentum I_p by the equations

$$J_p = \frac{I_p}{t^0 p_\infty}, \qquad I_p = \int_{t_*}^{t_k} (p^* - p_\infty)\, dt.$$

In addition, we can calculate the dimensionless momenta J_v of the velocity head from the equation

$$J_v = \int_{\tau_*}^{\tau_k} \frac{g^* f^{*2}}{2q}\, d\tau,$$

so that

$$J_v = \frac{I_v}{t^0 a_\infty^2 \rho_\infty}, \qquad I_v = \int_{t_*}^{t_k} \frac{\rho^* v^{*2}}{2}\, dt.$$

The quantity I_v is sometimes called the mass-transport momentum.

The time τ^* of action of the positive pressure phase is obtained from

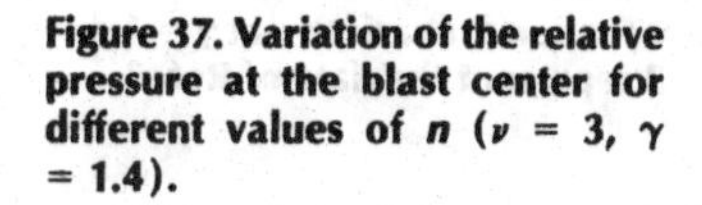
Figure 37. Variation of the relative pressure at the blast center for different values of n ($\nu = 3$, $\gamma = 1.4$).

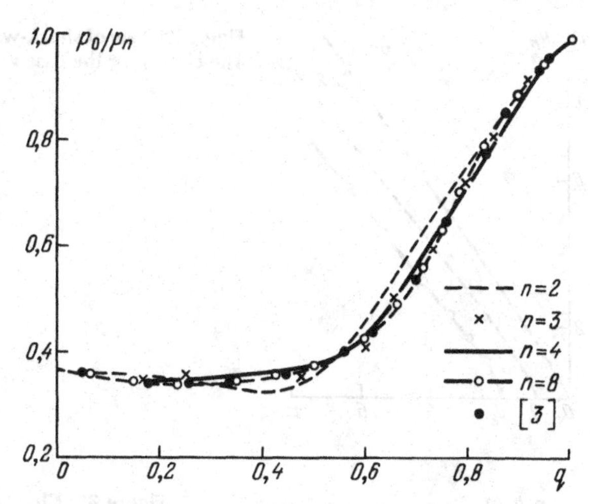

the relation $\tau^* = \bar{\tau} - \tau_*$, where $\bar{\tau}$ is the time when the excess pressure goes through zero.

Equation (4.36) can be used to determine the connection between the Lagrange coordinate η and the variables ξ and τ. The equations of the present section give the main quantities sought.

In the development of the method of calculation with Euler variables we have used the system of dimensionless variables (4.25). A similar method can naturally be developed also for other types of dimensionless variables. For example, it is possible to introduce the independent variables R, and the sought functions v/a_∞, p/p_∞, ρ/ρ_∞, and E/E_∞. The approximation is then with respect to R, and the approximating system contains differential equations in which the independent variable can be τ (or q). A calculation procedure using Lagrange variables is given in Ref. 31.

Note that in a treatment of the problem in either Euler or Lagrange variables the interpolation may not be with respect to the spatial coordinate but rather with respect to time (or with respect to q).

To solve the problem by the proposed method, E. Bishimov and K. V. Sharovatova have written computer programs for the integration of the approximating system of differential equations. These programs yield the basic functions φ_k, g_k, e_k, and ψ_k for $n = 1, \ldots, 4, 8$ using computation variants. An end-to-end approximation of the integrands by Lagrange polynomials was used, and piecewise-linear and parabolic approximations were considered also for $n = 8$.

The approximating system of ordinary differential equations was numerically integrated as a rule by the Runge–Kutta method with variable steps and by the Euler method with iteration.

In the main, blast cases were calculated for $\gamma = 1.3, 1.4, \frac{5}{3}$, and 2 and for different ν.

The influence of the number of bands n on the change of the pressure p_0/p_n at the center of the blast is shown in Fig. 37 for $\nu = 3$ and $\gamma = 1.4$.

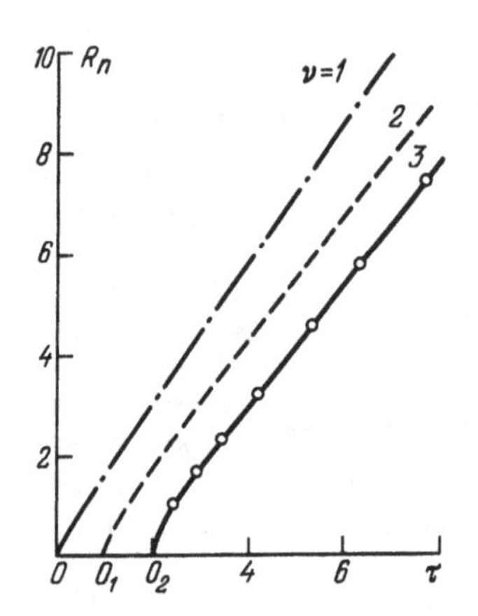

Figure 38. Law of shock-wave motion for various ν at γ = 1.4. The circles for the case ν = 3 represent the data of Ref. 3.

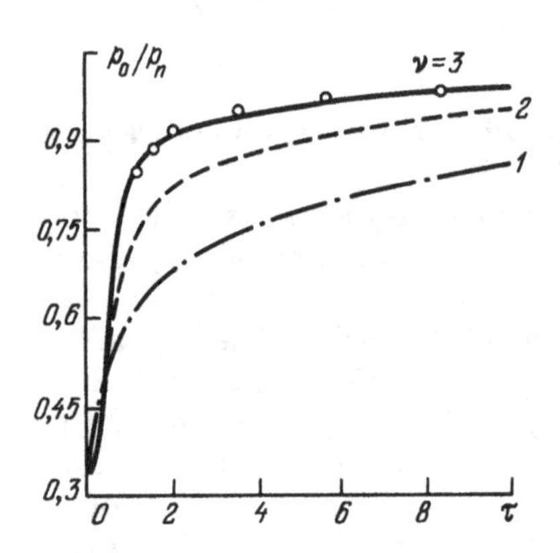

Figure 39. Change of relative pressure at the blast center for various ν at γ = 1.4. The circles for the case ν = 3 represent the data of Ref. 3.

These data were obtained for $q_0 = 0.05$ and $\xi_0(q_0) = 0.206$ for $n = 2$, 3, and 4 and $\xi_0(q_0) = 0.350$ for $n = 8$. No curve is shown in the figure for $n = 1$ in view of the low accuracy, and the points indicate data calculated by the finite-difference method.[3] Note that the solution converges much more rapidly for the law governing the shock-wave motion. Even the first-order approximation is acceptably accurate and differs little from the results for large n.

To illustrate the capabilities of the method let us dwell on some calculation results for $n = 4$. Figure 38 shows plots of $R_n(\tau)$ for $\nu = 1, 2, 3$ and for $\gamma = 1.4$. Since the $R_n(\tau)$ curves almost coalesce for large τ, they were plotted to different abscissa scales. Figure 39 shows plots of the dimensionless pressures p_0/p_n as functions of τ for ν = 1, 2, and 3 and for γ = 1.4. The circles in this figure and in Fig. 38 mark the data of Ref. 3 for ν = 3. The variations of p_0/p_n with increase of τ for ν = 2 and 3 and for various γ are shown in Figs. 40 and 41. The plots of $q(\tau)$ for different ν at γ = 1.4 are shown in Fig. 42. Calculations have shown that this function depends little on γ in the interval $1.4 \leq \gamma \leq 2$.

Note that the $\xi_0(q)$ curve in Fig. 36 corresponds to $\gamma = 2$.

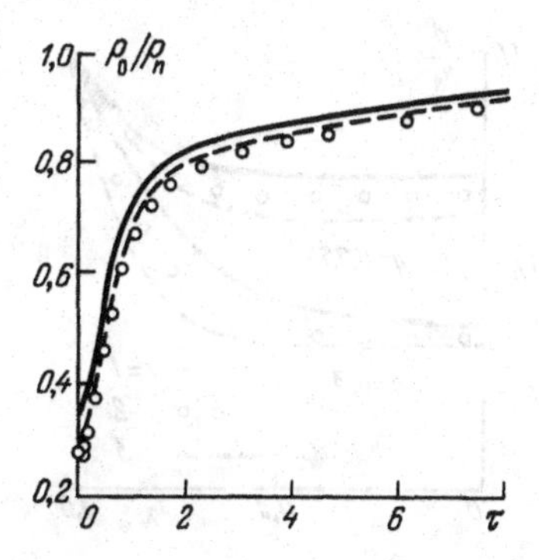

Figure 40. Change of relative pressure at the blast center for various γ, cylindrical case ($\nu = 2$).

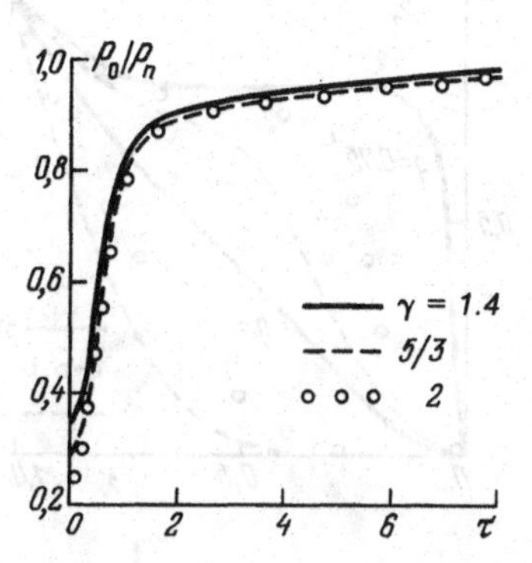

Figure 41. Change of relative pressure at the blast center for various γ, spherical case ($\nu = 3$).

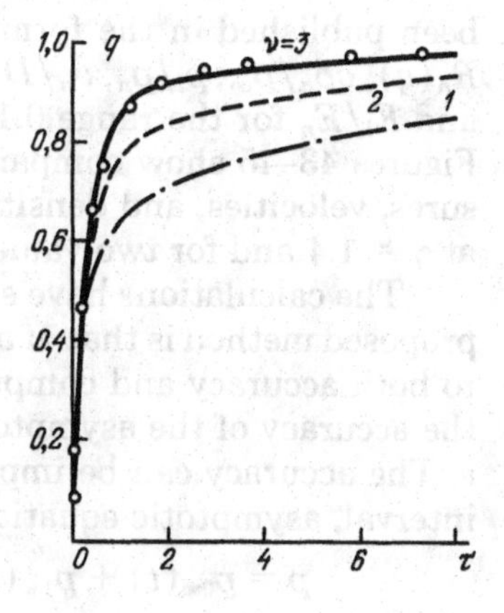

Figure 42. Plot of $q(\tau)$ for various ν at $\gamma = 1.4$. Circles represent data from Ref. 3.

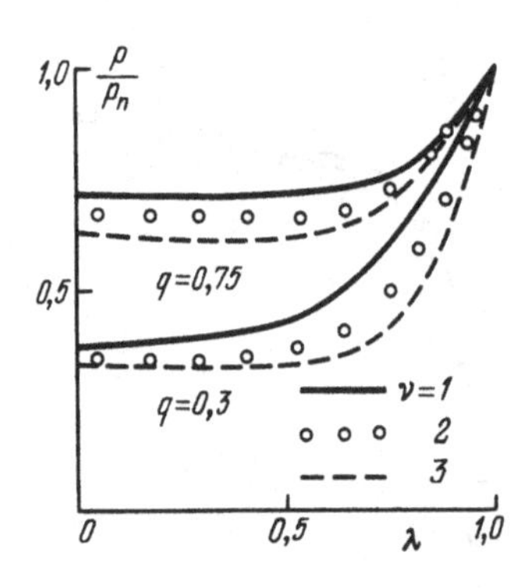

Figure 43. Distribution of relative pressure in space for fixed q and various ν ($\gamma = 1.4$, $n = 8$).

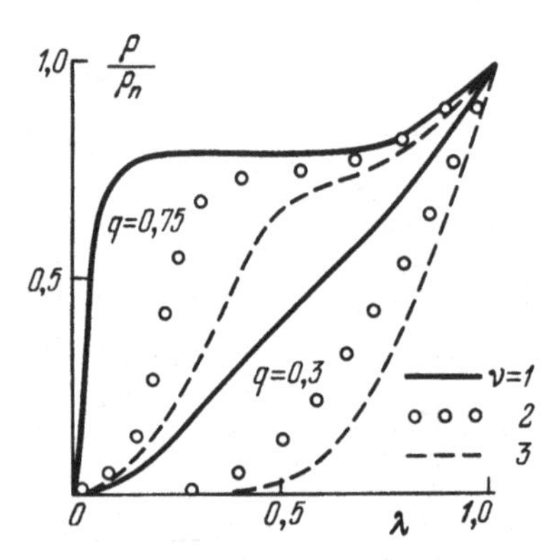

Figure 44. Distribution of relative density in space ($\gamma = 1.4$).

We cite also some calculation results for $n = 8$. The problem was solved at $n = 8$ for γ equal to 1.3, 1.4, and $\frac{5}{3}$. The main calculation results have been published in the form of special tables that list the functions $\tau(q)$, $R_n(q)$, qp_n/p_∞, p_l/p_n, v_n/D, v_l/D, ρ_n/ρ_∞, ρ_l/ρ_n, qT_n/T_∞, T_l/T_n, qE_n/E_∞, and E_l/E_n for the range $0.1 \leq q \leq 0.9$ (Ref. 32b) (see also Appendix II). Figures 43–45 show comparisons of the distributions of the relative pressures, velocities, and densities for spherical, cylindrical, and planar blasts at $\gamma = 1.4$ and for two values of q ($q = 0.3$ and 0.75).

The calculations have shown that the problem simplest to solve by the proposed method is that of a spherical blast. The worst results, with respect to both accuracy and computer time, are for the planar case. In addition, the accuracy of the asymptotic equations (4.37) decreases with decreasing ν. The accuracy can be improved by using for the pressure, in the central interval, asymptotic equations in the form

$$p = p_{0*}(t) + p_{1*}(t)r^d + o(r^d), \qquad d = \nu/(\gamma - 1) + 2.$$

3.4. Gasdynamic functions of spherical, cylindrical, and planar blasts in air ($\gamma = 1.4$)

In view of the appreciable interest from both the theoretical and applied viewpoints, we present a more detailed description of the behavior of the

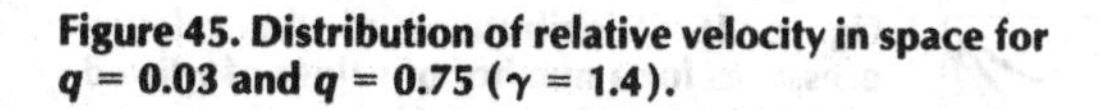

Figure 45. Distribution of relative velocity in space for q = 0.03 and q = 0.75 (γ = 1.4).

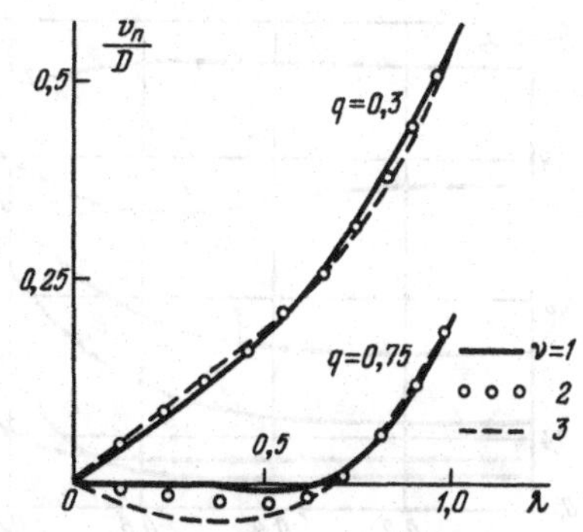

PBP in an ideal gas at γ = 1.4. It is known that this case corresponds approximately to air, if ionization and dissociation effects are disregarded.

We use the results of our calculations and the results of others published in the papers cited above.

We consider first the case ν = 3. Calculations have shown that the pressure in an appreciable vicinity of the center is constant for each fixed instant of time, i.e., a pressure plateau of sorts exists. The pressure in the vicinity of the center is not a monotonic function of the time t. During the initial stage of the blast, approximately up to τ = 0.054 (R_n = 0.334), the pressure decreases. Starting with τ = 0.054, the pressure in the vicinity of the center increases and approaches p_∞ asymptotically. At $\tau \approx 0.147$ ($R_n \approx 0.51$) the pressure at the center of the blast becomes lower than the initial p_∞ and a rarefaction phase sets in. For $R_n > 1.25$ we have for certain $R = R_*$ in the flow region $p(0, \tau) > p(R_*, \tau)$, and then the region of p smaller than $p(0, \tau)$ gradually shifts with increasing τ in the direction from the symmetry center towards the shockwave front.

The observed nonmonotonicity can lead to the onset of a weak pressure jump behind the front of the main shock wave, but in the numerical solution it can be "smeared out" in view of the appreciable procedural "viscosity." Plots of the distributions of the relative pressures for a number of values of τ (or R_n) are shown in Fig. 46, with the numbers of the curves and the values of τ and R_n, as well as of p_n/p_∞, listed in Table 4.

In this figure the spatial coordinate is referred to the shock-wave radius, i.e., $\sqrt[3]{\xi} = r/r_n = \lambda$ is assumed. The pressure is referred to the pressure on the front, i.e., p/p_n is plotted.

Let us dwell on the behavior of the solution at fixed points of space. Clearly, the maximum pressure at each point is equal to the pressure at the instant when the shock wave arrives at this point, i.e., it is equal to p_n. For all points of space, one can note, in the course of time, three characteristic phases of the pressure. The first, when $p > p_\infty$ (positive phase of excess pressures), the second, when $p < p_\infty$ (negative phase of excess pressures, $p/p_\infty - 1 < 0$), and third, when p differs insignificantly from p_∞ ($p \geq p_\infty$).

For points of space close to the blast center, the transition of the excess pressure into a negative phase is almost simultaneous, at $\tau \approx 0.14$.

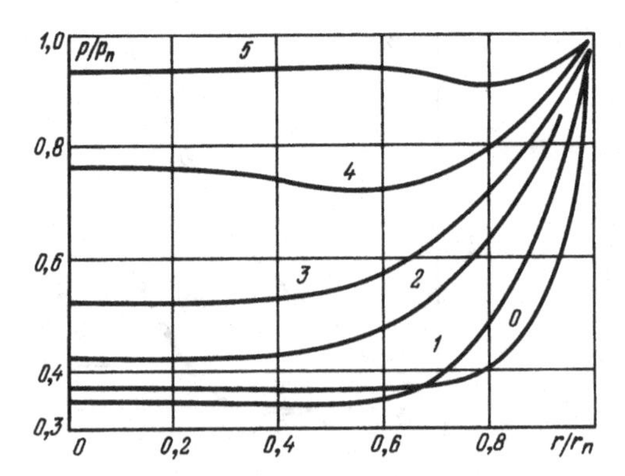

Figure 46. Distributions of the relative pressures for a number of values of τ listed in Table 4 ($\nu = 3$, $\gamma = 1.4$).

Table 4. Parameters of wave front.

N	τ	R_n	$\frac{p_n}{p_\infty}$	N	τ	R_n	$\frac{p_n}{p_\infty}$
0	0.0004	0.045	1740	3	0.432	0.980	1.521
1	0.123	0.489	3.056	4	0.830	1.517	1.255
2	0.281	0.757	1.835	5	2.770	3.633	1.066

A feature common to all points is that the duration $\Delta\tau_-$ of the negative phase of excess pressures in a given point is several times longer than the duration $\Delta\tau_+$ of the positive phase. Thus, at the point $R \approx 0.137$ the duration of the positive phase, in dimensionless units of the time τ, is 0.31, as opposed to 1.08 in the negative phase. For $R = 2.054$ we have $\Delta\tau_+ = 0.278$, and the negative-phase duration is $\Delta\tau_- = 1.024$. With increasing distance from the center, the relation between the duration of the negative and positive phases changes. We present also some data on the values of the pressure momenta and the velocity head.

Dimensionless equations for $J_p = J_p^+ - J_p^-$ and for J_p^+ were given above. Note that the negative momentum J_p^- is calculated for a time interval corresponding to the negative phase of the excess pressure, with $J_p^- > 0$ the absolute value of the negative momentum. Plots of J_p^+ and J_p^- are shown in Fig. 47.

We consider now the velocity distribution in the flow region. The preceding (Sec. 1) analysis of the solutions of the system of gasdynamic equations shows that near the symmetry center the dependence of the velocity v on the radius r is close to linear. Computations have shown that in the vicinity of the center, for small τ, the velocity is positive, i.e., the direction is from center (by virtue of the symmetry, the velocity at the center is zero). With time, the velocity near the center decreases, goes through zero, and becomes negative, after which it again goes through zero, and a small positive velocity pulsation is replaced by standstill. The negative phase for the particle velocities near the center, i.e., the reverse motion of the gas masses to

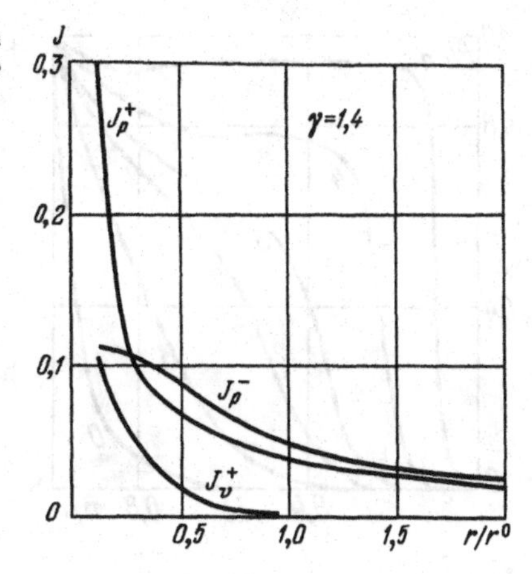

Figure 47. Distribution of the excess-pressure momenta and of the velocity head versus the distance to the blast center.

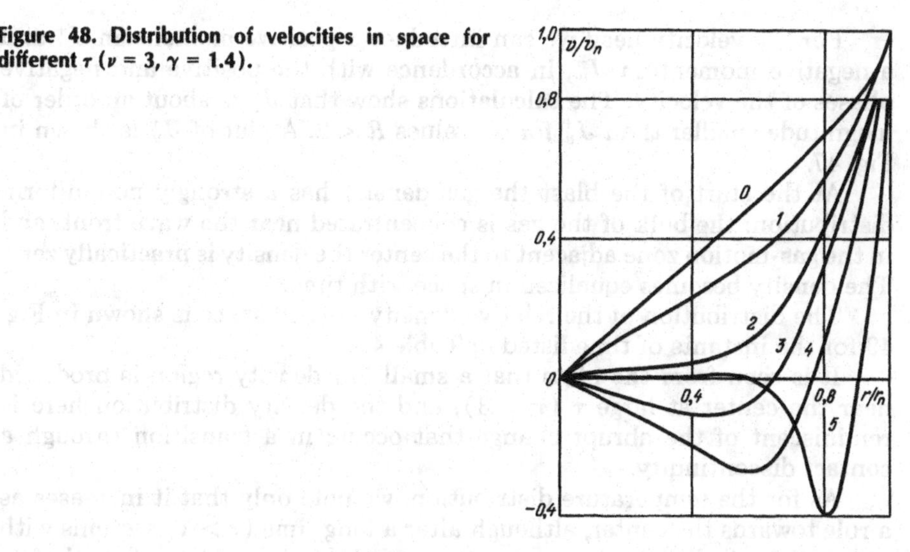

Figure 48. Distribution of velocities in space for different τ (ν = 3, γ = 1.4).

the center, sets in approximately at $R_n = 0.83$ ($\tau = 0.34$, $p_n/p_\infty = 1.68$). For $R_n > 2.4$ we have the following velocity distribution: near the center the gas is practically at rest, and on a certain section it has a velocity directed towards the center, while in the vicinity of the shock-wave front the velocity is directed away from the center.

Note that all the numerical methods used in the computations were such that it was impossible to observe the onset of the shock wave near the center when a converging flow was produced. Such a wave can arise in principle, and can alter somewhat the distribution of the gas-flow parameters, especially in the vicinity of the center.

Figure 48 shows the gas-velocity distribution in space (v/v_n as a function of r/r_n).

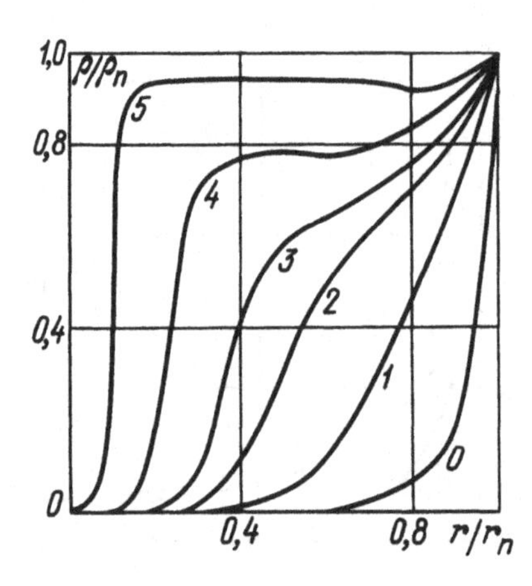

Figure 49. Distribution of the density in space for different τ ($\nu = 3$, $\gamma = 1.4$).

For the velocity head we can introduce a positive momentum J_v^+ and a negative momentum J_v^-, in accordance with the positive and negative phases of the velocity. The calculations show that J_v^- is about an order of magnitude smaller than J_v^+ for all values $R < 2$. A plot of J_v^+ is shown in Fig. 47.

At the start of the blast the gas density has a strongly nonuniform distribution: the bulk of the gas is concentrated near the wave front, and in the gas-motion zone adjacent to the center the density is practically zero. The density becomes equalized in space with time.

The distribution of the relative density ρ/ρ_n in space is shown in Fig. 49 for the instants of time listed in Table 4.

It is seen from the plots that a small low-density region is produced near the center at large τ ($\tau > 3$), and the density distribution here is reminiscent of the abrupt change that occurs in a transition through a contact discontinuity.

As for the temperature distribution, we note only that it increases as a rule towards the center, although after a long time ($\tau > 1$) sections with nonmonotonic distributions are encountered. In view of the adiabatic formulation of the problem, the temperature at the blast center remains infinite during the entire flow time. Naturally, allowance for radiation would influence strongly the character of the temperature distribution.

In PBP applications one encounters sometimes[36–38] the need for computing the momentum of the force acting on the walls of the cone at the apex of which a point blast took place. It turns out that the total momentum of the force can be computed in terms of J_p and J_v. We denote by θ the cone half-apex angle, and by l the length of its generatrix. Assume that we must determine

$$I_\infty = 2\pi \sin^2\theta \int_0^\infty \int_0^l (p - p_\infty)\, r\, dr\, dt.$$

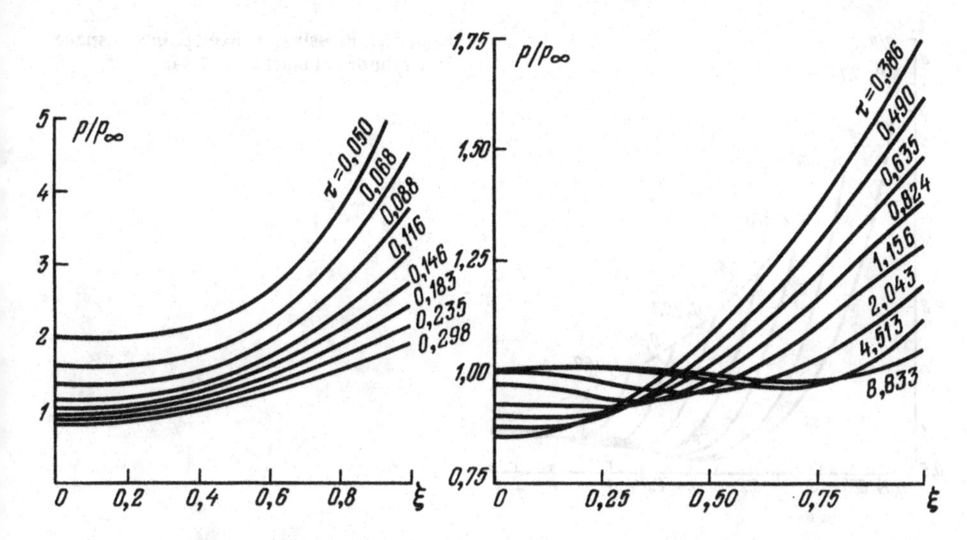

Figure 50. Pressure field in a cylindrical blast at τ from 0.05 to 8.833 ($\gamma = 1.4$).

From the momentum equations we obtain

$$r^{\nu-1}\frac{\partial}{\partial t}(\rho v) + \frac{\partial}{\partial r} r^{\nu-1}(\rho v^2) + \frac{\partial}{\partial r}(r^{\nu-1}\Delta p) - (\nu - 1) r^{\nu-2}\Delta p = 0,$$

$$\Delta p = p - p_\infty .$$

Integrating with respect to t from 0 to ∞ and then with respect to r from 0 to l, and assuming that $v \to 0$ and $\Delta p \to 0$ as $t \to \infty$, we obtain for $\nu = 3$ the following simple equation:

$$I_\infty = \pi l^2 \sin^2 \theta (I_p + 2I_v).$$

Note that the analogous equation for $\nu = 2$ is

$$I_\infty = l \sin \theta (I_p + 2I_v),$$

where θ is half the wedge angle.

We present now, for a cylindrical blast in air ($\nu = 2$, $\gamma = 1.4$), the results of computation by the integral-equations method. The distribution of the pressures p/p_0 in terms of the variable $\xi = (r/r_n)^2$ for a number of values τ are shown in Fig. 50. These distributions are qualitatively close to those for the case $\nu = 3$.

Figures 51–53 show the changes of p/p_∞ with increase of τ for several fixed points of space, and also the dependences of $\bar{\tau}$, τ^+, J_p^+, and J_v on R. The computation results for a cylindrical blast show that in the above dimensionless variables the flow pattern is close to that of the corresponding spherically symmetric flow. We shall therefore not describe in detail the velocity and density fields for different instants of time in the case $\nu = 2$.

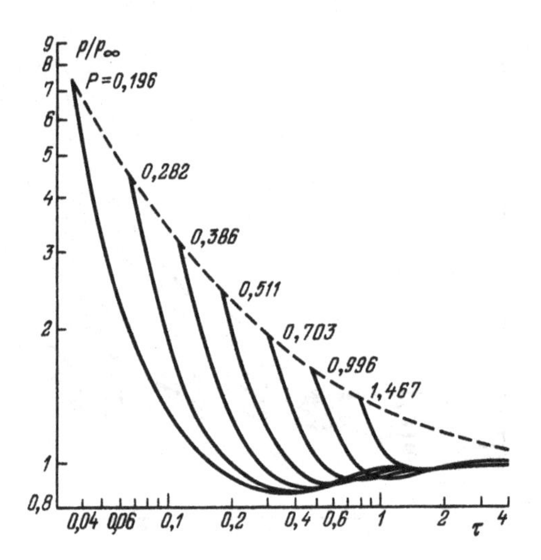

Figure 51. Pressures at fixed points of space in a cylindrical blast ($\gamma = 1.4$).

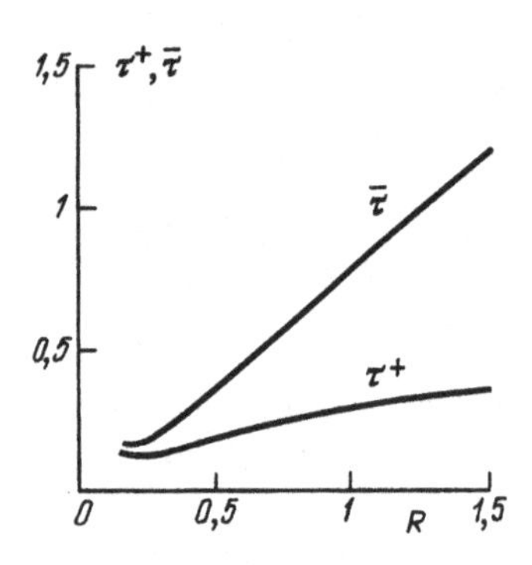

Figure 52. Positive-phase action time τ^+ and the time of transition of the excess pressure to the negative phase ($\nu = 2$, $\gamma = 1.4$).

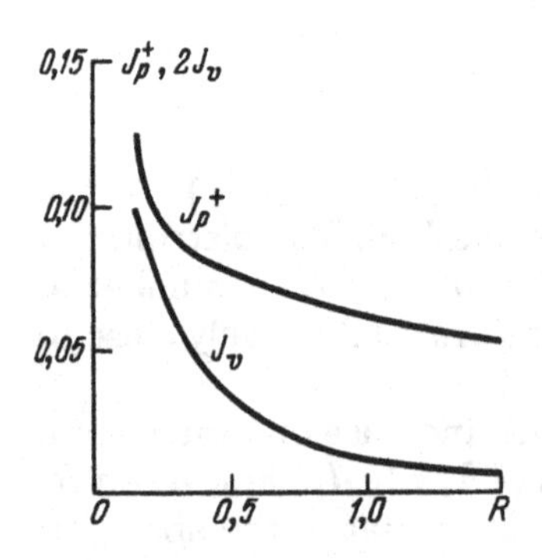

Figure 53. Positive pressure momentum J_p^+ and velocity-head momentum J_v ($\nu = 2$, $\gamma = 1.4$).

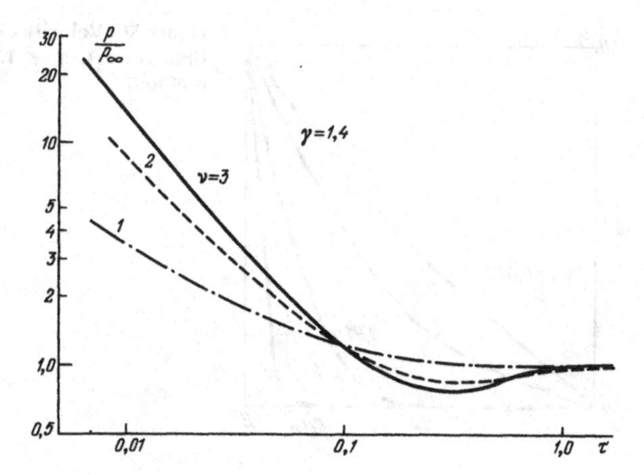

Figure 54. Comparison of the pressure variations at the symmetry center for planar, cylindrical, and spherical blasts.

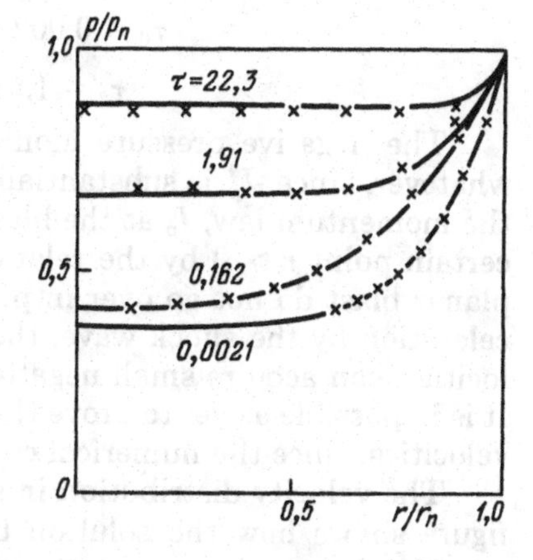

Figure 55. Pressure distribution in space for various instants of time; crosses—computed by the integral-equations method for $n = 8$, $\nu = 1$, and $\gamma = 1.4$.

The planar case ($\nu = 1$) differs substantially from the spherical. The relative pressure p/p_n in the central region likewise decreases first, but then increases with time and tends to unity. The negative-excess-pressures phase is quite weakly manifested. This phase sets in at the vicinity of the center at $\tau \approx 1$ and terminates at $\tau \approx 5$, while the absolute value of $p/p_\infty - 1$ does not exceed 0.01.

A comparative plot of the pressures at the symmetry center is shown in Fig. 54 for planar, cylindrical, and spherical explosions.

The pressure field is shown in Fig. 55. The solid curves correspond to computation by the large-particles method,[30] carried out specially for this study.* The points mark the results of computation by the integral-equations method ($n = 8$). The observed good agreement of the results confirms their accuracy. The curves correspond to the following instants of time:

* The computations were made by R. P. Agafonova with a BESM-6 computer using a program written in collaboration with L. V. Shidlovskaya.

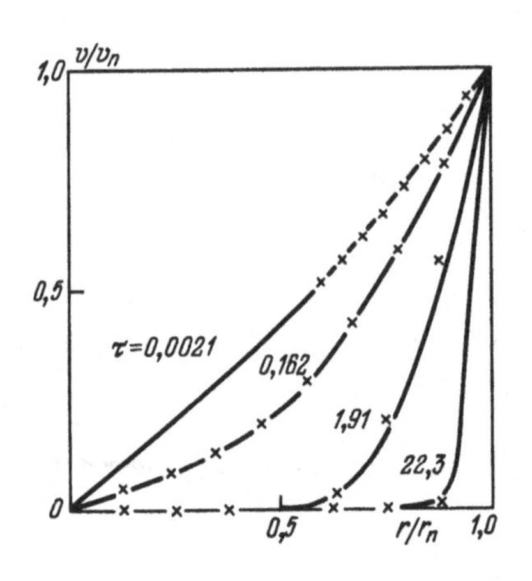

Figure 56. Velocity distributions at various instants of time ($\nu = 1$, $\gamma = 1.4$); crosses—integral-equations method.

$$\tau_0 = 0.002\,10, \qquad \tau_1 = 0.162,$$

$$\tau_2 = 1.91, \qquad \tau_3 = 22.3.$$

The negative pressure momentum in the planar case plays no role whatever, since J_p^+ is substantially larger than J_p^-. Note that, by virtue of the momentum law, I_p at the blast center is connected with I_p and I_v at a certain point $r > 0$ by the relation $I_p|_{r=0} = I_p + 2I_v$. The velocities in a planar blast do not go over in practice into the negative phase. After acceleration by the shock wave, the gas particles rapidly lose their own velocities, can acquire small negative velocities, and then become immobile. It is impossible as yet to prove the possibility of a total absence of negative velocities, since the numerical computations are not accurate enough.*

The velocity distribution in space is shown in Fig. 56. Note that this figure shows how the solution begins to acquire a triangular profile at $\tau > 20$.

The density distribution in space is smoother than in the cases $\nu = 2$ and 3.

Plots of this distribution, for the same instants of time $\tau_0 - \tau_3$, are shown in Fig. 57.

4. Parameters of shock-wave front

We use again the subscripts 1 and 2 to label parameters of the front and of the unperturbed region, respectively.

As already noted, the computed $R_2(\tau)$ dependences for $\nu = 1, 2, 3$ and $\gamma = 1.4$ are shown in Fig. 38.

* The absence of an extended zone of negative velocities in the case $\nu = 1$ and for long times ($p_n/p_\infty < 1.1$) follows from theoretical considerations. In fact, since the flow behind a weak planar shock wave can be approximated with high accuracy by a simple (Riemann) wave, and a simple wave can border only on a quiescent region or on a shock wave,[40] it is clear that the flow near the center must rapidly come to rest.

Figure 57. The density distribution for the various instants ($\nu = 1, \gamma = 1, 4$); cross-calculations by the integral ratio methods.

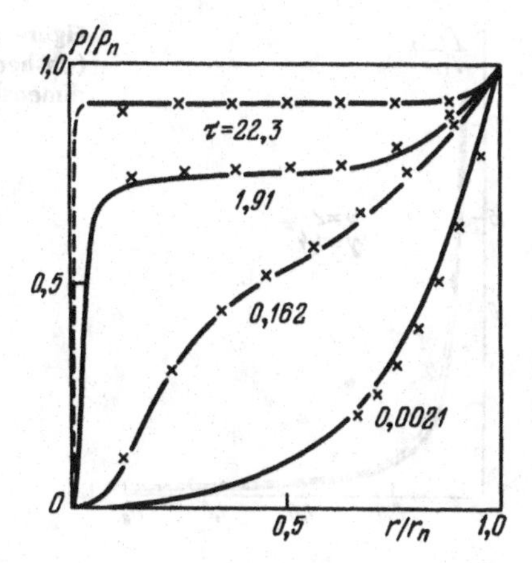

Figure 58. Comparison of shock-wave motion according to various theories: 1—calculated curve; 2—self-similar curve; 3—linearized theory.

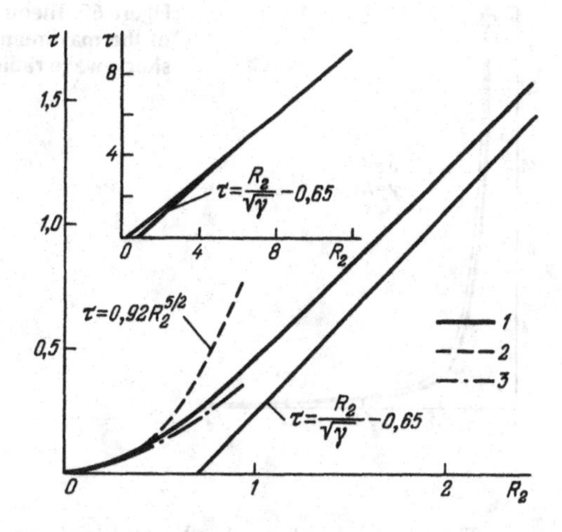

Figure 58 shows a comparison[5] of the laws of motion of a spherical shock wave ($\gamma = 1.4$), obtained from different theories. It shows also a straight line with slope equal to the speed of sound, passing through a point ($\tau = 9.527$, $R_2 = 12.04$) on the $R_2(\tau)$ curve.

A plot of $q(\tau)$ for various ν at $\gamma = 1.4$ is given in Fig. 42. The change of the excess pressure with an increase of R_2 is shown in Figs. 59 and 60 (solid curves) for $\nu = 2$ and 3, respectively.

To compare the results of the point-blast theory, we used the experimental data of M. A. Sadovskii,[41] M. A. Tsikulin,[42] and H. Shardin[43] on blasts of chemical explosives, and with the Los Alamos experimental and theoretical data on atomic explosions in Ref. 44. A comparison with other experimental data was made in Ref. 5. The experimental data were first reduced and converted to our dimensionless variables. From the experiments of M. A. Sadovskii and M. A. Tsikulin, we used the dependences of $p_2/p_1 - 1$ on R_2.

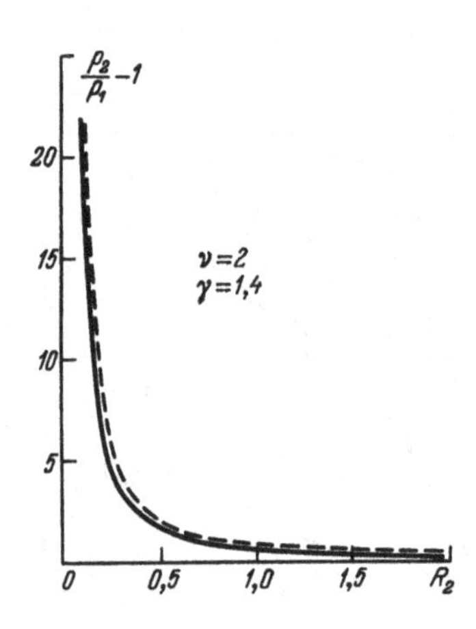

Figure 59. Theoretical (solid curve) and experimental (dashed) plots of the maximum excess pressures versus the dimensionless shock-wave radius for the cylindrical case.

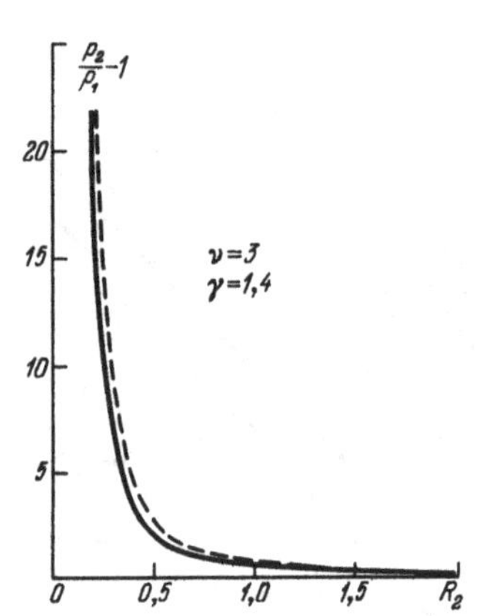

Figure 60. Theoretical (solid) and experimental (dashed) plots of the maximum excess pressures versus the dimensionless shock-wave radius for the spherical case.

For the case of a cylindrical blast, an equation that fits the experimental data is given in Ref. 42:

$$\frac{p_2}{p_1} - 1 = \frac{0.24}{R_2^2} + \frac{0.48}{R_2^{3/4}}. \tag{4.44}$$

The Sadovskii equation[45] for a spherical blast of a TNT charge, reduced in terms of our variables, takes the form

$$\frac{p_2}{p_1} - 1 = \frac{0.21}{R_2} + \frac{0.21}{R_2^2} + \frac{0.15}{R_2^3}. \tag{4.45}$$

Note that the Sadovskii formula (4.45) is given in Ref. 42 with coefficients that are somewhat larger, by a factor of 1.1. This difference can apparently be attributed to the use of initial (dimensional) equation coefficients that differ from the ones taken by us from Ref. 45 for TNT (see also Ref. 46).

Plots of $p_2/p_1 - 1$, corresponding to Eqs. (4.44) and (4.45), are shown in Figs. 59 and 60 (dashed lines).

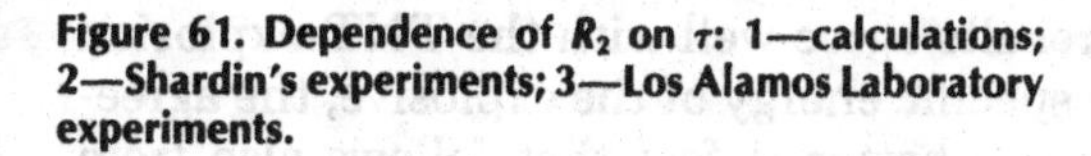

Figure 61. Dependence of R_2 on τ: 1—calculations; 2—Shardin's experiments; 3—Los Alamos Laboratory experiments.

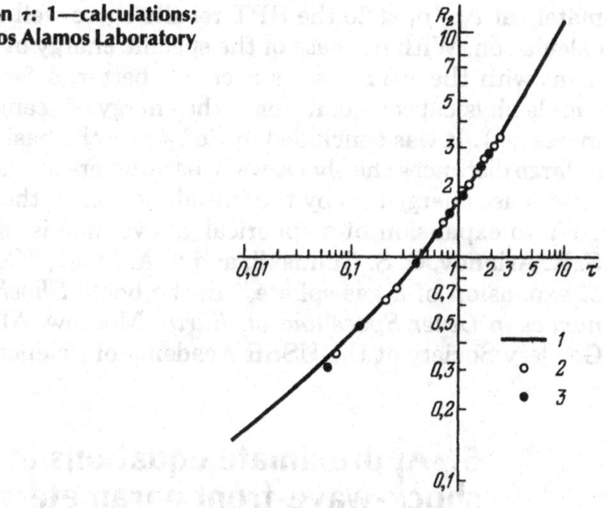

The experimental data on the motion of spherical waves were taken from Ref. 38. We used Shardin's results[43] for the experimental law of motion $r_2(t)$ of a shock wave produced by a lead-azide blast, assuming the lead azide to have 400 kcal/kg (360–380 kcal/kg is more accurate). From the weight of the charge we determined the blast energy E_0, after which we changed over from the dimensionless variables r_2 and t to the dimensionless R_2 and τ. These data are shown in Fig. 61, which also shows the Los Alamos data. In the reduction of these data to dimensionless form we have assumed $E_0 = 8.3 \times 10^{13}$ J and $p_1 = 1$ atm. The experimental data on $r_2(\tau)$ for a cylindrical wave[42] agree well with the calculation (the experimental values of R_2 for equal τ exceed the theoretical by 3–4%).

Comparison of the point-blast-theory (PBT) results given above (and in Ref. 5) with experiment leads to the following conclusions:

(1) Comparison with the experimental data on chemical-explosive blasts shows that the experimental excess pressures are higher than the theoretical and differ greatly from them near the blast center. At large distances from the blast location the experimental data agree better with the theoretical. The excess pressures obtained from the approximate equations (4.44) and (4.45) agree with the PBP results if the value of R_2 is decreased by a factor of 1.1–1.15; this is equivalent to decreasing the explosive charge.

(2) The experimental $R_2(\tau)$ dependences agree well with the theoretical ones.

(3) Atomic-explosion data lead to good agreement with the PBT at $\gamma = 1.4$, if the blast energy E_0 is assumed to equal 8.4×10^{13} J (this energy is close to the blast energy of the so-called nominal atomic bomb).

A. S. Fonarev and S. Yu. Chernyavskii[17] have compared the PBT data with calculations for the explosion of TNT charges in air taking account of the motion of the explosion products. Their results show that approximately at $\Delta p/p_1 < 20$ the BPT results agree well with the TNT-explosion calculation. With increase of the specific energy of the explosive, the agreement with the PBT results becomes better, a fact that follows also from simple physical considerations (the energy concentration in a given volume increases). It was concluded in Ref. 47 on the basis of the calculations that at large distances the shock-wave parameters are determined mainly by the total blast energy and by the initial motion in the air (the applicability of BPT to expansion of a spherical gas volume is discussed in an article by S. K. Aslanov, O. S. Golinskii, and S. A. Ivliev, "Contribution to the theory of expansion of a gas sphere," in the book: *Shock Physics and Wave Dynamics in Outer Space and on Earth,* Moscow, All-Union Astronomy and Geodesy Society at the USSR Academy of Sciences, 1983, pp. 102–113).

5. Approximate equations for shock-wave-front parameters

5.1. Approximate asymptotic equations for large distances

Analytic equations for the shock-wave front are important both for theoretical investigations and for practical applications. Since it is impossible to obtain exact relations, it becomes necessary to find effective approximate equations. This question is considered in the present section.

According to the results of Sec. 2, the parameters of the shock-wave front at large distances can be obtained by using asymptotic equations (4.10)–(4.13) or the more accurate Eqs. (4.14) and (4.15). The constants C_ν and r^* in these equations must be found from time calculations for $(\Delta p/p_1) \sim 0.1$–0.3 $(\Delta p = p_2 - p_1)$.

We consider by way of example the case $\nu = 3$, $\gamma = 1.4$. For p_2/p_1 at $\nu = 3$ we have

$$\frac{p_2}{p_1} = 1 + \frac{4\gamma}{5(\gamma+1)}\sqrt{\frac{E_0}{\alpha\gamma p_1}}\frac{k_3}{r_2\sqrt{\ln(r_2/r^*)}}.$$

In dimensionless variables, this equation takes the form

$$\frac{p_2}{p_1} = 1 + \frac{4\gamma}{5(\gamma+1)}\frac{1}{\sqrt{\alpha\gamma}}\frac{\bar{k}_3}{\sqrt{\ln(R_2/R^*)}}. \tag{4.46}$$

The constants $\bar{k}_3$ and R^* were obtained from the conditions that p_2/p_1 agree with the calculated values of Ref. 4 at certain chosen points for $R_2 > 2$. In the equations for ρ_2/ρ_1 and for v_2/a_1 these constants were assumed to be already known. Calculations for $\gamma = 1.4$ have yielded the asymptotic equations

$$\frac{p_2}{p_1} = 1 + \frac{0.227}{R_2\sqrt{\lg R_2 + 0.158}}, \tag{4.47}$$

$$\frac{\rho_2}{\rho_1} = 1 + \frac{0.163}{R_2\sqrt{\lg R_2 + 0.158}}, \tag{4.48}$$

$$\frac{v}{v_2} = \frac{0.163}{R_2\sqrt{\lg R_2 + 0.158}}. \tag{4.49}$$

These equations were first obtained by the author (Refs. 2 and 5). They yield sufficiently accurate values of the front parameters for values $R_2 > 1.5$ (see Ref. 5 for details).

An assessment of the accuracy of equations of type (4.47)–(4.49) and further development of the method proposed by us are contained in a paper by Miles.[48] He has proposed for the positive pressure momentum the equation

$$\frac{a_1}{p_1 r^0} I_p^+ = \frac{1}{4}\gamma(\gamma + 1)A^2R^{-1},$$

where the constant A is proportional to k_3 and yields the best results at $A = 0.23$.

Using the results of the calculations of the preceding section for planar and cylindrical blasts, we can determine the corresponding constants in the asymptotic relations (4.10)–(4.13) and obtain analogs of the approximate equations (4.47)–(4.49).

5.2. Coordinate dependence of the gas-particle velocity on a shock-wave front

Let us examine in detail the dependence of the particle velocity on the shock-wave front on the coordinate of a point-blast shock wave.

In the self-similar case we have from Eq. (2.12)

$$D_0 = \frac{2}{\nu + 2}\sqrt{\frac{E_0}{\alpha\rho_1}}\, r_{20}^{-\nu/2}. \tag{4.50}$$

From the conditions (1.62) on the shock wave we obtain for the function v_2

$$v_{20} = \frac{2}{\gamma + 1} D_0, \tag{4.51}$$

$$v_2 = \frac{2}{\gamma + 1}(1 - q)D. \tag{4.52}$$

If $r_2 \to \infty$, then $q \to 1$ and consequently $v_2 \to 0$ as $r \to \infty$. We see from Eqs. (4.50) and (4.51) that as $r_{20} \to \infty$ we also have $v_{20} \to 0$. The limiting values of v_2 as $r_2 \to \infty$ are thus the same for a strong blast as for a blast with back pressure. This limiting property does not hold for other characteristics of the motion. We compare next the dependence of v_2 on r_2 for a strong blast and for the case when a shock wave degenerates into a sound wave. For a strong blast we get from Eqs. (4.50) and (4.51)

$$v_{20} = \frac{4}{(\gamma+1)(\nu+2)} \sqrt{\frac{E_0}{\alpha \rho_1}}\, r_2^{-\nu/2}. \tag{4.53}$$

When the shock wave degenerates into a sound wave, we have the following asymptotic relations [see Eqs. (4.10) and (4.12)]:

$$v_{2\text{as}} = \frac{2a_1 C_\nu}{(\gamma+1) r_2^{(\nu+1)/4}} \qquad (\nu = 1, 2), \tag{4.54}$$

$$v_{2\text{as}} = \frac{2a_1}{(\gamma+1)} \frac{C_3}{r_2 \sqrt{\ln (r_2/r^*)}}.$$

Using the expression for the sound velocity $a_1 = (\gamma p_1 \rho_1)^{1/2}$ in Eq. (4.53) and replacing C_ν in Eqs. (4.54) by k_ν, we get

$$v_{20} = \frac{4a_1}{(\nu+2)(\gamma+1)} \sqrt{\frac{E_0}{\alpha \gamma p_1}}\, r_2^{-\nu/2},$$

$$v_{2\text{as}} = \frac{4a_1}{(\nu+2)(\gamma+1)} \sqrt{\frac{E_0}{\alpha \gamma p_1}} \frac{k\nu}{r_2^{(\nu+1)/4}} \qquad (\nu = 1, 2),$$

$$v_{20} = \frac{4a_1}{5(\gamma+1)} \sqrt{\frac{E_0}{\alpha \gamma p_1}}\, r_2^{-3/2},$$

$$v_{2\text{as}} = \frac{4a_1}{5(\gamma+1)} \sqrt{\frac{E_0}{\alpha \gamma p_1} \frac{k_3}{r_2^{3/2}}} \left(\frac{r_2}{\ln \dfrac{r_2}{r^*}} \right)^{1/2} \qquad (\gamma = 3),$$

where the constants k_ν are connected with C_ν by the relation

$$C_\nu = \frac{2}{\nu+2} \sqrt{\frac{E_0}{\alpha \gamma p_1}}\, k_\nu.$$

It is seen from the foregoing equations that the functional dependence of v_2 on r_2 changes little on going from small r_2 (strong blast) to large r_2, and it does not change at all in the case of plane waves.

Since the qualitative character of the change of v_2 with increase of r_2 remains the same as for v_{20} (it is clear from physical considerations that v_2 decreases when r_2 increases), and the asymptotic laws differ little from the self-similar ones, we can assume approximately that

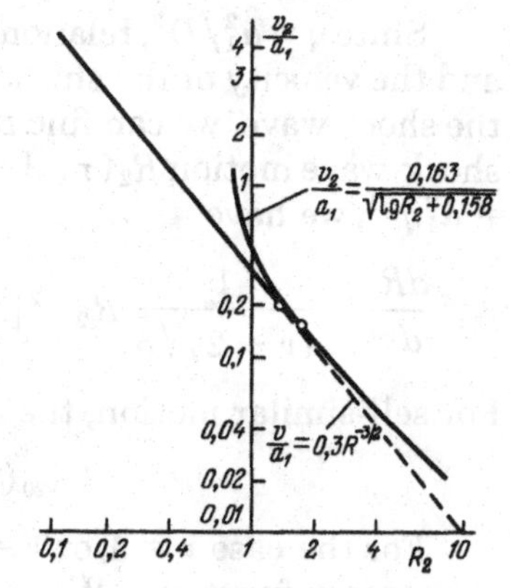

Figure 62. Dependence of v_2/a_1 on the shock-wave radius R_2.

$$v_2 = \frac{4}{(\gamma+1)(\nu+2)}\sqrt{\frac{E_0}{\alpha\gamma p_1}\frac{a_1}{r_2^{\nu/2}}} \tag{4.55}$$

also for sufficiently large values of r_2. Note that we do not take into account here the possible interactions of the main shock wave with secondary (weak) waves that can occur in the gas stream. A more accurate $v_2(r_2)$ dependence with a correct asymptotic behavior at infinity is obtained with the equations[44]

$$v_2 = \begin{cases} \dfrac{4a_1}{(\gamma+1)(\nu+2)}\sqrt{\dfrac{E_0}{\alpha\gamma p_1}}\, r_2^{-\nu/2}, & 0 \le r_2 \le r_2^*, \quad (4.56) \\ v_{2\mathrm{as}}, & r_2^* \le r_2 < \infty. \quad (4.57) \end{cases}$$

The quantity r_2^* and the constants k_ν and r^* will be chosen later. The above properties of the $v_2(r_2)$ dependence are confirmed by calculation and are illustrated in Fig. 62 for $\nu = 3$ and $\gamma = 1.4$.

5.3. Law of motion of a sufficiently strong shock wave and the wave parameters

Using Eq. (4.55), we can derive equations for all the shock-wave front parameters. From the condition for v_2 on the shock wave and from Eq. (4.55), we get

$$\frac{2}{\nu+2}\sqrt{\frac{E_0}{\alpha\gamma p_1}}\, r^{-\nu/2} = \frac{1-q}{\sqrt{q}}. \tag{4.58}$$

Introducing the dimensionless radius R_2, we get

$$R_2 = \left(\frac{2}{\nu+2}\right)^{2/\nu}\left(\frac{1}{\alpha\gamma}\right)^{1/\nu}\frac{q^{1/\nu}}{(1-q)^{1/\nu}}. \tag{4.59}$$

Since $q = a_1^2/D^2$, relation (4.59) gives the connection between the radius and the velocity of the shock wave. Using Eq. (4.59) and the conditions on the shock wave, we can find $p_2/p_1(R_2)$ and $\rho_2/\rho_1(R_2)$. We can also find the shock-wave motion $R_2(\tau)$. From Eq. (4.59) and from the relation $(dr_2/dt)^2 = a_1^2 q^{-1}$, we have

$$\frac{dR_2}{d\tau} = \frac{1}{(\nu+2)\sqrt{\alpha}} R_2^{-\nu/2}[1 + \sqrt{\theta_\nu}], \qquad \theta_\nu = 1 + \gamma\alpha(\nu+2)^2 R_2^\nu. \quad (4.60)$$

For self-similar motion, the $R_{20}(\tau)$ dependence is known to be

$$R_{20}(\tau) = \alpha^{-1/(\nu+2)}\tau^{2/(\nu+2)}. \quad (4.61)$$

For the case $\nu = 1$ or $\nu = 2$, Eqs. (4.60) can be integrated in terms of elementary functions. We consider these cases. From Eqs. (4.60) we get

$$d\tau = \frac{(\nu+2)\sqrt{\alpha}R_2^{\nu/2}\,dR_2}{1 + \sqrt{1 + \gamma\alpha(2+\nu)^2 R_2^\nu}}. \quad (4.62)$$

This yields $\tau(R_2)$ in quadrature. After integration we have

$$\tau = \frac{1}{4\gamma\sqrt{\alpha}}[\sqrt{\theta_2} - \ln(1 + \sqrt{\theta_2})] + C_2 \qquad (\nu = 2),$$

$$\tau = \frac{1}{3\gamma\sqrt{\alpha}}\left\{-2\sqrt{R_2} + \sqrt{R_2\theta_1} + \frac{1}{3\sqrt{\gamma\alpha}}\ln[\sqrt{\theta_1} + \sqrt{9\gamma\alpha R_2}]\right\} + C_1 \qquad (\nu = 1),$$

where C_2 and C_1 are integration constants. They can be chosen to satisfy the condition $R_2(\tau) = R_{20}$, where R_{20} is connected with τ_0 by the self-similar relation (4.62). Here τ_0 must be taken to be a small quantity corresponding to that stage of the blast in which the self-similar solution is valid.

The final equations relating R_2 and τ can be written in the form

$$\tau = \frac{1}{4\gamma\sqrt{\alpha}}\left[\sqrt{1 + 16\gamma\alpha R_2^2} - \sqrt{1 + 16\gamma\alpha R_{20}^2} - \ln\frac{1 + \sqrt{1 + 16\gamma\alpha R_2^2}}{1 + \sqrt{1 + 16\gamma\alpha R_{20}^2}}\right] + \tau_0 \qquad (\nu = 2),$$

$$\tau = \frac{1}{3\gamma\sqrt{\alpha}}\left[-2\sqrt{R_2} + 2\sqrt{R_{20}} + \sqrt{R_2(1 + 9\gamma\alpha R_2)} - \sqrt{R_{20}(1 + 9\gamma\alpha R_{20})}\right. \quad (4.63)$$

$$\left. + \sqrt{\frac{1}{9\gamma\alpha}}\ln\frac{\sqrt{1 + 9\gamma\alpha R_2} + \sqrt{9\gamma\alpha R_2}}{\sqrt{1 + 9\gamma\alpha R_{20}} + \sqrt{9\gamma\alpha R_{20}}}\right] + \tau_0 \qquad (\nu = 1).$$

In the case of spherical symmetry, we get from Eqs. (4.60)

$$\frac{dR_2}{d\tau} = \frac{R_2^{-3/2}}{5\sqrt{\alpha}}[1 + \sqrt{1 + 25\gamma\alpha R_2^3}]. \quad (4.64)$$

Figure 63. Dependence of R_2 on τ for various γ at $\nu = 1$.

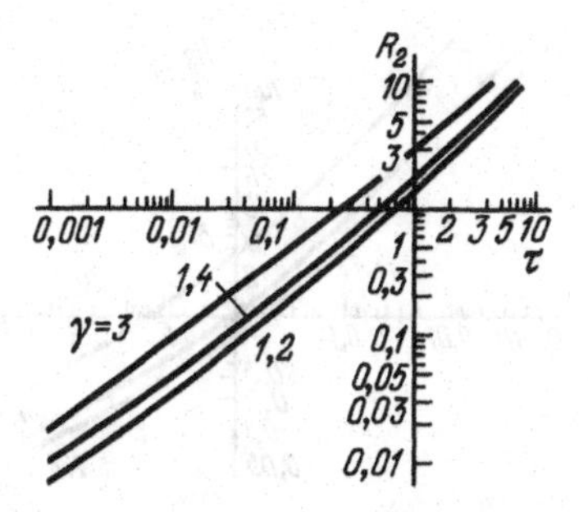

This equation can be integrated numerically after specifying the initial data from the self-similar solution. Tables of elliptic integrals can also be used.

Equations (4.59), (4.63), (4.64), and (4.22) [or Eqs. (1.62)] make it possible to determine with sufficient accuracy all the parameters of the shock wave while it is still quite strong ($r_2 \le r_2^*$). The question of deriving more accurate equations suitable for computation of the flow characteristics at arbitrary distances from the center, on the basis of relations (4.56) and (4.57), will be considered later.

All the relations given above remain valid also for $r_2 \le r_2^*$. To calculate the flow characteristics for $r_2 > r_2^*$ it is necessary to choose the quantity r_2^* and the constants k_ν and r^*.

5.4. Equations for the entire range of distances in the case of planar and cylindrical waves

For planar waves we choose $k_1 = 1$, for only under this condition does the $v_2(r_2)$ dependence for small distances go over continuously into this dependence for large ones.

Thus, by starting from our basic assumptions made at the beginning of the chapter, it is impossible to obtain for plane waves equations other than those of Sec. 5.3 for the calculation of the flow characteristics at large distances. Using Eq. (4.59) with $\nu = 1$ and the conditions on the shock wave, we obtain for the maximum excess pressures the simple equation

$$\frac{p_2}{p_1} - 1 = \frac{4\gamma}{(\gamma + 1)[-1 + \sqrt{1 + 9\gamma\alpha R_2}]}. \tag{4.65}$$

Plots of the shock-wave motion and of the flow characteristics on the shock-wave front for $\gamma = 1.2$, 1.4, and 3 are shown for planar symmetry in Figs. 63–66.

For cylindrical symmetry we can write in dimensionless variables, according to Eqs. (4.54) and (4.56),

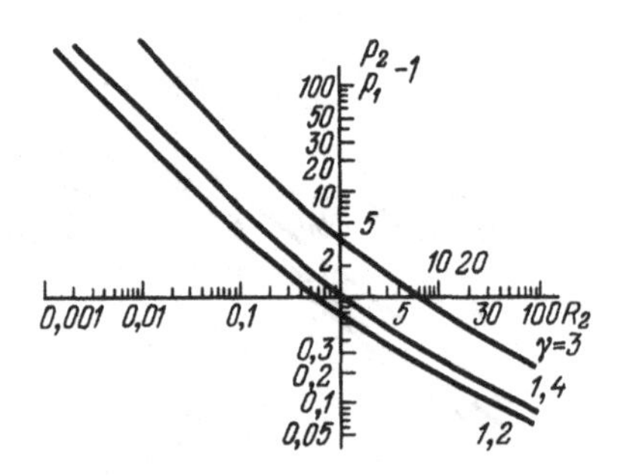

Figure 64. Maximum excess pressure at $\nu = 1$.

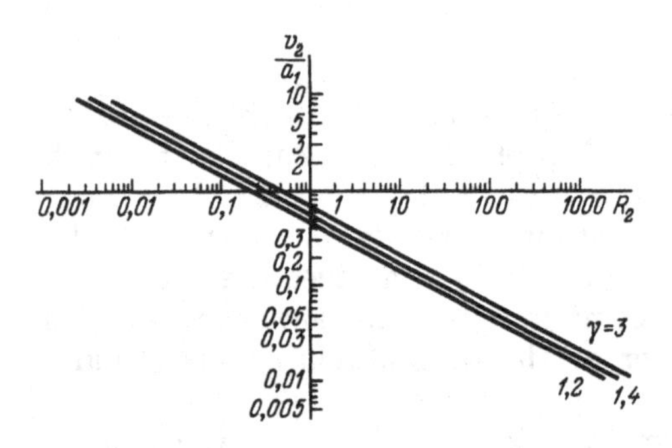

Figure 65. Dependence of the dimensionless velocity on R_2 for different γ at $\nu = 1$.

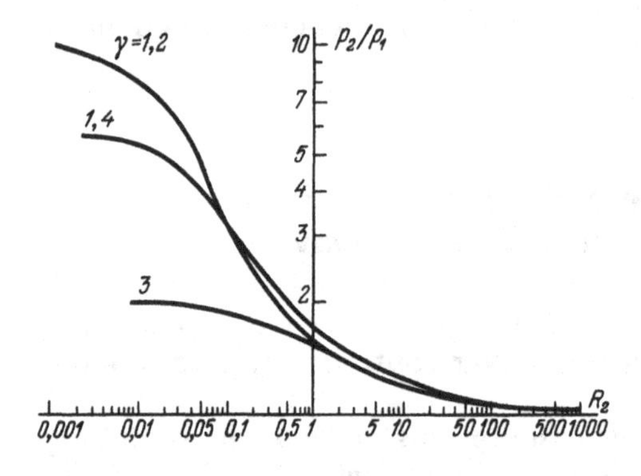

Figure 66. Variation of density on a shock wave at $\nu = 1$.

$$\frac{v_2}{a_1} = \begin{cases} \dfrac{(\gamma\alpha)^{-1/2}}{\gamma+1} R_2^{-1}, & 0 \leq R_2 \leq R_2^*, \\ [(\gamma\alpha)^{1/2}(\gamma+1)]^{-1}\bar{k}_2 R_2^{-3/4}, & R_2^* \leq R_2 < \infty, \end{cases} \tag{4.66}$$

where $\bar{k}_2 = k_2(E_0/p_1)^{1/8}$.

The requirement that v_2/a_1 be continuous at the point $R_2 = R_2^*$ leads to $\bar{k}_2 = (R_2^*)^{-1/4}$. We choose R_2^* such that the change from the equations corresponding to Eq. (4.56) to the asymptotic equations for shock-wave damping takes place at a_1/D approximately equal to 0.9. This value of a_1/D corresponds to an excess pressure $p_2/p_1 - 1 = 0.27$, i.e., the shock wave is quite weak and the use of asymptotic equations to calculate the shock-wave parameter does not introduce large errors.

Figure 67. Comparison of the calculated and approximate values of the excess pressures on a shock-wave front for the planar ($\nu = 1$) and cylindrical ($\nu = 2$) cases at $\gamma = 1.4$; the circles mark the calculated values.

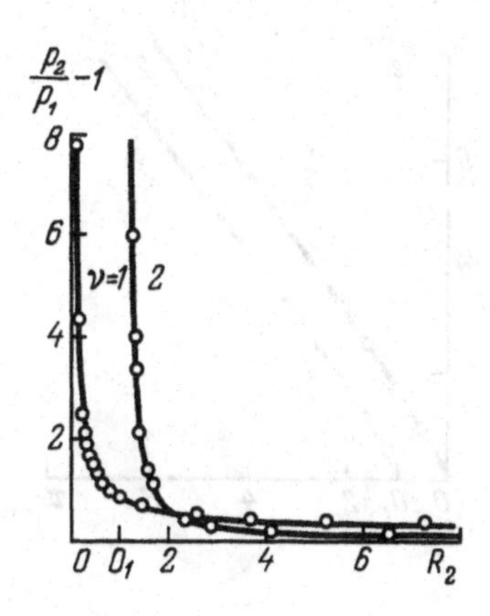

From Eqs. (1.62) and from the v_2/a_1 dependence for $R_2 < R_2^*$ we find that the value $a_1/D \approx 0.9$ corresponds to $R^* = 2$. We choose therefore on the basis of these premises $R_2^* = 2$.

Using Eqs. (4.66), the conditions (1.62), and the derived values of $\bar{k}_2$ and R_2^* we obtain equations for the maximum excess pressures

$$\frac{p_2}{p_1} - 1 = \frac{4\gamma}{(\gamma+1)(-1+\sqrt{16\gamma\alpha R_2^2+1})}, \qquad 0 < R_2 \le 2,$$
$$\frac{p_2}{p_1} - 1 = \frac{4\gamma}{(\gamma+1)(-1+\sqrt{16\sqrt{2}\gamma\alpha R_2^{3/2}+1})}, \qquad \infty > R_2 \ge 2. \tag{4.67}$$

We present next equations for the $R_2(\tau)$ dependence corresponding to Eqs. (4.66). For $R_2 \le 2$ the $R_2(\tau)$ dependence is given by Eqs. (4.63). If $R_2 \ge 2$ we choose for the shock-wave motion the asymptotic relation

$$a_1(t - t_*) = r_2\left[1 - \left(\frac{E_0}{\gamma\alpha p_1}\right)^{1/2} k_2 r_2^{-3/4} + \dots\right].$$

Changing to the dimensionless variables τ and R_2 and substituting the value of k_2, we get

$$\tau - \tau^* = \gamma^{-1/2} R_2\left[1 - \sqrt{\frac{1}{\gamma\alpha}}\, 2^{-1/4} R_2^{-3/4} + \dots\right], \tag{4.68}$$

where τ^* is a certain constant chosen such that the values of τ obtained from Eqs. (4.63) and (4.68) coincide at $R_2 = 2$. The calculations yielded for τ^* the values 0.17, 0.25, and 0.41 for γ equal to 1.2, 1.4, and 3, respectively.

Relations (4.64), (4.65), (4.66), and (1.62) made it possible to calculate all the shock-wave front parameters in the cylindrical case.

Plots of $R_2(\tau)$, $(p_2/p_1)(R_2) - 1$, $(v_2/v_a)(R_2)$, and (ρ_2/ρ_1) for different γ and for $\nu = 2$ are contained in Ref. 5 and are not reproduced here.

Figure 67 shows the dependences of $\Delta p/p_1$ on R_2 calculated by the

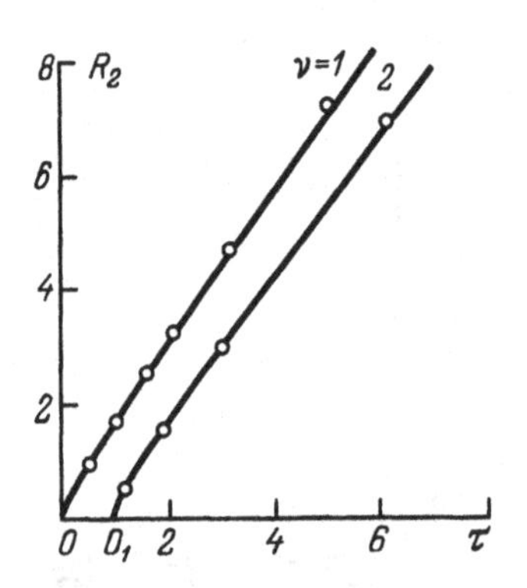

Figure 68. Comparison of the calculated and approximate laws of shock-wave motion for $\gamma = 1.4$.

method considered in Sec. 3 and using Eqs. (4.67). The calculated and approximate values of $R_2(\tau)$ are compared in Fig. 68, where the circles represent the data obtained by the numerical solution. The comparison shows the approximate equations to be fairly accurate.

5.5. Equations for the entire range of distances in the case of spherical waves

For a spherical shock wave we have according to Eqs. (4.56) and (4.57), in dimensionless coordinates,

$$\frac{v_2}{a_1} = \begin{cases} \dfrac{4}{5(\gamma+1)}\dfrac{1}{\sqrt{\alpha\gamma}}R_2^{-3/2}, & 0 < R_2 < R_2^*, \\[2ex] \dfrac{4}{5(\gamma+1)}\dfrac{1}{\sqrt{\gamma\alpha}}\dfrac{\bar{k}_3}{R_2\sqrt{\ln R_2 - \ln l^*}}, & \infty > R_2 \geq R_2^*. \end{cases} \tag{4.69}$$

Here $\bar{k}_3$ and l^* are certain dimensionless constants connected with k_3 and r^* by the relations $k_3 = \bar{k}_3(p_1/E_0)^{1/6}$ and $r^* = l^*(E_0/p_1)^{1/3}$. We stipulate that the function $v_1/a_a(R_2)$ and its first derivative be continuous at $R_2 = R_2^*$. This requirement leads to the relations $\bar{k}_3 = (R_2^*)^{-1/2}$ and $\ln l^* = \ln R_2^* - 1$. Just as in the cylindrical case, we choose $R_2^* = 2$. This choice of R_2 is natural, for at $R_2 \geq 2$ the values of v_2/a_1 obtained from the asymptotic relation (4.12) agree quite well with those obtained by numerical methods, as shown in Fig. 62. Approximately the same choice of R_2^* can result from reasoning similar to that used to choose R_2^* in the case $\nu = 2$.

For the chosen values of R_2^* and $\bar{k}_3$ we readily obtain from Eqs. (4.69) and (4.22) equations for the excess pressures

$$\frac{p_2}{p_1} - 1 = \begin{cases} \dfrac{4}{\gamma+1}\dfrac{1}{-1+\sqrt{1+25\gamma\alpha R_2^3}}, & 0 < R_2 \leq 2, \\[2ex] \dfrac{4}{\gamma+1}\dfrac{1}{-1+\sqrt{1+50\gamma R_2^2(\ln(R_2/2)+1)}}, & \infty > R_2 \geq 2. \end{cases}$$

The shock-wave motion for $R_2 \leq 2$ is determined by Eq. (4.64). To obtain the $R_2(\tau)$ dependence at $R_2 > 2$, we use, just as in cylindrical symmetry, an asymptotic equation that is valid for large distances. In dimensionless variables this equation takes the form

$$\tau = \tau^* + \frac{R_2}{\sqrt{\gamma}} - \frac{0.4}{\gamma\sqrt{\alpha}}\,\bar{k}_3\sqrt{\ln(R_2/l^*)}. \tag{4.70}$$

With the aid of Eq. (4.69) and the conditions (4.22) on the shock wave, we can also find the dependence of ρ_2/ρ_1 on R_2 for various adiabatic exponents γ. Reference 5 contains plots of the dimensionless excess pressure for particles directly behind the shock-wave front at $\gamma = 1.2$, 1.4, and 3 and for $\nu = 3$. The shock-wave motion $R_2(\tau)$ was obtained for these values of γ. Equation (4.64) was integrated in this case with initial conditions $\tau_0 = 10^{-5}$ and $R_{20} = \alpha^{-1/5} \times 10^{-2}$.

We changed over at $R_2 = 2$ from a calculation using Eq. (4.64) to a calculation using Eq. (4.70). Just as in the cylindrical case, the constant τ^* was determined from the condition that the values of τ coincide at $R_2 = 2$, and was found to be $\tau^* = -0.28$ for $\gamma = 1.2$, $\tau^* = -0.22$ for $\gamma = 1.4$, and $\tau^* = -0.19$ for $\gamma = 3$.

To assess the computational accuracy of the above approximate equations, we compared the excess pressures and the $R_2(\tau)$ obtained from the approximate equations with the computed values. It was found that in the spherical case with $p_2/p_1 - 1 \geq 0.05$ the errors of the hydrodynamic characteristics of the wave front do not exceed 5% if the computed values are taken to be "accurate."

For time instants corresponding to $p_2/p_1 - 1 \geq 0.05$, the error of $R_2(\tau)$ does not exceed 3% (Ref. 5).

Note that the equations obtained in this section are quite simple and highly accurate [they can be refined by using the damping laws (4.14) and (4.15)]. They have found various applications (see, e.g., Refs. 50 and 51). The equations of the present section can be generalized also to take dissociation and ionization into account. The constants in the asymptotic relations (4.10)–(4.13) can be obtained on the basis of the calculations of the relevant problems. These constants are given in Ref. 48 for a blast in air at spherical symmetry.

We have derived approximate equations based on approximating the $v_2(r)$ dependence for a perfect gas. The results are good if the energy parameter α is made more precise (e.g., by taking into account the effective value of the adiabatic exponent γ_0) and if γ is regarded as a variable that depends on density and pressure.

According to Eq. (4.53) we have for a strong blast

$$v_{20} = \frac{4}{(\gamma+1)(\nu+2)}\sqrt{\frac{E_0}{\alpha(\gamma_0)\rho_1}}\,r_2^{-\nu/2}.$$

Substituting this dependence into the conditions on the shock wave and assuming $\gamma = \gamma(p, \rho)$ we obtain implicit $p_2(r_2)$, $\rho_2(r_2)$, and $D(r_2)$ dependences.

In a simplified variant one can assume $\gamma = 1.4$ and refine only the value of α. This approach does not lead to large errors [particularly for the $r_2(t)$ dependence], since $(\gamma + 1)$ changes little (from 2.1 to 2.4).

Our computations and comparisons with the numerical results of H. Brode[52] have shown good agreement for the dependence of $\Delta p/p_1$ on R_2 (naturally, the agreement for the shock-wave motion will also be good). Note that if an approximation of $v_2(r_2)$ is used to calculate the pressures p_2, it is convenient to use Eqs. (2.68)–(2.70), which approximate N. M. Kuznetsov's tables (see Chap. 2).

6. Recalculation from dimensionless to dimensional quantities. Similarity law

6.1. Point blast in a perfect gas[5]

As already indicated, it is more convenient to use dimensionless quantities to solve point-blast problems numerically. We assume for simplicity an ideal gas with constant γ. A calculation performed for arbitrary fixed values of ν and γ yields the dependences of the sought dimensionless functions on the dimensionless variables r/r_2 and τ, and this makes it easy to obtain all the quantities that describe the flow (at the same values of ν and γ) for arbitrary values of E_0, ρ_1, and p_1. The calculations must be repeated for other values of ν and γ.

Let us examine the change from dimensionless to dimensional quantities. Assume that we have the independent variables $\lambda = r/r_2$, $\tau = t/t^0$, $t^0 = r^0(\rho_1/p_1)^{1/2} = (E_0/p_1)^{1/\nu}\gamma^{1/2}/a_1$, where r^0 is the dynamic linear dimension and t^0 is the dynamic time.

The energy released in the blast can be expressed in thermal or mechanical units (calories, ergs, foot-pounds). For greater clarity the energy can be described by the weight of some definite explosive, frequently TNT. (It is known in the explosion of the so-called nominal atomic bomb that the energy released in a short time interval is approximately the same as produced by explosion of 20 kilotons of TNT.)

Recognizing that

$$r = r_2\lambda, \qquad t = t^0\tau, \tag{4.71}$$

we can, having calculated the $R_2(\tau)$ dependence and assuming definite values for ρ_1, p_1, and E_0, use Eqs. (4.71) to convert the dimensionless λ and τ into the dimensional r and t.

We can similarly change to dimensional values of the pressure, density, velocity, etc. By way of example, we consider the following dimensionless variables:

$$P(\sigma, \tau) = \frac{p}{p_1}, \qquad G(\sigma, \tau) = \frac{\rho}{\rho_1}, \qquad V(\sigma, \tau) = \frac{v}{a_1},$$
$$R(\sigma, \tau) = \frac{r}{r^0}, \qquad R_2(\sigma, \tau) = \frac{r_2}{r^0}, \tag{4.72}$$

where $\sigma = \xi / r^0$ is the dimensionless Lagrange coordinate and $\tau = t/t^0$. The conversion to dimensional variables is by the equations

$$p(\xi, t) = p_1 P, \qquad \rho(\xi, t) = \rho_1 G, \qquad v(\xi, t) = a_1 V,$$
$$r(\xi, t) = r^0 R, \qquad r_2(t) = r^0 R_2, \qquad \xi = r^0 \sigma, \qquad t = t^0 \tau. \tag{4.73}$$

It may sometimes be necessary to change from dimensional quantities corresponding to specific values of the dimensional initial parameters E_0, p_1, and ρ_1 to dimensionless variables (for the same values of γ and ν). Consider the following example: the calculation is carried out for several values of γ and ν and for definite values E_{01}, p_{11}, and ρ_{11}. It is required to find a solution corresponding to the same values of ν and γ, but to different initial values E_{02}, p_{12}, and ρ_{12} of the energy, density, and pressure. In this case the procedure is as follows. Using Eqs. (4.72) and the known E_{01}, p_{11}, and ρ_{11} we change first to dimensionless variables. Next, using Eqs. (4.73) and the specified E_{02}, p_{12}, and ρ_{12} we obtain the relations of interest to us for the new system of dimensional variables. Thus, to recalculate the time, we find first τ and then $t_{(2)}$:

$$\tau = \frac{t_{(1)}}{t_1^0}, \qquad t_{(2)} = t_2^0 \tau = \frac{t_2^0}{t_1^0} t_{(1)},$$

where

$$t_i^0 = \left(\frac{E_{0i}}{p_{1i}}\right)^{1/\nu} \left(\frac{\rho_{1i}}{p_{1i}}\right)^{1/2} \qquad (i = 1, 2).$$

For the recalculation of the distances (the Euler coordinates of the particles), we have

$$R = \frac{r_{(1)}}{r_1^0}, \qquad r_{(2)} = r_2^0 R = \frac{r_2^0}{r_1^0} r_{(1)}, \qquad r_i^0 = \left(\frac{E_{0i}}{p_{1i}}\right)^{1/\nu} \qquad (i = 1, 2),$$

where $t_{(1)}$ and $r_{(1)}$ are the dimensional time and Euler coordinate, known from the calculation carried out with the parameters E_{01}, p_{11}, and ρ_{11}; similarly, $t_{(2)}$ and $r_{(2)}$ are the dimensional time and Euler coordinate recalculated from the parameters.

Similar equations can be rewritten for recalculation of the velocity, density, and other quantities. We assume here that all the quantities with subscripts 1 and 2 are specified in the same system of units. From these equations we easily find for $p_{11} = p_{12}$ and $\rho_{11} = \rho_{12}$ the relations

$$\frac{r_{(2)}}{r_{(1)}} = \left(\frac{E_{02}}{E_{01}}\right)^{1/\nu}, \qquad \frac{t_{(2)}}{t_{(1)}} = \left(\frac{E_{02}}{E_{01}}\right)^{1/\nu},$$

which are expressions of the so-called similarity law. For a spherical blast this law is frequently written in the form

$$\frac{r_{(2)}}{r_{(1)}} = \left(\frac{w_1}{w_2}\right)^{1/3}, \qquad \frac{t_{(2)}}{t_{(1)}} = \left(\frac{w_1}{w_2}\right)^{1/3},$$

where w is the weight of the charge (TNT equivalent for atomic explosions).

With the ratios of the distances from the blast center and times chosen in this manner, the pressures will be equal. This similarity law is valid for a large class of compressible media.

We consider further, by way of one specific example, the change from the variables used by Goldstine and von Neumann[4] to the dimensionless variables (4.72). The calculation for the problem of a spherical point blast ($\nu = 3$) is carried out in Ref. 4 using the following initial data taken from the solution of the self-similar problem:

$$\tau_{1N} = 0.018\,2575; \qquad P_{2N} = 100; \qquad G_{1N} = 1; \qquad R_{2N} = 0.5 \qquad (\gamma = 1.4), \tag{4.74}$$

where P_{2N} is the pressure on the shock-wave front, G_{1N} is the density of the unperturbed medium, R_{2N} is the shock-wave radius, and τ_N is the time. We assume here all the quantities to be dimensionless and connected with the dimensional ones by Eqs. (4.72):

$$P_N = p/p_1, \qquad G_N = \rho/\rho_1, \qquad \tau_N = t/t_N^0, \qquad R_{2N} = r_2 r_N^0, \qquad R_N = r/r_N^0, \qquad \sigma_N = \xi/r_N^0, \tag{4.75}$$

where t_N^0 is a constant with the dimension of time, r_N^0 is a constant with the dimension of length, r is a Euler coordinate, and ξ is a Lagrange coordinate.

In the self-similar strong-blast problem, the following condition holds on the shock wave:

$$p_2 = \frac{8\rho_1}{25(\gamma+1)} \left(\frac{r_2}{t}\right)^2. \tag{4.76}$$

The radius of the self-similar shock wave satisfies the relation

$$r_2 = \left(\frac{E_0}{\alpha\rho_1}\right)^{1/2} t^{2/5}. \tag{4.77}$$

Substituting Eq. (4.77) for r_2 in Eq. (4.76) and using Eqs. (4.75), we get

$$t_N^0 = \left(\frac{E_0}{\alpha p_1}\right)^{1/3} \left(\frac{\rho_1}{p_1}\right)^{1/2} \left[\frac{8}{25(\gamma+1)P_{2N}}\right]^{5/6} \frac{1}{\tau_N}.$$

For $\nu = 3$ and $\gamma = 1.4$, we have $\alpha = 0.851$. Consequently,

$$t_N^0 = \left(\frac{E_0}{0.851 p_1}\right)^{1/3} \left(\frac{\rho_1}{p_1}\right)^{1/2} \left[\frac{2}{15 P_{2N}}\right]^{5/6} \frac{1}{\tau_{1N}} = \left(\frac{E_0}{p_1}\right)^{1/3} \left(\frac{\rho_1}{p_1}\right)^{1/2} k_N.$$

Taking Eq. (4.74) into account, we obtain the numerical value of the constant $k_N = 0.232\,29$. Thus, $\tau_N = t/t_N^0 = t/k_N t^0$. Comparing this expression for the dimensionless time with Eq. (4.71), we have $k_N \tau_N = \tau$. We can next obtain $R_2 = k_N R_{2N}$, $R = k_N R_N$, $\sigma = k_N \sigma_N$, $P = P_N$, $G = G_N$, and $V = (\gamma)^{-1}(\partial R_N / \partial \tau_N)$. The equations obtained yield the relation between the dimensionless variables (4.75) assumed in Ref. 4 and the dimensionless variables (4.72).

6.2. Similarity parameters with allowance for high-temperature effects

It can be seen that the set (1.122) of gasdynamic equations, with allowance for radiation processes, contains the following functions that describe a two-parameter (thermodynamically speaking) medium:

$$p = (\gamma - 1)\rho\varepsilon, \qquad \gamma = \gamma_0 f_1(\rho/\rho_0, \varepsilon/\varepsilon_0, \gamma_1, \gamma_2, \ldots, \gamma_n),$$
$$T = T_0 f_2(\rho/\rho_0, \varepsilon/\varepsilon_0, \theta_1, \theta_2, \ldots, \theta_m), \tag{4.78}$$
$$k_\omega = k_{\omega_0} f_3(\rho/\rho_0, \varepsilon/\varepsilon_0, \omega/\omega_0, \varkappa_1, \varkappa_2, \ldots, \varkappa_l),$$

where T_0, ρ_0, ε_0, k_{ω_0}, and ω_0 are characteristic dimensionless constants whose meaning is clear from Eqs. (4.78); γ_i, θ_i, and $\varkappa_i$ are dimensionless constants. In addition, the transport equation contains the Planck function B_ω, which can be regarded as specified in the form

$$B_\omega = B_{\omega_0} f_4(\rho/\rho_0, \varepsilon/\varepsilon_0, \omega/\omega_0, b_1, b_2, \ldots, b_r), \tag{4.79}$$

where $B_{\omega_0} = 2\omega_0^3 h/(2\pi)^3 c^2$, and its dimensionality B_{ω_0} can be obtained from the relation $[B_{\omega_0} \times \omega_0] = [\sigma T_0^4]$.

The characteristic frequency can be taken to be the one corresponding to the maximum of the Planck distribution function at $T = T_0$,

$$\omega_0 = \frac{2\pi c}{\lambda_0} = bT_0, \qquad b = \frac{2\pi c}{b_*},$$

where $b_* = 0.29$ cm K is a known dimensional constant. We take the initial conditions, as above, for instantaneous release of a finite energy E_0 from a point, along a straight line, or along a plane in a gas with initial parameters ρ_1 and p_1, which we regard as constant. We have the following system of defining variables and constant parameters:

$$t,\ \mathbf{r},\ \boldsymbol{\Omega},\ \omega,\ B_{\omega_0},\ k_{\omega_0},\ \varepsilon_0,\ \rho_0,\ E_0,\ \rho_1,$$
$$p_1,\ \nu_0,\ \gamma_i,\ \varkappa_i,\ \theta_0,\ \theta_i,\ b_i. \tag{4.80}$$

Note that Eq. (4.80) includes the parameter $\theta_0 = T_0/\varepsilon_0$, which can be regarded as dimensionless. The set (4.80) contains 11 dimensional quantities, with only three having independent dimensionalities, since the basis of the system of units is assumed to be mechanical with the three fundamental dimensions of mass, length, and time. From dimensionality theory[2,53] we conclude that eight dimensionless combinations can be made up from these dimensional parameters. Five of them are constants. We choose the dimensionless combinations

$$\frac{r}{r^0},\quad \frac{t}{t_0},\quad \frac{\omega}{\omega_0},\quad \pi_\rho = \frac{\rho_1}{\rho_0},\quad \pi_k = k_{\omega_0} r^0,$$

$$\pi_B = \frac{\omega_0 B_{\omega_0}}{\rho_0 \varepsilon_0^{3/2}},\quad \pi_p = \frac{p_1}{(\gamma - 1)\rho_1 \varepsilon_0},\quad \pi_\omega = \frac{1}{\omega_0 t^0},\tag{4.81}$$

$$t^0 = r^0 \varepsilon_0^{-1/2},\quad r^0 = \left[\frac{E_0}{(\gamma_0 - 1)\rho_0 \varepsilon_0}\right]^{1/\nu},$$

where $\nu = 1, 2, 3$ in the case of planar, cylindrical, and spherical energy release, respectively.

Thus, any dimensionless function that yields a solution of the problem, for example

$$\rho/\rho_1 = g,$$

depends, by virtue of the π-theorem (Chap. 1), on the following variables and constant dimensionless parameters:

$$\frac{\rho}{\rho_1} = g\left(\frac{t}{t^0}, \frac{r}{r^0}, \Omega, \frac{\omega}{\omega_0}, \pi_i, \{m_i\}\right),\tag{4.82}$$

where $\{m_i\}$ denotes the set of dimensionless constants γ_i, $\varkappa_i$, θ_0, θ_i, and b_i which enter in Eq. (4.80), while π_i stands for the parameters (4.81). Since the function (4.82) includes more than three variable arguments and the set of constant parameters π_i and $\{m_i\}$, it follows, taking the complexity of the system (1.122) into account, that a complete solution of this problem is impossible at present analytically or numerically. We present a few particular cases. In most gasdynamics problems the parameter

$$\pi_\omega = \frac{1}{\omega_0 t^0}$$

is very small. Thus, for $t^0 = 10^{-3}$ s and $\omega = 10^{15}\ \text{s}^{-1}$ we get $\pi_\omega = 10^{-12}$. This parameter can be neglected in problems where there are neither high-frequency phenomena with very short characteristic times nor relativistic effects [the transport equation (1.110) does not contain the time explicitly].

We note furthermore that the constants $\{m_i\}$ represent a model of the medium but not the problem considered. The parameter π_B depends only on the constants of the model of the medium and on ω_0. It will therefore be automatically preserved if the model of the medium and the characteristic frequency ω_0 are preserved, in which case it can be included in $\{m_i\}$. If

$$p_1 \ll (\gamma - 1)\rho_0 \varepsilon_0$$

(strong blast) and the back pressure p_1 can be neglected, i.e., the dimensionless parameter π_p can be neglected, there remain two essential dimensionless parameters:

$$\pi_\rho = \frac{\rho_1}{\rho_0}, \qquad \pi_k = k_{\omega_0} r^0. \tag{4.83}$$

When π_ρ is preserved, there remains one essential dimensionless parameter on which the solution of the problem depends, namely,

$$\pi_k = \tau^0 = k_{\omega_0} r^0 = k_{\omega_0}\left(\frac{E_0}{(\gamma_0 - 1)\rho_0 \varepsilon_0}\right)^{1/\nu}. \tag{4.84}$$

This parameter is none other than the characteristic optical thickness. It follows from Eq. (4.84) that for a strong blast at fixed k_{ω_0}, ρ_0, ε_0, and γ_0 the calculations should be performed, strictly speaking, for all the energies E_0 of interest to us, and recalculation from one energy to another is impossible.[53] This is what makes the problem difficult even with the above simplifications. One should, however, hope that the actual dependence of the solution on τ^0 is weak and that interpolation with respect to the parameter τ^0 can be made if the number of necessary solutions for different τ^0 is small.

Note that in the diffusion approximation for a gray-body gas there is no significant simplification. In fact, in the model assumed above for the medium an essential parameter will be the characteristic optical thickness $k_0 r^0 = \tau^0$, where k_0 is the average characteristic absorption coefficient. Here, however, for $\tau^0 \ll 1$, we can use the approximation of an optically thin body (emission), and for $\tau^0 \gg 1$ we can use the optically thick approximation (radiant heat conduction). Assume that for $\tau^0 \sim 1$ we can write for the medium a model equation in the form

$$T = T_0(\varepsilon/\varepsilon_0)^{\alpha_2}(\rho/\rho_1)^{\alpha_1}, \qquad p = (\gamma_0 - 1)\rho\varepsilon, \qquad k = k_0 \rho^{\beta_1} \varepsilon^{\beta_2}.$$

It is easy to show then from dimensionality considerations that the solution of the strong-blast problem will depend on, besides γ_0, only one dimensionless parameter:

$$\pi_1 = k_0 \rho_1^{\delta_1} E_0^{\delta_2} \Sigma^{\delta_3}, \tag{4.85}$$

where δ_1, δ_2, and δ_3 are certain numbers expressed in terms of α_1, α_2, β_1, and β_2 and the symmetry parameter ν, and

$$\Sigma = 4\sigma T_0^4 \varepsilon_0^{-4\alpha_2} \rho_1^{-4\alpha_1}.$$

In these cases, for fixed k_0 and Σ, the calculation performed for one π_1 can be used for a number of E_0 and ρ_1 values connected by Eq. (4.85).

7. Use of the analogy between a blast and flow around thin blunted bodies

When thin bodies such as missiles and rockets move through a gas with hypersonic velocity, i.e., several times faster than sound in the surrounding medium, the front tips of the thin tapered bodies are blunted by the medium. Such moving bodies may also be initially constructed with blunted frontal parts.

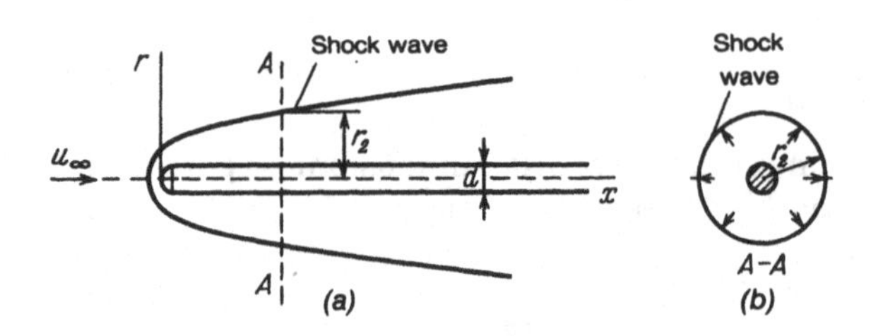

Figure 69. (a) Flow around cylinder with spherical head; (b) analogy with cylindrical blast.

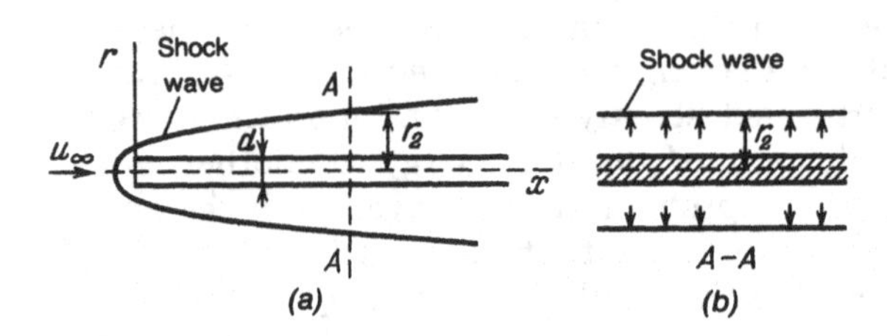

Figure 70. (a) Flow around a blunted plate; (b) analogy with blast of a planar charge.

Bodies flying at hypersonic speed are therefore almost always blunted and the solution of a problem involving the motion of blunted slender bodies is of great practical and theoretical significance. Supersonic motion of blunted bodies is accompanied by the onset, ahead of the body, of an outgoing shock wave that does not touch the surface of the body [Figs. 69(a) and 70(a)]. Methods of calculating the flow behind the outgoing shock wave were developed by O. M. Belotserkovskii, V. V. Lunev, G. F. Telenin, P. I. Chushkin, and others.[27,28,54–57]

There is a certain analogy between the blast problem and the problem of stationary hypersonic flow around a body. It was formulated in the papers by G. G. Chernyi[48] and others and is based on the use of the planar-sections principle.

The principles of this analogy are described also in Refs. 5 and 55. Investigations have shown that in planes perpendicular to the direction of unperturbed gas motion the pattern of the gas flow around a plate or a cylinder should be essentially the same as in propagation of a blast wave from a point blast for planar and cylindrical charges [Figs. 69(b) and 70(b)]. The resistance force X of a blunted body is approximately equal to the energy E_0, since the energy acquired per unit area by the gas layer on overcoming the blunted end is $E_0 = X \times 1$.

By using blast theory it is possible to obtain the basic data on flow around a blunted thin plate or a blunted thin cylinder (direct analogy). On the other hand, if we have the solution for hypersonic flow around a blunted plate or cylinder, we can determine approximately the parameters of the planar and cylindrical blasts (inverse analogy). After the foregoing analogy principle was formulated, principal attention was paid to the use of the strong-blast theory and of the linearized solution with allowance for back pressure (see, e.g., Refs. 5 and 54). The reason is that the corresponding non-self-similar non-linear problems were little investigated. M. A. Tsikulin used the analogy principle not for a strong blast, but in conjunction with experimental data for the front parameters for explosions of cylindrical charges.[42] A more complete study of the direct analogy became possible only after blast problems were solved by numerical and approximate methods.

To use the direct analogy, the parameters of the nonstationary problem are replaced by stationary ones, with t and E_0 replaced respectively by xu_∞^{-1} and $(1/2)c_x\rho_\infty u_\infty^2(\pi/4)^{\nu-1}\,d^\nu$.

We use here the following notation: x is the coordinate along the body axis, measured from its forward point; d is the characteristic transverse dimension of the blunted part; u_∞ is the velocity of the main gas stream, directed along the x axis; c_x is the blunt-nose resistance coefficient per unit cross section area of the body and referred to the velocity head.

In such a transition, the connection between the dimensionless quantities x/d, r/d and τ, R is given by the relations

$$\frac{x}{d} = \left(\frac{\pi}{4}\right)^{(\nu-1)/\nu}\left(\frac{c_x\gamma}{2}^{(\nu+2)/2}\right)^{1/\nu} \mathrm{M}_\infty^{(2+\nu)/\nu}\tau,$$
$$\frac{r}{d} = \left(\frac{\pi}{4}\right)^{(\nu-1)/\nu}\left(\frac{c_x\gamma}{2}\right)^{1/\nu} \mathrm{M}_\infty^{2/\nu} R, \tag{4.86}$$

where $\mathrm{M}_\infty = u_\infty/a_\infty$ is the Mach number of the incoming stream.

The results of the direct analogy are briefly the following.

First, the approximate equations considered in Sec. 5 for the shock-wave front make it possible to determine the shape of the shock wave and to write down analytic expressions for it. This was indicated in Ref. 5 and considered in our paper[19] in which we obtained also the following result: The shock-wave slope dr/dx for large x tends in the stationary case to the corresponding slope of the characteristics, i.e.,

$$\frac{dr}{dx} \to (\mathrm{M}_\infty^2 - 1)^{-1/2}.$$

On the other hand, the shock-wave velocity in the nonstationary problem tends to the sound velocity a_∞. Changing to the variables τ and R, according to 4.86, we obtain for large τ:

$$\frac{dR}{d\tau} \to \sqrt{\gamma}\,\frac{\mathrm{M}_\infty}{\sqrt{\mathrm{M}_\infty^2 - 1}}, \qquad \frac{dR}{d\tau} \to \sqrt{\gamma}$$

for streamline flow and blast, respectively. Thus, for streamlining, the shock-wave form at large distances will be closer to the form obtained from blast theory, and the larger the Mach number M_∞. This restricts the possibility of using nonstationary-problem data for the streamline problem (and vice versa).

The distributions of the pressures on the cylinder and of the forms of frontal shock waves are obtained by an approximate calculation for a cylindrical blast and by streamline calculations by the method of characteristics.[56] This comparison showed satisfactory agreement between the results of applying blast data to streamlining outside the blunted region. The higher-

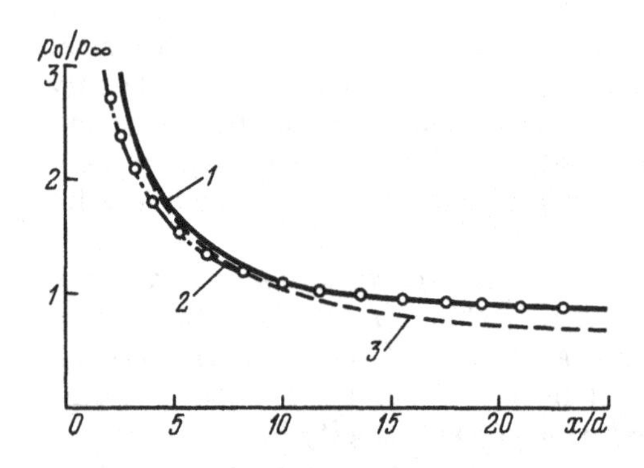

Figure 71. Comparison of the pressure distributions on a blunted cylinder, calculated by blast theory (curves 1 and 3) and by the method of characteristics ($M_\infty = 10$, $\nu = 2$).

accuracy calculations considered in Sec. 3 lead to more substantiated conclusions[32] concerning the accuracy of the analogy. It has turned out that for $M_\infty \sim 10$ the gasdynamic blast and streamlining functions, for both $\nu = 2$ and $\nu = 1$, are quite close (in the sense of the formulated analogy) outside a certain vicinity of the blunt edge ($x/d > 10$).

Figure 71 shows by way of illustration the distribution of the pressure on a cylinder with a spherically blunted end for $M_\infty = 10$, obtained from blast calculations (curve 1). Curve 2 corresponds to steady flow calculations by the method of characteristics[56] while curve 3 pertains to linearized blast theory. As seen from the figures, the first two curves practically overlap at a sufficient distance from the blunted edge.

Much progress was made in recent years in the calculations of hypersonic flow. This raises the question of using the inverse analogy. In fact, as shown using the variables R and τ, the dimensionless functions coincide outside a certain blunting zone, i.e., for sufficiently large τ and for not high p_2/p_1. For $\nu = 2$, $M_\infty = 10$, and $p_2/p_1 < 3$ the agreement is quite good. For approximate estimates of the blast parameters we can therefore use streamlining-calculation results for $M \geq 10$. To this end we can, for example, use tables[57] of thermodynamic functions for hypersonic flow around blunted cylinders for large M_∞. The data in these tables make it possible to obtain the pressure and density fields for blasts not only in an ideal gas, but also with allowance for dissociation and ionization. Naturally, the near zone of the blast (for small t) must be calculated separately (for example, for an ideal gas, using an effective value of γ).

A remark is in order here concerning a blast in a stationary translational gas stream.

Some applications may require calculations for a point blast in a stationary gas stream flowing with constant velocity $\mathbf{U}$. The case of a strong spherical blast was discussed in Ref. 58. If the blast evolution is considered in an immobile coordinate frame, it is necessary to recalculate the solution by using the Galileo–Newton transformation (see Sec. 1 of Chap. 1). Since the gasdynamic equations are invariant to this transformation, the solution will be transformed into some new solution.

Assume a certain fixed Cartesian rectangular frame $\hat{x}^1, \hat{x}^2, \hat{x}^3$ with origin at the center of the blast at $t = 0$. We denote by $J^j, j = 1, 2, 3$, the components of the initial velocity vector **U** on the coordinate axes, and by v^j the corresponding components of the gas velocity vector after the blast.

The gas-flow parameters in a fixed frame can then be calculated by using the transformation equations $\hat{x}^j = x^j + U^j t$, $\hat{v}^j = v^j + U^j$, $\hat{p} = p$, and $\hat{\rho} = \rho$. Here v^j, p, and ρ denote the values of the velocity, pressure, and density components in the x^j frame comoving with the gas. The solution of the problem in this moving system is assumed known. It can be obtained by the numerical method considered above, or taken from the tables of Ref. 32. Thus, the problem reduces to a simple recalculation of known data. In such a recalculation for cylindrical and spherical blasts, account must be taken of the geometric relations that follow from the connection between Cartesian and cylindrical or between Cartesian and spherical coordinates.

In the spherical case we have

$$x^1 = r \sin\theta \cos\varphi, \qquad x^2 = \sin\theta \sin\varphi, \qquad x^3 = r\cos\theta,$$

$$v^1 = v \sin\theta \cos\varphi, \qquad v^2 = v \sin\theta \sin\varphi, \qquad v^3 = v\cos\theta,$$

where r, θ, and φ are the spherical coordinates (of the moving system). It is clear from physical considerations that we have here a simple drift of the stream by a "constant wind." Unfortunately, for a blast in the atmosphere the wind velocity is rarely constant in magnitude or direction. Therefore the drift of the perturbation zone is more complicated. The foregoing equations can, however, sometimes be useful for determining the time of arrival of a shock wave at a given point of space, if the vector **U** varies slowly in space and in time. At high velocities U the drift of the perturbed-motion zone alters greatly (for an immobile observer) the picture of the shock-wave propagation.

8. Reflection of point-blast shock waves

8.1. Initial stage of reflection of a plane wave from a parallel plane wall and of cylindrical and spherical waves from concentric, cylindrical, or spherical walls, respectively

Reflections of point-blast shock waves have so far not been intensively investigated. The main reason is the rather high complexity of the problems. Naturally, to solve such problems successfully we must study the laws governing shock-wave propagation in unbounded space. Solution of these shock-wave reflection problems is essential for the development of a more complete PBT.

Consider the determination of the pressure, density, and velocity of the gas behind the front of a reflected shock wave at instants of time close to the instant t_* when the wave reaches the wall. By virtue of its proposed geometry, the wall reflects the shock normally, and the flow behind the shock-wave front can be regarded as one-dimensional with planar, cylindrical, or spherical symmetry. In this case the pressure p_*, the density ρ_*, and the temperature T_* in the reflected wave are obtained from the well-known gasdynamics equations

$$\frac{p_*}{p_1} = 2\left(\frac{p_2}{p_1} - 1\right) + \left[\left(\frac{p_2}{p_1} - 1\right)^2 \Big/ \left(\frac{\gamma - 1}{\gamma + 1}\frac{p_2}{p_1} + 1\right)\right], \tag{4.87}$$

$$\frac{\rho_*}{\rho_1} = \frac{\gamma p_2}{p_1}\frac{\rho_2}{\rho_1} \Big/ \left[(\gamma - 1)\frac{p_2}{p_1} + 1\right], \quad \frac{T_*}{T_1} = \frac{p_*}{p_1}\frac{\rho_1}{\rho_*}. \tag{4.88}$$

For the initial reflected-wave velocity we have the expression

$$\frac{D_*}{a_1} = \sqrt{\frac{2}{\gamma}}\left[(\gamma - 1)\frac{p_2}{p_1} + 1\right]\left[(\gamma + 1)\frac{p_2}{p_1} + (\gamma - 1)\right]^{-1/2}. \tag{4.89}$$

Since the wall is fixed, by virtue of the boundary condition on it the velocity v_* behind the reflected shock-wave front will be low at instants of time close to t_*.

If the distance r_* from the blast point to the wall and the characteristic length r^0 are known, we can, knowing the dimensionless coordinate R_*, obtain from the value of $R_n = R_*$, using the tables[32] or the approximate equations, the dimensionless time τ_* of arrival at the wall, and from it the time t_* as well as the corresponding values of q, p_2/p_1, and ρ_2/ρ_1. Equations (4.87) and (4.88) are used next to calculate the main parameters of the reflected shock wave for times close to t_*.

Note that Eqs. (4.87) and (4.88) are strictly speaking valid for the considered problem only at the instant of direct reflection of the shock wave from the wall since they have been obtained for a homogeneous flow behind the incident wave. These relations will, however, be approximately satisfied for instants of time when it is possible to neglect the variability of the flow parameters behind the shock wave.

Let us advance certain arguments concerning the character of the subsequent flow. In the case of reflection of a strong-blast wave we have in fact, from the physical viewpoint, a collision between a narrow gas layer and an immobile partition, followed by decay of an arbitrary discontinuity. It is possible to construct here various approximate models of the resultant flow. Note that certain aspects of strong-blast reflection (during its initial stages) were considered in Refs. 59 and 60. To solve the problem of reflection of waves of moderate intensity, numerical methods must be used. The gas flow in this reflection is one-dimensional, but a reflected shock wave will propagate towards the blast center through the moving gas. The solution obtained for instants of time close to t_* can be used to specify the initial data. Note that from the viewpoint of applications (blasts in artificial holes in the ground, cylindrical blasts in xenon tubes to produce light pulses, etc.)

greatest interest attaches to knowledge of the flow parameters in the immediate vicinity of the wall. This has to be taken into account when approximate theories are developed.

8.2. Initial stage of regular reflection of a planar, cylindrical, or spherical blast wave from a flat surface

We consider first the planar case. Let a plane blast wave be incident on an absolutely rigid flat wall in such a way that the angle between the shock-wave plane and the wall differs from zero and is equal to α. For instants of time close to a collision of the wave with the wall we can obtain for the gas velocity, from the relations on the shock wave and from the boundary condition on the wall, asymptotic relations between the parameters of the incident and reflected waves (see Ref. 61).

We denote by α_*, p_*, and ρ_*, respectively, the reflection angle, the pressure, and the density behind the shock wave at the instant of time immediately after the reflection, and introduce the following parameters:

$$\pi_1 = p_1/p_2, \qquad \pi_* = p_*/p_2, \qquad \omega = \operatorname{tg}\alpha, \qquad \omega_* = \operatorname{tg}\alpha_*.$$

If the angle α and the parameter π_1 are specified at the instant of encounter of the wave with the reflection plane, we have for ω_* the quadratic equation

$$\begin{aligned} m[(1-\mu)^2 - (\omega-\omega_*)^2 - (\mu+\omega\omega_*)^2] \\ + m^2(1-\mu)^2(\omega-\omega_*) + \omega_* - \omega = 0, \end{aligned} \tag{4.90}$$

where

$$\mu = \frac{\gamma-1}{\gamma+1}, \qquad m = \frac{(1-\pi_1)\omega}{1+\mu\pi_1+(\mu+\pi_1)\omega^2}.$$

The relative pressure π_* behind the reflected wave is given by

$$\pi_* = \frac{m(1+\mu\omega_*^2)+\omega_*}{\omega_* - m(\mu+\omega_*^2)}. \tag{4.91}$$

The density ratio ρ_*/ρ_2 and the temperature ratio T_*/T_2 can be obtained with the aid of Eqs. (4.88).

Since the point of intersection of the incident and reflected waves moves along the plane, we have for the velocity D_* of the reflected shock wave at the reflection point

$$D_* = D\,\frac{\sin\alpha_*}{\sin\alpha}. \tag{4.92}$$

Using the obtained values of D_* and ρ_*/ρ_2 and the law of mass conservation on passing through the discontinuity surface, we can determine the normal component of the gas velocity behind the reflected wave. The tangential

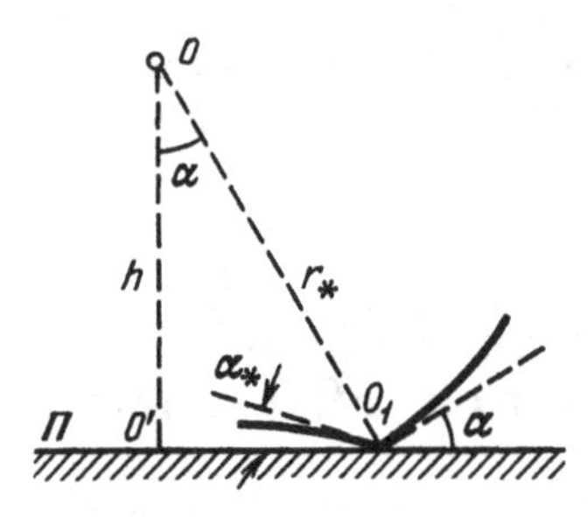

Figure 72. Diagram of regular reflection of a shock wave from the plane Π. The point O is the blast center, while α and α_* are the incidence and reflection angles.

component is obtained from the condition that it be continuous on going through the discontinuity. Note that the condition that Eq. (4.90) have no real roots yields the angles for which irregular (Mach) reflection sets in.

We consider now the reflection of a cylindrical or spherical wave. Let the blast occur at a distance h from the reflection plane Π (in the cylindrical case we assume for simplicity that the blast line is parallel to the plane Π). We denote by α the angle of incidence of the cylindrical or spherical wave on the Π plane at a certain instant t_* after the blast. It is obvious from geometric considerations that the angle α is equal to the angle between the normal to the plane Π and the radius vector drawn from the blast center O to the wave-reflection point O_1 at the considered instant of time t_* (Fig. 72). The gas flow upon reflection of the shock wave will have axial symmetry, with an axis passing through the blast point perpendicular to the plane Π. The solution will depend on the parameters h, E_0, p_1, ρ_1, and γ, while for the dimensionless functions p/p_1 we have

$$\frac{p}{p_1} = P\left(\frac{t}{t^0}, \frac{r}{r^0}, \frac{h}{r^0}, \theta, \gamma\right),$$

where θ is the angular coordinate. For various ratios h/r^0 the problem can be solved anew (we assume γ to be the same in all cases). The similarity law obtained in Sec. 6 is strictly speaking not satisfied here. If we vary r^0, say by varying the energy E_0, to recalculate the data for another h we must change r^0 in such a way that the ratio $(h/r^0) = H_*$ remains constant. In other words, the problem can be reformulated if

$$E_0 = (h/H_*)^\nu p_1. \tag{4.93}$$

For arbitrary E_0, reformulation by ordinary similarity can lead to errors. This must always be taken into account. Naturally, the similarity law can be approximately used also for energies that deviate little from Eq. (4.93). As already mentioned, reflection of a blast wave has an irregular character at certain values of the incidence angle (Refs. 44, 45, 46, 56–60, 62). At a given Δp an irregular (Mach) reflection takes place if the incidence angle exceeds a certain limiting value α_0. In irregular reflection, the pressure (on the plane) is substantially higher than the pressure expected for regular reflection. Generally speaking, the limiting angle α_0 increases with decreasing Δp. The fact that the problem is not one-dimensional, the limited validity of the similarity law, and the onset of Mach reflection all greatly complicate the theoretical solution of the problem. Consideration of this problem as a whole is a separate research project. We consider only some individual aspects of this problem.

By applying the solution of the reflected-plane-wave problem to reflection of an element of a cylindrical or spherical wave one can determine directly the gas parameters at the instant of reflection of the wave from the plane. Denoting by H_* the dimensionless quantity h/r^0, we have

$$\cos \alpha = \frac{h}{r_*} = \frac{H_*}{R_*} . \tag{4.94}$$

Assume that we are interested in the parameters of the reflected wave at a point O_1 of the plane π at a distance r_* from the blast center. If the parameter r^0 is known, we can, knowing $R_* = r_*/r_0$, calculate the angle α by using Eq. (4.94). With the aid of tables[32] or the equations of Sec. 5, we obtain from the value $R_2 = R_*$ the values q_* and τ_* and the ratios p_2/p_1 and ρ_2/ρ_1. If the angle α for the obtained p_2/p_1 corresponds to the regular reflection regime, we can, using the notation introduced, determine from Eqs. (4.88) and Eqs. (4.90)–(4.92) the parameters of the reflected wave at instants of time close to t_*.

Next, assume that we know the relation between p_2/p_1 and the limiting angle of regular reflection. On the basis of the theoretical analysis, these data follow from Eqs. (4.90)–(4.92); they are given, e.g., in Ref. 61. For a specified h we can then find the limit of the regular-reflection region on the Π plane. Note that for a spherical blast at $\gamma = 1.4$ the problem of the initial stage of reflection was solved by M. M. Vasil'ev.[63] A cylindrical blast is apparently considered here for the first time ever (experimental data are given in Ref. 42). The qualitative features of irregular reflection are described in many papers (Refs. 44, 46, 61, 62–66).

The question of determining the pressure on a plane at instants of time close to the instant of reflection, for the irregular reflection stage, might be solved if equations were known relating the pressures behind the incident and reflected waves at the threefold Mach configuration. Unfortunately, there is as yet no sufficiently accurate solution of this problem. Our arguments here have therefore only a qualitative and an approximate quantitative meaning.

Since the pressure on the plane does not vary (even during the stage of regular reflection) monotonically with increase of α, we are faced with the known problem of the most expedient (for a given value of E_0) choice of the blast height to obtain pressures $p_* \geq p_*^0$ on a certain area Σ (p_*^0 is a specified pressure). Let the blast be spherical. We can indicate here another such problem: find the value of h for which the total force

$$F = \int_{\Sigma_0} p_* \, d\Sigma$$

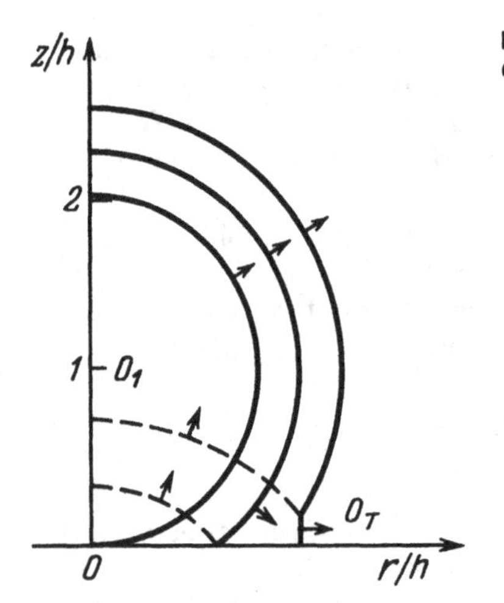

Figure 73. Reflection of a spherical wave from a plane. O_1 is the blast center and O_T is the triple point.

is a maximum (for all t) at a given value of Σ_0. The optimal choice of the height of a spherical blast has been considered in a number of papers (see Refs. 44, 45, 61). We are unaware of any complete theoretical solution of this problem for either a spherical or a cylindrical wave.

8.3. Results of calculation of the interaction of a spherical blast with a plane surface

The reflection of a spherical blast wave from a plane surface was recently investigated in great detail by A. S. Fonarev and his co-workers[67] (see also Ref. 68). We follow here their procedures. We introduce a cylindrical system of dimensionless coordinates z and r (Fig. 73) and take the length unit to be the distance h from the center of the blast to the plane Π, while the time t is reckoned from the instant when the shock wave reaches the origin. At $t<0$ the flow is a spherically symmetric solution of the point-blast problem. During the initial reflection stage it has a regular behavior. With time, however, the reflection becomes irregular or of the "Mach" type. In Fig. 73 the wave propagation directions are indicated by arrows. In the irregular case the reflected shock wave consists of two parts: an almost straight wave moving along the plane (called the "Mach stem") and a shock wave moving upward from the plane and propagating in a direction of the flow that is spherically symmetric about the point O_1. A shock-wave combination consisting of a spherical wave, a shock wave moving upwards, and a "Mach stem" is called a threefold configuration, and the point of intersection of the above-mentioned waves is called a triple point (point O_T in Fig. 73). The calculation was carried out by S. K. Godunov's method without explicitly distinguishing the fronts of the reflected waves.

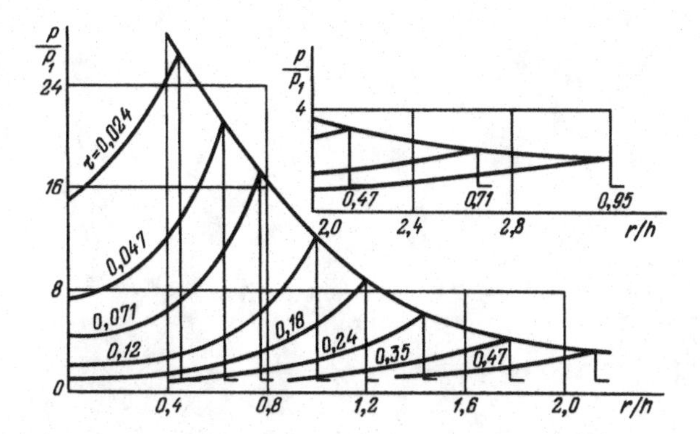

Figure 74. Pressure distribution over the plane at various instants of time $\tau = a_1t/h$ ($\Delta p_2 = 7.9p_1$, $\gamma = 1.4$).

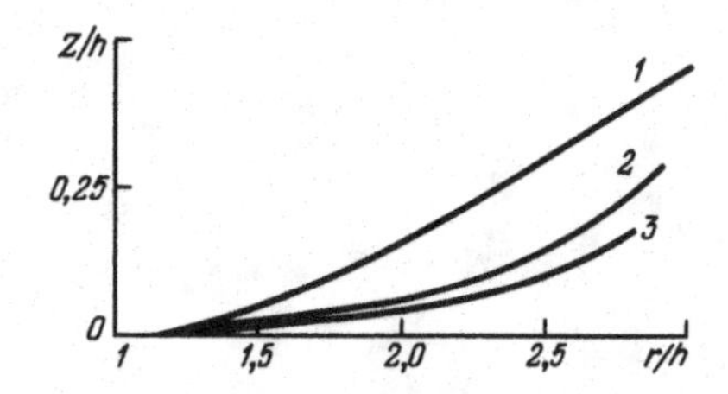

Figure 75. Trajectory of the irregular-reflection triple point for three values of the excess pressures of the incident wave. (1) $\Delta p_2 = 7.9p_1$; (2) $\Delta p_2 = 2.3p_1$; (3) $\Delta p_2 = 1.3p_1$.

Let us consider the case when the initial pressure jump Δp_2 in the incident wave is equal to $7.9p_1$. This means that a wave with pressure $p_2 = 8.9p_1$ passes through O at $t = 0$; the distribution of the pressure on the plane is shown in Fig. 74 for several values of the dimensionless time $\tau = a_1t/h$. The Mach reflection for this variant occurs at $r/h \approx 0.8$ ($\tau \approx 0.07$). Figure 75 shows also the trajectories of the triple points for three values of excess initial pressures of the incident wave (at $t - 0$), viz., $\Delta p_2 = 7.9p_1$, $\Delta p_2 = 2.3p_1$, and $\Delta p_2 = 1.3p_1$. It can be seen that for the first case z is almost linear in r if $r > h$. Note that specifying $\Delta p_2/p_1$ and the instant of wave arrival in a given point is equivalent to specifying the dimensionless height h/r_0 of the blast center. Let $\Delta p_2/p_1$ be specified at $t = 0$. This quantity is a known function of h/r^0 according to one-dimensional calculations or approximate equations, and by the same token h/r^0 is specified (for $\Delta p_2/p_1 = 7.9$ and $\gamma = 1.4$, namely $h/r^0 \approx 0.3$, where r^0 is the dynamic length of the spherical blast).

Chapter 5
Problems of point-blast theory in inhomogeneous media and with asymmetric energy release

1. One-dimensional flows with variable initial density

1.1. Exact solution for a special $\rho_1(r)$ dependence

We consider the point-blast problem for constant initial pressure p and for an initial gas density that varies with distance from the blast point in accordance with the law

$$\rho_1(r) = \frac{a}{\gamma}(\gamma-1)^2\left(\frac{\gamma+1}{2}\right)^{1-\beta}\left(\frac{r}{r^0}\right)^{-\omega}\left[\left(\frac{r}{r^0}\right)^{\nu} + \frac{\nu(\gamma^2-1)}{2\sigma_\nu\gamma}\right]^{-\beta}, \quad (5.1)$$

where γ is the ratio of the specific heats; a is a positive arbitrary constant; $\omega = [\nu(3-\gamma)+2\gamma-2]/[\gamma+1]$, $\beta = [3\nu\gamma+4-\nu]/[\nu(\gamma+1)]$, $r^0 = (E_0/p_1)^{1/\nu}$ is the dynamic length; $\nu = 3, 2, 1$. It is seen from Eq. (5.1) that ρ_1 depends parametrically on γ and on the dynamic length r^0.

The one-dimensional adiabatic motions of the gas behind the wave are described by the system (1.18)–(1.20) or by Eqs. (3.5). It is required to determine the dependences of the velocity, pressure, and density of the gas on the linear coordinate r and on the time t, and also the dependence of the shock-wave radius r_2 on the time.

The problem reduces to finding a solution of the system (3.5) with the above initial conditions, and also with the boundary condition $v(0,t) = 0$ at the symmetry center and conditions, which can be written in the form (1.62), on the front of the blast wave. It can be verified directly that the solution of our problem is given by the equations[1]

$$v = \frac{r}{kt}, \qquad k = \frac{\nu(\gamma-1)+2}{2},$$

$$\rho = \frac{2p_1[kt]^{-(\nu-1)/k}}{\nu(\gamma^2-1)r}\frac{d}{dx} \times\left\{\frac{4\gamma}{b(\gamma+1)}[f(x)]^{-(\gamma-1)/2} - (\gamma-1)[f(x)]^{-\gamma}\right\}, \quad (5.2)$$

$$p = \frac{p_1[kt]^{-\gamma\nu/k}}{\gamma+1}\left\{\frac{4\gamma}{b(\gamma+1)}[f(x)]^{-(\gamma-1)/2} - (\gamma-1)[f(x)]^{-\gamma}\right\},$$

$$\left(\frac{r_2}{r^0}\right)^\nu = \frac{(\gamma^2-1)}{2\gamma\sigma_\nu\left\{\dfrac{2}{(\gamma+1)b}[kt]^{-\nu(\gamma+1)/2k} - 1\right\}},$$

where $x = r[kt]^{-1/k}$, $b = [\nu^2 p_1/(\sigma_\nu r^0)^2 a]^{1/(\beta-1)}$, and $f(x) \geq 0$ is a function that takes on no negative values. It is determined from the equation

$$\left(\frac{x}{r^0}\right)^\nu + \frac{(\gamma^2-1)}{2\sigma_\nu\gamma} f - \frac{2}{(\gamma+1)b}\left(\frac{x}{r^0}\right)^\nu f^{(\nu+1)/2} = 0.$$

The change of pressure directly behind the shock-wave front is given by

$$p_2 = p_1\left[1 + \frac{\nu(\gamma-1)}{\sigma_\nu}\left(\frac{r^0}{r_2}\right)^\nu\right].$$

The above solution was obtained by us from L. I. Sedov's exact solution.[2] The method of constructing discontinuous solutions for this exact solution was developed in Ref. 3.

From our solution we obtain, for the particular case when $p_1 \to 0$ and $b \sim p_1 \to 0$, the known solution of the self-similar point-blast problem, when the initial density distribution is given by $\rho_1 = Ar_1^{-\omega}$ [see Eqs. (2.28)].

1.2. Integral estimates

In view of the variety of the possible initial values of $\rho_1(r)$ it is expedient to present some rather general integral estimates of the gasdynamic parameters. We follow hereafter for the most part Ref. 4. We use Lagrange variables and consider only the motion of an ideal gas with constant γ.

We denote by m the mass of a layer of radius r, where $r = r(m, t)$ describes the motion of the particle.

The integral equation for the energy of the gas layer between the surfaces $m = 0$ and $m = M_2(t)$ is

$$T + U = E_* + \int_0^t \left[\left(\frac{\dot r^2}{2} + \varepsilon\right)\dot M + S_\nu p\dot r\right]_0^{M_2} d\tau, \tag{5.3}$$

where

$$T = \frac{1}{2}\int_0^{M_2} \dot r^2 dm, \qquad U = \int_0^{M_2} \frac{f(m)\,dm}{(S_\nu r')^{\gamma-1}}, \qquad f(m) = \frac{p}{\rho^\gamma(\gamma-1)},$$

$$\dot r = \frac{\partial r}{\partial t}, \qquad r' = \frac{\partial r}{\partial m}, \qquad S_\nu = 1, 2\pi r, 4\pi r^2 \quad \text{for} \quad \nu = 1, 2, 3,$$

E_* is the initial value of the energy, T and U are the kinetic and internal energies of the layer, the integral in the right-hand side of Eqs. (5.3) has the meaning of the energy influx through the moving surface $M_2(t)$ and at the center of the blast, and $m = M(t)$ is a certain movable surface.

From the equation of motion we can determine also the integral momentum balance

$$\tfrac{1}{2}[(\dot{I} - r^2\dot{M})_0^{M_2}]^{\cdot} = 2T + \nu(\gamma - 1)U + r(\dot{r}\dot{M} - S_\nu p)_0^{M_2}, \qquad (5.4)$$

where $I = \int_0^{M_2} r^2 dm$ is the moment of inertia of the gas layer. For $\gamma = (\nu + 2)/\nu$, in particular, it follows from Eqs. (5.3) and (5.4) that

$$\tfrac{1}{2}(\dot{I} - r^2\dot{M})^{\cdot} = 2E_*.$$

This relation can be integrated twice to obtain

$$I = \int_0^R r^2 \rho_1 S_\nu dr + 2E_* t^2,$$

where $R = R(t)$ describes the shock-wave motion, $E_* = E_0$ is the blast energy, and $\dot{M} = \rho_1 S_\nu \dot{R}$.

Using the fact that the entropy increases on shock waves and is conserved in continuous motion, and also the known Holder inequality, the property of convex functions, and the relations on shock waves, we can obtain the following inequalities[4]:

$$F_2 \geq n^{1/\gamma}\left[\frac{X^{\gamma+2}}{t^2} - \frac{\gamma - 1}{2\gamma}Y^\gamma\right]^{1/\gamma}\left[1 + \frac{2}{\gamma - 1}\frac{Y^\gamma t^2}{X^{\gamma+2}}\right] \qquad (5.5)$$

for $\dot{R} \geq a_2$, $a_2^2 = (\gamma p/\rho)_{M-M_2}$;

$$F_2 = \int_0^{M_2}\left(\frac{na_2^2}{\rho_2^{\gamma-1}}\right)^{1/\gamma}\left(\frac{\dot{R}^2}{a_2^2} - \frac{\gamma - 1}{2\gamma}\right)^{1/\gamma}\left(1 + \frac{2}{\gamma - 1}\frac{a_2^2}{\dot{R}^2}\right)dm,$$

$$X = \int_0^R [\rho_2 S_\nu^\gamma]^{1/(\gamma+2)} dr, \qquad Y = \int_0^R (\rho_2 a_2^2)^{1/\gamma} S_\nu dr, \qquad (5.6a)$$

$$n = 2(\gamma - 1)^{\gamma-1}/(\gamma + 1)^{\gamma+1}, \qquad T \geq 2E_0^2 t^2 \bigg/ \left(\int_0^R r^2 \rho_1 S_\nu dr + 2E_0 t\right)$$

for $\gamma = (\nu + 2)/\nu$;

$$T \geq [\nu(\gamma - 1)E_0 t]^2 \bigg/ \left(2R^2 \int_0^R \rho_1 S_\nu dr\right) \qquad (5.6b)$$

for $\gamma < (\nu + 2)/\nu$; and

$$T \geq 2E_0^2 t^2 \bigg/ \left(R^2 \int_0^R \rho_1 S_\nu dr\right) \qquad (5.6c)$$

for $\gamma > (\nu + 2)/\nu$.

The inequalities (5.5) and (5.6) together with the energy equation (5.3) constitute a system of finite transcendental inequalities for the functions $R(t)$, and if $\rho_1(r)$ is actually specified they permit investigations of these functions with allowance for the conditions on the shock wave. These inequalities can be used, for example, for preliminary estimates of the function $R(t)$.

In addition, $R(t)$ can be estimated by using the less accurate inequalities

$$\gamma \leq \frac{\nu+2}{\nu}: \qquad R^2 \int_0^R \rho_1 S_\nu dr \geq \nu(\gamma-1) E_0 t^2 + \int_0^R r^2 \rho_1 S_\nu dr,$$

$$\gamma > \frac{\nu+2}{\nu}: \qquad R^2 \int_0^R \rho_1 S_\nu dr \geq 2 E_0 t^2 + \int_0^R r^2 \rho_1 S_\nu dr.$$

2. Blast on an interface of two media

2.1. Problem of planar, cylindrical, or spherical strong blast on the interface of two ideal compressible media

Assume two half-spaces filled with ideal compressible media whose internal energies depend on p and ρ in accordance with the equation

$$\varepsilon = p f(\rho/\rho_0) + \text{const}. \tag{5.7}$$

Following a point blast on the interface of these half-spaces, the flow is one-dimensional in the planar case and depends in the cylindrical and spherical cases on two spatial coordinates and the time. If the back pressure is assumed to be zero in both half-spaces, the system of the defining parameters will comprise E_0, ρ_0, ρ_{1-}, ρ_{1+}, r, and t, where ρ_{1-} and ρ_{1+} denote the constant densities in the two half-spaces. We assume these to be the upper and lower, respectively. Since the gasdynamic equations and the functions $f_-(\rho/\rho_0)$ and $f_+(\rho/\rho_0)$ contain no dimensional parameters, the formulated strong-blast problem is self-similar and depends in general on two dimensionless variables of form x^j/bt^δ or $\lambda = r/bt^\delta$ and θ, where θ is the polar angle. The choice of the coordinate frame depends on the solution method.

The boundary conditions are here the conditions on the shock waves in the upper and lower half-spaces, meaning equality of the pressures at the interface and equality of the normal gas velocities on the interface. All the necessary equations can be easily obtained from the equations and conditions on the surfaces of strong blasts, given in Chap. 1, and will not be considered in detail here.

For a planar blast on an interface of two ideal gases, the self-similar problem has an exact analytic solution.[5,6] The solution of the problem in fact calls for determining the partition of the energy E_0 among the two half-spaces from the conditions that the pressures be equal on the interface.

For cylindrical and spherical blasts, the problem has not yet been solved in detail. The simplest variants of this problem are the case of two ideal gases and the case when one of the half-spaces is incompressible. With a spherical blast as the example, we consider the approximate energy partition among two half-spaces occupied by different ideal gases.

We follow the paper by G. A. Ostroumov and the author.[7] For the radius $r_2(t)$ of the shock wave front we have in the one-dimensional case

$$r_2 = \left(\frac{E_0}{\alpha\rho_1}\right)^{1/5} t^{2/5}.$$

For the force F acting on the interface we have in the case of spherically symmetric flow, at arbitrary γ,

$$F = 2\pi p_2 \int_0^{r_2} \left(\frac{p}{p_2}\right) r dr = \frac{4\pi}{\gamma+1} \rho_1 D^2 r_2^2 K.$$

Substituting in the above equation

$$Dr_2 = \frac{2}{5}\left(\frac{E_0}{\alpha\rho_1}\right)^{2/5} t^{-1/5},$$

we obtain

$$F = \frac{16\pi}{25}\frac{1}{\gamma+1}\left(\frac{E_0}{\alpha}\right)^{4/5} \rho_1^{1/5} K t^{-2/5}. \tag{5.8}$$

We use in Eq. (5.8) the notation

$$\int_0^1 \left[\frac{p}{p_2}(\lambda)\right] \lambda d\lambda = K, \qquad \lambda = \frac{r}{r_2}.$$

Assume now that the total forces acting on the upper and lower half-spaces are approximately equal to the forces calculated for the corresponding spherically symmetric flows.[8] From this hypothesis that the forces F_- and F_+ acting on both media are constantly equal, we obtain the relation

$$\frac{E_+^0}{E_-^0} = \left(\frac{\gamma_+ + 1}{\gamma_- + 1}\right)^{5/4} \frac{\alpha_+}{\alpha_-} \left(\frac{\rho_{1-}}{\rho_{1+}}\right)^{1/4} \left(\frac{K_-}{K_+}\right)^{5/4}. \tag{5.9}$$

Let, for example, $\gamma_- = 7$, $\rho_{1-} = 1$, $\gamma_+ = 4$, and $\rho_{1+} = 2.65$. Using the results of Chap. 2 we obtain then $K_- = \frac{1}{5}$, $K_+ = 0.195$, $\alpha_- = 0.0279$, and $\alpha_+ = 0.0766$, E_+^0 is the energy for the upper half-space, and E_-^0 is the energy for the lower half-space (note that the chosen parameters are approximately equal to those for water and quartz). We obtain for these media $E_+^0 = 1.24E_-^0$, $E_+^0 = 0.55E_0$, and $E_0 = E_+^0 + E_-^0$, i.e., the energy is approximately equally divided among the half-spaces. An estimate of the energy partition among media such as water and air shows that the bulk of the energy goes into the air, as in the physical situation. We note further that in the planar case ($\nu = 1$) an equation such as Eq. (5.9) is exact. Calculations have shown that

K varies little with increasing γ. The value of K for $\nu = 3$ and $3 \leq \gamma \leq 7$ is close to 0.2.

Approximate solutions of the problem of strong spherical and cylindrical blasts under other assumptions and for certain special cases are considered in Refs. 9–18.

2.2. Determination of the flow parameters for a blast on a plane interface between a gas and a solid or liquid

Assume that the upper half-space is filled with gas and the lower with some other medium. We consider first an idealized case, where the lower half-space is occupied by an absolutely solid body. To determine, for example with the aid of tables,[19,20] the flow parameters following explosion of a charge on such an interface it is necessary to replace E_0 everywhere by $(1/2)E_0$.

The problem of a spherical blast at the vertex of an absolutely hard cone with vertex angle φ is solved similarly. It must only be taken into account that the energy goes into a solid angle $4\pi - \varphi$ rather than 4π. A blast in a conical cavity in an absolutely hard body can also be considered. This case has already been mentioned in Sec. 3 of Chap. 4. For a cylindrical charge it is necessary to choose corresponding wedge-shaped regions.

Let now the lower half-space be filled with a deformable medium, but let the flow be such that the processes in the lower half-space influence quite weakly the gas flow in the upper half-space. In addition, assume that we know (e.g., from experiment) the time of arrival of the shock wave at a certain fixed point of the upper half-space. In cylindrical and spherical cases it is convenient to choose this point somewhere on a straight line perpendicular to the interface and passing through the blast center.

We can then calculate approximately those fractions of the total blast energy E_0 that have gone into the upper and lower half-spaces, and find the gas-flow parameters in the upper half-space. In fact, from the fixed-point coordinate r_* and from the arrival time t_* of the shock wave, bearing in mind the relations $r_* = R_2 r^0$ and $t_* = \tau t^0$ and using the tabulated dependence of R_2 on τ, we can determine the energy E_+^0 released into the upper half-space. Obviously, the remaining energy E_-^0 goes to the lower half-space, since

$$E_+^0 = E_-^0 = E_0. \tag{5.10}$$

Knowing E_+^0 we can obtain from the tables all the gasdynamic parameters of interest and describe approximately the blast in the upper half-space.

Note that a similar but purely experimental approach to determining the energy partition among two half-spaces was considered in Ref. 21.

2.3. Planar blast on the interface of two like gases with equal initial pressures but different initial densities

Let us examine the use of the blast-point calculation result to determine the gasdynamic parameters for a planar blast on the interface between gases with initial parameters γ, p_1, ρ_{1+} and γ, p_1, ρ_{1-}.

Assume that the Lagrange coordinate of the interface is $\eta_* = 0$. From the equality of the pressures on this surface we have

$$P_1(0, \tau) = P_2(0, \tau), \tag{5.11}$$

where P_1 and P_2 are the dimensionless pressures ($P = p/p_1$) in the first and second half-spaces, respectively. We choose now for the functions P_1 and P_2 the relations obtained for a homogeneous medium. The equality (5.11) is then satisfied if $\tau_1 = \tau_2$ for all t. From this, since $\tau = t/t^0$, we obtain the connection between the energy fractions E_+^0 and E_-^0 going into the first and second half-space, namely

$$E_+^0 = E_-^0\sqrt{\rho_{1-}/\rho_{1+}}. \tag{5.12}$$

Naturally, Eq. (5.10) is valid here.

Thus, if a planar blast of energy E_0 takes place on the interface between like gases, the physical characteristics of the flow in the two half-spaces can be determined from the tables of Ref. 13 with allowance for relations (5.10)–(5.12). Note that in the case of a planar blast on an interface, with arbitrary ρ_1 and p_1 and with different values of γ, the problem can be solved numerically by generalizing the method described in Sec. 3 of the preceding chapter.

3. Approximate methods of determining shock-wave parameters for a blast in a layered inhomogeneous atmosphere

3.1. Formulation of problem and conclusions drawn from dimensionality analysis

It is of great interest to investigate the point-blast problem in a quiescent gas with a layered density distribution. Such a distribution obtains in the atmospheres of the earth and of other planets, with the gas density decreasing with altitude. The density variation with altitude influences the gas motion behind the shock wave and complicates the flow picture. The temperature and pressure in the earth's atmosphere also vary with altitude.

Assume variations of the initial gas density and pressure in the form

$$\rho = \rho_0\Omega_1(z/H), \quad \Omega_1(0) = 1, \quad p = p_0\Omega_2(z/H), \quad \Omega_2(0) = 1, \tag{5.13}$$

where z is the coordinate along which the density and pressure vary, ρ_0 and p_0 are the density and pressure at $z = 0$, and H is a constant with the dimension of length. For a spherical point blast the gas motion will be two-dimensional and will have axial symmetry. This point-blast problem can be solved only numerically and entails appreciable mathematical difficulties. Examples of numerical solutions for an inhomogeneous atmosphere were given by K. I. Babenko, V. V. Rusanov, and A. M. Molchanov.[25] We shall describe below several approaches to the solution of this problem.

The motion for a cylindrical blast is two-dimensional and nonstationary if the blast line is parallel or perpendicular to the z axis. For other configurations the motion is three-dimensional and the basic functions depend on x^j and t. In the planar case the motion is one-dimensional if the blast plane is perpendicular to the z axis. For other configurations the flow will be two- or three-dimensional.

Consider the case of a spherical blast (with account taken of the force of gravity). The system of gasdynamic equations for axial symmetry, in spherical coordinates, is given in Chap. 1 [see Eqs. (1.17)]. The conditions on the shock wave can be written in the form

$$\rho_1 D = \rho_2(D - v_2) = m, \qquad v_2 m = p_2 - p_1,$$

$$\epsilon_2 - \epsilon_1 = \frac{1}{2}(p_2 + p_1)\left(\frac{1}{\rho_1} - \frac{1}{\rho_2}\right). \tag{5.14}$$

Denoting by $r_2 = r_2(t, \theta)$ the law governing the change of the shock-wave front, we have from Eq. (1.45), in accordance with the definition of the velocity D (see Chap. 1, Sec. 2),

$$D = \frac{\partial r_2}{\partial t}\left[1 + \left(\frac{1}{r_2}\frac{\partial r_2}{\partial \theta}\right)^2\right]^{-1/2}. \tag{5.15}$$

The unknowns in this problem are v_r, v_θ, p, and ρ and depend on r, θ, and t. For an ideal gas the defining parameters of the problems are

$$r, \quad \theta, \quad t, \quad E_0, \quad \rho_0, \quad p_0, \quad H, \quad g, \quad \gamma, \tag{5.16}$$

where r is the length of the position vector and θ is the angle measured from the z axis. Out of the parameters (5.16) we can form the following dimensionless combinations:

$$R = r/r^0, \qquad \theta, r = t/t^0, \qquad h = H/r^0, \qquad \gamma, f = g(t^0)^2/r^0, \tag{5.17}$$

where $r^0 = (E_0/p_0)^{1/3}$ are the local dynamic length and $t^0 = r^0(\rho_0/p_0)^{1/2}$ is the local dynamic time. We conclude on the basis of the π-theorem that the dimensionless quantities sought, for example the pressure p/p_0, will depend on the dimensionless combinations indicated in Eqs. (5.17) (Refs. 26, 27):

$$p/p_0 = P(R, \tau, \theta, h, f, \gamma). \tag{5.18}$$

We disregard gravity, since it can be neglected for the initial stage of blast development.

It follows from Eq. (5.18) that a calculation for one energy E_0 and for fixed r^0, t^0, h, and γ cannot be used for other blast altitudes without varying H, p_0, and ρ_0 so as to keep h constant. If the blast energy is varied at fixed p_0 and ρ_0, then to use a calculation made for some fixed E_0 we must change the parameter H so as to leave h unchanged. It follows from this reasoning that the treatment of blasts in an inhomogeneous medium becomes very complicated, since it is necessary to perform calculations for sets of the parameters γ and h, and each solution is two-dimensional. The solution of the problems becomes even more complicated if the medium is not an ideal gas with constant heat capacities (e.g., air with allowance for dissociation and ionization).

During the later stage of blast development, an important role is assumed by the gravitational force (without allowance for this force the formulation of the problem is physically incorrect: the limiting solution for large t will not yield equilibrium conditions for the ambient medium).

Thus, "Froude's number" $f = g(t^0)^2/r^0$ becomes significant, i.e., one more dimensionless parameter is added.

If the aim is a numerical solution of the problem, it is clear that calculation of even one variant of the problem is quite laborious.[23–25] It becomes necessary therefore to find various approximate methods of solving the problem as a whole, and approximate methods of determining the parameters of the blast waves. Possible methods are: linearization, approximation of the quantities on the wave front, and use of exact solutions and hydrodynamics equations with approximate satisfaction of the conditions on the shock wave (one can use here L. V. Ovsyannikov's solution[28]). For this problem we shall consider here in detail only the second method.

Note also that, as indicated by L. I. Sedov, in the case of a strong blast of a spherically symmetric point charge ($p_1 = 0$) the problem is self-similar if the initial density distribution is given by

$$\rho_1 = A_0 \frac{\chi(\theta)}{r^\omega},$$

where χ depends only on the polar angle θ, and A_0 is a constant with dimensionality $[A_0] = ML^{3-\omega}$. Self-similar solutions can be obtained also for cylindrical and planar blasts in two-dimensional problems with variable initial density.

3.2. Determination of the form and parameters of the shock-wave front for a point blast

Let $\rho = \rho_0 e^{-z/H}$ and consider the case of a spherical point blast. For a strong blast, approximate methods for taking into account the variable density have been proposed in a number of papers (see Refs. 29–35). For sufficiently weak shock waves, asymptotic methods of nonlinear geometric acoustics have been developed[36–39] and asymptotic laws more general than those considered in Sec. 2 of Chap. 4 have been obtained for the shock-wave damping.

Practical interest attaches to a study of the blast evolution from its strong initial stage to a stage of near-degeneracy into sound waves. Note that K. E. Gubkin[37] presented one approximate method, based on the use of the asymptotic laws of shock-wave damping and on experimental data, of determining the pressure at large distances for a blast in an inhomogeneous atmosphere. We shall describe below approximate methods,[26–27] based on solution of gasdynamic equations and problems for strong and weak shock waves, of determining the blast-wave parameters.

We note first that if we specify, for a fixed medium, one of the quantities p_2, ρ_2, v_2, or D as a function of r_2 and θ, all the remaining quantities are obtained from relations (5.15). Assume that we are interested in determining the form of the wave and its variation with time. We introduce the dimensionless quantity $W = D/a_1$, where a_1 is the speed of sound in a medium at rest. Changing in Eq. (5.16) to dimensionless variables, we get

$$\beta \frac{\partial l}{\partial \tau} = W(l, \theta)\left[\left(\frac{1}{l}\frac{\partial l}{\partial \theta}\right)^2 + 1\right]^{1/2}, \qquad l = \frac{r_2}{r^0}, \qquad \beta = \left(\frac{p_0}{\rho_0}\right)^{1/2}\frac{1}{a_1}. \quad (5.19)$$

If we know from theory (or experiment) the $W(l, \theta)$ dependence, relation (5.19) can be regarded as the equation for $l(\tau, \theta)$.

Relation (5.19) is a first-order partial differential equation. If the function

$$l = l_0(\theta)$$

is specified at a certain instant of time $t = t_0$ $(\tau = \tau_0)$, we obtain, by solving the Cauchy problem for Eqs. (5.19), the dependence $l = l(\tau, \theta)$.

Let us consider the theoretical determination of the wave-front parameters from data on the strong stage of the blast wave and the asymptotic laws of wave damping at large distances, following in the main the work of the author jointly with V. P. Karlikov.[26,27]

We assume a point blast and an ideal gas with $\gamma = 1.4$. In this case we can find from Eq. (5.15) the values of $v_2(q)$ and $\rho_2(q)$ where $q = a_1^2/D^2$. These values are given by Eq. (1.63). Let

$$\Omega_1 = \Omega_2 = \exp(-z/H), \quad (5.20)$$

i.e., we are considering the case of an isothermal atmosphere. For small z/H we have

$$\Omega_1 = 1 - \frac{r}{H}\cos\theta. \quad (5.21)$$

For a strong shock wave, the point-blast problem with density variation given by Eqs. (5.13) and (5.21) was solved by V. P. Karlikov using a linearization method (see Refs. 29, 30). In accordance with these results, the equation for the shock-wave variation is

$$l = \alpha^{1/5}\tau^{2/5}(1 + \varkappa\alpha^{-1/5}\tau^{2/5}), \qquad \varkappa = \frac{0.1605}{h}\cos\theta, \quad (5.22)$$
$$\alpha = 0.851, \qquad \gamma = 1.4.$$

For the gas-particle velocity $v_2(r_2, \theta)$ behind the front we have for small $\varkappa$ the relation

$$v_{2\pi} = \frac{1}{3}\sqrt{\frac{E_0}{\alpha\rho_0}}\, r_2^{-3/2}(1 + 2\varkappa l)(1 + \varkappa l)^{3/2}. \tag{5.23}$$

In place of Eq. (5.23) we can write in the same approximation (for small $\varkappa$)

$$v_{2\pi} = \frac{1}{3}\sqrt{\frac{E_0}{\alpha\rho_0}}\, r_2^{-3/2}\left(1 + \frac{7}{2}\varkappa l\right). \tag{5.24}$$

It follows from the foregoing that Eqs. (5.23) and (5.24) give the $v_2(r_2, \theta)$ dependence at small values of r for a blast in an atmosphere in which the density varies in accordance with Eq. (5.20).

As follows from the results of Ref. 38, for large distances from the blast location we have the following asymptotic relation

$$v_{2\mathrm{ac}} = \frac{C_1 e^{ml}}{\rho_0 a_1 r_2}\left[\int_{r^*}^{r_2} \exp\left(\frac{r\cos\theta}{2H}\right)\frac{dr}{r}\right]^{-1/2}, \qquad m = \frac{\cos\theta}{2h}, \tag{5.25}$$

where C_1 and r^* are certain constants that depend on the form of the wave. Generalizing the method considered in Chap. 4, we assume that for an approximate determination of $v_2(r_2, \theta)$ one can use, up to appreciable distances, the following approximation. We assume that Eq. (5.24) [or Eq. (5.23) at $h > 1$] is valid up to a certain r_*, and that Eqs. (5.25) hold for $r > r_*$. We choose C_1, r^*, and r_* such that Eqs. (5.24) and (5.25) can be joined. It was shown in Chap. 4 that a similar approximation yields good results in the one-dimensional case. Since we have as yet no sufficiently accurate solution of the problem in an isothermal atmosphere with allowance for back pressure, it is impossible to assess the accuracy of the proposed approximation for a wide range of parameters by suitable comparisons. Note that for small m we have from Eqs. (5.25)

$$v_{2\mathrm{ac}} \sim \frac{1 + ml}{r_2^{3/2}\sqrt{m\left(1 - \dfrac{r^*}{r_2}\right) + \dfrac{1}{r_2}\ln\dfrac{r_2}{r^*}}},$$

which indicates that the functions (5.24) and (5.25) have qualitatively the same behavior at small ml. This also confirms the possibility of roughly approximating $v_2(r_2, \theta)$ by relations (5.24) and (5.25).

In dimensionless variables, Eqs. (5.24) and (5.25) take the form

$$V_\pi = \omega l^{-3/2}(1 + \tfrac{7}{2}\varkappa l) \qquad (V = v_2/a_1), \tag{5.26}$$

$$V_{\mathrm{ac}} = \sigma\omega e^{ml} l^{-1} I^{-1/2} \qquad [\omega = 1/(3\sqrt{\alpha})], \tag{5.27}$$

$$I = \int_{l^*}^{l} e^{m\xi}\frac{d\xi}{\xi} = E_i(ml) - E_i(ml^*), \tag{5.28}$$

where $E_i(x)$ is the integral exponential function and $l^* = r^*/r^0$. The quantity $l_*(\theta)$ can be chosen to satisfy the condition that q be close to unity ($0.5 < q < 1$). Let the transition to the asymptotic equations be made for $q = q_*$. Since $V = 2(1-q)/(\gamma+1)\sqrt{q}$, we easily obtain V from the known q and then determine $l_*(\theta)$ from Eq. (5.26).

By analogy with the one-dimensional case (Chap. 4), we choose σ and l^* that satisfy the condition of matching the functions V determined by Eqs. (5.26) and (5.27) and their derivatives with respect to l at $l = l_*$. For l^* and σ we obtain the equations

$$\sigma = l_*^{-1/2} e^{-ml_*} I_*^{1/2} (1 + \tfrac{7}{2}\varkappa l_*),$$

$$I_* = 0.5 e^{ml_*} \left[-\frac{1}{2} - ml_* + \frac{1}{1 + \frac{7}{2}\varkappa l_*} \right]^{-1}, \tag{5.29}$$

where $I_* = E_i(ml_*) - E_i(ml^*)$. We indicate also the connection between W and V:

$$W = \frac{\gamma+1}{4} V + \sqrt{\frac{(\gamma+1)^2}{16} V^2 + 1}. \tag{5.30}$$

For the above approximation of the function $V(l, \theta)$, and hence also of $W(l, \theta)$, we calculated the parameters and shapes of blast waves in an isothermal atmosphere. The pressures and densities were obtained from the conditions on the shock waves in terms of the function $V(l, \theta)$ known from the approximation. The change of the shock-wave form was determined from Eqs. (6.19) by numerically solving the Cauchy problem by the method of characteristics. The initial data for l were specified in accordance with Eq. (5.22) with $\tau_0 = 0.004$. Separate calculations down to small τ were made also with V approximated by Eq. (5.23). For $h > 1$ and $l < 2$ the shock-wave forms calculated in the approximation (5.23) or (5.24) differ little. This suggests that these approximations are valid.

Calculations by the described method were made for different values of $h = H/r^0$ ($h = 0.5, 1, 5, 8$). By way of example we note that for the earth's atmosphere ($H = 8$ km), when $h = 1$ and $p = 0.1$ atm (the pressure at an approximate altitude 16 km) the blast energy is $E_0 \sim 5 \times 10^{22}$ erg. Calculations have shown that the variability of the density is significant only for parameter values $h < 10$. The calculation results for $h = 8$ and $h = 1$ are shown in Figs. 76–78, where plots $V(l, \theta)$, $\rho_2/\rho_0(l, \theta)$, $p_2/p_0(l, \theta)$, and $p_2/p_1(l, \theta)$ are shown together with the shock-wave shape. Figure 78 shows the shock-wave shape for different τ at $h = 1$ (the wave shape is close to a sphere at $h = 8$). The dashed lines also show circles with radii close to the distances from the center to the point where the shock wave and the trajectory of an element of the wave (beam) intersect at an initial angle $\theta_0 = \pi/2$. For $\tau = 5$ the figure (dotted line) shows the form of the wave at $h = 0.5$. The plots can be used to follow the deviation of the wave shape from a sphere.

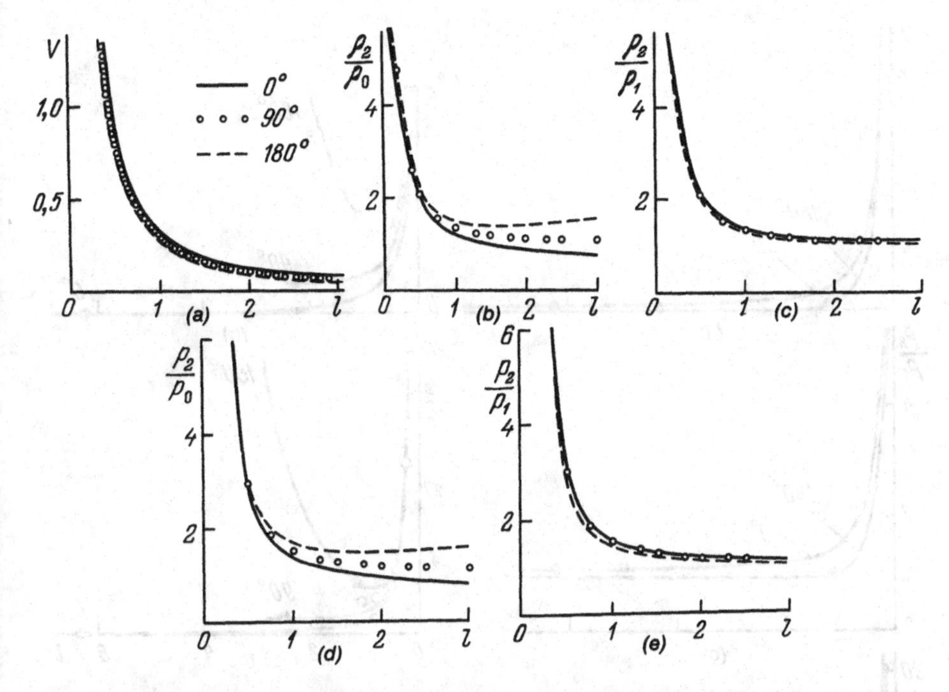

Figure 76. Variation of blast parameters for three values of the angles θ ($h = 8$). (a) Velocity behind the shock front; (b) relative density; (c) ratio of the density ρ_2 behind the shock front to the local initial density ρ_1 of the medium; (d) relative pressure; (e) ratio of pressure p_2 behind the shock wave to the local pressure p_1.

The circles in Fig. 77(d) mark the dependence of p_2/p_0 on l in the linearized theory, with allowance for the variability of the density and for a constant back pressure. Equations for p_2/p_0 and $l(\tau, \theta)$ can be obtained by the linearization method by using the principle of superposition of the linear corrections (see Chap. 3). They take the form

$$\frac{p_2}{p_0} = \frac{0.157}{l^3}\left[1 - 0.36\,\frac{l}{h}\cos\theta\right] + 2.407,$$

$$l = \alpha^{-1/5}\tau^{2/5}\left[1 + \varkappa\alpha^{-1/5}\tau^{2/5} + \frac{0.7\cdot 25}{6}A_1\alpha^{2/5}\tau^{6/5}\right], \qquad (5.31)$$

$$A_1 = \text{const} = 1.92 \qquad (\gamma = 1,4).$$

We note here one more interesting property of the approximate solutions obtained. It is possible to use as the abscissa in place of $l = r_2/r^0$ the quantity $\tilde{l} = r_2/\tilde{r}^0$, where $\tilde{r}^0$ is the local dynamic length at the given point, and for an exponential atmosphere we have

$$\tilde{l} = l\exp(-z/3H).$$

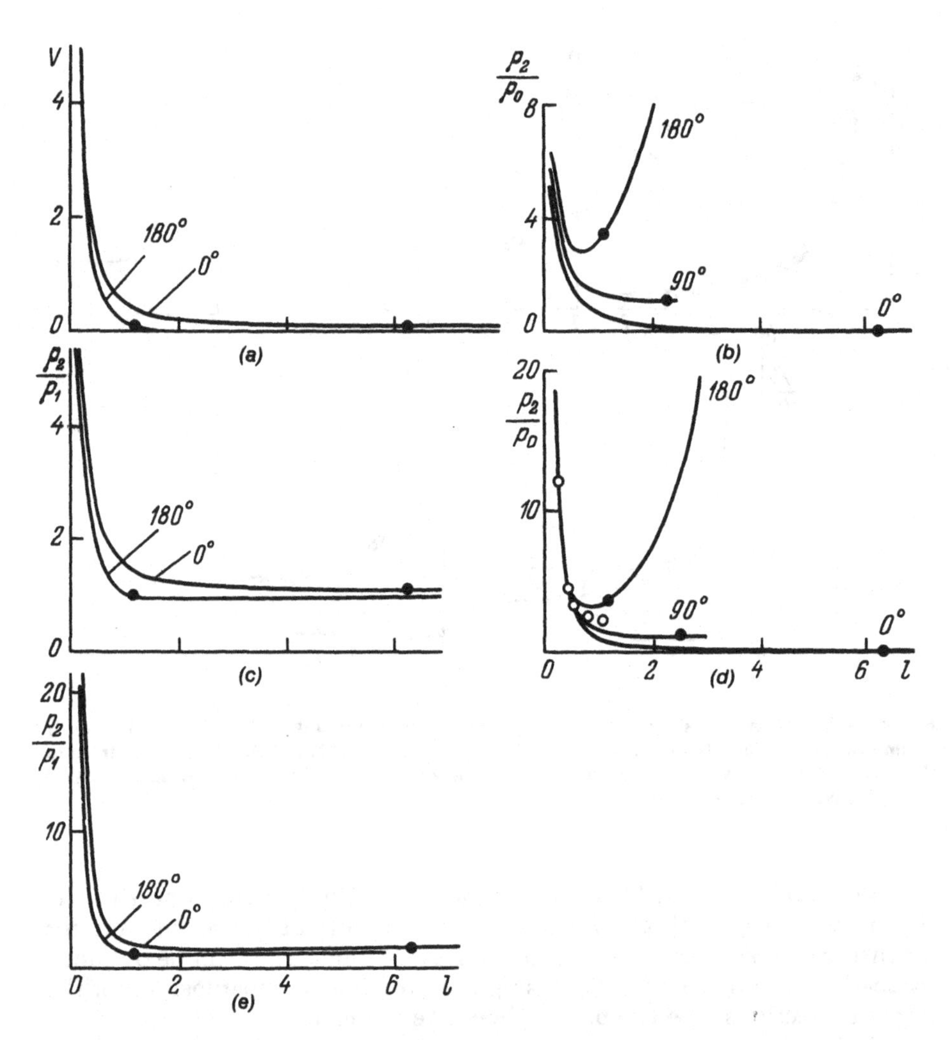

Figure 77. Variation of dimensionless parameters of a blast wave for $\theta = 0°, 90°$, and $180°$ ($h = 1$). The circles mark the dependence of p_2/p_0 on l, determined from the linearized theory. The points mark the values of l starting with which the asymptotic equations for the damping of weak shock waves were used.

It turns out that with this change of the variable coordinate l the curves corresponding to different θ on Figs. 76 and 77 come closer together.

The following conclusions of Ref. 27 follow from the investigation of the problem and from the presented results.

1. For $h < 1$ the shock waves differ from a sphere and become "egg-shaped," somewhat elongated above ($\theta = 0$) and flattened below ($\theta = \pi$). This is attributed to the decrease of the density with increasing z. For $h > 5$ the change of the shock-wave form is negligible. This conclusion agrees with the results of a complete calculation of the blast for initial parameters corresponding to a standard atmosphere.[23,24]

Figure 78. Form of shock wave $l(\theta)$ on an isothermal atmosphere for certain values of the time τ (τ = 0.5–5, cases h = 1 and 0.5). Circles corresponding to a one-dimensional wave propagation are shown for comparison.

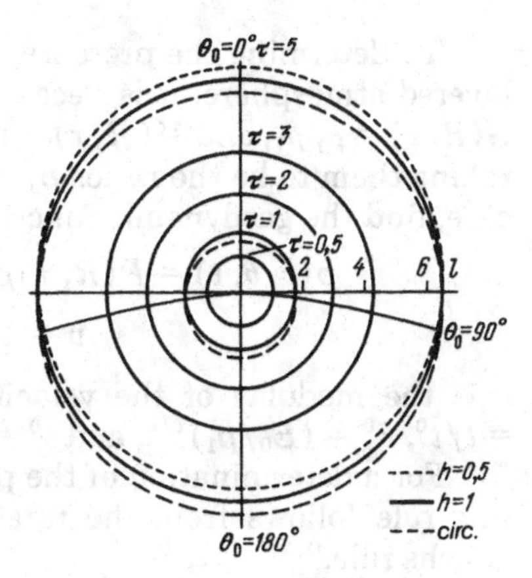

2. The relative pressure p/p_1 depends rather weakly on θ, whereas the dimensionless pressure p/p_0 varies greatly when the polar angle θ is changed. This leads to the rule, considered below, for an approximate determination of the pressure p_2.

Calculations of some other variants by the proposed method were published in Ref. 40. The authors compared calculations by our method with other approximate approaches. It was shown that its accuracy is perfectly satisfactory for sufficiently strong waves. In the direction $\theta = \pi$ the accuracy of the method is quite high also for the weak wave-propagation stage.

It is possible to study similarly the cases of cylindrical and planar blasts, and if the medium is weakly inhomogeneous (e.g., the distance H over which the density of the medium varies is substantially smaller than the characteristic dynamic length r^0) it is possible to use for an approximate calculation of the shock-wave parameters the hypothesis of locally one-dimensional flows (the analog of the known hypothesis of "flat cross sections"), neglecting the local change of the form of the wave along the blast axis or plane.

Approximate methods of this type for determining the parameters of spherical and cylindrical shock waves were developed in Refs. 35 and 40–42. Detailed calculations were made in Refs. 40 and 41 by the method of locally one-dimensional flows, which is sometimes called the sector-approximation method (all of space is subdivided into a number of sectors within which the dependence on the angle θ is disregarded and the exchange of energy, mass, or momentum between sectors is neglected).

3.3. Recalculation rule for an approximate determination of the flow

The foregoing analysis shows that for $h \geq 1$ it is possible, with accuracy sufficient for practical needs, to calculate the gasdynamic functions by using the following rule.

To determine the pressure, density, and velocity in a point blast in a layered atmosphere, it is necessary to use the functions $P(R, \tau) = p/p_1$, $G(R,\tau) = \rho/\rho_1$, and $V(R,\tau) = v/a$ for the one-dimensional case and then, taking them to be the ratios p/p_1, ρ/ρ_1, and v/a_1 for the two-dimensional case, find the gasdynamic functions in accordance with the rule

$$p(r, \theta, t) = P(R, \tau)p_1(r, \theta), \qquad \rho = G(R, \tau)\rho_1(r, \theta),$$

$$v = V(R, \tau)a_1(r, \theta).$$

v is the modulus of the velocity or its radial component $R = r/\tilde{r}^0$, $\tau = t/\tilde{t}^0$, $\tilde{r}^0 = (E_0/p_1)^{1/\nu}$, and $\tilde{t}^0 = \tilde{r}^0(\rho_1/p_1)^{1/2}$.

For a determination of the parameters directly behind the wave front, this rule follows from the foregoing results. It is sometimes called the "Sachs rule."

As shown by Lutzky and Lehto,[43] this rule leads to a difference of only 10%–20% from the corresponding calculations in the sector approximation with the force of gravity taken into account.

Comparison with more exact numerical calculations that will be considered below shows that the maximum error in the calculation of the pressures behind the front does not exceed 20% up to an excess pressure $\Delta p_2 = 0.01p_1$. Clearly, the effects of the "floating up" of the central part, and also the transverse component of the velocity v_θ, will not be determined by this approximate recalculation, and this is its major drawback. But these effects are important mainly during the weak stage of the waves.

For additional analytic substantiation, we can state the following:

(1) If Eqs. (1.17) are transformed to the variables P, G, and V and then expanded in terms of a small parameter proportional to $1/H$, the first terms of the expansion will obey the equations of one-dimensional flow with spherical symmetry.

(2) Comparing the asymptotic equations obtained for Δp_2 from the "Sachs rule," and the equations for Δp_2 in the one-dimensional case, with the equations for $\Delta p_2 = v_2\rho_1 a_1$ which follow from Eq. (5.27), we obtain by simple calculation for large r the estimate

$$\Delta p_{2as}/\Delta p_{2s} \approx e^{-z/4H},$$

where Δp_{2as} stands for Δp_2 determined from an equation such as Eq. (5.27), while Δp_{2s} is the pressure obtained by the Sachs rule. Thus, at $z \leq H$ the difference between these equations is small.

(3) For the propagation of one-dimensional waves in a medium with variable parameters, recalculation laws of the Sachs-rule type flow from the results obtained by L. V. Ovsyannikov.[44]

3.4. Results of numerical calculation of a strong blast in an exponential atmosphere

A strong blast in an exponential atmosphere was numerically investigated by L. A. Chudov and his co-workers.[25,40] Their results are followed here.

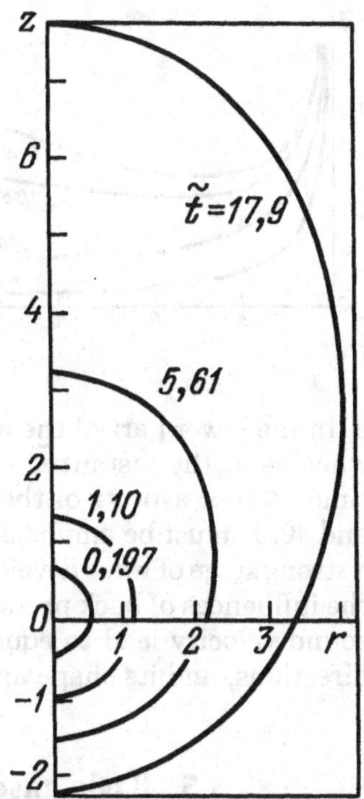

Figure 79. Form of shock-wave front for a blast in an exponential atmosphere at various instants of time.

The strong stage of axisymmetric flow following a blast in an exponential atmosphere was calculated by a finite-difference method. The specified initial data were obtained by a self-similar solution. The calculations were made for two specific-heat ratios, $\gamma = 1.3$ and 1.4.

We present certain calculation results for the case $\gamma = 1.4$, $h < 0.1$. We introduce the dimensionless cylindrical coordinates r/H and z/H and the dimensionless time $\tilde{t} = tE_0^{1/2}(\rho_0 H^5 \alpha)^{-1/2}$. For the initial state we choose L. I. Sedov's solution for a strong blast at $\tilde{t} = 0.37 \times 10^{-3}$, $p_2 = 1750 E_0/\alpha H^3$, and $r_2 = 0.042H$. The calculated forms of the shock waves for various instants of time are shown in Fig. 79. The shock-wave form is seen here to differ greatly from spherical. The shock wave is strongly elongated upwards. Owing to the inhomogeneity and to failure to take the back pressure into account, the wave front velocity is larger in the upper part than in the lower. Calculations show that the wave velocity D first decreases because of the wave damping, but then, owing to the decrease of the density, it begins to increase strongly in the upward direction, whereas in the downward part of the shock wave this velocity always decreases.

The velocity of the upper point of the front ($\theta = 0°$) reaches a minimum at $z = 3.5H$ at the instant of time $\tilde{t} = 7.1$. Plots of the variation of the velocity D along a number of rays are shown in Fig. 80.

In the above case the spatial inhomogeneity of the density influences strongly all the gas-flow parameters. Thus, at the instant $\tilde{t} = 17$ the pressure

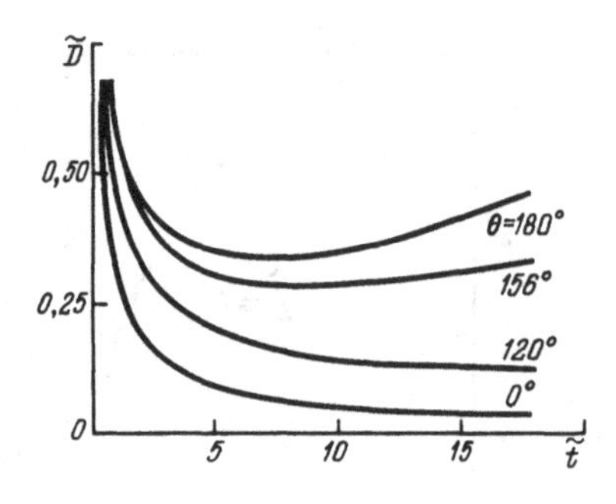

Figure 80. Change of the relative wave velocity with time for various directions ($D = D/D_0$, D_0 is the characteristic velocity).

p_2 in the lower part of the wave front is 100 times larger than in the upper, whereas at the instant $\tilde{t} = 1.1$ these pressures differ by only two or three times. Other aspects of the calculations are discussed in detail in Refs. 25 and 40. It must be emphasized that the solution considered corresponds to a strong stage of wave development. For an "unbounded" layered atmosphere the influences of back pressure, gravitation, and weak inhomogeneity of the sound velocity lead to equalization of the shock-wave velocity in various directions, and its shape approaches that of a sphere.

3.5. Basic gasdynamic functions for axisymmetric flow in an inhomogeneous atmosphere with allowance for back pressure and gravitation

We examine here the flow pattern for blast-wave propagation in an inhomogeneous atmosphere, using the result of Refs. 45 and 46.

We consider the case of an isothermal (exponential) atmosphere, and choose for the initial conditions a homogeneous sphere of radius r_0 filled with compressed gas of density low enough to make the total initial energy of the sphere equal to E_0, and make the initial pressure in it considerably higher than the pressure at the specified blast height. An investigation of this problem for one-dimensional flows shows that its solution approaches rapidly the asymptote of a point blast, at least outside the region $r > r_0$.

It should be expected that, starting with certain instants of time [e.g., when $r_2(\theta, t) > 5r_0$], the flow produced in an exponential atmosphere by expansion of the hot compressed sphere would correspond well to a point blast and would unconditionally exhibit all the qualitative features of the evolution of a point blast.

The problem was solved numerically by S. K. Godunov, who used a difference method modified for this problem by L. V. Shurshalov.[45,47] The center of the blast was placed at a height z_0, and we shall assume that $z = 0$ corresponds to the earth's surface.

Let us consider the following example: $E_0 = 6.5 \times 10^{22}$ erg, $z_0 = 6.5$ km, $h \approx 1$, $p_1 = p_0 \exp(-z/H)$, $H = 8$ km, $p_0 = 1$ kg/cm^2, and $\rho_0 = 1.25 \times 10^{-3}$ g/cm^3. This example corresponds to the parameters of a blast simulation

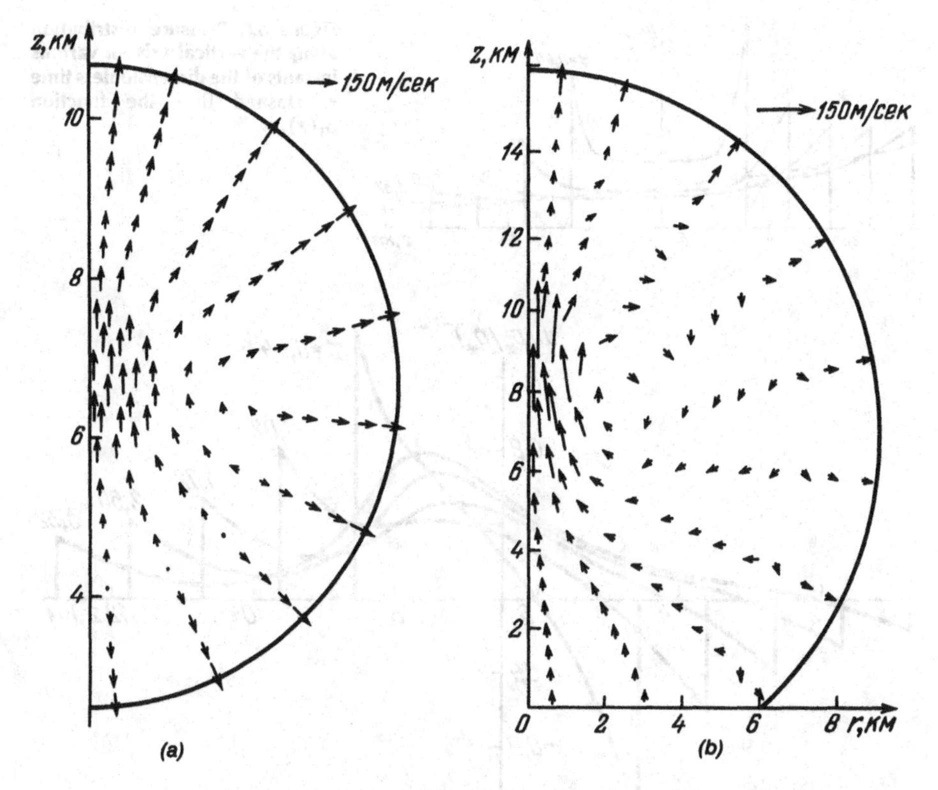

Figure 81. Velocity field for two instants of time: (a) t = 5.4 s, (b) t = 19 s.

of the Tunguska catastrophe (this will be discussed further below; see Ref. 46 for details).

Figures 81(a) and 81(b) show the velocity fields for two instants of time. The arrows are directed along the gas velocity, and their length is indicative of the absolute velocity. It can be seen that the lighter central zone floats up and a vortex flow is produced (an annular vortex is produced in space). Figure 81(b) shows also the effect of reflection of the shock wave from the plane $z = 0$.

The distribution of the pressure p/p_* along the z axis at various instants of the dimensionless time is shown in Fig. 82, while the distribution of the dimensionless vertical velocity $w/\sqrt{p_*/\rho_*}$ is given in Fig. 83 [$p_* = p_0$, $\rho_* = 10^{-3}$ g/cm^3, $\tau = t/t_*$, $t_* = l_*(p_*/\rho_*)^{-1/2}$, $l_* = 1$ km].

It can be seen from Fig. 82 that the pressure in the central zone changes little (just as in a point blast in a homogeneous atmosphere). The absolute values of the pressure in the lower part of the wave are highest, although their relative value p_2/p_1 is here the lowest. At the instant of arrival of the wave at the earth's surface, the relative excess pressure is 0.21 (note that the Sachs rule yields 0.23, i.e., 10% higher).

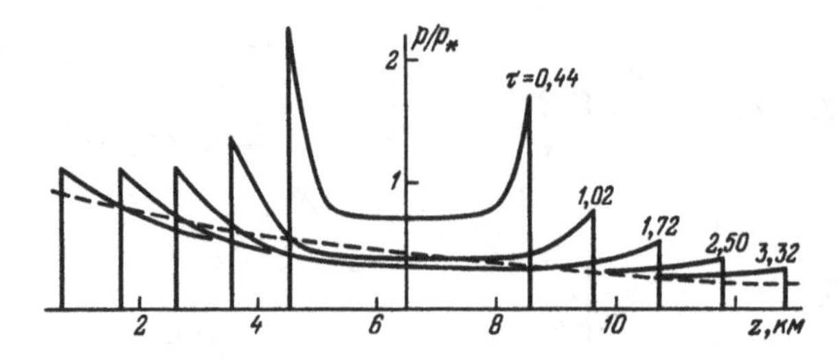

Figure 82. Pressure distribution along the vertical axis for various instants of the dimensionless time τ. Dashed line—the function $p_1(z)$.

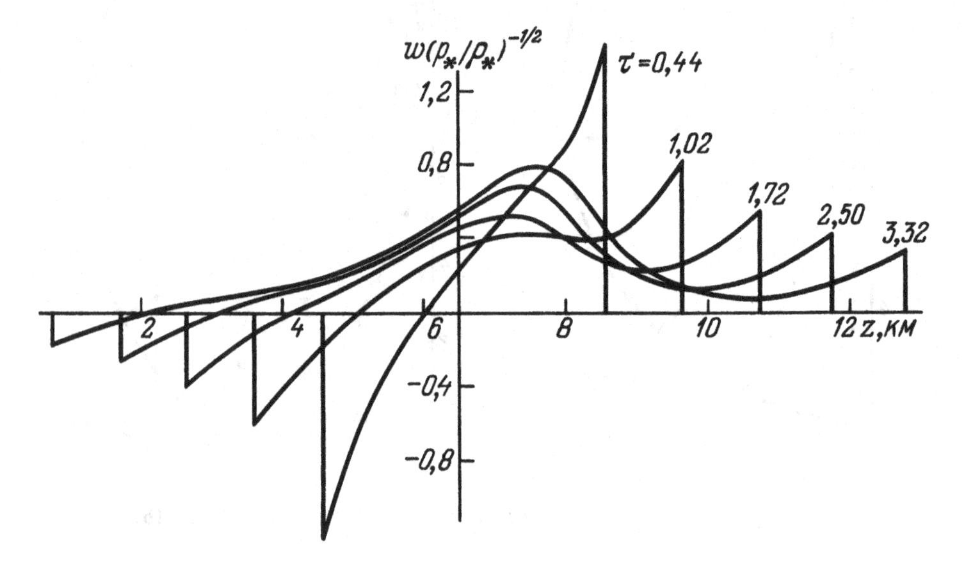

Figure 83. Altitude distribution of the dimensionless vertical velocity.

So long as the wave is strong, the velocity in the central region is almost linearly distributed. At later times the distribution of the vertical velocity changes strongly. The velocity distribution has a characteristic form with a maximum at the central zone and with a gently sloping minimum between this zone and the upward-propagating wave.

At the instant $t \approx 19$ s ($\tau = 5.94$) the velocity in the central zone, which shifts gradually upward, reaches 230 m/s [see Fig. 81(a)], and an intensive outflow of air from the earth's surface layer takes place.

To estimate the validity of the various quasi-one-dimensional models, it is important to know the energy distribution along the rays.

We introduce the integral of the total excess energy

$$E = \iiint\limits_{\Omega} \left[\frac{p - p_1}{\gamma - 1} + \frac{\rho v^2}{2} + (\rho - \rho_1) g z \right] d\Omega, \tag{5.32}$$

where Ω is the volume of the sector along some ray, and $E = E_0$ if Ω is equal to the total perturbed region. Let us see how energy is exchanged between

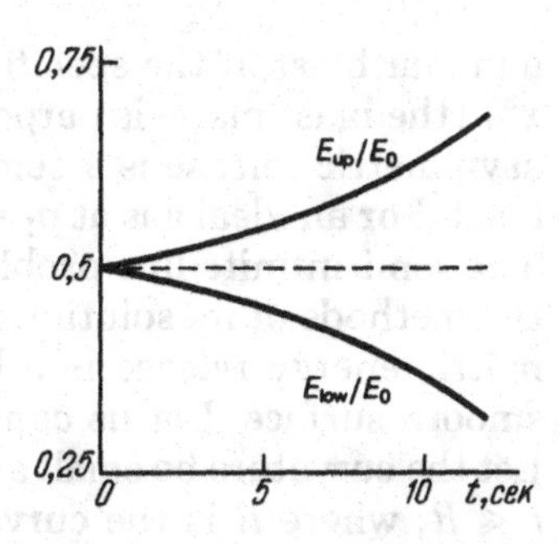

Figure 84. Energy distributions over the upper and lower half-spaces.

the upper and lower parts of the perturbed region. To this end we calculate integrals of the type (5.32) over the upper and lower parts of the perturbed zone and refer the results to the value of E_0. The results for a blast at an altitude $z_0 = 6.5$ km ($E_0 \approx 10^{23}$ erg) are shown in Fig. 84, where the dashed line corresponds to a blast in a homogeneous atmosphere, when $E_{up}/E_0 = E_{low}/E_0 = \frac{1}{2}$. It is seen from this figure that the redistribution process changes with time. As expected from physical considerations, the greater part of the energy goes upwards.

The foregoing circumstances must be taken into account when using the results of the previously published[40,41] quasi-one-dimensional calculations made in the sector approximation. L. V. Shurshalov[47] also carried out blast calculations for the case when the initial medium parameters are those of a standard atmosphere. It turned out that by suitable choice of the parameters ρ_0 and H of the exponential medium (as was done, for example, above) the difference between the flow parameters is negligible. This difference can be disregarded for blasts of energy 10^{23} erg and less, and at altitudes lower than 8 km.

Allowance for the real properties of the air and of the radiation will influence the flow more substantially. Some results have already been obtained along these lines.[48–50]

4. Asymmetric supply of energy or momentum

4.1. Asymmetric energy release

It was noted in Chap. 2 that in the study of the action of a laser beam on a medium we encounter blast problems when the energy is released along a certain straight line, and the specific energy is variable, $E_0 = E_0(s)$ if s is the coordinate along the blast line. If the $E_0(s)$ distribution is such that dE_0/ds is small we can, using the planar-cross-sections hypothesis, calculate the gas flow by the locally one-dimensional theory. An explicit correction to the solution might be obtained by the linearization method if $E_0 = E_0^0 + \mu f(s)$, where μ is a small parameter. A similar remark holds true also for a planar blast, if the specific energy E_0 along the plane varies, $E_0 = E_0(x^2, x^3)$ (the blast plane is perpendicular to the x^1 axis). A classical example of asymmetric release is a semi-infinite straight line or a semi-infinite plane blast. For an ideal gas at $p_1 = 0$ the corresponding problems are self-similar. The semi-infinite line problem was formulated by L. I. Sedov,[51] and one of the methods of its solution is considered in Ref. 52. Another case of asymmetric energy release is a blast along a certain curve or along a certain smooth surface. Let us consider the case of a blast along a certain curve. Let the curvature be small and the specific energy E_0 constant. For distances $r \ll R$, where R is the curvature radius of the blast line, we can use (approximately) the theory of a blast along a straight line. Combined variants of the considered asymmetric cases are also possible. Thus, a semi-infinite straight line can have a variable specific energy $E_0(s)$. These cases are realized in practice (laser beam, cosmic bodies in the earth's atmosphere).

For linear equations of the heat-conduction or wave type, the above problems can be solved by using the superposition principle for the elementary solutions considered in Chap. 1. We know of no exact solution of these problems for the nonlinear case.

The foregoing problems can be solved by numerical methods. By way of example we consider the blast of a semi-infinite cylindrical charge in a homogeneous unbounded medium, assumed to be an ideal gas with an adiabatic exponent $\gamma = 1.4$. The blast will be simulated by a scattering, at the initial instant of time, of a semi-infinite cylindrical charge.

Let a semi-infinite high-pressure gas cylinder of radius r_0 and energy E_0 per unit length ($E_0 = \text{const}$) be located along the x axis in a cylindrical (x, r) coordinate system. At the initial instant $t = 0$ this cylindrical semi-infinite gas column begins to expand into the surrounding gas medium having a pressure p_1 and a density ρ_1 ($p_1 \ll p_0$, where p_0 is the initial pressure in the cylinder). It is required to determine the character of the resultant flow. It follows from physical considerations that at an appreciable distance from the end of the charge the flow will have cylindrical symmetry. Two-dimensional axisymmetric flow will take place near the end of the charge.

A solution of the above problem was obtained by S. K. Godunov using a numerical method and was reported in Refs. 53 and 54.

Without dwelling on the details of the initial stage of the gas expansion, we consider only the pattern of the flow at an instant when the radius of the cylindrical part of the shock wave significantly exceeds the initial radius r_0.

We introduce the dimensionless ratios x/r_0, r/r_0, $\tau = t/t_0$, $t_0 = r_0\sqrt{p_1/\rho_1}$, $e = E_0(\gamma - 1)/\pi r_0^2 p_*$, and p/p_1, where p_* is the characteristic pressure, equal to 1 atm ($p_* = 1\ \text{kg/cm}^2$). Some results of the calculation for the case $e = 10^4$ are shown in Fig. 85. It shows the form of the shock wave for a number of instants of the dimensionless time τ. The values p/p_1 on the shock wave are given for two characteristic points A and B. The point A is the forward point of the shock front, and the point B was taken quite far from the frontal part O of the charge, so that the end-point

Figure 85. Form of shock wave at a number of instants of dimensionless time $\tau = t/t_0$. The values of p/p_1 for these instants are 5.7, 2.1, 1.7, and 1.3 at point A and 28, 5.6, 2.5, and 1.7 at point B.

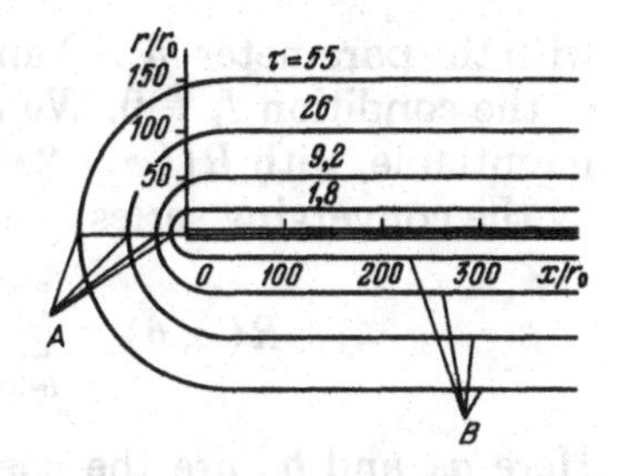

perturbations do not reach it. The pressure decreases in the direction from B to A along the shock wave.

It is seen from Fig. 85 that the shock wave takes the form of a cylinder with a spherical blunted end. Analysis of the behavior of the solution of this problem shows that at $\tau > 10$ the pressures at the points A and B can be calculated by using the asymptote of a spherical point blast in the forward part, with a specially chosen value of the energy, and the asymptote of a cylindrical blast for the semi-infinite part.

The cited papers[53,54] contain also calculations of the blast of a semi-infinite charge with variable energy along the symmetry axis. The results of these calculations were used to estimate the parameters of the shock waves produced by the falling Tunguska meteorite on 30 June 1908.

4.2. Strong point blast with action of a finite momentum

Let now the nonstationary flow be produced by a blast, but let the gas acquire, on top of the energy E_0, a momentum I_z directed along the z axis, from which we measure the polar angle θ of a spherical coordinate frame (r, θ, φ). The unperturbed gas is at rest and has a constant density ρ_1, we neglect the back pressure p_1 (i.e., we put $p_1 = T_1 = 0$), assume absence of mass forces, disregard viscosity and heat conduction, and assume the gas to be ideal with constant γ $(1 < \gamma < 2)$. The foregoing assumptions concerning the initial state of the medium allow us to conclude that the energy E_0 and the momentum I_z will be conserved in the course of the motion. In the absence of momentum, $I_z = 0$, the solution of the problem is self-similar and was considered in Chap. 2. We consider hereafter a solution for sufficiently long instants of time $t > 0$ and use the method of linearization about a self-similar solution, regarding the perturbation due to the application of the momentum as small. This solution was obtained by E. D. Terent'ev[55] and is reported below.

Assume that in the presence of momentum the position of the shock wave at sufficiently large t can be represented in the form

$$r_2(\theta, t) = (bt)^{2/5}[1 + t^{-2m/5}R(\varphi, \theta) + \ldots], \tag{5.33}$$

with the parameter $m > 0$ and the function R chosen to ensure satisfaction of the condition $I_z \neq 0$. We assume the function $R(\varphi, \theta)$ to be doubly differentiable, with $R[(\varphi + 2\pi), \theta] = R(\varphi, \theta)$, and we can represent $R(\varphi, \theta)$ by the converging series

$$R(\varphi, \theta) = \sum_{l=0}^{\infty} \sum_{k=0}^{l} a_{lk} P_l^k(\cos\theta) \cos(k\varphi + b_{lk}).$$

Here a_{ik} and b_{ik} are the coefficients of the expansion of the function R, $P_l^k(\cos\theta)$ are adjoint Legendre functions, and $Y_l^k = P_l^k(\cos\theta)\cos(k\varphi + b_{lk})$ are spherical harmonics.

Since the perturbation problem is linear, we choose in the spherical-harmonic series a term of arbitrary number and put $R = Y_l^k(\varphi, \theta)$. We seek the gas parameters corresponding to this shock-wave perturbation in the form

$$v_r = \frac{4}{5(\gamma+1)} b^{2/5} t^{-3/5} [f(\lambda) + t^{-2m/5} f_m(\lambda) Y_l^k + \ldots],$$

$$v_\varphi = -\frac{4}{5(\gamma+1)} b^{2/5} t^{-3/5-2m/5} u_m(\lambda) \frac{1}{\sin\theta} \frac{\partial Y_l^k}{\partial\varphi} + \ldots,$$

$$v_\theta = -\frac{4}{5(\gamma+1)} b^{2/5} t^{-3/5-2m/5} w_m(\lambda) \frac{\partial Y_l^k}{\partial\theta} + \ldots, \tag{5.34}$$

$$\rho = \frac{\gamma+1}{\gamma-1} \rho_1 [g(\lambda) + t^{-2m/5} g_m(\lambda) Y_l^k + \ldots],$$

$$p = \frac{8\rho_1}{25(\gamma+1)} b^{4/5} t^{-6/5} [h(\lambda) + t^{-2m/5} h_m(\lambda) Y_l^k + \ldots],$$

$$\lambda = r/(bt)^{2/5}.$$

The first-approximation functions f, g, and h specify the self-similar motion that describes a strong point blast. Let us dwell on the second-approximation functions with index m. The Rankine–Hugoniot conditions on the shock-wave front yield the initial velocity at the point $\lambda = 1$:

$$f_m = \frac{\gamma-7}{2(\gamma+1)} - m, \qquad u_m = 1, \qquad w_m = 1,$$

$$\rho_m = -\frac{5\gamma+13}{\gamma^2-1}, \qquad h_m = \frac{1-7\gamma}{\gamma^2-1} - 2m. \tag{5.35}$$

Substituting the expansion (5.34) in the set (1.16) of the gasdynamics equations (Chap. 1) and retaining terms of like powers of t, we obtain for the second-approximation functions a set of five equations. The functions u_m and w_m, however, are solutions of one and the same ordinary differential equation. Taking into account the equality of the initial conditions (5.35) for the functions w_n and u_m, we obtain

$$u_m(\lambda) = w_m(\lambda).$$

Retaining hereafter only $w_m(\lambda)$ and taking into account the properties of Y_l^k, we write down the system of equations

$$gf'_m + \left(f - \frac{\gamma+1}{2}\lambda\right)g'_m + \left(g' + \frac{2g}{\lambda}\right)f_m + \frac{l(l+1)}{\lambda}gw_m$$
$$+ \left[f' + \frac{2f}{\lambda} - \frac{m(\gamma+1)}{2}\right]g_m = 0,$$
$$\left(f - \frac{\gamma+1}{2}\lambda\right)gf'_m + \frac{\gamma-1}{2}h' + \left[f' - \frac{(3+2m)(\gamma+1)}{4}\right]gf_m$$
$$+\left[\left(f - \frac{\gamma+1}{2}\lambda\right)f' - \frac{3}{4}(\gamma+1)f\right]g = 0 \qquad \left(f'_m = \frac{df_m}{d\lambda}\ \text{etc.}\right),$$
$$\left(f - \frac{\gamma+1}{2}\lambda\right)gw'_m \tag{5.36}$$
$$+\left[\frac{f}{\lambda} - \frac{(3+2m)(\gamma+1)}{4}\right]gw_m - \frac{\gamma-1}{2\lambda}h_m = 0,$$
$$\gamma hf'_m + \left(f - \frac{\gamma+1}{2}\lambda\right)h'_m + \left(h' + \frac{2\gamma}{\lambda}h\right)f_m + \frac{\gamma(l+1)l}{\lambda}hw_m$$
$$+\left[\gamma f'_\lambda + \frac{2\gamma}{\lambda}f - \frac{(\gamma+1)(3+m)}{2}\right]h_m = 0.$$

From among the perturbations of the shock-wave front, we take those that make a finite contribution at large t to the gas momentum along the z axis. The expression for I_z is

$$I_z = \lim_{r\to 0}\int_0^{2\pi}\int_0^{\pi}\int_0^{r_s}\rho v_z r^2 \sin\theta\, dr\, d\theta\, d\varphi;$$

where $v_z = v_r\cos\theta - v_\theta\sin\theta$. Taking into account the expansion (5.34) as well as the expression $Y_l^k = p_l^k(\cos\theta)\cos(k\varphi + b_{kl})$ we have

$$I_z = \frac{4}{5(\gamma-1)}\rho_1 b^{8/5} t^{3/5-2m/5}\int_0^{2\pi}\cos(k\varphi + b_{lk})\,d\varphi$$
$$\times\int_0^{\pi} P_l^k(\cos\theta)\cos\theta\sin\theta\, d\theta[1 + \lim_{\lambda\to 0} I_{z1}(\lambda)], \tag{5.37}$$
$$I_{z1}(\lambda) = \int_\lambda^1 (gf_m + fg_m - 2gw_m)\lambda_1^2 d\lambda_1.$$

To separate the perturbations that determine the momentum I_z, we must put $k = 0$, for only in this case will the integral with respect to the variable φ differ from zero. Without loss of generality we can set the constant

b_{lk} equal to zero. Let us examine the integral with respect to θ. For $k = 0$, the associated Legendre functions go over into Legendre polynomials. Recognizing that $P_1(\cos\theta) = \cos\theta$, and also the orthogonality of the Legendre polynomials, we conclude that the integral with respect to θ will differ from zero only if $l = 1$. We choose m such that I_z is independent of time, so that $m = \frac{3}{2}$. Note that for $l = 1$ and $m = \frac{3}{2}$ the system (5.36) exhibits the momentum integral obtained in Ref. 56:

$$\lambda(g f_{3/2} + f g_{3/2} - 2g w_{3/2})$$
$$- \frac{1}{(\gamma+1)} [4 f g f_{3/2} + 2 f^2 g_{3/2} - 4 f g w_{3/2} + (\gamma - 1) h_{3/2}] = C_1/\lambda^2. \quad (5.38)$$

Determining the constant C_1 from the conditions (5.35), we obtain $C_1 = 0$. Excluding from the system (5.36) the function $w_{3/2}$ with the aid of Eq. (5.38) and discarding the third equation of Eqs. (5.36), we obtain a system of three equations for the functions $f_{3/2}$, $g_{3/2}$, and $h_{3/2}$. The solution of the system for $\gamma = 1.4$ was obtained numerically, and plots of the functions are given in Ref. 55. As $\lambda \to 0$ all the functions are oscillatory but, whereas $q_{3/2}$ and $h_{3/2}$ tend to zero, $f_{3/2}$ and $w_{3/2}$ increase without limit. Nonetheless, the integral I_{z1} tends to a finite value -1.052.

Consequently, for $\gamma = 1.4$ perturbation of a centrosymmetric shock front with $t^{-2m/5} R(\varphi, \theta) = C_0 t^{-3/5} \cos\theta$ corresponds to motion with a momentum directed along the z axis:

$$I_z = -0.140 C_0 \rho_1 b^{8/5}.$$

Let us prove the convergence of the integral $I_{z1}(\lambda_1)$ as $\lambda_1 \to 0$ for arbitrary γ $(1 < \gamma < 2)$. We study for this purpose the asymptotes of the second-approximation function as $\lambda_1 \to 0$; since these functions satisfy a system of third-order differential equations, a complete solution of this system is a sum of three linearly independent solutions. The asymptotes of two linearly independent solutions are oscillatory:

$$f_{3/2} = \frac{(\gamma^2-1)(\gamma-1)}{6\gamma k_1} \left\{ \left[\alpha_1 + \frac{\gamma+2}{2(\gamma-1)} \right] C_{20}(\lambda) + \alpha_2 C_{30}(\lambda) \right\} \lambda^{\alpha_1} + \ldots,$$

$$w^{3/2} = -\frac{(\gamma^2-1)(\gamma-1)}{12\gamma k_1} \left\{ \left[\alpha_1^2 - \alpha_2^2 + \frac{5\gamma-2}{2(\gamma-1)} \alpha_1 + \frac{\gamma+2}{\gamma-1} \right] C_{20}(\lambda) + \alpha_2 \left[2\alpha_1 + \frac{5\gamma-2}{2(\gamma-1)} \right] C_{30}(\lambda) \right\} \lambda^{\alpha_1} + \ldots,$$

$$g_{3/2} = C_{20}(\lambda) \lambda^{\alpha_1 + (4-\gamma)/(\gamma-1)} + \ldots, \quad (5.39)$$

$$h_{3/2} = \frac{(\gamma^2-1)^2}{6\gamma^2} \left\{ \left[\alpha_1^2 - \alpha_2^2 + \frac{7\gamma-6}{2(\gamma-1)} \alpha_1 + \frac{3\gamma^2+2\gamma-2}{2(\gamma-1)^2} \right] C_{20}(\lambda) + \alpha_2 \left[2\alpha_1 + \frac{7\gamma-6}{2(\gamma-1)} \right] C_{30}(\lambda) \right\} \lambda^{\alpha_1 + (2+\gamma)/(\gamma-1)} + \ldots,$$

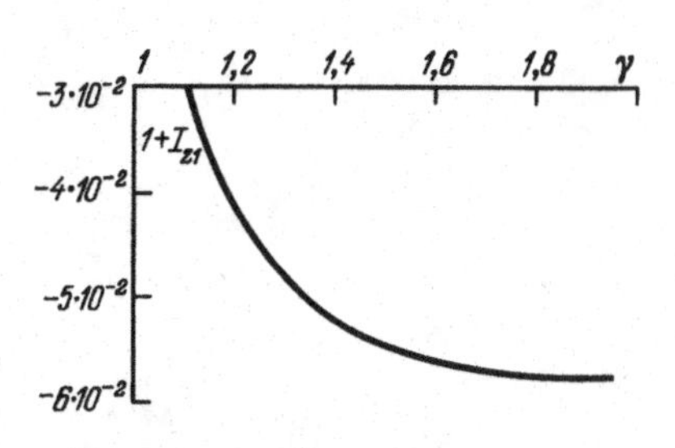

Figure 86. Plot of dimensionless momentum as a function of γ.

$$C_{20}(\lambda) = C_3 \cos(\alpha_2 \ln \lambda) + C_3 \sin(\alpha_2 \ln \lambda),$$

$$C_{30}(\lambda) = -C_2 \sin(\alpha_2 \ln \lambda) + C_3 \cos(\alpha_2 \ln \lambda),$$

$$f_{3/2} = C_4 \frac{(\gamma^2-1)(\gamma-1)}{6\gamma k_1}\left[\alpha_3 + \frac{\gamma+2}{2(\gamma-1)}\right]\lambda^{\alpha_3} + \ldots,$$

$$w_{3/2} = -C_4 \frac{(\gamma^2-1)(\gamma-1)}{12\gamma k_1}\left[\alpha_3^2 + \frac{5\gamma-2}{2(\gamma-1)}\alpha_3 + \frac{\gamma+2}{\gamma-1}\right]\lambda^{\alpha_3} + \ldots, \quad (5.40)$$

$$g_{3/2} = C_4 \lambda^{\alpha_3 + (4-\gamma)/(\gamma-1)} + \ldots,$$

$$h_{3/2} = C_4 \frac{(\gamma^2-1)^2}{6\gamma^2}\left[\alpha_3^2 + \frac{7\gamma-6}{2(\gamma-1)}\alpha_3 + \frac{3\gamma^2+2\gamma-2}{2(\gamma-1)^2}\right]\lambda^{\alpha_3+(2+\gamma)/(\gamma-1)} + \ldots.$$

Here C_2, C_3, and C_4 are arbitrary constants determined in the actual solution by the initial data (5.35); $\alpha_1 + i\alpha_2$, $\alpha_1 - i\alpha_2$, and α_3 ($i = \sqrt{-1}$) are roots of the cubic equation

$$\alpha^3 + \frac{7\gamma}{2(\gamma-1)}\alpha^2 + \frac{3\gamma^2+17\gamma-14}{2(\gamma-1)^2}\alpha + \frac{3\gamma}{(\gamma-1)^2} = 0.$$

Plots of α_1, α_2, and α_3 vs γ are given in Ref. 55. For $\gamma = 1.4$ we have $\alpha_1 = -5.810$, $\alpha_2 = 2.815$, and $\alpha_3 = -0.630$. From the asymptotes (5.39) we can conclude that the integrand in Eqs. (5.37) tends to zero as $\lambda \to 0$, so that the total integral (5.37) converges, and $I_{z1}(0)$ exists for all γ. A perturbation (5.33) in the form $C_0 t^{-3/5} \cos\theta$ corresponds thus to motion having a momentum

$$I_z = \frac{16 C_0 \rho_1 b^{8/5}}{15(\gamma-1)}[1 + I_{z1}(0)].$$

Figure 86 shows the dependence of $1 + I_{z1}(0)$ on γ. Since $1 + I_{z1}(0) < 0$, the momentum I_z is directed opposite to the relative displacement of the shock front along the z axis.

We note one more property of the solution: as $\lambda \to 0$ the perturbed pressure, according to Eq. (5.38), also tends to zero.

The problem considered has a bearing on the blast of a moving charge. Blasts of this type were recently investigated in the paper "Blast in flight" by L. V. Shurshalov, Izv. AN SSSR, Mek. Zhidk. i Gaza No. 5, 1984.

Chapter 6
Blast in a combustible gas mixture

1. Statement of problem

This chapter is devoted to the basic problem for a quiescent gas in which exothermal chemical reactions can take place. A blast initiates in the gas propagation of a strong shock wave that heats to a state in which combustion is possible. Problems of this type occur when a detonation is set off in a combustible gas mixture and in a thermonuclear detonation, and their study is both of theoretical and practical interest.

We assume the gas motion to be one-dimensional. Let us examine the overall picture of the flow of the medium. It follows from physical considerations that during the first instants after the blast the gas motion will follow the usual point-blast laws,[1,2] since the contribution of the combustion energy to the total energy balance is still small. We can estimate the distances up to which it is expedient to disregard the effect of the detonation energy on the motion of the medium. Let Q be the calorific capacity per unit volume of the combustible mixture. We consider by way of example a constant initial density ρ_1 and a constant energy Q. If the width of the combustion band is assumed to be negligibly small compared with the wave radius r_2, the energy released by the burning gas is

$$U = \frac{1}{\nu}\,\sigma_\nu \rho_1 Q r_2^\nu, \qquad \sigma_\nu = 2(\nu-1)\pi + (\nu-2)(\nu-3).$$

Assuming $E_0 > U$, we obtain the condition under which the detonation energy has little influence on the flow:

$$r_2 < r_*, \qquad r_*^\nu = \nu E_0/\sigma_\nu Q\rho_1. \tag{6.1}$$

The condition (6.1) will be met also when account is taken of the reaction zone for incomplete combustion of the combustible mixture, for in these cases the combustion energy will be less than U.

If the charge has a finite radius r_0, to use the standard point-blast theory to describe the motion we must use the estimate

$$r_0 < r < r_*. \tag{6.2}$$

The conditions (6.1) and (6.2) greatly restrict the range of applicability of the point-blast laws in an inert gas to flows of a detonating medium. If, however, the energy E_0 is high and is released in a small volume, the flow at $r_2 < r_*$ will be mainly the same as for an ordinary point blast. On the other hand, for instants of time when $E_0 < U$ the principal role will be assumed by combustion processes, and the gas flow will have the main features of detonation combustion.

At values of r_2 close to r_* the gas flow will assume a more complicated intermediate character.

The distinctive features of the considered flow are that the shock-wave front is adjacent to a region in which chemical reactions take place, and that near the blast center there will be a certain region in which (just as in an ordinary blast) the gas particles will have a high entropy.

From the mathematical standpoint the solution of the problem reduces to integration of a system of gasdynamics and chemical-kinetics equations with boundary conditions on the surfaces of the strong blast and in the vicinity of the blast center. In addition it must be recognized that at the initial instant $t = 0$ a finite energy is released and initial values of the velocity, density, pressure, and other parameters of the medium are specified. For a complete description of the gas motion account must be taken of the kinetics of all the basic chemical reactions of the combustion, and also the ionization and dissociation processes in the high-temperature zones.

When the kinetics of the chemical reactions is taken into account, the solution of the problem above meets with appreciable mathematical difficulties, even if viscosity and heat conduction are neglected. The gasdynamics equations will contain, besides Q, a number of additional parameters (activation energy, the coefficients in the expressions for the reaction rates, and others). The presence of an appreciable number of constant parameters and large gradients of the functions in the chemical-reaction zone greatly hinders the analytic and numerical solutions of the problems. Fairly simple approaches to the solution are possible in a number of cases. The following gas-flow models can be introduced.

(1) The detonation-wave model, in which it is assumed that the material burns up in the immediate vicinity of the shock-wave front and the shock wave, together with the chemical-reaction zone behind it, is regarded as a single strong-discontinuity surface.

(2) Simple model with kinetics, i.e., gas-flow model with chemical reactions, when the reaction is described by one variable quantity—the mass density of the non-combusted substance.[3,8]

(3) A flow model with two fronts, when a shock wave propagates through the unperturbed gas, at some distance behind which follows a combustion wave in which all the heat is released, i.e., an induction zone or an ignition zone is introduced (see, e.g., Refs. 4–7 and 9).

(4) A model including a chemical reaction with heat release in the stream behind the shock-wave front after the termination of the induction period. The reaction is also described by a single variable, the mass density of the non-combusted matter; an inverse "recombination" reaction may also be taken into account here.

(5) All important chemical reactions including behind shock wave front.

In the case of the detonation-wave model, the problem was investigated in detail for a perfect gas with constant heat capacities. It should be noted here that the first to solve the problem for this model, in certain special cases, was I. S. Shikin.[13] The parameters of a strong detonation wave were determined approximately by V. A. Levin,[14] who also formulated a condition similar to Eq. (6.1).

2. Solution of problem in the detonation-wave mode

In view of the above peculiarities of the flow during the initial blast stage, the detonation will not follow the Chapman–Jouguet (C–J) rule. Let us consider this question in somewhat greater detail for an ideal gas with constant heat-capacity ratio. Let D be the shock-wave velocity, v_2 the gas velocity behind its front, and a_2 the corresponding sound velocity. From the conditions on the shock wave we have then for a strong blast

$$v_2 + a_2 = \frac{D}{\gamma+1}\left[2 + \sqrt{2\gamma(\gamma-1)}\,\right].$$

It follows from this expression that for $\gamma > 1$ we have $v_2 + a_2 > D$, i.e., the detonation during the initial stage of the process is overcompressed. We know that $v_1 + a_2 = D$ for a C–J wave. As the shock wave propagates (in the detonating gas) $v_2 + a_2$ decreases, and a C–J detonation regime can be reached. For a blast in a chemically inert gas, a condition of the C–J type is met in fact only asymptotically: as $t \to \infty$, $v_2 \to 0$, $a_2 \to a_1$, and $D \to a_1$.

Let us consider the mathematical formulation of the problem for the detonation-wave model. We choose the initial system of gasdynamics equations for one-dimensional adiabatic motions of the gas in the form (1.18)–(1.20). To solve the point-blast problem in an immobile detonating medium we must find the solution of the system (1.18)–(1.20) with the boundary conditions on the shock-wave front, i.e., for $r = r_2$ (see Sec. 2 of Chap. 1):

$$\rho_2(D - v_2) = \rho_1 D, \qquad p_2 = p_1 + \rho_1 D v_2,$$
$$\varepsilon_1 + \frac{p_1}{\rho_1} + \frac{D^2}{2} + Q = \varepsilon_2 + \frac{p_2}{\rho_2} + \frac{1}{2}(D - v_2)^2. \tag{6.3}$$

In addition, we have at the symmetry center the boundary condition

$$v(0, t) = 0. \tag{6.4}$$

At the instant $t = 0$ there is released at the symmetry center a finite energy E_0, and the following conditions are specified:

$$v(r, 0) = 0, \qquad p(r, 0) = p_1 = \text{const}, \qquad \rho(r, 0) = \rho_1(r)$$
$$(r > 0), \qquad r_2(0) = 0. \tag{6.5}$$

Consider the solution of this problem for the case of an ideal gas and assume that the initial density is variable: $\rho_1 = Ar^{-\omega}$. The system of the defining parameters of the problem contains the constants $Q, A, p_1, E_0, \omega, \gamma, \nu$, and in addition the solution will depend on the coordinate r and on the time t. We assume for simplicity that the detonation wave is strong and neglect the initial gas pressure p_1. We introduce an effective detonation pressure in the form

$$p_* = \frac{1}{\gamma} Q\rho_{10},$$

where ρ_{10} is a constant with the dimension of density, which can be written as $\rho_{10} = A(r^0)^{-\omega}$, $r^0 = (E/p_*)^{*1/\nu}$. We can write for p_* the more explicit expression

$$p_* = \left(\frac{QA}{\gamma}\right)^{\nu/(\nu-\omega)} E_0^{-\omega/(\nu-\omega)}.$$

The parameter p_* can be included in the system of defining parameters in place of Q.

The simplest premises of dimensionality theory allow us to verify that the problem obeys in the case of a strong detonation wave a similarity law analogous to that for an ordinary blast (see Sec. 6, Chap. 4). The role of the initial gas pressure is assumed here by p_*.

From the system of the defining parameters of the problem it follows that in terms of the dimensionless functions

$$f = \frac{v}{D}, \qquad g = \frac{\rho}{\rho_1}, \qquad h = \frac{q}{p_*}\left(\frac{r_2}{r^0}\right)^{\omega} p, \tag{6.6}$$

the problem depends on several constant parameters and two dimensionless variables, which we choose to be

$$\lambda = r/r_2, \qquad q = Q/D^2. \tag{6.7}$$

We introduce also the dimensionless shock-wave radius $R_2(q) = r_2/r_1$. Transforming Eq. (1.20) of the initial system to take the adiabaticity of the flow into account and changing to the dimensionless variables (6.6) and (6.7) we obtain the system

$$\begin{gathered}
D_{\lambda,q} f + \frac{1}{\gamma g}\frac{\partial h}{\partial \lambda} - \eta\frac{f}{2q} = 0, \qquad \eta = R_2\frac{dq}{dR_2}, \\
D_{\lambda,q} g + g\frac{\partial f}{\partial \lambda} - \omega g + \frac{(\nu-1)fg}{\lambda} = 0, \\
D_{\lambda,q} h + \gamma h\frac{\partial f}{\partial \lambda} - \omega h - h\left(\frac{\eta}{q} - \gamma(\nu-1)\frac{f}{\lambda}\right) = 0, \\
D_{\lambda,q} = (f-\lambda)\frac{\partial}{\partial \lambda} + \eta\frac{\partial}{\partial q}.
\end{gathered} \tag{6.8}$$

From Eqs. (6.3), putting $p_1 = 0$, we get

$$(f_2 - 1)g_2 = -1, \qquad h_2 = \gamma f_2, \qquad \frac{1}{2} + q = \frac{1}{\gamma - 1}\frac{h_2}{g_2} + \frac{1}{2}(f_2 - 1)^2. \quad (6.9)$$

Note that at $q = 0$ we get the self-similar strong-blast problem (see Chap. 2). At $E_0 = 0$ we obtain the self-similar detonation-wave propagation problem,[1,3] which can also be expediently treated as a point-blast problem. In the complete solution of the problem we must track the transition from one self-similar flow regime (strong point blast in an inert gas) to another self-similar regime (detonation wave or point blast in a detonating gas at negligibly low blast energy). Note also that if we reverse the sign of Q in Eqs. (6.3) and (6.7) (assuming that $Q > 0$) we have a point-blast problem with energy absorption on the wave front (e.g., with evaporation). The wave will then be additionally damped. The solution methods developed below will be valid also for this case. The blast problem with allowance for the influence of the detonation energy can be solved by numerical methods.

For the initial stage of the blast, when

$$(p_2 - p_*)/p_* > 1, \qquad q < q_j, \qquad r_2 < r_*,$$

a simpler method can be proposed, based on linearization of the initial equations with respect to the self-similar solution for a strong blast. We denote here by q_j the value of q for the C–J wave. The solution of the linearized problem in the variables assumed can be obtained by analogy with the case considered in Chap. 3 with back pressure taken into account. We represent the sought functions in the form

$$f(\lambda, q) = f_0(\lambda) + qf_1(\lambda) + O(q^2), \qquad g(\lambda, q) = g_0(\lambda) + qg_1(\lambda) + O(q^2),$$
$$h(\lambda, q) = h_0(\lambda) + qh_1(\lambda) + O(q^2), \quad (6.10)$$
$$\frac{R_2}{q}\frac{dq}{dR_2} = \frac{\nu - \omega}{1 + A_1 q + O(q^2)},$$

where $O(q^2)$ denotes terms of order q^2 for small q. The solution (6.10) must be substituted in the system (6.8). Neglecting terms of order q^2 and higher, we obtain then a system of linear ordinary equations, with coefficients that depend on the self-similar solution f_0, g_0, h_0. It is necessary to determine from this system the functions f_1, g_1, and h_1. In addition, it is necessary to find the constant A_1 that determines the non-self-similar correction to the motion of the shock wave.

From the conditions (6.4) and (6.9) we obtain the boundary values for the functions f_1, g_1, and h_1:

$$f_1(1) = -(\gamma - 1), \; h_1(1) = -\gamma(\gamma - 1),$$
$$g_1(1) = -\frac{(\gamma + 1)^2}{\gamma - 1}, \qquad f_1(0) = 0. \quad (6.11)$$

Let now $\omega = 0$. We shall dwell on this case in greater detail, following Sec. 3 of Chap. 3, where a linearized problem with back pressure was con-

Table 5. Values of the constant A_1.

ν	γ						
	8/7	1.2	1.3	1.4	5/3	2	3
1	0.5091	0.7246	1.121	1.545	2.811	4.685	12.28
2	0.4695	0.6694	1.037	1.428	2.592	4.296	11.10
3	0.4551	0.6489	1.005	1.383	2.502	4.132	10.59

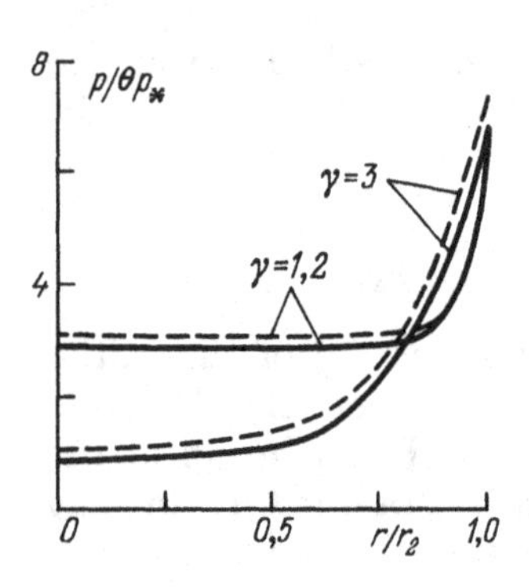

Figure 87. Distribution of the relative pressures in space for two values of γ. The dashed lines correspond to $h = h_0$.

sidered. It is easily seen that by virtue of the introduced dimensionless parameters the system of linear equations for f_1, g_1, and h_1 coincides with the corresponding system in Sec. 4 of Chap. 3. This system must be solved with the boundary conditions (6.11). For arbitrary ν and γ the problem can be solved by numerical integration. Calculations for a number of values were performed for $\nu = 1$, 2, and 3 by the procedure developed in Chap. 3.[10]

Table 5 lists the values of the constant A_1. Figure 87 shows plots (for $\nu = 3$) of $p/\theta p_*$ for the case $\gamma = 1.2$, $q = 0.15$, and $\theta = 1$, and plots of $p/\theta p_*$ for the case $\gamma = 3$, $q = 0.04$, and $\theta = 5$. The dashed lines correspond to a blast without allowance for detonation ($h = h_0$).

We present also equations that describe, in parametric form, the shock-wave motion $R_2(\tau)$ [see Eqs. (3.47)]:

$$R_2^\nu = \frac{\sigma^2}{\alpha\gamma}\, q \exp(A, q), \qquad \delta = \frac{2}{\nu + 2}, \qquad R_{20}^\nu = \frac{\delta^2}{\alpha\gamma}\, q,$$

$$\tau = \delta\left(\frac{q}{\gamma}\right)^{1/2} R_{20}\left[1 + \frac{\nu\delta + 2}{2\nu\delta(\nu\delta + 1)}\, A_1 q\right], \tag{6.12}$$

$$\tau = \frac{t}{t^0}, \qquad t^0 = r^0\left(\frac{\rho_1}{p_*}\right)^{1/2}, \qquad \alpha = \alpha(\gamma, \nu).$$

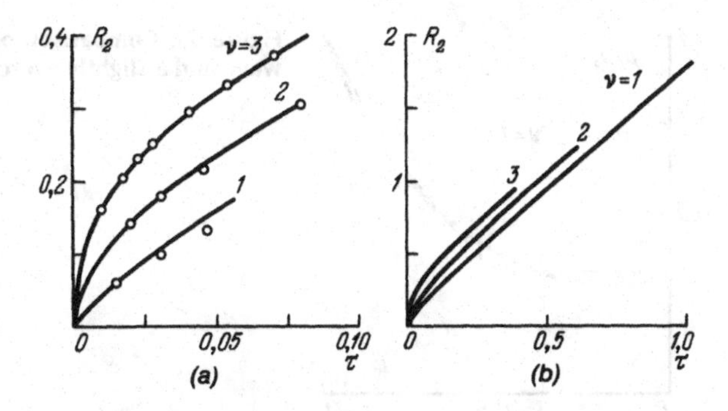

Figure 88. Dependence of the motion $R_2(\tau)$ of an overcompressed detonation wave on the time: (a) from the linearized solution (the circles show the results of the numerical solution of the nonlinear problem); (b) from the numerical solution (in the cases $\nu = 2$ and 3 the end points of the curves correspond to a transition to the Chapman–Jouguet regime).

Figure 88(a) shows plots of $R_2(\tau)$ for different ν at $\gamma = 1.4$. Note that for $\nu = 3$ and $\gamma = 7$ the linearized problem has an exact analytic solution, in analogy with the situation when backpressure is taken into account (here $A_1 = 60$). We can similarly investigate the linearized problem also for the case of a variable initial density $\omega \neq 0$.

Since the validity of the linear solution is confined to small q, numerical methods must be used to solve the problem for large q (and correspondingly large instants of time t), when a transition of the wave to the C–J regime takes place. It is possible to use for this purpose the method developed in Sec. 3 of Chap. 4 for the solution of blast problems.[15] When this method is used, the calculation procedure is different only if the shock-wave front parameters are to be calculated. The initial data can be specified for a certain value of τ or $q < q_j$, using the solution of the linearized problem considered above. This method was used to obtain a numerical solution of a constant initial density. The calculations were performed only up to the instants of time when the shock-wave parameters became close to those of the C–J wave parameters ($q_j/q - 1 < 0.01$). When the calculation method developed in Chap. 4 and based on a numerical integral-equation method is used, the integration region between the detonation wave and the blast center was broken up into an even number n of space bands. The calculations were performed for $n = 2$ and $n = 4$. The results that follow pertain to the case $n = 4$, i.e., the functions sought were approximated by fourth-degree polynomials (the problem was programmed by E. Bishimov).

The computations have shown that in the case of a blast in a perfect gas with a constant value of γ, the spherical and cylindrical shock waves approach quite rapidly the parameters of the C–J wave. In the planar case, the C–J regime was approached after much longer time intervals (theoretically, the C–J regime is reached after an infinite time[16]). Using the computation results for a planar blast, one can determine the constants in the asymptotic equations, obtained by G. G. Chernyi,[16] for the damping of overcompressed detonation waves, and find analytic relations for the detonation-wave front parameters as the C–J regime is reached. Thus, at $\gamma = 1.4$ we have for the time dependence of the coordinate of a strong planar detonation wave

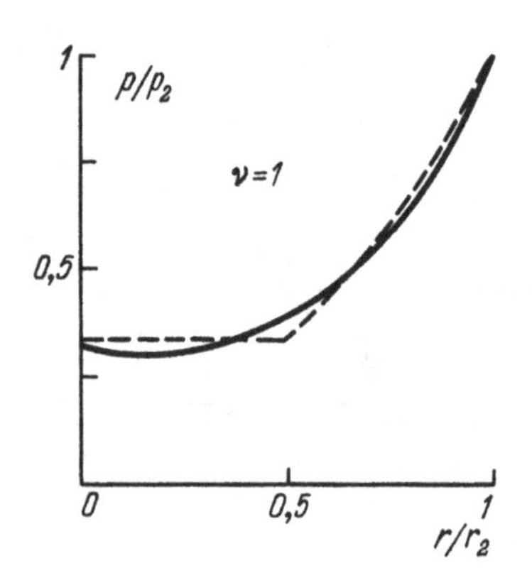

Figure 89. Comparison of the pressure profiles for a C–J wave and a slightly overcompressed shock wave ($\tau \sim 1$).

$$\tau = \tau_0 + \frac{R_2}{k}\left(1 + \frac{m^2}{R_2^2}\right), \qquad \tau_0 = -0.0835, \qquad k = 1.64,$$
$$m = 0.0741, \qquad t = E_0 p_*^{-3/2} \rho_1^{1/2} \tau, \qquad R_2 = \frac{p_*}{E_0} r_2. \tag{6.13}$$

Comparison of the values of τ obtained from this equation and by numerically solving the problem reveals good accuracy even at rather small values of R_2 ($R_2 \sim 0.5$).

Figure 88(b) shows a numerically computed plot of $R_2(\tau)$ for $\nu = 1$, 2, and 3.

In the cases $\nu = 2$ and 3 the curves terminate at values close to the instants when the C–J regime is reached. The $R_2(\tau)$ curve for the planar case can be continued for larger instants by using Eqs. (6.13).

Denoting by R_j the distance at which the strong wave has reached the C–J regime, we have in the spherical and cylindrical cases

$$R_j = \varkappa (E_0/p_*)^{1/\nu}, \qquad p_* = \rho_1 Q/\gamma, \tag{6.14}$$

where $\varkappa$ depends on γ and ν (and is close to unity for $\gamma = 1.4$, see Ref. 11).

It follows from Eqs. (6.13) that in blasts produced in combustible mixtures with equal and low initial temperature and equal E_0, but with different pressures, the value of R_j increases with decreasing pressure and hence with decreasing initial density ρ_1.

Note that the similarity law mentioned above is reflected in Eqs. (6.14).

Figure 89 shows the spatial distribution of the relative pressure p/p_2 for a planar blast in a gas ($\gamma = 1.4$) at large values of τ ($\tau = 0.65$). The figure shows also a curve (dashed) based on the known solution of the problem of propagation of a strong self-similar planar C–J detonation wave (neglecting the initial blast energy E_0). It is seen here that in the vicinity of the wave front the gas flow is close enough to the flow in a C–J detonation.

The distances at which an overcompressed detonation wave reaches the C–J regime were determined by V. V. Markov[13] using a finite-difference method, for $\nu = 2$ and 3 and for various γ ($1.2 \le \gamma \le 3$). These computations have shown that the C–J regime is reached at $U > 2.5E_0$, i.e., the total combustion energy is more than double the energy of the initiating blast ($U = 5E_0$ when the C–J regime is reached at $\nu = 3$ and $\gamma = 1.3$).

The same problem with allowance for back pressure was solved by E. Bishimov. He has studied also the case of variable heat release $Q = Q(\xi_0)$, where ξ_0 is the initial coordinate of the gas particle.[17,18] The results, as well as those of investigation of the asymptotic shock-wave behavior,[11,16] lead to the conclusion that in the detonation-wave model an overcompressed wave reaches the C–J regime at a finite distance from the blast center in the case of spherical and cylindrical blasts, but for a planar blast the C–J wave is approached only at appreciable distances.

Thus, for explosive gas mixtures (or for liquid and solid explosives in the ideal-gas approximation), for which model 1 is applicable, we can neglect at certain distances the initiation energies of detonation cylindrical and spherical waves. These distances are substantially larger for plane waves.

3. Some properties of gas flow with chemical-reaction kinetics taken into account

Allowance for the kinetics of the chemical reactions can influence strongly, in a number of cases, the picture of the gas flow in the case of a point blast.[11,19,20]

(1) Consider the simplest model of such flows (model 2). We neglect the influence of viscosity and heat conduction on the gas flow and assume that the chemical reaction can be described by a single chemical variable, the mass density β of the molecules that have not yet reacted. A unit mass of the combustible mixture contains then a chemical energy βQ. In the system (1.18)–(1.20) the energy equation is changed[3] and its right-hand side is no longer zero but

$$-\rho r^{\nu-1}\frac{d}{dt}(\beta Q).$$

For an ideal gas with constant γ, the energy equation can be transformed into[8]

$$\frac{dp}{dt}-\frac{\gamma p}{\rho}\frac{d\rho}{dt}+(\gamma-1)\rho\frac{d(\beta Q)}{dt}=0. \tag{6.15}$$

We choose the chemical-kinetics equation in the form[8,11,19]

$$\frac{d\beta}{dt}=-L\beta^m p^n \rho^l \exp\left(-\frac{u_0}{\mu}\frac{\rho}{p}\right). \tag{6.16}$$

Here u_0 is the activation energy, μ the average molecular weight, m the order of the reaction, n and l constants, L a positive constant, $\beta = 1$ at the start of the reaction, and $\beta = 0$ at the end.

The complete set of equations contains thus Eqs. (1.18), (1.19), (6.15), and (6.16). The boundary conditions are here the usual conditions on the shock wave and the condition (6.4).

We can assume in this problem $\beta = 1$ directly behind the shock-wave front. It follows from Eq. (6.16) that in a gas particle β can only decrease.

We write down also the integral energy-conservation law, which takes the form

$$E_0 = \sigma_\nu \int_0^{r_2} \left(\frac{\rho v^2}{2} + \varepsilon\rho - Q(1-\beta)\rho \right) r^{\nu-1}\, dr + \frac{\sigma_\nu}{\nu} \rho_1 \varepsilon_1 r_2^\nu. \tag{6.17}$$

This relation is quite useful for the treatment of a number of questions. For $\beta = 0$ and $\varepsilon_1 = 0$ we obtain from Eq. (6.17) the integral conservation law for a strong detonation wave (Sec. 2).

A solution of the point-blast problem for the introduced model of flow with chemical reactions can in general be obtained only by approximate or numerical methods. An attempt can be made to find a class of self-similar solutions of the gasdynamics and chemical-kinetics equations. We assume that

$$\frac{u_0}{\mu} = Bp^{\alpha_1}\rho^{\alpha_2}, \qquad \rho_1 = Ar^{-\omega}. \tag{6.18}$$

Neglecting the energy E_0, i.e., consider the problem of propagation of a strong shock wave that is self-sustained through chemical energy, the system of defining parameters of the problem comprises $Q, L, B, A, m, l, n, \omega, \alpha_1, \alpha_2, \gamma$, and ν. It is easy to see from dimensionality considerations that for $\alpha_1 = \alpha_2 = 0$ the problem is self-similar if the condition $l + n = 1/\omega$ is satisfied.

We conclude thus that for a strong shock wave the detonation–propagation problem is self-similar if the initial density varies as $\rho_1 = Ar^{-1/(l+n)}$. If $E_0 \neq 0$, the problem of a blast in a chemically active gas is self-similar only if Q depends on the gas parameters (or on the gas parameters, the coordinates, and the time).

Thus, if we have under the condition (6.18) the relation $Q = Q_0 p^{\alpha_1}\rho^{\alpha_2}$, the problem of a strong point blast is self-similar, provided that

$$\begin{gathered} 2(1-\alpha_1)(\delta-1) + \delta\omega(\alpha_1+\alpha_2) = 0, \qquad 2 = \delta(\nu+2-\omega), \\ \delta[3l + n + (\nu-1)(l+n)] - 1 - 2l = 0. \end{gathered} \tag{6.19}$$

Other classes of self-similar solutions can be indicated[19] if it is assumed that L, B, and Q_0 depend on certain powers of r and t. For self-similar solutions, if special (nondimensional) variables are used, the considered system of gasdynamic and chemical-kinetics equations reduces to a system of four ordinary differential equations

$$
\begin{aligned}
&\varphi(\lambda)\frac{dg}{d\lambda}+\frac{2}{\gamma+1}g\frac{df}{d\lambda}-\omega g+\frac{2(\nu-1)gf}{(\gamma+1)\lambda}=0,\\
&g\left(\frac{\delta-1}{\delta}f+\varphi(\lambda)\frac{df}{d\lambda}\right)+\frac{\gamma-1}{\gamma+1}\frac{dh}{d\lambda}=0,\\
&\left[2\frac{\delta-1}{\delta}+\omega(\gamma-1)\right]h+\varphi(\lambda)\left(\frac{dh}{d\lambda}-\frac{\gamma h}{g}\frac{dg}{d\lambda}\right)\\
&\quad=-\sigma_2 g\left[2\frac{\delta-1}{\delta}\beta h^{\alpha_1}g^{\alpha_2}+\varphi(\lambda)\frac{d}{d\lambda}(\beta h^{\alpha_1}g^{\alpha_2})\right],\\
&\varphi(\lambda)\frac{d\beta}{d\lambda}=-\sigma_0 h^n g^l\beta^m\exp(-\sigma_1 h^{\alpha_1-1}g^{\alpha_2+1}),
\end{aligned}
\tag{6.20}
$$

where

$$
\varphi(\lambda)=\frac{2f}{\gamma+1}-\lambda,\quad f=\frac{v}{v_2},\quad h=\frac{p}{p_2},\quad g=\frac{\rho}{\rho_2},\quad \lambda=\frac{r}{r_2}.
$$

v_2, ρ_2, and p_2 are the velocity, density, and pressure behind the strong shock wave.

For the self-similar strong-blast case, the function $r_2(t)$ takes the form

$$
r_2=\left(\frac{E_0}{\alpha_* A}\right)^{\delta/2}t^{\delta},\quad \delta=\frac{2}{2+\nu-\omega}.
$$

Here α_* is a quantity that depends on the dimensionless parameters of the problem and is determined (as in Chap. 2) from the energy-conservation law, viz.,

$$
\alpha_*=2\sigma_\nu\frac{\sigma^2}{\gamma^2-1}\int_0^1[gf^2+h-\sigma_2 g^{\alpha_2+1}h^{\alpha_1}(1-\beta)]\lambda^{\nu-1}\,d\lambda. \tag{6.21}
$$

It is required to find the solution of the system (6.20) with the boundary conditions $f(1)=g(1)=h(1)=1$ on the shock wave and with account taken of the condition (6.21) or of the condition $f(0)=0$ at the blast center.

In contrast to the usual self-similar gasdynamics problems,[1] allowance for chemical reactions complicates the solution greatly. Thus, the system of self-similar equations for the considered cases has no classical adiabaticity integral and does not reduce to investigation of one ordinary first-order differential equation. Complicated singularities can be encountered here in the vicinity of the center. Approximate and computer methods must therefore be used also to investigate self-similar problems involving shock-wave propagation in a combustible mixture with allowance for kinetic reactions.

To monitor the computational accuracy one can use the integral mass-conservation law, which takes in the assumed dimensionless variables the form

$$\int_0^1 g\lambda^{\nu-1}\,d\lambda = \frac{\gamma-1}{\gamma+1}\frac{1}{\nu-\omega}.$$

If $\alpha_2 = 0$, the conditions of self-similarity at $\omega = 0$ lead to $n = (\nu+2)/2\nu$.

Another interesting case is $\omega = \nu - 1$. As noted in Chap. 3, this case corresponds to a point blast or to propagation of a detonation wave in a gas stream produced in a wedge-shaped ($\nu = 2$) or conical ($\nu = 3$) nozzle. In the case of a blast we have the relation $\omega\alpha_2 = \nu - \alpha - \nu\alpha_1$. Examples of solutions for a self-sustaining wave ($\alpha_* = 0$) at $\omega = 1$ and $\alpha_1 = \alpha_2 = 0$ are given in Ref. 19.*

(2) Blast phenomena in a chemically active gas mixture can sometimes be additionally described using a flow model, by introducing a discontinuity surface such as a combustion wave behind the front of the propagating shock wave (model 3).

It is assumed in this model that after the gas particle passes through the shock-wave front instantaneous combustion of the substance sets in after a certain time t_{ind} and an energy Q is released on the combustion-wave front. The time t_{ind} is usually called the ignition-delay time or the induction time. It is known[4-7,9] that the induction time depends on the thermodynamic parameters of the gas. Let t_{ind} be given by the equation

$$t_{ind} = Lp^{-n_1}\rho^{-l_1}\exp(K/T), \tag{6.22}$$

where n_1, l_1, K, and L are certain constants ($K > 0$, $L > 0$).

Thus, for a stoichiometric mixture of hydrogen with air it can be assumed that[22,23]

$$t_{ind} = 4.5 \times 10^{-9}\frac{1}{p}\exp(10^4/T), \tag{6.23}$$

where t_{ind} is in seconds, the pressure p is in atmospheres, and T is the temperature (K).

Disregarding for the time being the detailed properties of the gas stream behind the shock-wave front, let us trace approximately the pattern of the variation of t_{ind} behind the shock-wave front. It follows from Eqs. (6.22) and (6.23) that t_{ind} is small for large p and T. As the shock wave attenuates and p and T decrease, the value of t_{ind} in the gas particle increases. Consequently, the distance between the shock-wave front and the combustion wave also increases and the combustion front is detached from the shock-wave front. Thus, starting with certain instants of time the shock wave and the combustion zone can no longer be regarded as one discontinuity surface (the detonation wave). After sufficiently long t intense combustion can take place only outside a certain vicinity of the points of the shock-wave front (in the region of sufficiently high temperatures), and practically no

* This reference deals also with the case $\omega = 0$, $E_0 \neq 0$, $\alpha_1 = 1$, and $\alpha_2 = 0$. The extrapolation of the solution to the region $\lambda < 0.2$ is not correct. A corrected solution and its application to gas-liquid systems is given in Ref. 21.

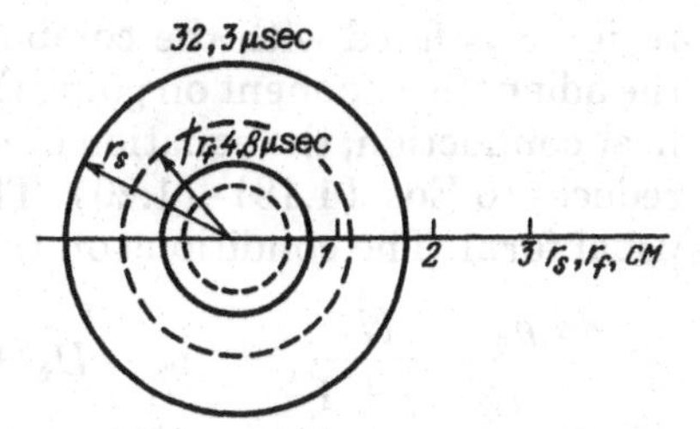

Figure 90. Locations of shock wave and ignition front at two instants of time. Ignition by a laser spark; stoichiometric hydrogen–acetylene mixture (p_1 = 80 Torr, E_0 = 0.275 J).

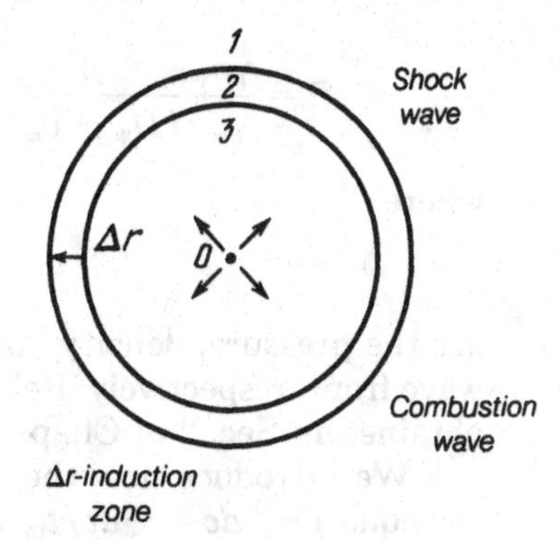

Figure 91. Gas flow in the two-front model. Key: (1) Shock wave; (2) Combustion wave; (3) Δr = induction zone.

combustion will take place in the vicinity of the front of a strongly attenuating shock wave.

The above conclusion that a detonation wave can become separated under point-blast conditions was first reached by us in 1967 (Ref. 11). The picture described was also indirectly confirmed by experimental and theoretical data on hypersonic flow of a combustible mixture around a body.[22,23] Using the inverse analogy of hypersonic flow around blunted bodies and a cylindrical blast (see Chap. 4) we can qualitatively describe a cylindrical blast by using the stationary flow-around picture. Using this analogy and carrying over the data on flow around a blunted cylinder, we can naturally expect, in the case of a cylindrical blast in a mixture of hydrogen with air (or oxygen), a flow with two fronts—a shock-wave front followed by, in accordance with Eq. (6.23), a combustion front. Simultaneously with our results, a similar conclusion was reached by R. Soloukhin, J. Lee, and A. Oppenheim[24] from experiments with a laser beam in a combustible mixture. Diffraction was also found to cause detonation waves to break up.[25]

Figure 90 shows the wave fronts for a concentrated energy supply to a stoichiometric mixture of acetylene and oxygen by a laser spark (E_0 = 0.275 J, p_1 = 80 Torr), and agrees with the experimental data of Ref. 33.

Some results of recent[12,20,27,28] attempts at a theoretical corroboration of shock-wave splitting during the initial stage of a point blast will be described below. Figure 91 shows the gas-flow scheme in the two-front model. In region 1 the gas is at rest, and in region 2 it moves behind an ordinary shock wave. The gas burns up on passage of the combustion wave, and region 3 is filled with the combustion products. Neglecting the change of the adiabatic exponent on going through the discontinuity and disregarding heat conduction, the equation of motion of the gas in the continuity regions reduces to Eqs. (1.18)–(1.20). The conditions on the shock wave have the usual form. The conditions on the combustion front can be expressed as

$$\frac{\rho_*}{\rho_3} = \frac{N_1}{\gamma+1}, \qquad v_3 = D_* + \frac{v_* - D_*}{\gamma+1} N_1, \qquad p_3 = \frac{p_* N}{(\gamma+1)\Omega},$$
$$N = \gamma + \Omega + \gamma\sqrt{(\Omega-1)^2 - 2(\gamma^2-1)Q(D_* - v_*)^{-2}}, \tag{6.24}$$
$$\Omega = \frac{\gamma p_*}{\rho_*}\frac{1}{(D_* - v_*)^2}, \qquad N_1 = \gamma + \Omega + \frac{1}{\gamma}(\gamma + \Omega - N),$$

where

$$p_*,\ \rho_*,\ v_*,\ p_3,\ \rho_3,\ v_3$$

are the pressure, density, and velocity ahead and behind the combustion-wave front, respectively. Relations (6.24) follow from the general conditions obtained in Sec. 2 of Chap. 1 for strong discontinuities.

We introduce now the induction-time fraction dc in accordance with the equation[7] $dc = -dt/t_{\text{ind}}$. This yields, taking Eq. (6.22) into account,

$$\frac{dc}{dt} = -k_1 p^{n_1} \rho^{l_1} e^{-E_1\rho/p}. \tag{6.25}$$

We shall assume that $c = 1$ on the shock-wave front. The instant t at which $c = 0$ is then the time of termination of the induction period for the gas particle in a variable pressure and density field. If t_* is the instant when the particle crosses the shock wave, we have $t_{\text{ind}} = t - t_*$.

To solve the problem we must thus integrate Eqs. (1.18)–(1.20) and Eq. (6.25) in the region 2, and Eqs. (1.18)–(1.20) in region 3, with appropriate initial conditions and with the aforementioned boundary conditions. The right-hand side of Eq. (6.25) is a total derivative. From the computational viewpoint it may seem more expedient to use an equation in Lagrange variables. For the initial instant of time, when the estimate (6.1) holds, we can neglect the effect of the combustion wave on the dynamics of the gas. Equation (6.25) enables us then to determine the instant t by quadrature, since p and ρ can be regarded as known functions of Lagrange coordinates and the time from the solution of the gasdynamic problem. For a strong blast this solution is given in Chap. 2. Such calculations with Lagrange coordinates (see also the more complicated reaction model below) have confirmed fully the qualitative arguments favoring the splitting of the front by an overcompressed detonation wave.

Figure 92 shows plots of the lines $c = 1$, $R_2(\tau) = r_2/r^0$ (shock wave), and $c = 0$ (combustion front), $R_1 = r_f/r^0$ for the case $\nu = 2$, $\gamma = 1.3$, $E_0 = 10^{10}$ erg/cm^2, $p_1 = 10^6$ dyn/cm^2, and $\rho_1 = 10^{-3}$ g/cm^2 for constants

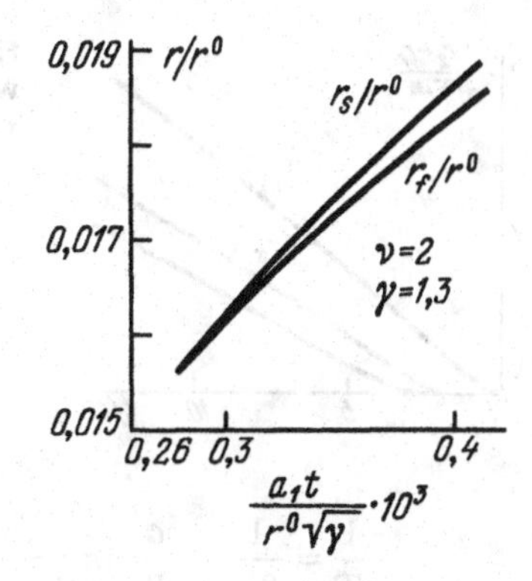

Figure 92. Plots of the motions of the shock-wave front and the ignition front for the model mixture ($r_s = r_2$).

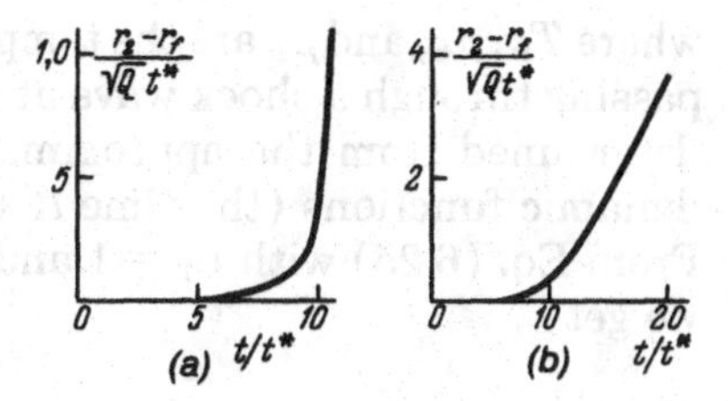

Figure 93. Growth of induction-zone width with time (model reaction, $\gamma = 1.4$, $\nu = 2$, $E_0 = 10^{10}$ erg/cm, $t^* = 10^{-7}$ s): (a) for a gas particle behind the wave front as a function of the end of the induction period; (b) in space.

$(1/5)E_1$ and k_1 close to those for a hydrogen–oxygen mixture ($Q \approx 7 \times 10^{10}$ erg/g).

It follows from the plots in Figs. 92 and 93 that the combustion front and the shock wave almost coincide at small τ ($\tau = t/t^0$, with r^0 and t^0 the dynamic length and the time, determined from E_0, ρ_1, and p_1). As τ increases, the detonation wave breaks up into a shock wave and a flame front. This circumstance takes place also in other symmetries. Calculations have shown that the instants of time t, starting with which a substantial growth of the distances between r_2 and r_f sets in, increase as E_0 increases.

It was assumed in the calculation that $n_1 = 1$ and $l_1 = 0$, i.e., the pre-exponential factor in Eq. (6.25) depended only on the pressure. R. I. Soloukhin has called to the author's attention that for a hydrogen–oxygen mixture it is better to make this factor dependent on the density ρ. Subsequent calculations by V. V. Markov have shown that the qualitative picture remains the same also for the case $n_1 = 0$ and $l_1 = 1$, and only the numerical values of $c(\tau)$ and $R_f(\tau)$ change. Typical calculations of the induction zones for $\nu = 1, 2$, and 3 are shown in Fig. 94 (stoichiometric mixture of acetylene with oxygen, $\rho_1 = 0.5 \times 10^{-3}$ g/cm^3, $p_1 = 10^6$ dyn/cm^2, $\gamma = 1.4$, $E_0 = 3 \times 10^6$ erg/cm$^{3-\nu}$, $n_1 = 0$, $l_1 = 1$). No account was taken here of the possible onset of combustion in the vicinity of the center, where the density is low.

A rough qualitative estimate of the induction time can be obtained by using the self-similar-solution approximation proposed in Ref. 29. Indeed, assume that

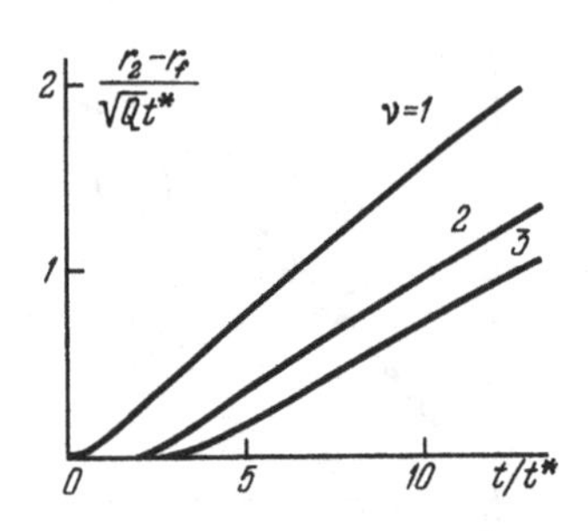

Figure 94. Change of width of induction zone behind a shock wave for an acetylene–oxygen mixture ($E_0 = 3 \times 10^6$ erg/cm$^{3-\nu}$, $\gamma = 1.4$).

$$\frac{1}{T} = \frac{1}{T_*} + \frac{a}{T_*} \ln \frac{t}{t_*}, \qquad \rho = \rho_* \left(\frac{t_*}{t}\right)^{2b}, \qquad p = p_* \left(\frac{t_*}{t}\right)^{2b\gamma}, \tag{6.26}$$

where T_*, ρ_*, and p_* are the temperature, density, and pressure in a particle passing through a shock wave at the instant t_*, while a and b are constants determined from the approximation conditions for the strong-blast gas-dynamic functions (the time t_* can be used as the Lagrange coordinate). From Eq. (6.25) with $n_1 = 1$ and $l_1 = 0$ (recognizing that $c = 1$ for $t = t_*$) we get

$$c = 1 - \frac{t_*}{t_{0\text{ind}}\,\Omega}\left[1 - \left(\frac{t_*}{t}\right)^{\Omega}\right],$$

$$t_{0\text{ind}} = \frac{1}{k_1 p_*} e^{E_1 \rho_k / p_*},$$

$$\Omega = 2b\gamma + \frac{a E_1 \rho_*}{p_*} - 1.$$

From this we have for the instant $\bar{t}$ of termination of the induction period

$$\bar{t} = t_* \left[1 - \frac{t_{0\text{ind}}}{t_*}\,\Omega\right]^{-1/\Omega}. \tag{6.27}$$

Using for the induction time the relation $t_{\text{ind}} = \bar{t} - t_*$ we get from Eq. (6.27)

$$t_{\text{ind}} = t_* \left[1 - \frac{t_{0\text{ind}}}{t_*}\,\Omega\right]^{-1/\Omega} - t_*. \tag{6.28}$$

In the spherical case, the values $a = 0.44$ and $b = 0.75$ were obtained[29] for $\gamma = 1.3$. Calculations made by T. Baiteliev at our request, using the constants for an oxygen–hydrogen mixture, show that Eq. (6.28) describes correctly the splitting of the detonation wave, although for small t it yields t_{ind} with large errors. Note that approximate methods of estimating t_{ind} were developed also in Refs. 30 and 31.

The results of the theory were compared with experiments[32,33] in an acetylene–oxygen mixture. It was assumed in Ref. 33 that

Figure 95. Comparison of the theoretical and experimental laws of front motion. Acetylene–oxygen mixture, p_1 = 50 Torr, E_0 = 0.3 J. Dashed curve—shock wave (experiment); dots—ignition front (experiment); solid lines—theoretical curves.

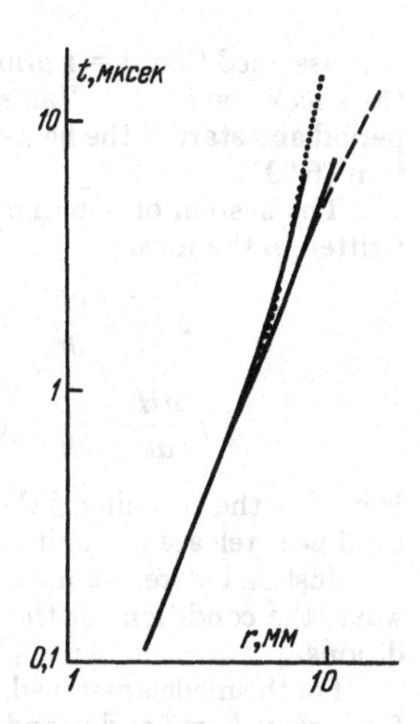

$$t_{\text{ind}}^{-1} = 2.62 \times 10^{12} \rho \exp{(-2500/RT)},$$

and the values $p_1 = 50$ Torr, $E_0 = 3 \times 10^6$ erg, and $\nu = 3$ were used. Comparisons for the trajectories of the shock-wave and ignition fronts (Fig. 95) show good agreement between theory and experiment.

The foregoing analysis pertains only to the strong shock-wave stage. The gasdynamic functions for long times differ from those for a strong blast, and the problem can be solved either by approximating the density in Lagrange variables, by an approximate integral-equations method, or by the end-to-end calculation method.[28,34] Note also that the use of the two-front model is limited by the onset of instabilities.

(3) Let us examine certain features of the solution for a model in which a chemical reaction is initiated and heat is released into the stream behind the shock-wave front after the end of the induction period (model 4). We discuss a strong blast in a quiescent gas.

The equations that describe the evolution of the chemical reactions are taken in the Arrhenius form

$$\frac{dc}{dt} = -\frac{1}{t_{\text{ind}}} = -k_1 p^{n_1} \rho^{l_1} \exp\left(-\frac{E_1 \rho}{p}\right), \tag{6.29}$$

$$\frac{d\beta}{dt} = -k_2 \beta^{m_1} p^{n_2} \rho^{l_2} \exp\left(-\frac{E_2 \rho}{p}\right) + k_3 (1-\beta)^{m_2} p^{n_3} \rho^{l_3} \exp\left(-\frac{E_3 \rho}{p}\right) \tag{6.30}$$

$$(E_1, E_2 > 0, E_3 \geq 0).$$

It is assumed that $\beta = 1$ prior to the start of reaction (6.30) while $c = 1$ on the shock-wave front. Vanishing of c means termination of the induction period and start of the heat-releasing reaction. No heat is released in reaction (6.29).

The system of equations describing one-dimensional gas motion are written in the form

$$\rho \frac{dv}{dt} + \frac{\partial p}{\partial r} = 0, \qquad \frac{\partial \rho}{\partial t} + \frac{\partial \rho v}{\partial r} + \frac{(\nu - 1)\rho v}{r} = 0,$$
$$\rho \frac{dH}{dt} - \frac{dp}{dt} = 0, \qquad H = \frac{\gamma}{\gamma - 1}\frac{p}{\rho} + \beta Q. \tag{6.31}$$

Here H is the enthalpy, β the fraction of non-reacting molecules, and Q the total heat release per unit gas mass.

Just as before, we must take into account the condition on the shock wave, the condition for the release of the energy E_0, and the initial conditions.

For the models assumed, the conditions for self-similarity of the problem for various l_i, m_i, and n_i and for the dependences of E_i and the heat release Q on the pressure and density (as well as on the coordinates and time) will in fact be the same as for model 2. When a strong blast wave propagates in a quiescent gas of constant density, the absolute width of the induction zone increases in the self-similar case in proportion to $t^{2/(\nu+2)}$. The relative thickness of this zone, however, equal to the ratio of the difference between the shock-wave and combustion-wave coordinates to the shock-wave coordinate, will remain constant. Note that for this self-similar motion the quantities E_i and Q are proportional to the pressure. Allowance for the finite rate of the chemical reaction leads here too to a substantial change of the qualitative picture of the gas flow.

If E_i and Q are constant, the point-blast problem is not self-similar. Consider a blast in a homogeneous quiescent medium. Close to the initial instant of time, the total energy released by combustion in the volume bounded by the shock wave is much lower than the blast energy, i.e.,

$$E_0 > U_4 = \sigma_\nu \int_0^{r_2} Q(1 - \beta)\rho r^{\nu-1}\, dr, \qquad \sigma_\nu = 2(\nu - 1)\pi + (\nu - 2)(\nu - 3), \tag{6.32}$$

and the combustion therefore has little influence on the gasdynamic flow.

For the initial blast stage, when inequality (6.32) is valid, one can seek a solution by linearization in a small parameter μ proportional to the ratio $Q\rho_1/E_0$. We have then for any sought function f

$$f = f_{(0)} + \mu f_{(1)} + o(\mu). \tag{6.33}$$

After substituting a function in the form (6.33) in Eqs. (6.29)–(6.31) (and linearizing them) we obtain for f_0 and f_1 a system of partial differential equations, one of which is nonlinear and contains only the principal terms of the expansions, i.e., the quantities $v_{(0)}$, $\rho_{(0)}$, and $p_{(0)}$, and the other nonlinear $(v_{(1)}, \rho_{(1)}, p_{(1)})$. The system for the principal terms also breaks up in turn into two. The solution of the gasdynamic equations consists of self-similar functions describing the flow from the strong point blast considered in Chap. 2 (the influence of the chemical reactions is negligible). The chemical reactions take place in this case against a specified flow field and are described by Eqs. (6.20)–(6.30), in which each function should be labeled by an index (0). We assume that $m_1 = m_2 = 2$, $n_1 = 1$ and $n_2 = 2$.

We introduce the dimensionless quantities

$$G = \frac{\rho_{(0)}}{\rho_1}, \qquad P = \frac{p_{(0)}}{\rho_1 Q}, \qquad V = \frac{v_{(0)}}{\sqrt{Q}}, \qquad \tau = \frac{t}{t^*}, \qquad y = \frac{r}{\sqrt{Q}t^*},$$

$$\delta_0 = t^* k_1 Q \rho_1, \qquad \delta_1 = \frac{E_1}{Q}, \qquad \delta_2 = t^* k_2 Q^2 \rho_1^2, \qquad \delta_3 = \frac{E_2}{Q},$$

where t^* is the characteristic time.

The equations for c and β take then in Euler variables the form

$$\frac{\partial c}{\partial \tau} + V\frac{\partial c}{\partial y} = -\delta_0 P \exp\left(-\delta_1 \frac{G}{P}\right), \tag{6.34}$$

$$\frac{\partial \beta}{\partial t} + V\frac{\partial \beta}{\partial y} = -\delta_2 P^2\left\{\beta^2 \exp\left(-\delta_3 \frac{G}{P}\right) - (1-\beta)^2 \exp\left[-(\delta_3 + 1)\frac{G}{P}\right]\right\}. \tag{6.35}$$

This system is a model for the kinetics of total chemical reactions of second order ($m = 2$). Since the functions V, G, and P are assumed in the considered approximation to be given, this system consists of two independent partial differential equations of first order. The initial data for them are specified at $r = r_2$, $t = t_*$ ($c = 1$, $\beta = 1$) and at $r = r_f$, $t = \bar{t}$ ($c = 0$, $\beta = 1$), with β = const in the region between r_2 and r_f.

We have thus for Eqs. (6.34) and (6.35) a Cauchy problem that can be solved by the standard method of characteristics. The ordinary differential equations that determine the variation of the functions along the characteristics were solved numerically by the Runge–Kutta method. We shall not dwell on the details of these calculations. The method developed by E. Bishimov was used in the computations with the parameters $t^* = 10^{-7}$ s, $\delta_0 = 10^3$, $\delta_1 = 5.1$, and $\delta_2 = 20$. These constants were chosen to make the kinetics equations correspond to the induction times and to the resultant reaction for mixtures of the hydrogen–oxygen type. The data obtained on the induction time and on the width of the induction zone agreed with the corresponding values for model 3 (see Figs. 92 and 93). Note that Fig. 93(a) shows the variation of the relative difference $(r_2 - r_f)/Q^{1/2}t^* = \Delta y$ along the particle trajectory as a function of the relative time $\bar{t}/t^*$ of the end of the induction period. These calculations show also that allowance for the finite chemical-reaction rates leads to a qualitatively different pattern of the flow development compared with the model of an infinitesimally thin

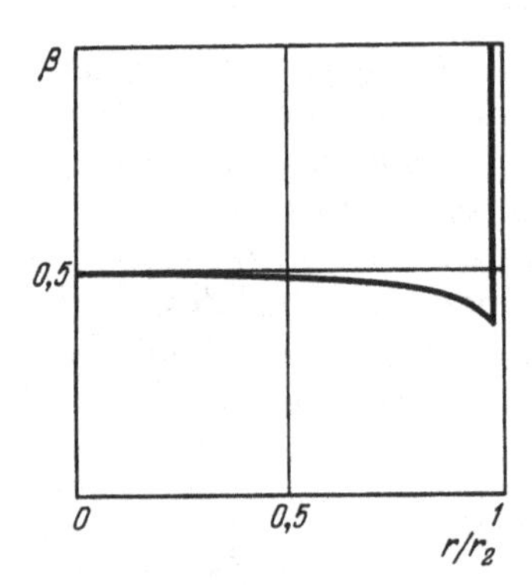

Figure 96. Change of concentration β behind a strong shock wave (model reaction, $\gamma = 1.4$, $\nu = 2$, $E_0 = 10^{10}$ erg/cm).

detonation wave. Since the temperature, pressure, and density of the gas behind the wave had large negative gradients in the region of strong stream spreading, the ignition zone breaks away from the shock front. The ignition delay time increases strongly as the wave propagates, even though the blast wave is still quite strong: $p_2 > 50p_1$. This leads to a breakup of the detonation wave into an ordinary shock and a flame front. Calculations have shown that for the assumed model of the direct and inverse reaction (6.30) the inverse reaction, which is usually neglected in the model description of the gas motion behind detonation waves, plays a large role. The distribution of the concentration β in space is shown in Fig. 96. Owing to the high temperature in the vicinity of the blast center, the mixture is not completely burned and only a fraction of the energy is released. The ratio of the released energy of model 4 to that which would be released in model 3 is approximately one-half for $\tau = 10$. By that time the ignition zone has already been separated from the blast wave and the energy released in the entire perturbed region is still negligibly small compared with the blast energy ($U'_4/E_0 \sim 0.02$).

In this approximation it is not difficult in principle to solve the problem, with account taken of a complete kinetic theory of the chemical reactions. For blast and combustion this theory has been developed in depth in the papers of the school of N. N. Semenov (see Ref. 35 for a survey of these papers) and others. Their results permit calculation of the chemical reactions accompanying a strong blast in a gas.

4. Calculation results for a model kinetic theory

We have analyzed above flows in the case of short times, when the energy released in the chemical reactions is much lower than E_0. Even in this approximation, the calculations have shown that when E_0 is increased the induction zone decreases (for equal values of the shock-wave radius), and the time for a noticeable splitting of the wave increases. In addition, if

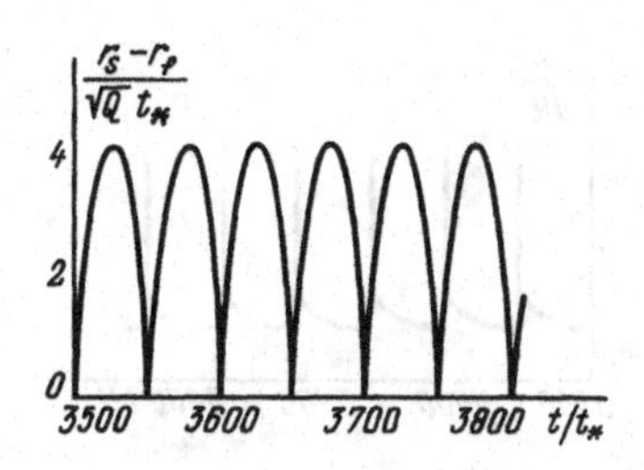

Figure 97. Variation of induction zone with time.

appreciable combustion energy is supplied, the flow structure can be greatly altered. The flow behind the wave front can become quasistationary and similar to a Chapman–Jouguet wave.

To answer these questions, a number of point-blast-problem calculations for Eqs. (6.29)–(6.31) were undertaken and reported in Refs. 1, 28, 36, and 37. Below, we describe briefly the more recent results. The calculations were performed by a modification of S. K. Godunov's numerical method.[38] This modification consisted mainly of a special numerical procedure for the calculation of the kinetic equations (6.29) and (6.30), for which this method was not used before.

We consider first the results of calculation of the one-dimensional flows for planar, cylindrical, and spherical symmetries. The employed parameters of the medium and of the combustion processes were:

$$Q = 5.25 \times 10^{10}\ \text{erg/g}, \qquad \rho_0 = 4.87 \times 10^{-4}\ \text{g/cm}^3,$$

$$p_0 = 1\ \text{atm}, \qquad E_1/R = 9850\ \text{K}, \qquad E_2/R = 2000\ \text{K},$$

$$\gamma = 1.3, \qquad t_* = 10^{-7}\ \text{s}, \qquad t_* k_1 \rho_0 = 15.15,$$

$$t_* k_2 \rho_0^2 Q^2 = 3.16, \qquad t_* k_3 \rho_0^2 Q = 3.16 \times 10^{-2},$$

$$E_1/\mu Q = 1.36, \qquad E_2/\mu Q = 0.35, \qquad m_1 = m_2 = n_2 = n_3 = 2,$$

$$l_1 = 1, \qquad n_1 = l_2 = l_3 = 0.$$

These parameters correspond to combustion of a stoichiometric hydrogen–oxygen mixture and are based on experimental data.[39–43] It was assumed that $k_3 = k_2 \times 10^{-2}$ to make the combustion more complete; μ is the molecular weight.

The calculations have shown that in the case of spherical, cylindrical, and planar symmetry there exists a certain critical blast energy E_{0*}, which is the lowest value of E_0 for which a process reminiscent of quasistationary detonation is realized. Thus, for spherical symmetry the detonation–combustion regime sets in at a blast energy $E_0 = 10^{13}$ erg. At this energy there is first produced an overcompressed detonation wave that does not split up as it attenuates into a shock wave and a combustion front. It ceases gradually to be monotonic and changes into a pulsating detonation that propagates on the average at constant velocity. The ignition front executes strong nonlinear oscillations towards and away from the shock wave (see Fig. 97). The shock-wave and flame-front velocities oscillate about certain constant

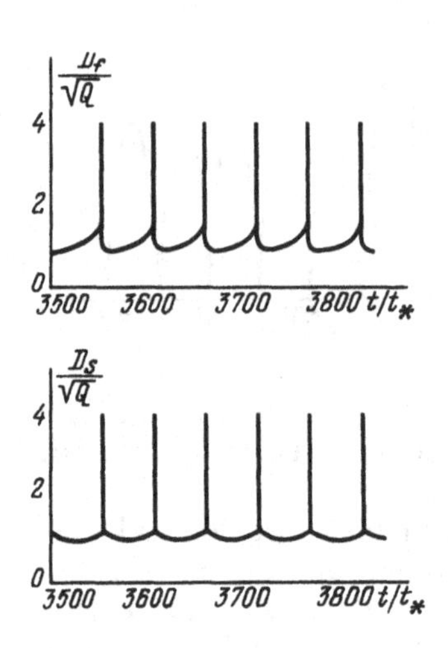

Figure 98. Oscillations of shock-wave velocity D_s and the combustion-front velocity D_f.

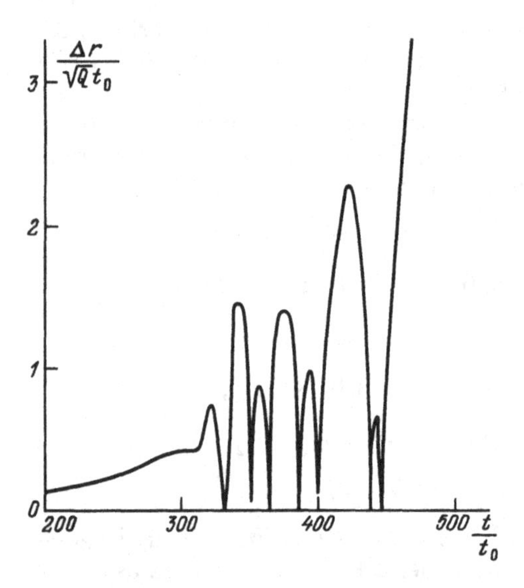

Figure 99. Growth of induction zone and the splitting of the spherical detonation wave, $E_0 = 10^{12}$ erg, $t_0 = t_*$.

values (Fig. 98). The gas parameters directly behind the shock wave also vary periodically. A certain intermediate regime sets in at a blast energy $E_0 = 10^{12}$ erg, wherein pulsating detonation combustion sets in temporarily after the flame front has moved away some distance from the shock wave. The ignition front approaches the shock wave, near the front of which the flow executes several oscillations, followed by a strong growth of the induction zone and by disruption of the detonation. The variation of the induction zone with time is shown for this case in Fig. 99. The temporary restoration of the detonation and its subsequent decay was recorded experimentally by R. I. Soloukhin and his colleagues working on detonation initiation.[42]

At a blast energy $E_0 = 10^{11}$ erg, the ignition front moves monotonically away from the shock wave and the overcompressed wave breaks up into a shock wave and a slow combustion front. It was thus established for the considered model that the critical energy E_{0*} is in the range 10^{12}–10^{13} erg in the spherical case, 10^{10}–10^{11} erg/cm in the cylindrical case, and 10^{8}–10^{9} erg/cm^2 in the planar case. Calculations show also that in the case of detonation regimes the monotonic departure of the ignition front terminates when the released chemical energy is approximately $E_0/2$, while the flame-front velocity exceeded by approximately 15% the Champan–Jouguet wave velocity calculated from the total heat release in the mixture ($D_J = 1.217 \sqrt{Q}$ in the case considered). After the pulsating detonation regime set in ($E_0 \geq E_{0*}$), the average detonation wave velocity was 10% lower than the velocity D_J, with insignificant differences between the symmetries.

The main conclusion drawn from the results is that it is possible theoretically to obtain analogs of all the experimentally observed main detonation–combustion regimes. Subsequent calculations with actual kinetic data have confirmed both the good approximation of the chemical processes by Eqs. (6.29) and (6.30) and the existence of the described types of flow. On the other hand, the question of quantitative agreement of the experimental results with the theory calls for additional research. According to data by J. H. Lee (Annual Review of Fluid Mechanics, 1984, pp. 311–316) the experimental estimate of E_{0*} at $\nu = 3$ for hydrogen–air mixtures is 5×10^{10}–5×10^{12} erg.

Calculations in accord with the one-dimensional scheme have shown that E_{0*} depends substantially on the constants of the kinetic equations (6.29) and (6.39), and also on the initiation method.[12]

Thus, when the initiation is by condensed explosives, the piston effect of the detonation product and the onset of secondary shock waves can decrease E_{0*} by several times. It has also been shown that E_{0*} decreases by more than an order when the constant k_1 is increased tenfold. The establishment of the detonation combustion regime depends strongly on the value of E_1 in the expression for t_{ind}. Thus, if E_1 is decreased by a factor of 100, then E_{0*} for the cylindrical case decreases to 3×10^6 erg/cm.

The value of the critical energy and the entire process of gas flow behind the wave front is strongly affected by instability and by the deviation of the flow from one-dimensional. A non-one-dimensional multifront detonation was observed in many experiments. It was possible to model it theoretically.

The following two-dimensional nonstationary problem was considered in Refs. 12, 36, and 37. Assume a planar point blast in a homogeneous gas in a channel with plane walls. We solve this one-dimensional problem by a two-dimensional computation scheme. If this scheme is stable, the results agree with a one-dimensional calculation. Trial computations have demonstrated the stability of the scheme in the indicated sense. Assume $E_0 = 10^9$ erg/cm^2 and a channel width $l_* = l/t_*, \sqrt{Q} = 10$. Using two-dimensional calculations, we can answer the following question: what happens if we

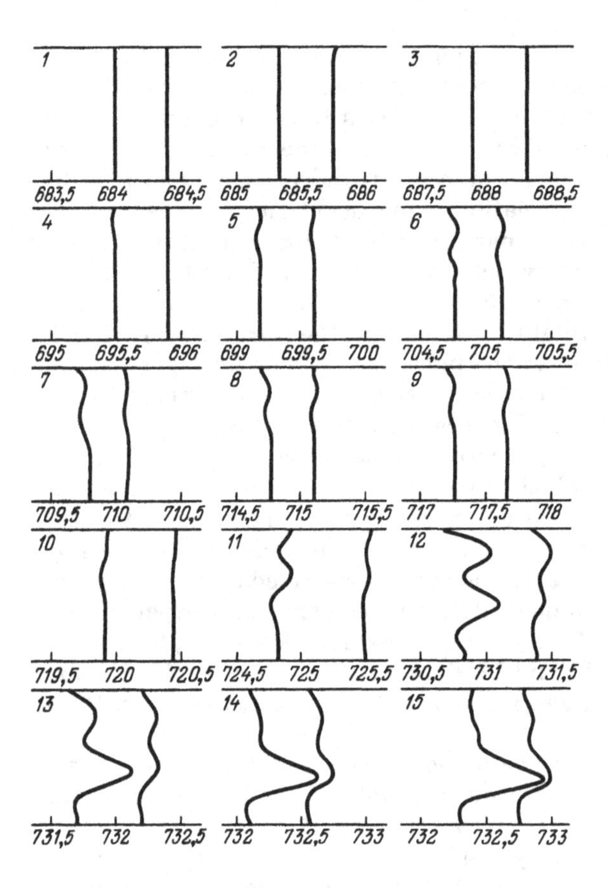

Figure 100. Shapes of shock wave and ignition front at various instants of time. Planar non-one-dimensional stream; the numbers on the horizontal axis show the values of the dimensionless coordinate $r/\sqrt{Q}t_*$.

introduce a small inhomogeneity in the initial density, i.e., if we introduce inhomogeneity of the initial gas temperature at an initial constant pressure?

It turns out that although the local change of the density is only a few percent, the gas flow rapidly loses its one-dimensional character, the wave fronts become deformed, and transverse waves are produced.

The result of a calculation by V. V. Markov is shown in Fig. 100. The perturbation in the stream was simulated here by introducing regions with higher temperature near one of the channel walls, and the initial density differed by 1% from the density of the ambient medium. The place of inhomogeneity was reached by the shock wave propagating in the channel after longitudinal oscillations of the ignition front have already developed in the stream, and the released chemical energy amounted to 75% of the blast energy. The perturbation development in this case is illustrated in Fig. 100, which shows the front shapes of the shock waves and of the ignition waves for various instants of time. It can be seen from this figure that small perturbations of the initial conditions lead to substantial distortions of the one-dimensional flow, i.e., to nonlinear effects. These effects act as instabilities and lead to a multifront picture of the combustion zone. Calculations have shown that after a certain transient period there are produced in the

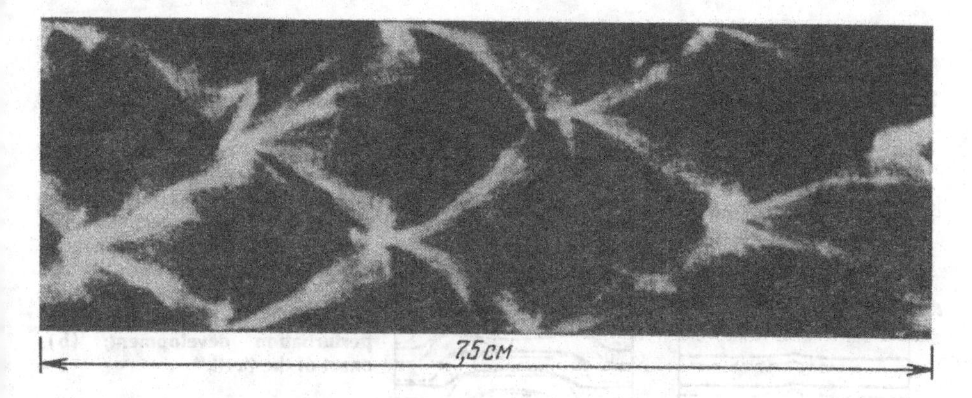

Figure 101. Cellular structure of detonation wave; experimental data of Ref. 24, acetylene–oxygen mixture; p_1 = 0.1 atm, channel width 0.3 cm.

flow non-one-dimensional pulsations having a certain cellular structure.[43] This structure is the result of interaction of longitudinal (one-dimensional) oscillations with transverse (two-dimensional) waves. The calculated trajectories[37,44] of the intersection points of the shock waves were found to agree qualitatively with the experimentally observed picture of the wave front. An example of such a picture for detonation propagation in a channel with transverse dimensions 0.3 × 2.5 cm at an initial pressure 0.1 atm in an equimolar mixture of acetylene and oxygen is shown in Fig. 101, based on experimental results (see Ref. 24). The lines correspond here to points where intense combustion takes place, i.e., to trajectories of the "triple points" in Mach collisions of shock waves.

The calculation results for a mixture of oxygen and hydrogen are shown in Figs. 102(a) and 102(b).

V. V. Markov has shown also that a similar instability and subsequent multifront structure are produced by local perturbation of the constants k_1 and E_1 of the kinetic equations; this can simulate the presence of concentration inhomogeneities of the gas-mixture components. Naturally, the detonation instability and the onset of multifront combustion can influence the critical energy E_{0*} if the latter is now taken to mean the blast energy needed to excite multifront detonation. The quantitative measure of this influence is a question that can be answered by additional calculations and experiments. Thus, the considered detonation model, taking account of the kinetics of the chemical reactions, describes the principal experimental data. Moreover, it points to the possibility of describing other gasdynamic processes far from equilibrium, such as detonation absorption waves and ionization fronts.

We note in conclusion that a similar analysis could also be carried out for thermonuclear detonation reactions. There are no differences in principle with respect to the role of the kinetic reactions.[35,45] As noted in Ref. 45, however, an important role in the structure of a thermonuclear detonation

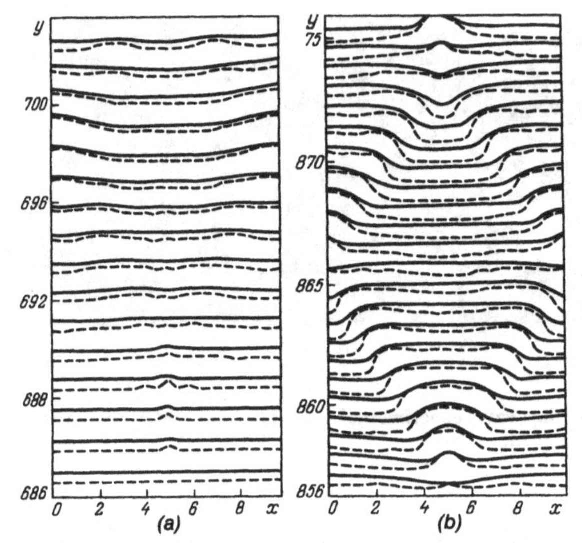

Figure 102. Results of calculation of two-dimensional nonstationary flow in a planar tube; y—relative coordinate along the tube; x—relative coordinate across the tube ($E_0 = 10^{10}$ erg/cm^2; the linear dimensions are relative to $t_*\sqrt{Q}$, and the initial density perturbation $\Delta\rho/\rho_1 = -0.1$ is located at $x = 5$ and $y = 684$. The continuous and dashed lines correspond to the shock wave and to the ignition front, respectively. (a) Initial stage of perturbation development; (b) onset of the "cell."

wave is played by radiation effects, since thermonuclear reactions occur at very high temperatures (see also Ref. 46).

5. Creation of inverted population behind a shock wave

5.1. Population inversion

Any macroscopic volume of gas consists of a multitude of molecules or atoms. We assume for the sake of argument that the gas consists of molecules. It is known from statistical mechanics and quantum mechanics that a gas molecule can be located at any particular instant of time only on some perfectly defined energy level. The number N of molecules per unit volume on some particular energy level is called the population of this level.

If the gas is in disequilibrium, a situation may arise where the number N_2 of molecules on a higher energy level is higher than the number N_1 on a lower level. This is called population inversion. A medium with an inverted population has the remarkable ability of amplifying a small optical signal of definite wavelength, i.e., it can serve as an active laser medium.

One of the important characteristics of an active medium is its gain. The gain is exactly equal to the absorption coefficient k_ω of Chap. 1, but of opposite sign.

For an optical signal propagating along the z axis we have in the stationary case the transport equation $dI/dz = kI$, where $k > 0$ is the gain. The amplification process is physically connected with the so-called laser transitions, i.e., transitions of the molecules from an upper energy state ε_2 to a lower ε_1.

These transitions are accompanied by emission of photons of frequency ν and energy $\varepsilon_2 - \varepsilon_1 = \Delta\varepsilon = h\nu$, where h is Planck's constant.

In an external electromagnetic field, a transition of a particle from an upper energy level is more intense than in the absence of a field, i.e., an electromagnetic field is capable of amplifying the probability of emission of an energy quantum by a particle. This additional field-induced radiation is called stimulated or induced.

A feature of an active medium is that the number of photons emitted in it by passage of a beam in the z direction is larger than the number of absorbed photons.

The gain is proportional to the excess inverted population (i.e., to the difference $N_2 - N_1$ of the number of excited particles on the upper ε_2 and lower ε_1 energy levels). A weak optical signal becomes amplified if $k > 0$.

In the absence of inversion, (e.g., in an equilibrium state), the signal is absorbed. Thus, there is no amplification of an optical signal in the absence of an inverted population.

In 1963, N. G. Basov and A. N. Oraevskii[47] advanced the idea that population inversion can be produced in molecular systems by rapid heating or cooling of a gas.

This was followed by the advent of gasdynamic lasers based on rapid expansion of pre-heated gas flowing out of a nozzle at supersonic velocity.

A similar situation arises when a strong shock wave propagates in a gas mixture: the gas particles are strongly heated by the passing shock wave and can then be abruptly cooled in a rarefaction wave behind the shock-wave front. The question is: is population inversion of molecules behind point-blast shock waves theoretically feasible? The answer is in the affirmative, and furthermore by two different methods. The first is population inversion by rapid heating in the shock wave, and the second is by rapid cooling behind the wave front.

5.2. Nonequilibrium state in heating by a shock wave

As already mentioned, it was demonstrated in Ref. 47 that an inverted population can be produced in polyatomic gas molecules by rapid heating. This heating can be effected in a shock wave.[48] Investigations[49] have shown that when a $CO_2 + N_2 + He$ gas mixture passes through a stationary wave, inversion is produced, under certain conditions, between the vibrational levels 04^00 and 00^01 or 20^00 and 00^01 levels of the CO_2 molecule.

Here 00^01 denotes the upper vibrational energy level, and 04^00 and 20^00 are certain lower vibrational levels of the CO_2 molecule (see Refs. 50 and 51 for details). When a gas particle passes through a shock wave, vibrational degrees of freedom of the molecules are excited and behind the front there is produced a relaxation region in which the population is inverted.

Following Ref. 52, we examine the onset of vibrational-level populations in $CO_2 + N_2 + He$ mixtures behind a shock wave produced in a nonequilibrium zone by motion of a gas expelled by a piston whose motion is described by $r_p = \lambda_p t^\delta$ with λ_p and δ constant. The gasdynamic piston problem has been thoroughly investigated, and at δ close to $2/(\nu + 2)$ the gasdynamic functions in the vicinity of the wave behave in the same manner as for a point blast.[1,2]

For a theoretical investigation of the problem we must study the solution of the gasdynamics equations jointly with the system of kinetic equations describing the vibrational relaxation of the working-gas molecules.

The equations of motion for an ideal gas can be taken in the form of Eqs. (6.31), with only a change in the dependence of the enthalpy in the gas parameters:

$$H = \frac{\gamma}{\gamma - 1}\frac{p}{\rho} + \sum_i \varepsilon_i, \tag{6.36}$$

where ε_i is the energy of the ith vibrational degree of freedom.

To analyze the vibrational relaxation, we use Anderson's approximate kinetic model,[49,53] according to which the vibrational levels of the CO_2 and N_2 are grouped into two "modes." This is based on a characteristic property of the CO_2 molecule, manifested in the fact that the deformation vibrations of frequency ω_2, and the symmetric vibrations ω_1 which enter rapidly into equilibrium with them, have a shorter relaxation time than the asymmetric vibrations ω_3. On the other hand, the asymmetric ω_3 vibrations exchange energy rapidly with the ω vibrational level of the N_2 molecule. This property of the CO_2 and N_2 molecules makes it possible to combine the ω_1 and ω_2 vibrations into a "mode" I, and ω_3 and ω into a "mode" II.

Since principal interest attaches to the lower vibrational levels, the harmonic oscillator model,

$$\frac{d\varepsilon_{\mathrm{I}}}{dt} = \frac{\varepsilon_{\mathrm{I}}(T) - \varepsilon_{\mathrm{II}}}{\tau_{\mathrm{I}}}, \tag{6.37}$$

remains valid in the calculations, and analogously for $\varepsilon_{\mathrm{II}}$, where τ_{I} and τ_{II} are the characteristic relaxation times. These times are certain effective values determined in accordance with the "parallel resistance" rule.[49,53] The vibrational temperatures T_{I} and T_{II} can be obtained by using the relations

$$\varepsilon_{\mathrm{I}} = c_{CO_2} R_{CO_2} \left\{ \frac{\theta_1}{\exp(\theta_1/T_{\mathrm{I}}) - 1} + \frac{2\theta_2}{\exp(\theta_2/T_{\mathrm{I}}) - 1} \right\},$$

$$\varepsilon_{\mathrm{II}} = c_{CO_2} R_{CO_2} \frac{\theta_3}{\exp(\theta_3/T_{\mathrm{II}}) - 1} + c_{N_2} R_{N_2} \frac{\theta}{\exp(\theta/T_{\mathrm{II}}) - 1},$$

where c_{CO_2} and R_{CO_2} are the CO_2 mass fraction and gas constant; c_{N_2} and R_{N_2} are the analogous quantities for N_2; $\theta_i = h\omega/2\pi k$ are the respective characteristic temperatures. The vibrational temperatures T_{I} and T_{II} are

Figure 103. Relative-temperature profiles for $\nu = 3$ and $\delta = 1$.

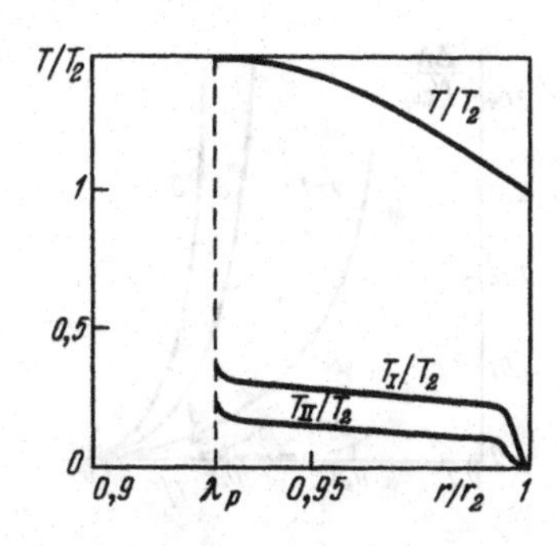

used to calculate the level populations in "modes" I and II, assuming a Boltzmann distribution over the levels inside the vibrational modes.[49,50,53]

We shall consider vibrational relaxation under the condition that the energy contained in the vibrational degrees of freedom is less than the work performed by the piston:

$$A > U = \sigma_\nu \int_{r_p}^{r_2} (\varepsilon_{\rm I} + \varepsilon_{\rm II})\, \rho r^{\nu-1}\, dr,$$

where the piston work is $A = \sigma_\nu \int_0^t p v r^{\nu-1} dt$. Here r_2 is the shock-wave coordinate and ν denotes the symmetry parameter. For instants of time when the inequality $A > U$ is valid we can seek the solution by linearization in the small parameter $\mu = U/A$. After linearizing Eqs. (6.31), (6.36), and (6.37) we find that the system for the principal terms breaks up into two. The solutions of the gasdynamic equations will be self-similar functions that describe the flow away from the piston.

The relaxation of the vibrational energies can be considered in this case against a specified flow field. The situation here is perfectly analogous to that in the initial stage of a blast in a combustible mixture [see Sec. 3 of this chapter, Eqs. (6.32)–(6.35)].

The following conditions are met on the front of a shock wave ($r = r_2$ for a strong wave):

$$\rho_2 = \frac{\gamma+1}{\gamma-1}\,\rho_1, \qquad u_2 = \frac{2}{\gamma+1}\,D, \qquad p_2 = \frac{2}{\gamma+1}\,\rho_1 D^2,$$

where D is the shock-wave velocity. The functions $\varepsilon_{\rm I}$ and $\varepsilon_{\rm II}$ retain directly behind the shock-wave front their values corresponding to the state of the mixture ahead of the shock front.

Figures 103 and 104 show results of a typical calculation for $\delta = 1$ and $\nu = 3$. The shock-wave velocity is $D = 2100$ m/s, and the temperature behind the discontinuity is $T_2 = 1200$ K, while the molar fractions of the mixture are $\alpha_{CO_2} = 0.02$, $\alpha_{N_2} = 0.38$, and $\alpha_{He} = 0.6$. Note that the expression assumed for the adiabatic exponent γ of the mixture is

$$\gamma = (1+\alpha)/\alpha, \qquad \alpha = \tfrac{3}{2}\alpha_{He} + \tfrac{5}{2}(\alpha_{CO_2} + \alpha_{N_2}).$$

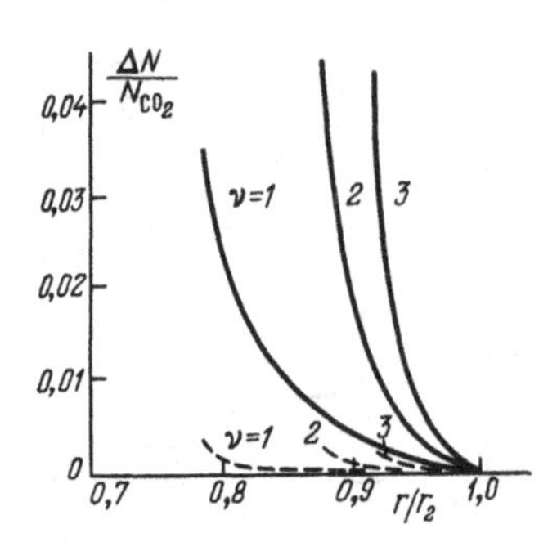

Figure 104. Inverted populations at $\delta = 1$ for various symmetries: the solid curves correspond to $\Delta N = N_{04^00} - N_{00^01}$, and the dashed ones to $N_{20^00} - N_{00^01}$.

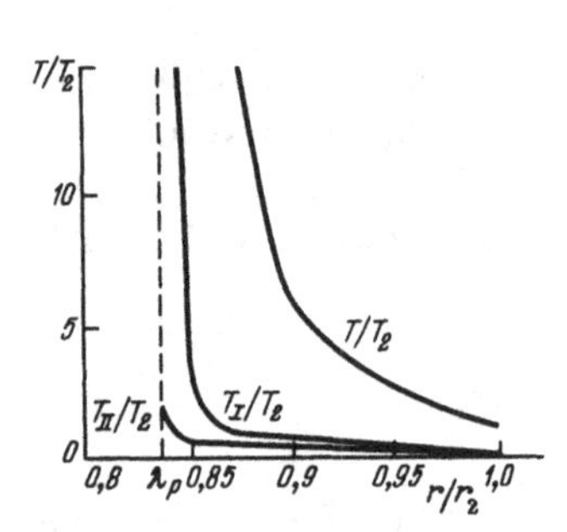

Figure 105. Temperature profiles for $\nu = 3$ and $\delta = 0.55$.

Figure 103 shows the variation of the translational temperature T and of the vibrational temperatures T_{I} and T_{II} behind the shock-wave front. A rapid increase of T_{I} is clearly seen, while T_{II} increases more slowly. This indicates a rapid population of the vibrational levels in the "mode" I. Figure 104 shows the inversions between the levels $(04^00\text{–}00^01)$ and $(20^00\text{–}00^01)$ of the CO_2 molecule.

If the ratio $\alpha_{CO_2}/\alpha_{He}$ is increased without changing the He content, a somewhat stronger rise of the temperatures T_{I} and T_{II} is observed. Decreasing the He content without changing the shock-wave velocity increases the inversion. Note also that larger values of γ correspond to smaller inversions. Calculations were also performed for damped shock waves, when $\delta < 1$. The results of one of the calculations are shown in Figs. 105 and 106, where $\delta = 0.55$. Changes of the mixture composition and the shock-wave acceleration do not affect the $(04^00\text{–}00^01)$ level inversion, while no $(20^00\text{–}00^01)$ inversion sets in.

Similar calculations were made for cylindrical and planar shock waves. The character of the variation inversion does not change in this case, but the maxima of the inverted populations increase: the maxima of the $(04^00\text{–}00^01)$ and $(20^00\text{–}10^01)$ inversions lie in the ranges $(0.3\text{–}1.0) \times 10^{-1}$ and $(0.1\text{–}1.0) \times 10^{-2}$, respectively.

Note also that the $(20^00\text{–}00^01)$ and $(04^00\text{–}00^01)$ inversions correspond to approximate laser transitions at 22 and 50 μm, respectively. Flow of $CO_2 + N_2 + He$ mixtures through a supersonic nozzle produces a $00^01\text{–}10^00$ level inversion in CO_2, with emission at 10.6 μm. The basic characteristic of a

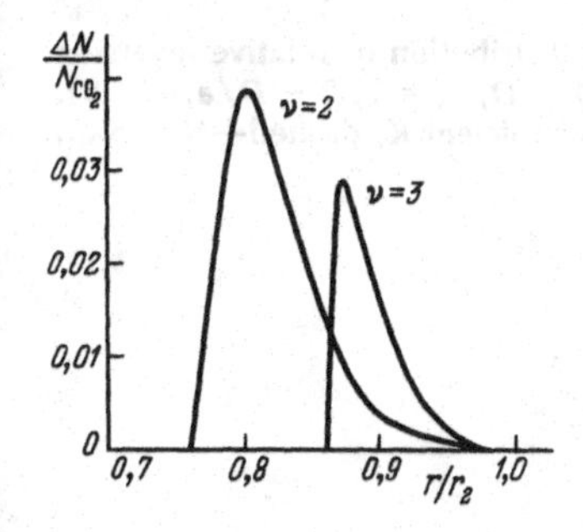

Figure 106. Typical inverted populations for $\nu = 2$ and 3 and $\delta = 0.55$; $\Delta N = N_{04^00} - N_{00^01}$.

gasdynamic laser is the gain. It can be shown for various inversions[49,51,53] that the gain for the 22 and 50 μm transitions is lower by three orders than the gain for the 10.6 μm transition.

The accuracy of the foregoing calculations is governed for the most part by the reliability of the constants employed. Unfortunately, there are as yet no reliable experimental data on the vibrational relaxation of CO_2 at temperatures $T > 1000$ K. The optimal conditions for shock-induced inversions were not found in Ref. 52. However, optimization of the $CO_2 + N_2 + He$ mixture is not likely to increase the inversion order above the foregoing results.

5.3. Production of active medium by expansion behind a shock wave

The possibility of obtaining population inversion and hence amplification of emission in the gas mixture $CO_2 + N_2 + He$ by expansion behind a cylindrical blast wave produced by a point blast was first demonstrated in Ref. 54.

The physical picture of the phenomenon is the following. Passage of a shock wave perturbs the gas particles strongly, the gas molecules are excited, and the gas is in a nonequilibrium state. The gas particles are heated and if the pressures are high enough they rapidly reach an equilibrium state. In the rarefaction region behind the shock-wave front, the translational temperature in the gas particles decreases rapidly. This can lead, just as in expansion in a supersonic nozzle, to inversion.

Here, just as in the initial stage of blast development in a combustible gas mixture, the equations of the relaxation kinetics of a gas mixture can be integrated, using an approximate approach with neglect of the effect of the disequilibrium on all the gas characteristics. This means that, just as above, nonequilibrium processes can be investigated against a known gasdynamic background. An approximate kinetic scheme was used in Ref. 54 to calculate the populations and to determine the gain of a medium.

The kinetic equations for this case can be written in the form

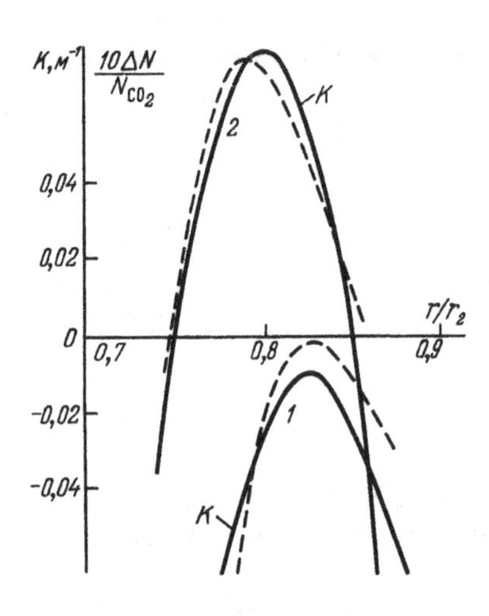

Figure 107. Spatial distribution of relative inversion and of the gains ($1 - D/a_1 = 2$; $2 - D/a_1 = 1.8$). Solid curves—the coefficient K, dashed—$N = N_{00^01} - N_{10^01}$.

$$\frac{d\varepsilon_i}{dt} = F_i(c_i, \varepsilon_i, \tau_i, p, \rho), \tag{6.38}$$

$$\varepsilon_1 = \varepsilon_2^2/(2\varepsilon_2 + 1), \qquad i = 2, \ldots, 5,$$

where F_i are known functions of their arguments (they are given in Refs. 51 and 53), c_i are the concentrations, ε_i are the "internal energies" of the ith vibrational mode of the molecule, and τ_i are the relaxation times of the molecules of species i.

We know from statistical mechanics that ε_i is connected with the so-called vibrational temperatures T_i:

$$\varepsilon_i = [\exp(\theta_i/T_i) - 1]^{-1},$$

where θ_i = const is the characteristic temperature.

It is necessary to add to these equations also the equation of state of the gas mixture, the relations for the concentrations, and other needed thermodynamic relations and constants. From the values of ε_i and T_i one can determine the number N_i of particles in the ith energy state [$N_i \sim \exp(\varepsilon_i/RT_i)$ at equilibrium]. Gasdynamic CO_2 lasers amplify and generate radiation obtained by inversion between the first level of the antisymmetric vibration mode of the triatomic CO_2 molecule (this level is designated 00^01 as above) and the first level of the symmetric vibrational mode (designated 10^00).

The solution procedure is perfectly analogous to that considered above for the kinetic calculations on a specified flow field. Without dwelling on the calculation details, we present only the results for the inverted population and for the gain. The calculations were performed in Refs. 55 and 56.

Figure 107 shows the spatial distribution of the relative inversion $\Delta N/N_{CO_2} = (N_{00^01} - N_{10^00})/N_{CO_2}$ and the gains for two cases. The first corresponds to the instant of time when the Mach number of the shock wave is $D/a_1 = 2$, and the second to $D/a_1 = 1.8$. It can be seen that in the first case the inversion has not yet set in, and in the second it is already substantial. The gasdynamic functions were taken here from cylindrical-blast theory, and the value of γ was determined in accordance with the composition of the $CO_2 + N_2 + He$ mixture and was approximately equal to 1.4. (The approximate concentrations were $c_{CO_2} = 0.1$, $C_{N_2} = 0.4$, and $c_{He} = 0.5$). The initial temperature was $T_1 = 293$ K, and the initial gas mixture density corresponded to $p_1 = 0.2$ atm.

The maximum gain obtained in these calculations was 7.6×10^{-2} m^{-1}. Naturally, the question of the feasibility of using the results in experimental devices calls for further analysis, since an idealized formulation of the problem is used here. However, the very fact that a substantial inversion is obtained behind cylindrical (and spherical) shock waves is of interest.

This problem was extended in Refs. 55–58 to include more realistic situations: a finite charge radius, allowance for the influence of the detonation products of a condensed explosive with simultaneous calculation of the gasdynamics and kinetics of the relaxation, and others. Recent calculations have pointed to a number of situations in which appreciable inversion behind the fronts of diverging waves can be obtained. Inversions on ionic laser transitions, which have so far been little investigated, are also possible here.

We conclude this chapter with the following remark. Phase transitions, disequilibrium states, and chemical reactions take place in a stream behind a passing shock wave in the case of a moving gas mixture containing minute solid particles. In the case of explosions in dust-laden air (e.g., in coal mines) we have a two-phase flow and these effects take place. Little progress has been made in the theory of blasts in two-phase media. The problem of a strong point blast in a dust-laden gas has already been investigated[1,21,59] for simple models of the motion of the medium. It was observed, in particular, that thin layers with increased dust-particle density (ρ layers) are produced behind the front of a strong wave. In the case of small particle dimensions and outside the relaxation zones, one can use the equilibrium approximation (the temperatures and velocities of the phases are equal). If the volume fraction of the solid phase is small, the problem reduces in this approximation to a blast in an ideal gas with an effective γ close to unity.[1] To estimate the flow parameters in this case we present the approximate equations[60]

$$p = 0.5p_2(1 - 0.5\lambda^b)^{-1}, \qquad \rho = 0.25\rho_2\lambda^b(1 - 0.5\lambda^b)^{-2}, \qquad v = v_2\lambda,$$

$$b = \nu/(\gamma - 1), \qquad \lambda = r/r_2,$$

where p_2, ρ_2, and v_2 are the parameters behind the strong-wave front (see Secs. 1 and 2 of Chap. 2).

Chapter 7
Point blast in an electrically conducting gas with allowance for the influence of a magnetic field

1. Formulation of problems involving magnetohydrodynamic equations

The problems considered in the preceding chapters can be extended to include an electrically conducting gas capable of interacting with applied magnetic fields. This generalization of the point-blast problem is of interest for astrophysical applications, blast-wave propagation in the ionosphere, laser-beam focusing, detonation blasts, electric discharge in a high-temperature laboratory plasma, investigation of pulsed-current sources for explosive magnetohydrodynamic (MHD) generators, etc. Let us consider the formulations of the problems and the basic equations that describe the gas motion.

Assume that instantaneous energy has been released, i.e., a blast was produced, at a point along a line or along a plane in a quiescent electrically conducting gas with an initial magnetic field $\mathbf{H}_1$. A shock wave will propagate in the gas and perturb the initial magnetic field, the initial pressure p_1, and the initial density ρ_1.

We denote by E_0 the energy released by the blast. Assuming that the resultant motion is described by a system of MHD equations, we arrive at the problem of integrating an MHD system with boundary conditions at the center of the blast, on the shock-wave front, and at infinity.

Since the magnetic field introduces anisotropy into the motion, an exact formulation of the problem, in contrast to the corresponding problem for the gasdynamic equations, does not reduce in the spherical case to one-dimensional even at constant p_1 and ρ_1, while in the planar and cylindrical cases it can be reduced to one-dimensional only for a special magnetic field configuration. The system of MHD equations for the motions of a non-viscous and non-heat-conducting ideal gas is of the form (1.69)–(1.73). The electric field $\mathbf{E}$ and the current density $\mathbf{j}$ are calculated from the equations

$$\mathbf{E} = \frac{\mathbf{j}}{\sigma} - \frac{1}{c}\mathbf{v} \times \mathbf{H}, \qquad \mathbf{j} = \frac{c}{4\pi} \operatorname{rot} \mathbf{H}, \tag{7.1}$$

where σ is the conductivity and c the speed of light.

If the magnetic Reynolds numbers are large ($R_m \gg 1$) we can neglect the dissipative terms in the system of MHD equations. For small magnetic Reynolds numbers ($R_m < 1$) we can neglect the change of the initial electromagnetic field in the course of the gas motion. For a moving gas having a finite electric conductivity we shall assume that the gas conductivity as a function of pressure and density is given by

$$\sigma = \sigma_1 \rho^n p^m, \tag{7.2}$$

where σ_1, n, and m are constants. These three additional parameters appear thus in solutions of problems with finite electric conductivity. This also complicates the investigation.

We consider the case of infinite conductivity and of motion of a gas with finite Reynolds numbers R_m.

2. One-dimensional motion of an ideally conducting gas with velocity perpendicular to the field vector

2.1. Formulation of problem

Consider one-dimensional flow of an infinitely conducting gas, a case in which only cylindrical and planar point blasts are meaningful. Assume that the vector **H** is perpendicular to the gas-particle trajectories. In the cylindrical case the field vector **H** can, in particular, be directed along the symmetry axis, along tangents to concentric circles with centers on the symmetry axis, and produces in the general case a helical field with components H_φ and H_z (the magnetic force lines are helices; see the gas-motion diagram in Fig. 108). A planar blast can be one-dimensional also if the field vector is inclined to the blast plane. We shall consider this case in the next section.

From the system of MHD equations we can obtain one-dimensional flow equations in the form

$$\rho \frac{dv}{dt} = -\frac{\partial}{\partial r}(p + h) - \frac{2(\nu - 1)}{r} h_\varphi, \qquad \frac{d\rho}{dt} = -\rho\left(\frac{\partial v}{\partial r} + \frac{(\nu - 1)v}{r}\right),$$

$$\frac{d}{dr}\left(\frac{p}{\rho\gamma}\right) = 0, \qquad \frac{d}{dt}\left(\frac{h_\varphi}{r^2\rho^2}\right) = 0, \qquad \frac{d}{dt}\left(\frac{h_z}{\rho^2}\right) = 0, \tag{7.3}$$

$$\frac{d}{dt} = \frac{\partial}{\partial t} + v\frac{\partial}{\partial r}, \qquad h = h_z + (\nu - 1)h_\varphi, \qquad h_z = H_z^2/8\pi, \qquad h_\varphi = H_\varphi/8\pi.$$

We consider shock waves of two types: magnetohydrodynamic, and those having a conductivity jump and ionizing the gas. For magnetohydrodynamic

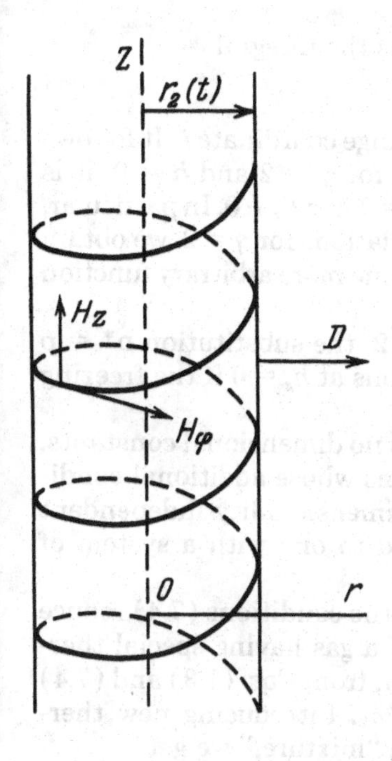

Figure 108. Propagation of a cylindrical shock wave in a helical field.

shock waves in a medium of infinite conductivity we have from Eqs. (1.100)–(1.104)

$$[\rho(v-D)]=0, \qquad [\rho v(v-D)+p+h]=0,$$
$$\left[(v-D)\left(\frac{p}{\gamma-1}+\frac{\rho v^2}{2}+h\right)+(p+h)v\right]=0, \tag{7.4}$$
$$[h_\varphi(v-D)^2]=0, \qquad [h_z(v-D)^2]=0.$$

For a shock wave with a conductivity jump we assume that **H** remains continuous on going through the discontinuity surface. This means that the magnetic viscosity in the shock layer is assumed to be larger than the other dissipative effects.[1] The discontinuities of the other quantities satisfy the gasdynamic conditions

$$[\rho(v-D)]=0, \qquad [\rho v(v-D)+p]=0,$$
$$\left[(v-D)\left(\frac{p}{\gamma-1}+\frac{1}{2}\rho v^2\right)+pv\right]=0. \tag{7.5}$$

From the system (7.3) we can obtain an energy-conservation law in the form

$$\frac{\partial}{\partial t}\left(\frac{\rho v^2}{2}+\frac{p}{\gamma-1}+\frac{H^2}{8\pi}\right)+r^{1-\nu}\frac{\partial}{\partial r}\left[r^{\nu-1}v\left(\frac{\rho v^2}{2}+\frac{\gamma p}{\gamma-1}+\frac{H^2}{4\pi}\right)\right]=0. \tag{7.6}$$

Let us note some known properties of the system (7.3).

Property 1. For $\gamma = 2$, the system (7.3) has the integral

$$p = \Phi_1(\xi)h_z, \tag{7.7}$$

where $\Phi_1(\xi)$ is an arbitrary function of the Lagrange coordinate ξ. It follows hence that if a solution of Eqs. (7.3) is known for $\gamma = 2$ and $h = 0$, it is easy to obtain a solution of this system also for $\gamma = 2$ and $h_z \neq 0$. In particular, knowing the solution of ordinary gasdynamic equations for $\gamma = 2$, we obtain for $h_\varphi = 0$ a solution of Eqs. (7.3) which contains one more arbitrary function of ξ.

In the case of MHD shock waves and $\gamma = 2$, the substitution $p^* = p + h$ reduces Eq. (7.4) to the form of the conditions at $h_z = 0$ if the freezing condition for h_z is disregarded.

Property 2. Since the system (7.3) contains no dimensional constants, the motion will be self-similar for those problems whose additional conditions include not more than one constant of dimensionality independent of E_0. In this case the problem can be reduced to one with a system of ordinary differential equations.

Property 3. At $H_\varphi = 0$ the system (7.3) and the conditions (7.4) reduce to the gasdynamic conditions for the motion of a gas having special thermodynamic properties (see also Chap. 1). In fact, from Eqs. (7.3) and (7.4) there follows the freezing integral $H_z = H_{z1}\rho/\rho_1$. Introducing new thermodynamic quantities p^*, ε^* for the gas + field "mixture," we get

$$p^*(S,\rho) = p + h_z = f(S)\rho^\gamma + \frac{1}{8\pi}\left(\frac{H_{z1}}{\rho_1}\right)^2 \rho^2, \tag{7.8}$$

$$\varepsilon^* = \frac{p}{\rho(\gamma - 1)} + \frac{H^2}{8\pi\rho} = \frac{p^*}{\rho(\gamma - 1)} + \frac{\gamma - 2}{\gamma - 1}\left(\frac{H_{z1}}{\rho_1}\right)^2 \frac{\rho}{8\pi}, \tag{7.9}$$

where S is the entropy and $f(S)$ is an arbitrary function of the entropy.

Equations (7.8) and (7.9) define thermodynamic functions of a gas whose flow can be considered in investigations of MHD problems. It follows also from property 3 that for $\gamma = 2$ the problem reduces to a gasdynamic one with equations of state for an ideal gas in accordance with property 1. These properties of the equations and boundary conditions give grounds for introducing changes in the formulation of the point-blast problem compared with the gasdynamic case. Let us proceed to specific problems for Eqs. (7.3)–(7.6).

2.2. Constant initial magnetic field parallel to the shock-wave front

Consider the point-blast problem with $\mathbf{H}_1 = \mathbf{H}_{z1}$ for the case of planar and cylindrical blasts. It follows from Eq. (7.9) that in this case the problem is not self-similar, even for a strong blast ($p_1 = 0$). If the magnetic field is weak, i.e., $H_{z1}^2 < 8\pi p_1$, the initial distribution of the field H_z can be obtained from the freezing integral $H_z = \rho H_{z1}/\rho_1$.

To calculate blast problems for long times, allowing for the influence of the field H_z and the back pressure, we must use numerical methods similar to those in gasdynamics problems.

We consider here[2] a method based on the use of the integral-equation method, with the initial equations in divergent form:

$$\frac{1}{\nu}\frac{\partial \rho}{\partial t} + \frac{\partial}{\partial \zeta}(\zeta \rho u) = 0, \qquad \frac{1}{\nu}\frac{\partial \varepsilon}{\partial t} + \frac{\partial}{\partial \zeta}[\zeta u(\varepsilon + p^*)] = 0,$$

$$\frac{1}{\nu}\frac{\partial p^{1/\gamma}}{\partial t} + \frac{\partial}{\partial \zeta}(\zeta u p^{1/\gamma}) = 0, \quad \frac{\partial}{\partial t}\frac{\rho u}{r^{\nu-2}} + \nu\frac{\partial}{\partial \zeta}(\rho v^2 + p^*) + \frac{\nu - 1}{\zeta}\rho v^2 = 0, \tag{7.10}$$

$$\frac{H_z}{H_{z\infty}} = \frac{\rho}{\rho_\infty}, \qquad \rho_\infty = \rho_1, \qquad H_{z\infty} = H_{z1},$$

$$\varepsilon = \frac{\rho v^2}{2} + \frac{p}{\gamma - 1} + \frac{H^2}{8\pi}, \qquad \zeta = r^\nu, \qquad u = \frac{v}{r}.$$

We introduce new independent variables

$$h^* = \frac{p^* q}{p_\infty}, \quad G = \frac{H_z}{H_{z\infty}}, \quad \xi = \frac{\zeta}{\zeta_n}, \quad q = \frac{\gamma p_\infty}{\rho_\infty D^2}, \quad \psi = \left(\frac{q}{p_\infty} p\right)^{1/\gamma},$$
$$\varphi = \frac{u r_n}{D}, \qquad g = \frac{\rho}{\rho_\infty}, \qquad e = \frac{\varepsilon q}{p_\infty}. \tag{7.11}$$

The subscript n labels in the present section quantities directly behind the wave front, while ∞ labels the initial parameters.

In the new variables, we can rewrite Eq. (7.10) in the form

$$\frac{\partial g}{\partial q} + \mu\left[\frac{\partial}{\partial \xi}(g\varphi\xi) - \frac{\partial}{\partial \xi}(g\xi) + g\right] = 0,$$

$$\frac{\partial e}{\partial q} + \mu\left[\frac{\partial}{\partial \xi}(e\varphi\xi) - \frac{\partial}{\partial \xi}(e\xi) + \frac{\partial}{\partial \xi}(h^*\varphi\xi) + e\right] - \frac{e}{q} = 0, \tag{7.12}$$

$$\frac{\partial \psi}{\partial q} + \mu\left[\frac{\partial}{\partial \xi}(\psi\varphi\xi) - \frac{\partial}{\partial \xi}(\psi\xi) + \psi\right] - \frac{\psi}{\gamma q} = 0,$$

$$\frac{\partial m}{\partial q} + \frac{\mu}{\nu}\left[m + \nu\frac{\partial}{\partial \xi}\left(g f^2 + \frac{h^*}{\gamma} - m\xi\right) + \frac{\nu - 1}{\xi} g f^2\right] - \frac{m}{2q} = 0, \qquad G = g.$$

From the kinematic condition $D = dr_n/dt$ there follows a connection between the dimensionless time $\tau = t/t^0$ and q:

$$\tau = R_n \frac{\mu}{\nu}\sqrt{\frac{q}{\gamma}}.$$

Here and elsewhere in this section a prime denotes a derivative with respect to q. In addition, we introduce the notation

$$\mu = \frac{\delta'}{\delta} + \frac{1}{q}, \qquad m = \varphi g \xi^{(2-\nu)/\nu}, \qquad h^* = (\psi^*)^\gamma, \qquad f = \frac{v}{D}, \qquad R = \frac{r}{r^0},$$

$$\delta = \frac{\sigma_\nu R_n^\nu}{\nu q}, \qquad r^0 = \left(\frac{E_0}{p_\infty}\right)^{1/\gamma}, \qquad t^0 = r^0\left(\frac{\rho_\infty}{p_\infty}\right)^{1/2},$$

$$\sigma_\nu = 2(\nu - 1)\pi + (\nu - 2)(\nu - 3).$$

Note that not all the equations in the systems (7.10) and (7.12) are independent. One of the equations from these systems can be regarded as a consequence of the others when account is taken of the connection between ε, v, ρ, and H. The "redundant" relations between the functions sought can be used as an additional check on the accuracy of the solution. In view of the presence of the freezing integral $G = g$, only the three functions g, ψ, and φ suffice to solve the problem.

The conditions on the shock wave take in the new variables the form

$$g_n(\varphi_n - 1) = -1, \qquad g_n = G_n, \qquad m_n(\varphi_n - 1) + \frac{h_n^*}{\gamma} = \frac{1}{\gamma} q\left(1 + \frac{\gamma}{2}\beta\right),$$

$$\beta = \frac{H_\infty^2}{4\pi p_\infty}, \qquad (\varphi_n - 1)e_n + h_n^* \varphi_n = q\left(\frac{1}{\gamma - 1} + \frac{\gamma}{2}\beta\right). \tag{7.13}$$

We must thus solve the system (7.12) with boundary conditions (7.13) and with the condition that symmetry obtain at the blast center.

Integrating the system (7.12) with respect to ξ from a certain $\xi_n = \xi_l(q)$ to $\xi_n = 1$ and taking Eqs. (7.13) into account, we obtain the system of integral equations

$$\mathcal{J}'_{1l} + g_l \xi'_l - \mu[1 + g_l \xi_l(\varphi_l - 1) - \mathcal{J}_{1l}] = 0,$$

$$\mathcal{J}'_{2l} + e_l \xi'_l - \mu\left[e_l \varphi_l \xi_l - e_l \xi_l + h_l^* \varphi_l \xi_l - \mathcal{J}_{2l} + q\left(\frac{1}{\gamma - 1} + \frac{\gamma}{2}\beta\right)\right] - \frac{1}{q}\mathcal{J}_{2l} = 0,$$

$$\mathcal{J}'_{3l} + \psi_l \xi'_l - \mu\left[\psi_l \xi_l(\varphi_l - 1) - \mathcal{J}_{3l} + \frac{\psi_n}{g_n}\right] - \frac{1}{\gamma q}\mathcal{J}_{3l} = 0, \tag{7.14}$$

$$\mathcal{J}'_{4l} + m_l \xi'_l + \mu\left[\frac{1}{\nu}\mathcal{J}_{4l} + m_l \xi_l - \frac{h_l^*}{\gamma} - g_l f_l^2 + \frac{\nu - 1}{\nu}\mathcal{J}_{5l}\right.$$

$$\left. + \frac{q}{\gamma}\left(1 + \frac{\gamma}{2}\beta\right)\right] - \frac{1}{2q}\mathcal{J}_{4l} = 0,$$

where the following relations hold:

$$e_l = \frac{\gamma}{2} g_l \varphi_l^2 \xi_l^{2/\nu} + \frac{\psi_l^\gamma}{\gamma - 1} + q \frac{\gamma\beta}{2} g_l^2, \qquad f_l = \varphi_l \xi_l^{1/\nu},$$
$$m_l = g_l \varphi_l \xi^{(2-\nu)/\nu}, \qquad h_l^* = \psi_l^\gamma + \gamma q \frac{\beta}{2} g^2. \tag{7.14a}$$

We use in Eqs. (7.14) the notation

$$\mathcal{I}_{1l} = \int_{\xi_l}^1 g\,d\xi, \qquad \mathcal{I}_{2l} = \int_{\xi_l}^1 e\,d\xi, \qquad \mathcal{I}_{3l} = \int_{\xi_l}^1 \psi\,d\xi,$$

$$\mathcal{I}_{4l} = \int_{\xi_l}^1 m\,d\xi, \qquad \mathcal{I}_{5l} = \int_{\xi_l}^1 \frac{gf^2}{\xi}\,d\xi.$$

The system (7.14) jointly with Eqs. (7.14a) can serve as the basis for the development of programs for the numerical solution of the problem. If $\beta < 1$, the calculation for arbitrary γ is similar to the gasdynamic calculation. For small q, the influence of the field is weak and the initial data can be taken to be the numerically obtained gasdynamic functions of Chap. 3. By analogy with the results of Chap. 4 we introduce in the integration range $0 \le \xi \le 1$ a central interval $\xi_0(q)$ by fixing the Lagrange coordinate in accordance with the initial data. We introduce, just as in the corresponding gasdynamic problem (see Chap. 4), into the central interval a calculation based on special asymptotic equations.

We determine the solution in the region between the boundary of the central interval $\xi = \xi_0(q)$ and the shock wave $\xi = \xi_n = 1$ by the integral-equations method.

If the integrands are approximated by interpolating Lagrange polynomials with interpolation nodes on the lines $\xi = \xi_k$ ($k = 0, 1, \ldots, n$), we can obtain an approximating system of ordinary differential equations. This approximating system will differ from the analogous system of Sec. 3 of Chap. 4 only by the coefficients of the equations obtained from the second and last integral equations in Eqs. (7.14). For the approximating system of equations there will exist a Cauchy problem whose numerical solution can yield all the functions sought. The overall accuracy can be checked against the satisfaction of the integral energy- and mass-conservation laws. By way of example, the problem was calculated for the cases $n = 4$, $\nu = 2$, $\gamma = 1.4$, $\gamma = \frac{5}{3}$, and various $\beta < 1$. Calculation for the planar case was carried out for $\gamma = 1.4$ and $\beta = 0.2$. Figure 109 shows the calculated $R_n(\tau)$ dependence for $\nu = 2$. For comparison, it shows also the corresponding dependence for $\beta = 0$. Figure 110 shows p_0/p_n as a function of τ [$p_0 = p(0, t)$] for $\gamma = 1.4$, $\nu = 2$, $\beta = 0$, and $\beta = 0.2$. The functions $P(\xi) = p/p_\infty$ for $q = 0.43$, $\beta = 0.2$, and $\beta = 0$ are shown in Fig. 111. The integral-equation method can be used also for $\beta > 1$. In this case, however, it is desirable to solve exactly

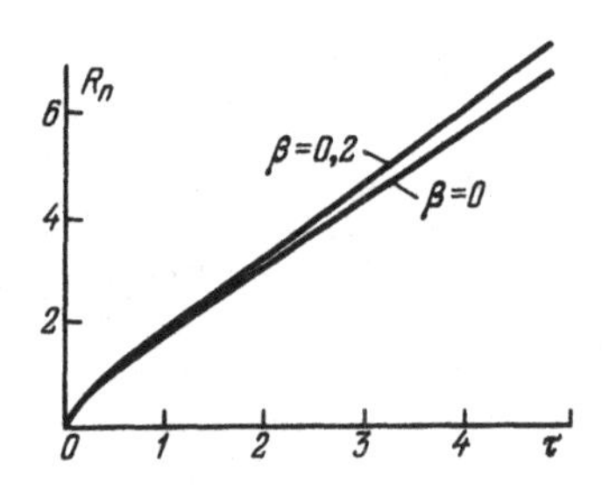

Figure 109. Motion of a cylindrical shock wave in a parallel field.

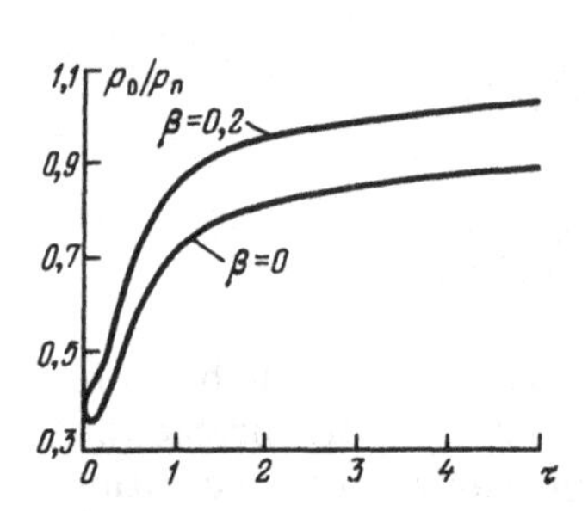

Figure 110. Change of relative pressure at the blast center ($\nu = 2$, $\gamma = 1.4$).

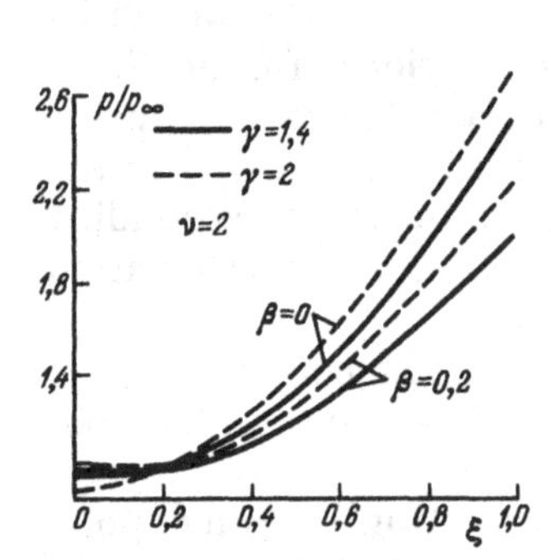

Figure 111. Pressure distribution in ξ for two values of γ.

the linearized blast problem, taking account of the back pressure and the magnetic field. An approximate solution was considered in Ref. 3.

Since the problem reduces at $\gamma = 2$ to the corresponding gasdynamic problem, to obtain the solution for arbitrary β it suffices in this case to use the results of the calculation of the gasdynamic point-blast problem by the method considered in Chap. 4.

Denoting by g_2, P_2, and f_2 ($P = p/p_\infty$) the sought functions of the gasdynamics problem, and by P, g, and f the corresponding functions of the MHD problem, we have the equalities

$$P^* = p^*/p^*_\infty = P_2, \qquad g = G = g_2, \qquad f = f_2,$$

$$P = P_2(1+\beta) - \beta g^2, \qquad h_z = \Phi(\xi)\rho^2.$$

Let us dwell particularly on the case of a strong wave.[4-6]

Neglecting p^*_∞ in the conditions on the shock waves, we obtain

$$v_n = \tfrac{2}{3}D, \qquad \rho_n = 3\rho_\infty, \qquad p_n^* = \tfrac{2}{3}\rho_\infty D^2, \qquad h_{zn} = 9h_{z\infty}.$$

We assume that $h_{z\infty}$ is constant. A solution for $p^*(r, t)$, $\rho(r, t)$, and $v(r, t)$ is known (see Chap. 2).

Let us find the arbitrary function $\Phi(\xi)$ contained in $h_z(r, t)$. From the condition on the discontinuity we get

$$\frac{h_{zn}}{\rho_n^2} = \frac{h_{z\infty}}{\rho_\infty^2}, \qquad \Phi(\xi) = \frac{h_{z\infty}}{\rho_\infty^2}.$$

It follows hence that the MHD equations and the conditions (7.4) on the shock wave can be satisfied by putting

$$h_z = \frac{h_{z\infty}}{\rho_\infty^2}\rho^2.$$

Taking into account the form of the solution of the analogous problem on ordinary gasdynamics, we get

$$r = r_n\mu^{-\delta}\left[2\left(\frac{3}{2} - \mu\right)\right]^{-\delta\nu/2}\left[4\left(\mu - \frac{3}{4}\right)\right]^{\delta/2}, \qquad v = v_n\lambda\mu,$$

$$\rho = \rho_n\left[4\left(\mu - \frac{3}{4}\right)\right]^{\nu\delta/2}\left[\left(2\left(\frac{3}{2} - \mu\right)\right)^{(\nu-2)/(\nu+2)} \exp\left(-3\nu\delta\,\frac{1-\mu}{\frac{3}{2} - \mu}\right)\right], \tag{7.15}$$

$$p = p_n\mu^{\delta\nu}\left[\left(2\left(\frac{3}{2} - \mu\right)\right)^{\delta(\nu-2)} \exp\left(-3\nu\delta\,\frac{1-\mu}{\frac{3}{2} - \mu}\right)\right] - h_z,$$

$$h_z = \frac{h_{z\infty}}{\rho_\infty^2}\rho^2, \qquad \frac{3}{4} \le \mu \le 1.$$

Here r_n is the shock-wave coordinate. The $r_n(t)$ dependence coincides with the gasdynamic one.

For a strong blast in a nonconducting medium with conductivity discontinuities $\sigma_\infty = 0$ and $\sigma_n = \infty$ we obtain from the boundary condition $h(r_n, t) = h_{z\infty}$, in analogy with the preceding case,

$$\Phi(\xi) = \frac{h_{z\infty}}{9\rho_\infty^2}.$$

The solution of the problem is given in this case by Eqs. (7.15) if it is assumed that $h_z = h_{z\infty}\rho^2/9\rho_\infty^2$. It follows from Eqs. (7.15) that, high temperatures are maintained at the center and the pressure in the vicinity of the shock-wave front is lower than the corresponding pressure for a blast without a magnetic field.

It follows also from the equations for h_z that the magnetic pressure for a blast in an infinitely conducting medium is nine times stronger than the

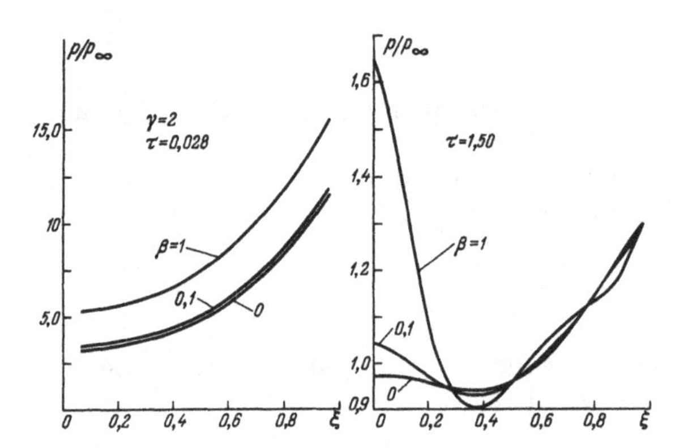

Figure 112. Pressure distribution for different parameters β at a fixed time, γ = 2.

magnetic pressure for a blast in a nonconducting medium. Note that no account was taken, for a blast with a conductivity discontinuity on a shock wave, of the change of the initial field H_∞ by the radiated electromagnetic wave.[1] This effect may turn out to be significant if the characteristic gas-particle velocities are high and $H_{z\infty}$ is not small.

To take the back pressure and the magnetic field into account at $\gamma = 2$ we use the numerical solution considered for the problem in Chap. 4. Plots of the pressure are presented in Figs. 111 and 112. They show that at $q > 0.5$ the pressure distribution differs greatly from the gasdynamic one, especially near the center. This difference is not only quantitative but also qualitative: substantial pressure gradients are present in the vicinity of the blast center.

Calculations have shown that the accuracy of the calculation of the considered problem decreases with increasing q. Just as in gasdynamic problems, to improve the accuracy it is expedient to use for the calculation of the blast-wave parameters, at large distances from the blast axis [as $q \to q_\infty = 1/(1+\beta)$], the asymptotic shock-wave damping laws.[2,7] We shall not consider this question in detail here.

2.3. Self-similar problem for an annular initial field

If the magnetic field has only the component h_φ, with $h_{\varphi 1} = Br^{-2}$ and $\rho_1 = Ar^{-\omega}$, the problem of cylindrical strong blast in an infinitely conducting gas will be self-similar (we use here again the subscript 1). In fact, since the dimensionality of the constant B coincides with that of the energy E_0', the problem contains only one additional constant A with a dimensionality independent of E_0'. This leads to self-similarity in accordance with property 2. The field-variation law $h_{\varphi 1} = Br^{-2}$ shows that the initial field is produced by flow of a constant linear current whose return path is a concentric cylindrical conductor of very large radius. This problem was first investigated for the system (7.3) in Refs. 5, 8, and 9. Let us consider the method of its solution and the results for the case $\omega = 0$.

We introduce dimensionless variables using the equations

$$v = \frac{r}{t} V(\lambda), \quad p = \rho_1 \frac{r^2}{t^2} P(\lambda), \quad \rho = \rho_1 R(\lambda), \quad h_\varphi = \rho_1 \frac{r^2}{t^2} G(\lambda),$$

$$r_2 = \left(\frac{E}{\rho_1}\right)^{1/4} t^{1/2}, \quad \lambda = \frac{r}{r_2},$$

where $E = \alpha E_0'$, α is a constant to be determined, E_0' is the finite part of the energy E_0, and $r_2(t)$ the law of the shock-wave motion. The corresponding system of four ordinary equations for self-similar motions is

$$R\lambda(V - 0.5)V' + \lambda G' + \lambda P' = R(V - V^2) - 2(P + G) - 2G,$$

$$\lambda\left[(V - \delta)\frac{R'}{R} + V'\right] = -2V,$$

$$\lambda\left[(V - \delta) + \frac{P'}{P} + \gamma V'\right] = 2 - 2(1 + \gamma)V, \tag{7.16}$$

$$\lambda\left[(V - \delta)\frac{G'}{G} + 2V'\right] = 2 - 4V.$$

The boundary conditions of the problem follow from the need to satisfy the conservation laws on the front of the shock wave (7.4) and from the symmetry condition on the blast axis $v(0, t) = 0$. The system of differential equations has three first algebraic integrals[5,8]: adiabaticity, energy, and freezing (see Sec. 6 of Chap. 1):

$$P = R^{\gamma-1}(V - 0.5)^{-1}\lambda^{-4}\varkappa_t, \qquad G = (V - \delta)^{-2}\lambda^{-4}\varkappa_2,$$

$$(P + G)V + (V - 0.5)\left(0.5RV^2 + \frac{1}{\gamma - 1}P + G\right) = \lambda^{-4}\varkappa_4.$$

The arbitrary constants in the analytic expressions for the integrals are obtained from the boundary conditions $\varkappa_2 = 0.25G_1$ and $\varkappa_4 = -0.5G_1$ on the shock-wave front. With the aid of these integrals the problem reduces to an investigation of one ordinary differential equation that can be obtained from Eqs. (7.16) and expressed in the form

$$\frac{d \ln y}{dV} = \frac{2(\gamma V - 1)[\psi_1(y, V) + (\gamma - 2)/\gamma - y(V - 0.5)^2] + (\gamma - 1)\psi_2}{(V - 0.5)\psi_2}, \tag{7.17}$$

where

$$\psi_1 = (V - 0.5)V^{-2}\left[\frac{1}{\gamma}\left(2V - \frac{1}{\gamma - 1}\right) - 0.5yV^2\right],$$

$$\psi_2 = V(V - 1)(V - 0.5)y + \frac{1}{\gamma} - 2V, \qquad y = \frac{R}{\gamma P}.$$

The main difficulty of the problem lies in the integration of Eq. (7.17) inasmuch as, given the function $y(V)$, all the functions sought are obtained with the aid of first integrals and quadratures. Labeling by the subscript 1 quantities pertaining to the unperturbed state of the gas and by 2 quantities directly behind the shock-wave front, we can obtain, using the conditions on the shock wave, the dependences of y_2 on V_2 and γ:

$$y_2^{-1} = \gamma(\gamma-1)\left[0.5V_2^2 + \left(1 + \frac{0.5}{V_2 - 0.5} + 2V_2\right)G_1\right],$$

$$G_1 = \frac{0.5V_2\left[(0.5 - V_2)\dfrac{1}{\gamma-1} - 0.5V_2\right]}{\dfrac{1}{\gamma-1}(V_2 - 0.5) + \dfrac{\gamma-2}{4(\gamma-1)}\dfrac{1}{V_2-0.5} + V_2 + 0.5}.$$

From the condition $D^2 \geq H_{\varphi 1}^2/4\pi\rho_1$ and from the equation for $y_2(v_2)$ it follows that

$$\frac{(\gamma+1)^2}{0.5\gamma(\gamma-1)} < y_2 < \infty, \qquad \frac{1}{\gamma+1} > V_2 > 0 \qquad \left(G_1 < \frac{1}{8}\right).$$

Note that in the self-similar gasdynamic point-blast problem the connection between V_2 and y_2 at fixed γ was mapped on the (V, y) plane by a point, whereas in our problem the presence of a magnetic field leads to a $y_2(V_2)$ dependence given by the relations on the shock wave. A qualitative analysis of the behavior of the integral curves was carried out for Eq. (7.17). It has been shown that the symmetry center of the flow of the blast axis corresponds to the point $y = 0$, $V = 0$. A qualitative analysis of the behavior of the solutions of Eq. (7.17) has also shown that corresponding to the solution of the problem of a strong blast in the V, y plane are integral curves that start out from points on the $y_2(V_2)$ curve and enter, tangent to the abscissa axis, the singular point (0, 0). [The presence of a set of curves entering the point (0, 0) along the abscissa axis can be proved on the basis of the Perron criterion.[10]] For small V, Eq. (7.17) has the asymptotic solution[5]

$$y = c_1 V^{-2(\gamma-2)/(\gamma-1)}[V - 0.5]^{-(3-2\gamma)/(\gamma-1)}[1 - 2\gamma V]^{-\gamma/(\gamma-1)} \times \exp\left(\frac{-2}{(\gamma-1)V}\right), \tag{7.18}$$

where c_1 is an arbitrary constant. Examples of numerically obtained integral $y(V)$ curves are shown in Fig. 113. The system of self-similar equations was numerically integrated for a wide range of G_1 and for a number of values of γ. The calculation in the vicinity of the blast center was calculated with special asymptotic equations that follow from Eq. (7.18). Calculations corresponding to $\gamma = 1.4$ and $\gamma = \frac{5}{3}$ were later performed in Ref. 11. Some calculation results for $\gamma = \frac{5}{3}$ and $\gamma = 2$ are shown in Fig. 114 for $V_2 = 0.3$. Note the following differences between the obtained solution and the corresponding gasdynamic-problem solution considered in Chap. 2:

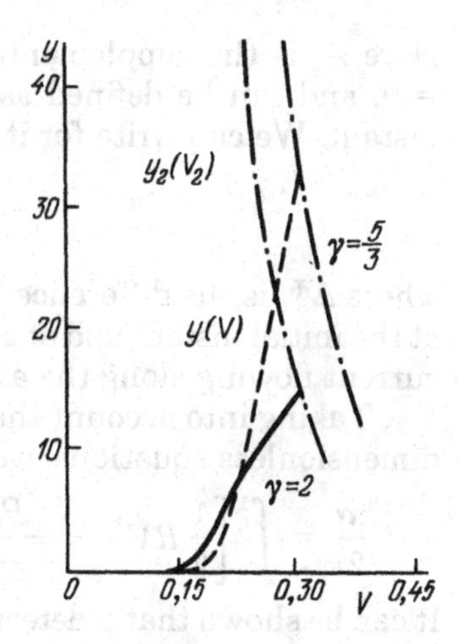

Figure 113. Examples of integral curves in the (y, V) plane.

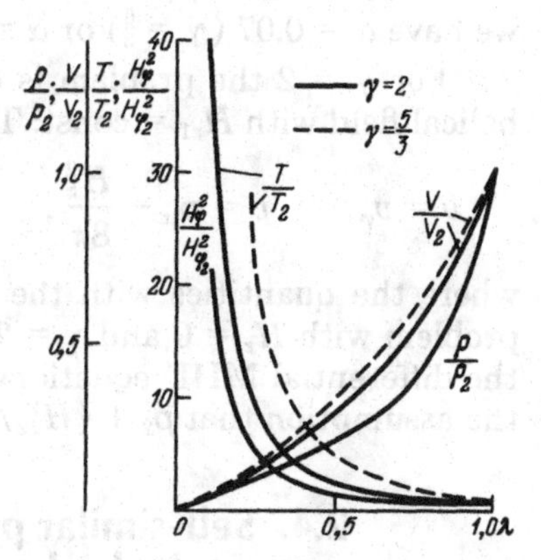

Figure 114. Distributions of the relative densities, velocities, temperatures, and magnetic field for a strong cylindrical blast in an annular field.

(1) the shock-wave intensity is lower than in the gasdynamic case;

(2) $H_\varphi \neq 0$, viz., $H_\varphi \sim 1/r$ near the center, i.e., a constant concentrated current flows along the symmetry axis;

(3) the temperature in the flow region decreases when the influence of the magnetic field is taken into account;

(4) the equation for the integral energy balance has a more complicated form.

Let us examine in greater detail the integral energy-conservation law. Analysis of the solution of the problem shows that the total energy of the moving gas remains infinite if the initial (infinite) energy is subtracted from it, and the corresponding integral diverges as $\ln|\ln \lambda|$ as $\lambda \to 0$. This indicates that to obtain the considered solution we must admit an infinite energy release at the initial instant of time on the $r = 0$ axis. A possible way of determining this energy is given in Ref. 11.

We choose the energy balance in the form

$$E_0' = 2\pi \int_0^{r_2} \left(\frac{\rho v^2}{2} + \frac{p}{\gamma - 1} + \frac{H^2}{8\pi} - \frac{H_1^2}{8\pi} \right) r dr - A_c.$$

Here A_e is the supplementary (infinite) part of the energy released at $t = 0$, and can be defined as the electromagnetic energy fed at the initial instant. We can write for it[11]

$$A_e = \frac{I_0}{c} \Delta\Phi_e,$$

where $\Delta\Phi_e$ is the difference between the field induction flux per unit length at the initial instant and at any arbitrary instant t, and I_0 is the concentrated current flowing along the axis $r = 0$.

Taking into account the dependence of $\Delta\Phi_e$ on H_φ and changing to the dimensionless equations, we get

$$\frac{\alpha}{2\pi} = \int_0^1 \left[\frac{1}{2} RV^2 + \frac{P}{\gamma - 1} + G - G_1\left(\frac{2V}{0.5 - V} + 1\right)\lambda^{-4}\right]\lambda^3\, d\lambda. \quad (7.19)$$

It can be shown that α determined from Eq. (7.19) is finite. Its actual values for various γ and G_1 (or V_2) were obtained numerically. Thus, for $V_2 = 0.3$ we have $\alpha = 0.07$ ($\gamma = \frac{5}{3}$) or $\alpha = 0.05$ ($\gamma = 2$).

For $\gamma = 2$ the problem is easily generalized to include the case of a helical field with $H_{z1} = \text{const}$. To this end it suffices to choose

$$v = v_\varphi, \qquad p = p_\varphi - \frac{H_z^2}{8\pi}, \qquad \rho = \rho_\varphi, \qquad H = H_\varphi, \qquad H_z = H_{z1}\frac{\rho}{\rho_1},$$

where the quantities with the subscript φ correspond to the considered problem with $H_z = 0$ and $\gamma = 2$. This follows from the cited properties of the differential MHD equations and from the boundary conditions, under the assumption that $p_2 + (H_{z2}^2/8\pi) \gg p_1 + (H_{z1}/8\pi)$.

2.4. Self-similar problem for the magnetohydrodynamic equations of a rarefied plasma

The problem of a strong cylindrical blast in a quiescent medium, considered in the preceding section, can be investigated also for the MHD equations of a tenuous plasma (Chew–Goldberger–Low equations) given in Sec. 3 of Chap. 1. We neglect outside the perturbed-motion region the stress tensor components $p_{\perp 1}$ and $p_{\parallel 2}$. We assume for simplicity that the initial density is constant, $B_{\varphi 1} \sim 1/r$, and $B_{z1} = 0$. From the system (1.81)–(1.85) we have for the case of cylindrical symmetry

$$\rho \frac{dv}{dt} = -\frac{\partial}{\partial r}(p_\perp + h) + \frac{p_\parallel - p_\perp}{r} - \frac{2h}{r},$$

$$\frac{d\rho}{dt} = -\rho\left(\frac{\partial v}{\partial r} + \frac{v}{r}\right), \qquad \frac{d}{dt}\frac{h}{r^2\rho^2} = 0, \qquad h = \frac{H_\varphi^2}{8\pi}, \quad (7.20)$$

$$\frac{d}{dt}\frac{p_\perp}{h^{1/2}\rho} = 0, \qquad \frac{d}{dt}\frac{p_\parallel h}{\rho^3} = 0.$$

Assuming that the flow is accompanied by the onset of a shock wave, the following condition must be met on the latter:

$$\rho_2(v_2 - D) = -\rho_1 D, \qquad \rho_2 v_2(v_2 - D) + p_{\perp 2} + h_2 = h_1,$$
$$(v_2 - D)(p_{\perp 2} + 0.5p_{\parallel 2} + \tfrac{1}{2}\rho_2 v_2^2 + h_2) + (p_{\perp 2} + h_2)v_2 = -Dh_1,$$
$$h_2(v_2 - D)^2 = h_1 D^2, \qquad p_{\perp 2} = 0.$$

These conditions follow from Eqs. (1.106). The problem considered is self-similar. In the dimensionless variables λ, V, $P_{\parallel}$, $P_{\perp}$, R, and G we obtain from Eqs. (7.20) a self-similar system of five equations:

$$R\lambda(V - 0.5)V' + \lambda P'_{\perp} + \lambda G' = R(V - V^2) - 3P_{\perp} - 4G + P_{\parallel},$$
$$\lambda\left[(V - \delta)\frac{R'}{R} + V'\right] = -2V, \qquad \lambda\left[(V - \delta)\frac{P'_{\perp}}{P_{\perp}} + 2V'\right] = 2 - 5V, \quad (7.21)$$
$$\lambda\left[(V - \delta)\frac{P'_{\perp}}{P_{\parallel}} + V\right] = 2 - 6V, \qquad \lambda\left[(V - \delta)\frac{G'}{G} + 2V'\right] = 2 - 4V.$$

The system (7.21) has four algebraic integrals (see Sec. 4 of Chap. 1): the energy integral (1.154), the freezing integral (1.149), which we express in the form

$$G = (V - 0.5)^{-2}\lambda^{-4}\varkappa_2, \qquad (7.22)$$

and the adiabatic-invariant integrals (1.151) and (1.152), which take the form

$$P_{\perp} = G^{1/2}R^{1/2}(V - 0.5)^{-1/2}\lambda^{-2}\varkappa_3, \qquad (7.23)$$
$$P_{\parallel} = G^{-1}R(V - \delta)^{-2}\lambda^{-8}\varkappa_4. \qquad (7.24)$$

The presence of these integrals permits the system (7.21) to be reduced, for arbitrary constants $\varkappa_2$, $\varkappa_3$, $\varkappa_4$, and $\varkappa_7$, to a single differential equation in the variables R and V, of the form

$$dR/dV = f(R, V, \varkappa_7, \varkappa_1, \varkappa_3, \varkappa_4).$$

It is more expedient, however, to reduce the system (7.21) to a single equation of different form, in the (V, λ) plane. It follows from the boundary condition $P_{\perp 2}$ and from the integral (7.23) that $\varkappa_3 = 0$, i.e., $P_{\perp} = 0$ everywhere in the stream. The conditions on the shock wave take in dimensionless variables the form

$$R_2(V_2 - 0.5) = -0.5, \qquad R_2(V_2 - 0.5)V_2 + G_2 = G_1,$$
$$G_2V_2 + (V_2 - 0.5)(\tfrac{1}{2}R_2V_2^2 + 0.5P_{\parallel} + G_2) = -0.5G_1, \qquad (7.25)$$
$$G_2(V_2 - 0.5)^2 = G_1(0.5)^2,$$

where $G_1 = B/\alpha\varepsilon'_0$ and ε'_0 is the specific energy.

From the conditions (7.25) we get $\varkappa_7 = -0.5G_1$ and $\varkappa_2 = (0.5)^2G_1$, while the constant $\varkappa_4$ is obtained from Eqs. (7.24) and (7.25). Using the algebraic integrals from the system (7.21), we obtain

$$\lambda\frac{dV}{d\lambda} = \frac{V^2(V - V^2 + (\varkappa_4/\varkappa_2)\lambda^{-4})}{V^2(V - 0.5) + 0.5(V^2 + (\varkappa_4/\varkappa_2)\lambda^{-4})}. \qquad (7.26)$$

Once this equation is integrated, the $R(\lambda)$ dependence is obtained from the continuity equation, while $G(\lambda)$ and $P_{\parallel}(\lambda)$ are obtained from the integrals (7.22) and (7.24). Near the symmetry center, i.e., for small λ, neglecting in the right-hand side of Eq. (7.26) terms of order $\lambda^4 V^2$ and higher, we obtain the asymptotic equation $V = c/\ln \lambda$ (c = const). Corresponding asymptotic equations can be obtained also for the other functions.

The energy parameter α in the relation $r_2(t) = (\alpha\varepsilon_0'/\rho_1)^{1/4}t^{1/2}$ is obtained from the integral energy-conservation law

$$\alpha = 2\pi \int_0^1 (0.5RV^2 + 0.5P_{\parallel} + G - G_1\lambda^{-4})\lambda^3\, d\lambda - A_0,$$

where, as before, A_0 is the (dimensionless) work performed on the field by the currents maintaining the current I_0 at the center of the blast (see the preceding section). Note that if instead of the condition that $p_{\perp}$ be continuous we assumed the condition $[p_{\parallel}] = 0$, we would reach the conclusion that $p_{\parallel} = 0$ everywhere and the problem reduces to the one for the MHD equations at $\gamma = 2$.

For nonzero initial $p_{\parallel 1}$ and $p_{\perp}$ the problem is no longer self-similar. The back pressures $p_{\parallel 1}$ and $p_{\perp 1}$ can be taken approximately into account by the linearization method discussed for the gasdynamic equations in Chap. 3.

3. Planar blast in an oblique magnetic field

Assume that energy has been instantaneously released in an unbounded ideally conducting gas along a plane, i.e., a planar blast was produced. We designate the energy released per unit area by E_0. We assume that in the initial state the gas was at rest; the pressure p_∞ and the initial density ρ_∞ are constant. The initial magnetic field H_∞ is constant in magnitude and makes a certain angle α_0 with the blast plane. Without loss of generality it can be assumed that the field has components H_x and H_y, where the y axis is parallel to the blast plane and the x axis is perpendicular to it. It follows from the symmetry of the problem that it will be one-dimensional and that all the initial characteristics of the flow will depend only on the time t and the coordinate x. The resultant planar shock wave propagates in the x direction. A nonstationary flow is produced behind the shock-wave front and will be described by using the equation for a non-viscous non-heat-conducting ideal gas. For the considered flows we have the system of equations

$$\frac{\partial \rho v_x}{\partial t} = -\frac{\partial}{\partial x}(\rho v_x^2 + p^*), \qquad \frac{\partial \rho v_y}{\partial t} = -\frac{\partial}{\partial x}\left(\rho v_y v_x - \frac{H_x}{4\pi} H_y\right),$$

$$\frac{\partial \rho}{\partial t} = -\frac{\partial}{\partial x}(\rho v_x), \qquad \frac{\partial p^{1/\gamma}}{\partial t} = -\frac{\partial}{\partial x}(v_x p^{1/\gamma}), \tag{7.27}$$

$$\frac{\partial H_y}{\partial t} = -\frac{\partial}{\partial x}(v_x H_y - H_x v_y), \qquad H_x = H_{x\infty} = \text{const.}$$

Here v_x and v_y are the gas-velocity components, and $p^* = p + H_y^2/8\pi$.

From Eqs. (7.27) and the thermodynamic relations we can obtain the energy equation in the form

$$\frac{\partial \varepsilon^*}{\partial t} = -\frac{\partial}{\partial x}\left[v_x(\varepsilon^* + p^*) - \frac{1}{4\pi} H_x v_y H_y\right], \qquad \varepsilon^* = \frac{1}{2}\rho v^2 + \frac{p}{\gamma - 1} + \frac{H_y^2}{8\pi}.$$

The conditions (1.100)–(1.104) on the shock wave will take the form

$$(v_{xn} - D) H_{yn} - H_x v_{yn} = -D H_{y\infty},$$

$$-D\rho_\infty v_{yn} - \frac{1}{4\pi} H_x H_{yn} = -\frac{1}{4\pi} H_x H_{y\infty},$$

$$(v_{xn} - D)\rho_n = -D\rho_\infty, \qquad -D\rho_\infty v_{xn} + p_n^* = p_\infty^*, \tag{7.28}$$

$$(v_{xn} - D)\varepsilon_n^* + v_{xn} p^* - \frac{H_x}{4\pi} H_{yn} v_{yn} = -D\left(\frac{p_\infty}{\gamma - 1} + \frac{H_{y\infty}^2}{8\pi}\right), \qquad D = \frac{dx_n}{dt}.$$

The subscript n labels here the parameters of the shock-wave front, and D is the shock-wave velocity. In addition to the conditions (7.28) we have the condition at the symmetry center of the flow:

$$v_x(0, t) = 0. \tag{7.29}$$

To solve the problem we must find the functions v_x, v_y, p, ρ, H_y, and $x_n(t)$ that satisfy the system (7.27) and the boundary conditions (7.28) and (7.29).

For $H_x = 0$ the magnetic field is parallel to the wave front. A problem corresponding to this case was considered in Sec. 2. The solution of the problem for $H_x \neq 0$ is made complicated by the presence of a new unknown function v_y and by the larger number of differential equations. A complete solution requires the use of numerical and special approximate methods.

Note certain general features of the solution of the problem.[12] The system (7.27) and the conditions (7.28) are invariant to the transformation $v_y' = -v_y$, $v_x' = -v_x$, and $x' = -x$. This means, in particular, that the function v_y is odd in x, and if the solution is continuous in the vicinity of $x = 0$ we have $v_y(0, t) = 0$. Thus, at $x = 0$ the gas velocity is zero.

We consider next the law of the shock-wave damping at large distances from the blast point. In the course of its propagation from the blast center the shock wave will be damped, and the overall region of perturbed motion will increase. A gas flow will set in between the blast center and the leading front of the wave, accompanied by a complicated interaction between the magnetohydrodynamic compression and rarefaction waves. At large distances the shock wave becomes weak. Since the entropy changes little in a passage through a weak shock wave, the flow can be regarded as isentropic in a certain vicinity of the wave front. By analogy with gasdynamics, we can assume that in a certain vicinity behind the magnetohydrodynamic shock-wave front the flow is an isentropic Riemann wave.[13] Then, using for example the methods considered in Refs. 7 and 14 of obtaining asymptotic laws for shock-wave damping, we can find the asymptotic variation of v_{xn}

as a function of x_n. In analogy with hydrodynamics or with the case $H_x = 0$ we have[7]

$$v_{xn} = Cx_n^{-1/2}, \tag{7.30}$$

where C is a certain constant.

Using relations (7.28), we can derive the corresponding asymptotic equations also for the remaining functions sought.

Let us examine several methods of solving the problem. For a numerical solution we can use the method of characteristics. This method would yield a sufficiently accurate description of the system of magnetohydrodynamic waves in the flow region, at least up to the instant when secondary shock waves are produced in the stream. This method, however, is quite laborious and does not lend itself to computer calculations. We shall indicate below simpler calculation methods.

For $H_{x\infty} < H_{y\infty}$, i.e., when the angle α_0 is small, we find, neglecting in Eqs. (7.27) and (7.28) terms of order of $v_y H_x$ compared with $H_y v_x$, that the problem of determining v_x, p, and ρ becomes separable from the problem of determining v_y. If the planar-blast problem is solved for the case $H_x = 0$ and $\alpha_0 = 0$, we can find v_y from this solution by integrating the second equation of the system (7.27). In Sec. 2 we have calculated H_y, v_x, p, and ρ by the method of integral equations. If the calculation results obtained by this method are used, it is expedient to apply this solution also to the second equation of the system (7.27) so as to determine v_y in the case of small α_0.

To calculate the problem for arbitrary α_0, it is expedient to use either the method of straight lines, or an integral-equations method similar to the methods developed above. By the method of straight lines we mean here the integral-equation variant in which the integrand containing the functions sought is approximated by linear functions between the points at which the spatial coordinate is broken up into strips, i.e., the integrals over space are in fact calculated by the trapezoid rule. The advantage of the use of the straight-line method is that the system of approximating differential equations has a simple structure.

The important characteristics of the flow are v_n, p_n, ρ_n, H_{yn}, and the motion $x_n(t)$ of the shock wave. Let us consider an approximate method of determining these quantities independently of the calculation of the entire flow field.

Since the finite magnetic field has little influence on the gas flow at the initial instants of time, the $v_n(x_n)$ dependence for small x_n is close to gasdynamic and is of the form

$$V_{xn} = \frac{4}{3}\frac{1}{\gamma+1}\frac{1}{\sqrt{\gamma\alpha R_n}}, \tag{7.31}$$

where $V_{xn} = v_{xn}/a_\infty$, $R_n = x_n p_\infty / E_0$, $a_\infty^2 = \gamma P_\infty / \rho_\infty$, and $\alpha = \alpha(\gamma)$.

The relation (7.31) is similar to Eq. (7.30), and it is natural to assume (see Chap. 4) that $V_{xn}(R_n)$ can be approximated by Eq. (7.31) for an ar-

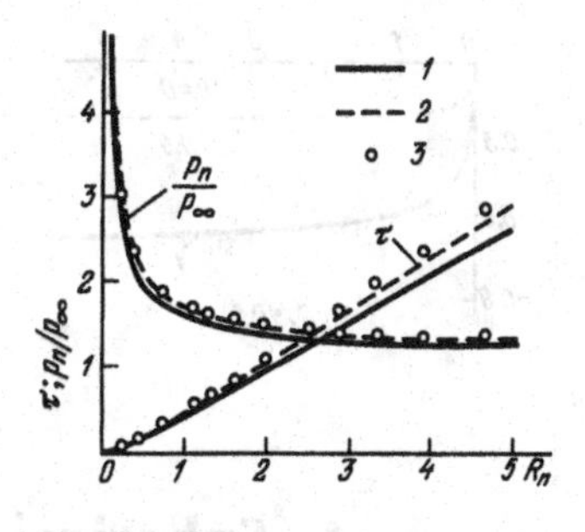

Figure 115. Dimensionless time τ and pressure p_n/p on a planar shock-wave front versus the front coordinate R_n. Curves 1–3 correspond to β = 1, 0.5, and 0, respectively.

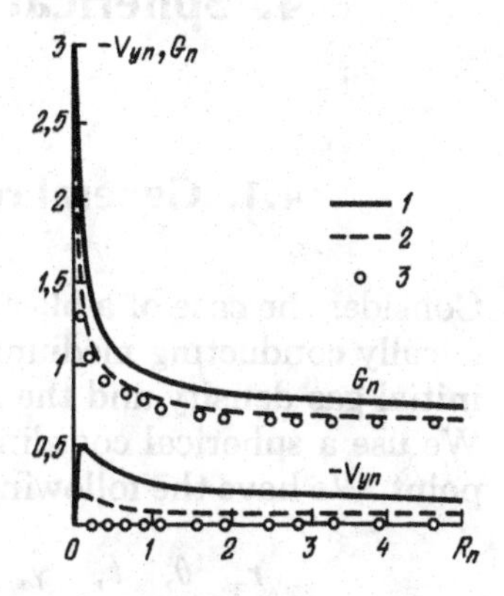

Figure 116. Dependences of V_{yn} and G_n on R_n. Curves 1, 2, and 3 correspond to β = 1, 0.5, and 0, respectively.

bitrary range of x_n. Using Eqs. (7.28) and (7.31) we can determine all the sought dimensionless quantities on the wave front as functions of R_n or of the dimensionless time $\tau = E_0^{-1} \rho_\infty^{-1/2} p_\infty^{3/2} t$. Note that an investigation of this method as applied to the problem with $H_x = 0$ (and to the gasdynamics problem, see Chap. 4) has shown its accuracy to be high enough.

The results of calculations are shown in Figs. 115–117. The problem considered has, besides γ, two dimensionless independent parameters, viz., $\beta^2 = H_x^2/4\pi\gamma p$ and $G_\infty^2 = H_{y\infty}^2/4\pi\gamma p_\infty$.

Figure 115 shows plots of $\tau(R_n)$ and $p_n/p_\infty(R_n)$ for $\gamma = 1.4$, $G_\infty = 0.5$, and various β. Figure 116 shows plots of V_{yn} vs R_n and G_n vs R_n for the same values of G_∞ and β, with $V_y = v_y/a_\infty$ and $G^2 = H_y^2/4\pi\gamma p_\infty$.

Figure 117 shows the velocity-component ratio v_{yn}/v_{xn}, as a function of R_n.

It follows from the foregoing results that the component v_{ny} behind the shock wave is not monotonic but increases in absolute value from zero to a certain maximum at the start of the blast and drops to zero in a later stage. This means that the gas particles close to the blast center acquire velocities almost parallel to x. On the other hand, the initial velocities of particles located outside a certain vicinity of the center will be directed at a certain angle to the x axis. The gasdynamic parameters p_n and ρ_n and the velocity component v_{xn} will behave qualitatively in the same way as the corresponding quantities in a blast without allowance for the influence of the magnetic field.

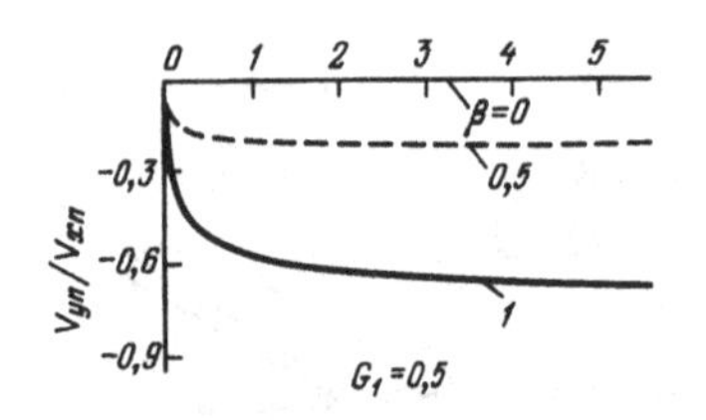

Figure 117. Ratio of tangential and normal velocity components behind a shock wave vs R_n ($G_1 = G_\infty = 0.5$).

4. Spherical point blast in a constant field

4.1. General remarks

Consider the case of a blast of a spherical point charge in a quiescent electrically conducting medium. We assume that $H_1 = |\mathbf{H}_1| = \text{const}$. Let the initial gas density and the initial pressure be constant for the time being. We use a spherical coordinate frame (r, θ, φ) with its center at the blast point. We have the following system of defining parameters of the problem:

$$r, \quad \theta, \quad t, \quad \gamma, \quad p_1, \quad \rho_1, \quad H_1, \quad \sigma_1, \quad n, \quad m, \quad E_0. \qquad (7.32)$$

Since the gas flow will have axial symmetry, the solution is independent of the other coordinate angle φ. The following dimensionless combinations can be made up from the defining parameters (7.32):

$$R = \frac{r}{r^0}, \quad \theta, \quad \tau = \frac{t}{t^0}, \quad \beta = \frac{H_1^2}{8\pi p_1}, \quad \tilde{\sigma}_1 = \rho_1^n p_1^m \sigma_1 / t^0, \quad \gamma, \quad n, \quad m$$
$$\left[r^0 = \left(\frac{p_1}{E_0}\right)^{1/3}, \qquad t^0 = r^0 \left(\frac{\rho_1}{p_1}\right)^{1/2}, \qquad ct^0 \gg r^0 \right]. \qquad (7.33)$$

It follows from Eqs. (7.33) that even in this simplified formulation of the blast problem we have four additional dimensionless parameters β, $\tilde{\sigma}_1$, n, and m compared with the gasdynamics problem for a homogeneous medium. This complicates significantly the general solution of the problem, since a solution obtained for a specific system of parameters γ, $\tilde{\sigma}_1$, β, n, and m cannot be recalculated to other values of these parameters.

For infinite conductivity, $\sigma = \infty$, an additional dimensionless parameter remains compared with the analogous gasdynamics problem. Furthermore, the flow remains axisymmetric and the solution depends on the angle θ.

Let us examine the problem in greater detail for the case of infinite conductivity of the gas.[15,16] We choose the system of MHD equations in a spherical frame in the form

$$\frac{\partial}{\partial t}(r^2\rho v_r \sin\theta) = r\sin\theta(\pi_{\theta\theta}+\pi_{\varphi\varphi}) - \frac{\partial}{\partial r}(r^2\sin\theta\pi_{rr}) - \frac{\partial}{\partial\theta}(r\sin\theta\pi_{r\theta}),$$

$$\pi_{rr} = \rho v_r^2 + p^* - \frac{H_r^2}{4\pi}, \qquad \pi_{\theta\theta} = \rho v_\theta^2 + p^* - \frac{H_\theta^2}{4\pi},$$

$$\pi_{\varphi\varphi} = p^* = p + \frac{H^2}{8\pi}, \qquad \pi_{r\theta} = \rho v_r v_\theta - \frac{1}{4\pi}H_r H_\theta = \pi_{\theta r},$$

$$\sin\theta\,\frac{\partial r^3\rho v_\theta}{\partial t} = -\pi_{\varphi\varphi}r^2\cos\theta + \frac{\partial}{\partial\theta}(r^2\sin\theta\pi_{\theta\theta}) + \frac{\partial}{\partial r}(r^3\sin\theta\pi_{\theta r}),$$

$$\frac{\partial r^2\rho\sin\theta}{\partial t} = -\frac{\partial}{\partial r}(r^2\rho v_r\sin\theta) - \frac{\partial}{\partial\theta}(r\rho v_\theta\sin\theta), \tag{7.34}$$

$$\frac{\partial}{\partial t}(r\sin\theta H_r) = \frac{\partial}{\partial\theta}[\sin\theta(v_r H_\theta - v_\theta H_r)],$$

$$\frac{\partial(rH_\theta)}{\partial t} = -\frac{\partial}{\partial r}[r(v_r H_\theta - v_\theta H_r)],$$

$$\frac{\partial}{\partial t}\left[r^2\sin\theta\left(\frac{\rho v^2}{2} + \frac{p}{\gamma-1} + \frac{H^2}{8\pi}\right)\right] = -\frac{\partial}{\partial r}(r^2\sin\theta S_r) - \frac{\partial}{\partial\theta}(r\sin\theta S_\theta),$$

$$\frac{\partial}{\partial r}(r^2\sin\theta H_r) + \frac{\partial}{\partial\theta}(rH_\theta\sin\theta) = 0,$$

where r is the radius, θ the latitude measured from the direction of the vector $\mathbf{H}_1$ (see Fig. 118), S_r and S_θ the energy-flux vector components, H_r and H_θ the radial and transverse magnetic field components, and v_r and v_θ the velocity components.

The conditions on the shock wave are

$$[\rho(v_N - D)] = 0, \qquad \left[\rho(v_N - D)v_\tau - \frac{1}{4\pi}H_N H_\tau\right] = 0,$$

$$\left[p + \rho(v_N - D)^2 + \frac{1}{8\pi}(H_\tau^2 - H_N^2)\right] = 0, \tag{7.35}$$

$$[S] = \left[\rho(v_N - D)\left(\frac{v^2}{2} + \frac{\gamma p}{\rho(\gamma-1)}\right) + \frac{1}{4\pi}\{(v_N - D)H^2 - H_N(\mathbf{vH})\}\right] = 0,$$

$$[H_N\mathbf{v}_\tau - \mathbf{H}_\tau(v_N - D)] = 0, \qquad [H_N] = 0.$$

Here $\mathbf{H}_\tau$ and H_N are the tangential and normal components of the magnetic field, square brackets denote differences of quantities on opposite sides of the discontinuity surface, and D is the shock-wave velocity. Quantities behind the wave front will be labeled by a subscript 2.

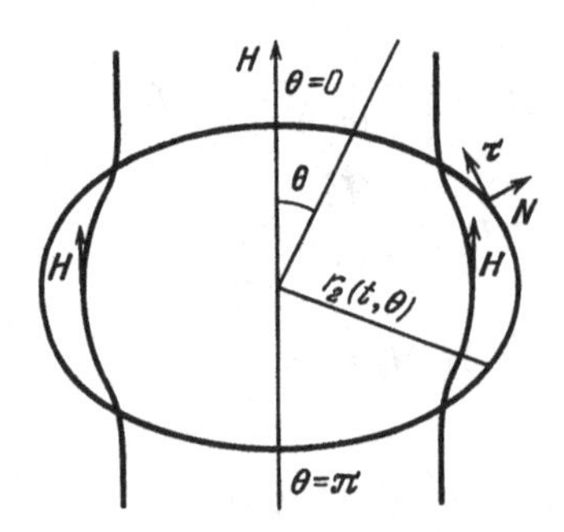

Figure 118. The deformation scheme of the magnetic field and the front shock wave.

If $r_2(\theta, t)$ is the shock-wave radius-vector length, we have

$$D = W\frac{\partial r_2}{\partial t}, \qquad W = \left[1 + \left(\frac{1}{r_2}\frac{\partial r_2}{\partial \theta}\right)^2\right]^{-1/2}.$$

In our case H_{N2} and $H_{\tau 2}$ are connected with H_{r2} and $H_{\theta 2}$ by the relations

$$H_{N2} = H_{\theta 2}\sin\psi + H_{r2}\cos\psi,$$

$$H_{\tau 2} = H_{\theta 2}\cos\psi - H_{r2}\sin\psi,$$

$$\sin\psi = -\frac{W}{r_2}\frac{\partial r_2}{\partial \theta}, \qquad \cos\psi = W.$$

Similar relations hold also for the components of the velocity $\mathbf{v}$.

From the requirement that the motion be symmetric we have the following condition for v_θ:

$$v_\theta(r, 0, t) = v_\theta(r, \pi, t) = 0. \tag{7.36}$$

From the system (7.34) and from the initial and boundary conditions there follow integral energy- and mass-conservation laws:

$$E_0 = 2\pi\int_0^\pi\int_0^{r_2}\left(\frac{\rho v^2}{2} + \frac{p}{\gamma - 1} + \frac{H^2}{8\pi}\right)r^2\sin\theta\,dr\,d\theta$$
$$- \frac{2\pi}{3}\left(\frac{p_1}{\gamma - 1} + \frac{H_1^2}{8\pi}\right)\int_0^\pi r_2^3(\theta, t)\sin\theta\,d\theta, \tag{7.37}$$

$$\int_0^\pi\int_0^{r_2}\rho r^2\sin\theta\,dr\,d\theta = \frac{1}{3}\rho_1\int_0^\pi r_2^3(\theta, t)\sin\theta\,d\theta. \tag{7.38}$$

The solution of the problem reduces to integration of the system (7.34) with conditions (7.35) and (7.36). Its solution for a large time range can be obtained by numerical methods and by using the asymptotic laws governing the shock-wave damping at large distances.

The initial data for the solution of the problem can be specified by using the solution of the problem for small t, which we shall consider in detail below.

4.2. Determination of magnetic field during the initial flow stage

For short enough t, the shock wave is quite strong. In moderate magnetic fields we can disregard, for the initial stage of the blast, the influence of the field on the gas motion, and determine only the magnetic field perturbation by a shock-wave field. For a rough estimate of the limits of the validity of this approach, we use the fact that a magnetic field has little influence on the motion if the total magnetic energy contained in the perturbed gas is low compared with the blast energy.

If r_* is the admissible shock-wave radius, the following condition must be met if the energies are to be equal:

$$E_0 = 4\pi \int_0^{r_*} \frac{H_1^2}{8\pi} r^2 \, dr \tag{7.39}$$

or

$$E_0 = \frac{H_1^2}{6} r_*^3. \tag{7.40}$$

From this we have the condition for a weak influence of the field:

$$r < r_*, \quad r_* = \left(\frac{6E_0}{H_1^2}\right)^{1/3}. \tag{7.41}$$

From similar reasoning we have an estimate of the condition of neglecting the back pressure p_1:

$$r < r_{**} = \left(\frac{3(\gamma - 1)}{4\pi}\right)^{1/3} r^0. \tag{7.42}$$

If $r > r_*$ and $r > r_{**}$ we can disregard the back pressure and the effect of the field on the motion. We assume both inequalities (7.41) and (7.42) to be satisfied.

For a strong wave we can neglect in the conditions for the shock wave the quantity p_1 compared with $\rho_2 v_2^2$, and neglect H_1 in the conditions that follow from the momentum and energy conservation law. For v_2, p_2, and ρ_2 we have then the usual hydrodynamic conditions, and the field discontinuity is determined from the relation $[H_r(v_2 - D)] = 0$. Since in our case v is independent of θ and H_1 and is known from the solution of the gasdynamic problem, the induction equations of the system (7.34) are integrable and the dependences of the magnetic field components on r, θ, and t are given by

$$H_r = H_1 \cos\theta \left[\frac{\gamma+1}{\gamma-1}\left(1 - \frac{5}{2}V\right)\right]^{\beta_3} \left[\frac{\gamma+1}{\gamma-1}\left(\frac{5}{2}\gamma V - 1\right)\right]^{\beta_4}$$

$$\times \left[\frac{5(\gamma+1)}{7-\gamma}\left(1 - \frac{3\gamma-1}{2}V\right)\right]^{\beta_5},$$

$$H_\theta = -\frac{\gamma+1}{\gamma-1} H_1 \sin\theta \left[\frac{\gamma+1}{\gamma-1}\left(1-\frac{5}{2}V\right)\right]^{\beta_3-1}\left[\frac{\gamma+1}{\gamma-1}\left(\frac{5}{2}\gamma V-1\right)\right]^{\beta_4}$$
$$\times\left[\frac{5(\gamma+1)}{7-\gamma}\left(1-\frac{3\gamma-1}{2}V\right)\right]^{\beta_5},$$

$$\lambda = \frac{r}{r_{20}} \tag{7.43}$$
$$= \left[\frac{5(\gamma+1)}{4}V\right]^{2/5}\left[\frac{\gamma+1}{\gamma-1}\left(\frac{5}{2}\gamma V-1\right)\right]^{\beta_2}\left[\frac{5(\gamma+1)}{7-\gamma}\left(1+\frac{3\gamma-1}{2}V\right)\right]^{\beta_1},$$

$$\beta_1 = \beta_2 + \frac{\gamma+1}{3\gamma-1} - \frac{2}{5}, \qquad \beta_2 = \frac{\gamma-1}{2\gamma-1}, \qquad \beta_3 = \frac{2\gamma}{3(\gamma-2)},$$

$$\beta_4 = \frac{2}{2\gamma+1}, \qquad \beta_5 = -\frac{2}{3}\frac{5}{(\gamma-2)}\beta_1,$$

where

$$r_{20} = \left(\frac{E_0}{\alpha\rho_1}\right)^{1/5} t^{2/5}, \qquad \alpha = \text{const}, \qquad \frac{2}{5\gamma} \le V \le \frac{4}{5(\gamma+1)}. \tag{7.44}$$

Equations (7.43) can be used directly to calculate the distributions of H and H_θ if $\gamma \neq 2$. The case $\gamma = 2$ is special (see Chap. 2) and must be treated separately.

Since, by assumption, the field does not affect the motion, the solution for the sought functions v_r, ρ, and p coincides with the gasdynamic solution (2.5)–(2.10).

Solution (7.43) was first obtained in Ref. 15. This solution was generalized by E. V. Ryazanov[16] to include a blast with variable initial density $\rho_1 = Ar^{-\omega}$. He found, in particular, that at $\omega = (7-\gamma)/(\gamma+1)$ the solution for the field takes the simple form

$$H_r = H_1\lambda^{4/(\gamma-1)}\cos\theta,$$
$$H_\theta = -\frac{\gamma+1}{\gamma-1}H_1\lambda^{4/(\gamma-1)}\sin\theta. \tag{7.45}$$

It follows from Eqs. (7.41) and (7.43) that the components j_r and j_θ are zero, and the component j_φ is proportional to $\sin\theta$. The dependence of H/H_1 on $\lambda = r/r_{20}$ is shown for $\theta = 0°$, 45°, 90°, and 180° at $\gamma = 1.4$ in Fig. 119.

Using Eqs. (7.43) and the differential equations for the force lines, we can determine the deformation of the force lines by the gas stream. For the case considered the differential equation of the force lines is

$$\frac{dr}{H_r} = \frac{r\,d\theta}{H_\theta}.$$

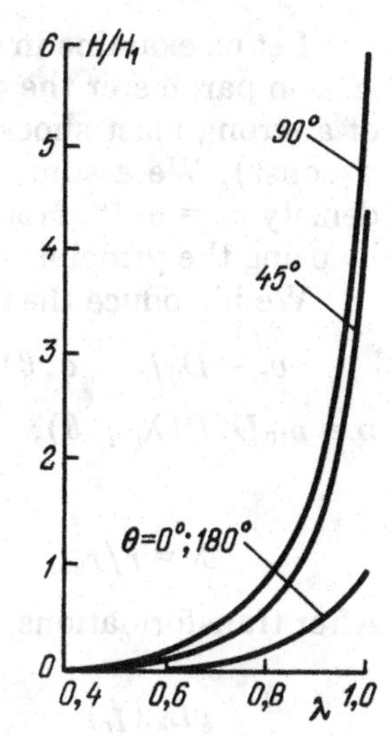

Figure 119. Distribution of field intensity behind wave front during the initial stage of a spherical blast.

Using Eqs. (7.43) for H_r and H_θ we get

$$\frac{d\lambda}{\lambda(1-\frac{5}{2}V)} = -\text{ctg}\,\theta. \tag{7.46}$$

This equation can be integrated by taking the $\lambda(V)$ dependence into account. As a result we obtain, in the $\lambda(\partial)$, $V(\theta)$ parametric form, the equation for the force lines. Since this expression is quite unwieldy, we consider a simple way of solving Eq. (7.46) approximately.

It follows from the inequality (7.44) that the parameter V changes little when λ ranges from 0 to 1. Putting $V = V_*$ in Eq. (7.46) we get

$$r = r_{20}C(\sin\theta)^{-a}, \tag{7.47}$$

where C is the integration constant and $a = 1 - (5/2)V_* > 0$. Equation (7.47) yields the approximate form of the force lines of the magnetic field behind the shock-wave front. An approximate picture of the variation of the force lines is shown in Fig. 118.

4.3. Allowance for the influence of the magnetic field on the gas motion

To take into account the influence of the field on the considered flow we can use a linearization method. It is noted in Ref. 16 that the influence of the magnetic field can be obtained by linearization, using the small parameter $\mu = (\alpha/E_0)^{1/2}r_{20}^{3/2}H_1$, where r_{20} is the radius of the shock wave in the corresponding problem without a magnetic field, and α = const. A preliminary investigation of the problem by this method was carried out by E. V. Ryazanov.[16]

Let us examine in detail the linearization idea, choosing as the linearization parameter the quantity $q = H_1^2/4\pi D_0^2\rho_{10}$, where D_0 is the velocity of a strong-blast shock wave in the absence of a field, and $\rho_{10} = ar_{20}^{-\omega}$ (ω = const). We assume zero back pressure ($p_1 = 0$) and a variable initial density $\rho_1 = ar^{-\omega}$. Note that the back pressure can be taken into account by using the principle of superposition of the linear increments.

We introduce the dimensionless variables

$$v_r = D_0 f_r(\lambda, q, \theta), \qquad v_\theta = D_0 f_\theta(\lambda, q, \theta), \qquad \rho = \rho_{10} g(\lambda, q, \theta),$$
$$p = \rho_{10} D_0^2 P(\lambda, q, \theta), \qquad H_r = H_1 G_r(\lambda, q, \theta),$$
$$H_\theta = H_1 G_\theta(\lambda, q, \theta), \tag{7.48}$$
$$\lambda = r/r_{20}, \qquad r_{20} = (E_0/\alpha a)^{\delta/2} t^\delta, \qquad \delta = 2/(5-\omega).$$

After transformations, the MHD equation (7.43) system takes the form

$$gL_{\lambda\theta}(f_r) - \frac{gf_\theta^2}{\lambda} = -\frac{\partial P}{\partial\lambda} + q\left[\frac{G_\theta}{\lambda}\frac{\partial G_r}{\partial\theta} - \frac{1}{2\lambda^2}\frac{\partial}{\partial\lambda}(G_\theta^2\lambda^2)\right],$$
$$gL_{\lambda\theta}(f_\theta) + g\frac{f_r f_\theta}{\lambda} = -\frac{1}{\lambda}\frac{\partial P}{\partial\theta} + q\left[G_r\frac{\partial G_\theta}{\partial\lambda} + \frac{G_r G_\theta}{\lambda} - \frac{1}{2\lambda}\frac{\partial G_r^2}{d\theta}\right],$$
$$-\omega g + L_\lambda(g) + \frac{1}{\lambda^2}\frac{\partial}{\partial\lambda}(gf_r\lambda^2) + \frac{1}{\lambda\sin\theta}\frac{\partial}{\partial\theta}(gf_\theta\sin\theta) = 0,$$
$$\left(2 - \omega - \frac{2}{\delta}\right)P + L_\lambda(P) + f_r\frac{\partial P}{\partial\lambda} + \frac{f_\theta}{\lambda}\frac{\partial P}{\partial\theta} \tag{7.49}$$
$$+ \gamma P\left[\frac{1}{\lambda^2}\frac{\partial}{\partial\lambda}(f_r\lambda^2) + \frac{1}{\lambda\sin\theta}\frac{\partial}{\partial\theta}(f_\theta\sin\theta)\right] = 0,$$
$$L_\lambda(G_r) - \frac{1}{\lambda\sin\theta}\frac{\partial}{\partial\theta}[\sin\theta(f_r G_\theta - G_r f_\theta)] = 0,$$
$$L_\lambda(G_\theta) + \frac{1}{\lambda}\frac{\partial}{\partial\lambda}[\lambda(f_r G_\theta - G_r f_\theta)] = 0.$$

Here $L_{\lambda\theta}$ and L_λ are operators:

$$L_{\lambda\theta} = \varkappa + L_\lambda + f_r\frac{\partial}{\partial\lambda} + \frac{f_\theta}{\lambda}\frac{\partial}{\partial\theta}, \qquad L_\lambda = -\lambda\frac{\partial}{\partial\lambda} + q\frac{b}{\delta}\frac{\partial}{\partial q},$$
$$\varkappa = 1 - 1/\delta, \qquad b = 2 + \delta(\omega - 2).$$

In addition, we have from the equation div $\mathbf{H} = 0$

$$\frac{\partial}{\partial\lambda}(G_r\lambda^2) + \frac{\lambda}{\sin\theta}\frac{\partial}{\partial\theta}(G_\theta\sin\theta) = 0. \tag{7.50}$$

We introduce next the notation

$$R = r_2(t, \theta)/r_{20}, \qquad U = D/D_0. \tag{7.51}$$

The conditions (7.35) on the shock waves take in terms of the dimensionless variables (7.48) and (7.51) the form

$$P_2 - Uf_{N2}R^{-\omega} + \frac{q}{2}(G_{\tau 2}^2 - G_{\tau 1}^2) = 0,$$

$$R^{-\omega}Uf_{\tau 2} + q(G_{N1}G_{\tau 2} - G_{N1}G_{\tau 1}) = 0,$$

$$g_2(f_{N2} - U) = -UR^{-\omega}, \qquad G_{N1}f_{\tau 2} - G_{\tau 2}(f_{N2} - U) = G_{\tau 1}U, \quad (7.52)$$

$$\frac{P_2}{\gamma - 1} + \frac{P_2}{2}(1 - g_2R^{\omega}) + \frac{q}{4}(1 - g_2R^{\omega})(G_{\tau 2} - G_{\tau 1})^2 = 0, \quad G_{N2} = G_{N1}.$$

Note that the next-to-last condition in Eq. (7.52) is the shock adiabat. We use it in place of the energy equation. In the approximation considered, the shock wave will differ from a sphere (see Fig. 118). With the aid of the above kinematic and geometric relations we obtain for the shock-wave front

$$U = \frac{[R + (b/\delta)qR'_q]R}{\Delta}, \qquad f_{N2} = f_{r2}\frac{R}{\Delta} - f_{\theta 2}\frac{R'_\theta}{\Delta},$$

$$f_{\tau 2} = f_{r2}\frac{R'_\theta}{\Delta} + f_{\theta 2}\frac{R}{\Delta}, \qquad G_{Ni} = G_{ri}\frac{R}{\Delta} - G_{\theta i}\frac{R'_\theta}{\Delta}, \quad (7.53)$$

$$G_{\tau i} = G_{ri}\frac{R'_\theta}{\Delta} + G_{\theta i}\frac{R}{\Delta}, \qquad \Delta = \sqrt{R^2 + (R'_\theta)^2},$$

$$R'_q = \frac{\partial R}{\partial q}, \qquad R'_\theta = \frac{\partial R}{\partial \theta}, \qquad i = 1, 2.$$

We represent any one of the sought functions $F(\lambda, q, \theta)$ in the form $F(\lambda, q, \theta) = F_0(\lambda) + qF_1(\lambda, \theta) + o(q)$ and put $f_{r0}(\lambda) = f_0(\lambda)$.

Linearization of the MHD system (7.49) and (7.50) with respect to the parameter q yields equations for the functions F_1:

$$g_0\left[(k + \varkappa)f_{r1} - \lambda\frac{\partial f_{r1}}{\partial \lambda} + kf_{r1}\frac{\partial f_0}{\partial \lambda} + f_0\frac{\partial f_{r1}}{\partial \lambda}\right] + g_1\left(\varkappa f_0 - \lambda\frac{df_0}{d\lambda} + f_0\frac{\partial f_0}{\partial \lambda}\right)$$

$$= -\frac{\partial P_1}{\partial \lambda} + \frac{G_{\theta 0}}{\lambda}\frac{\partial G_{r0}}{\partial \theta} - \frac{1}{2\lambda^2}\frac{\partial}{\partial \lambda}(G_{\theta 0}^2\lambda^2) \qquad (k = b/\delta), \quad (7.54)$$

$$g_0\left[\varkappa f_{\theta 1} - \lambda\frac{\partial f_{\theta 1}}{\partial \lambda} + kf_{\theta 1} + f_0\frac{\partial f_{\theta 1}}{\partial \lambda} + \frac{f_0 f_{\theta 1}}{\lambda}\right]$$

$$= -\frac{1}{\lambda}\frac{\partial P_1}{\partial \theta} + G_{r0}\frac{\partial G_{\theta 0}}{\partial \lambda} + \frac{G_{r0}G_{\theta 0}}{\lambda} - \frac{1}{2\lambda}\frac{\partial G_{\theta 0}^3}{\partial \theta}, \quad (7.55)$$

$$-\omega g_1 - \lambda\frac{\partial g_1}{\partial \lambda} + kg_1 + \frac{1}{\lambda^2}\frac{\partial}{\partial \lambda}\lambda^2[g_0 f_{r1} + g_1 f_0]$$

$$+ \frac{1}{\lambda \sin\theta}\frac{\partial}{\partial \theta}(g_0 f_{\theta 1}\sin\theta) = 0, \quad (7.56)$$

$$-\lambda \frac{\partial P_1}{\partial \lambda} + kP_1 + f_0 \frac{\partial P_1}{\partial \lambda} + f_{r1} \frac{\partial P_0}{\partial \lambda}$$

$$+ \gamma P_0 \left[\frac{1}{\lambda^2} \frac{\partial}{\partial \lambda} (f_{r1} \lambda^2) + \frac{1}{\lambda \sin\theta} \frac{\partial}{\partial \theta} (f_{\theta 1} \sin\theta) \right] \quad (7.57)$$

$$+ \gamma P_1 \frac{1}{\lambda^2} \frac{d}{d\lambda} (\lambda^2 f_0) = 0,$$

$$-\lambda \frac{\partial G_{r1}}{\partial \lambda} + kG_{r1} = \frac{1}{\lambda \sin\theta} \frac{\partial}{\partial \theta} [\sin\theta (f_0 G_{\theta 1} + f_{r1} G_{\theta 0} - G_{r0} f_{\theta 1})], \quad (7.58)$$

$$-\lambda \frac{\partial G_{\theta 1}}{\partial \lambda} + kG_{\theta 1} = -\frac{1}{\lambda} \frac{\partial}{\partial \lambda} [\lambda (f_0 G_{\theta 1} - f_{r1} G_{\theta 0} - G_{r0} f_{\theta 1})], \quad (7.59)$$

$$\frac{\partial}{\partial \lambda} (G_{r1} \lambda^2) + \frac{\lambda}{\sin\theta} \frac{\partial}{\partial \theta} (G_{\theta 1} \sin\theta) = 0. \quad (7.60)$$

The last equation was obtained from Eq. (7.50) and can be used in place of either Eq. (7.58) or Eq. (7.59) (see the note in Sec. 3 of Chap. 1 on this subject).

This MHD system has a remarkable property, namely, that the equations that determine the gasdynamic functions f_{r1}, $f_{\theta 1}$, P_1, and g_1 are separable from the equations for G_{r1} and $G_{\theta 1}$. For the functions with subscript 0 the solution of the gasdynamic system is known and is given in Chap. 2, while G_{r0} and $G_{\theta 0}$ satisfy the equations

$$\frac{\partial G_{r0}}{\partial \lambda} = \frac{f_0}{\lambda^3} \frac{\partial}{\partial \lambda} (G_{r0} \lambda^2), \qquad \frac{\partial G_{\theta 0}}{\partial \lambda} = \frac{1}{\lambda^2} \frac{\partial}{\partial \lambda} (\lambda f_0 G_{\theta 0}). \quad (7.61)$$

Equations (7.61) were solved in the preceding section [see Eqs. (7.43)], where we have investigated the problem for weak fields. It can be formally verified that the functions (7.43) satisfy Eqs. (7.61).

If the gasdynamic functions f_{r1} and $f_{\theta 1}$ are known, Eqs. (7.58)–(7.60) make it possible to determine the field corrections G_{r1} and $G_{\theta 1}$. The main problem is thus to determine the gasdynamic increments f_{r1}, $f_{\theta 1}$, g_1, and P_1 from the system of inhomogeneous linear equations (7.54)–(7.57).

For the shock-wave coordinate R and for its velocity U we assume

$$R = 1 + qR_1, \qquad U = 1 + qU_1, \quad (7.62)$$

so that Eqs. (7.53) yield the following connection between U_1 and R_1 (k = const):

$$U_1 = (1 + k) R_1. \quad (7.63)$$

Since the shock wave is not a sphere, we have in the linear approximation, for any function $F(\lambda, q, \theta)$,

$$F_2 = F_0(1) + qF_1(1, \theta) + \left(\frac{\partial F}{\lambda} \right)_{\substack{q=0 \\ \lambda=1}} R_1 q \qquad (F_2 = F|_{r=r_2}). \quad (7.64)$$

Linearization of the conditions on the shock wave taking account of Eqs. (7.62)–(7.64) yields the relations

$$G_{r02} = G_{r01} = \cos\theta, \quad G_{\theta 02}(f_{02} - 1) = -G_{\theta 1}, \quad -\frac{\gamma+1}{\gamma-1}\sin\theta = G_{\theta 02}, \quad (7.65)$$

$$P_{12} - f_{r12} + [(P'_0)_2 - (f'_0)_2 - \omega f_{02}]R_1 - u_1 f_{02} + \tfrac{1}{2}(G_{\theta 02}^2 - \sin^2\theta) = 0,$$

$$\left((F'_0)_2 = \frac{\partial F}{\partial\lambda}\bigg|_{q=0,\lambda=1}\right), \quad f_{02}\frac{\partial R_1}{\partial\theta} + f_{\theta 12} + \cos\theta[G_{\theta 02} + \sin\theta] = 0,$$

$$g_{02} f_{r12} + g_{12}(f_{02} - 1) + \{(f'_0)_2 g_{20} + [(g'_0)_2 + \omega g_{02}](f_{02} - 1) - \omega\}R_1 + (1 - g_{02})U_1 = 0, \quad (7.66)$$

$$\cos\theta\left[f_{02}\frac{\partial R_1}{\partial\theta} + f_{\theta 12}\right] - G_{\theta 02}[f_{r12} + (f'_0)_2 R_1 - U_1] - \left[G_{r02}\frac{\partial R_1}{\partial\theta} + G_{\theta 12} + (G'_{\theta 0})_2 R_1\right](f_{02} - 1) = -\sin\theta U_1 + \cos\theta\frac{\partial R_1}{\partial\theta},$$

$$\left[-\frac{1}{2}P_{02}g_{12} - \frac{1}{2}(g'_0)_2 P_{02} - \omega\frac{P_{02}}{2}g_{02}\right]R_1 + \frac{1 - g_0}{4}(G_{\theta 02} + \sin\theta)^2 = 0,$$

$$G_{r12} + (G'_{r0})_2 R_1 - G_{\theta 02}\frac{\partial R_1}{\partial\theta} = \sin\theta\frac{\partial R_1}{\partial\theta}.$$

Linearization of the integral energy-conservation and mass-conservation laws yields

$$\int_0^\pi\left[\int_0^1\left(\frac{g_1 f_0^2 + 2f_1 f_0 g_0}{2} + \frac{P_1}{\gamma-1} + \frac{1}{2}G_0^2\right)\lambda^2\,d\lambda + R_1\left(\frac{f_{02}^2 g_{02}}{2} + \frac{P_{02}}{\gamma-1}\right)\right]\sin\theta\,d\theta = \frac{1}{3}, \quad (7.67)$$

$$\int_0^\pi\left[\int_0^1 g_1\lambda^2\,d\lambda + R_1(g_{20} - 1)\right]\sin\theta\,d\theta = 0.$$

The problem consists of solving the system of linear partial differential equations (7.54)–(7.57) with boundary conditions (7.66) on the shock wave (for $\lambda = 1$). In addition, we have the symmetry condition $f_{\theta 1}(\lambda, \pi) = f_{\theta 1}(\lambda, 0) = 0$. With the aid of this condition (and the integral relations) we can determine the shape of the shock wave in the approximation considered.

Analysis of the equations and of the boundary conditions has shown that the solution of the problem should be sought in the form

$$g_1 = M_0(\lambda) + M_1(\lambda)\sin^2\theta, \qquad P_1 = \pi_0(\lambda) + \pi_1(\lambda)\sin^2\theta,$$
$$f_{r1} = \varphi_{r0}(\lambda) + \varphi_{r1}(\lambda)\sin^2\theta, \quad (7.68)$$
$$f_{\theta 1} = F_{\theta 1}(\lambda)\sin 2\theta, \qquad R_1 = R_{10} + R_{11}\sin^2\theta.$$

Note also that we have, in accord with Eqs. (7.43),

$$G_{r0} = \psi_r(\lambda)\cos\theta, \qquad G_0 = -\psi_\theta(\lambda)\sin\theta. \quad (7.69)$$

If the solution (7.68), with allowance for Eqs. (7.69), is substituted in Eqs. (7.54)–(7.57), the variables separate and we obtain for the seven unknown functions in Eqs. (7.68) systems of ordinary differential equations and corresponding boundary conditions at $\lambda = 1$. These linear systems can be solved numerically by the methods developed in Chap. 3 for one-dimensional problems. In the solution of the linear systems one must use the integral laws (7.7) both to monitor the computational accuracy and to determine the constants R_{10} and R_{11}. Note that for the special solution (2.28) at $\omega = \omega_1$ the linear systems have constant coefficients. In this case (just as for one-dimensional problems), the task of finding the solution becomes purely algebraic in the sense of finding the eigenvalues and the arbitrary constants contained in the solutions.

We can conclude thus from the foregoing analysis that the solution of the considered problem for small q is given by

$$
\begin{aligned}
p &= ar_{20}^{-\omega} D_0^2 \{P_0(\lambda) + q[\pi_0(\lambda) + \pi_1(\lambda)\sin^2\theta] + o(q)\},\\
\rho &= ar_{20}^{-\omega}\{g_0(\lambda) + q[M_0(\lambda) + M_1(\lambda)\sin^2\theta] + o(q)\},\\
v_r &= D_0\{f_0(\lambda) + q[\varphi_{r0}(\lambda) + \varphi_{r1}(\lambda)\sin^2\theta] + o(q)\},\\
v_\theta &= D_0[q\varphi_{\theta 1}(\lambda)\sin 2\theta + o(q)],\\
H_r &= H_1[\psi_r(\lambda)\cos\theta + qG_{r1}(\lambda,\theta) + o(q)],\\
H_\theta &= H_1[-\psi_\theta(\lambda)\sin\theta + qG_{\theta 1}(\lambda,\theta) + o(q)],\\
r_2 &= r_{20}(t)[1 + q(R_{10} + R_{11}\sin^2\theta) + o(q)],\\
q &= H_1^2/4\pi D_0^2\rho_{10}.
\end{aligned}
$$

Note the analogy in this solution with the case where the density varies with altitude (without a magnetic field), where the density distribution along z is symmetric about $z = 0$ (see Ref. 17). Note also that the case of piston motion in an infinitely conducting gas with a magnetic field (both a self-similar and a non-self-similar linearized solution) was considered by a number of workers in the U.S. (see, e.g., Ref. 18).

To calculate the wave-front parameters at large distances from the blast point one can use the asymptotic laws of weak-wave damping, investigated by us in Ref. 7. That reference contains asymptotic equations that make it possible to develop an approximate theory of the shock-wave front, based on a method similar to that described in Chaps. 4 and 5 and in Sec. 3 of the present chapter. We shall not dwell on them in detail here.

5. Perturbation of an arbitrary weak magnetic field

The above treatment of perturbations of a constant weak magnetic field $\mathbf{H}_1$ by propagation of a strong shock wave from a spherical point blast in an infinitely conducting quiescent gas can be generalized to include the case of any weak field $\mathbf{H}_1$ for spherical as well as planar and cylindrical blasts.[19] We shall also assume the shock wave to be strong enough and put $\rho_2 v_2^2 > H_1^2/8\pi$ and $r < r_*$, where r_* is calculated from some average value for the field $\mathbf{H}_1$. To first order, we neglect the influence of the field $\mathbf{H}_1$ on the perturbed motion of the gas. We follow Ref. 19.

In the case of a spherical shock wave ($\nu = 3$) we consider the problem in spherical coordinates r, θ, and φ. For a cylindrical wave ($\nu = 2$) we use a cylindrical (r, φ, z) frame, and for a plane wave ($\nu = 1$) we use the Cartesian coordinates x, y, and z.

To determine the perturbed field $\mathbf{H}$ we can use the induction equation (1.76). Here, however, it is convenient to follow a simpler path.

If account is taken of the continuity equation and a change is made to Lagrange variables (we take the Lagrange coordinate to be the initial coordinate of the particle), the induction equation takes the form (1.78) (see Sec. 3 of Chap. 1 for details)

$$\frac{\mathbf{H}}{\rho} = \left(\frac{\mathbf{H}_1}{\rho_1} \operatorname{grad}_0\right)\mathbf{r}.$$

Here $\mathbf{H}_0 = \mathbf{H}_1$, $\rho_0 = \rho_1$, and r is the particle radius vector. Since the gasdynamic variables can be regarded as depending only on the Lagrange coordinate r_0 and on t, we obtain[19] from Eq. (1.78), with allowance for the gasdynamic continuity equation, the following equations for the magnetic field components in the perturbed-motion region:

$$H_1 = H_{11}\left(\frac{r_0}{r}\right)^{\nu-1}, \qquad H_2 = H_{21}\frac{\rho}{\rho_0}\left(\frac{r}{r_0}\right)^{\omega_*}, \qquad H_3 = H_{31}\frac{\rho}{\rho_0}\left(\frac{r}{r_0}\right)^{(\nu-2)(\nu-1)/2}. \tag{7.70}$$

Here $\omega_* = 1$ for $\nu = 2$ or 3, $\omega_* = 0$ for $\nu = 1$, and H_i are the components of the magnetic field vector in the spherical, cylindrical, and Cartesian frames. It follows from Eqs. (7.70) that the corresponding conditions on the shock wave are met for the field.

We assume hereafter for simplicity that the initial density ρ_0 is constant. In this case we have for r/r_0 (see Chap. 2)

$$\left(\frac{r}{r_0}\right)^{\nu} = \frac{p}{p_2}\left(\frac{\rho}{\rho_2}\right)^{-\gamma}\left(\frac{r}{r_2}\right)^{\nu}.$$

Here r_2 is the shock-wave coordinate, p_2 the pressure behind the wave front, and the dependences of p/p_2, ρ/ρ_2, and ρ/ρ_0 on r/r_2 are known from the analytic solution of the strong-blast problem (see Chap. 2).

Consider the particular case of a constant initial magnetic field $\mathbf{H}_1$. Let the initial field be directed along the z axis in the spherical case, make an angle α with the z axis in the cylindrical case, and have Cartesian components H_{r1}, H_{y1}, and H_{z1}, respectively. For the spherical and cylindrical cases we have

$$H_{11} = H_1 \cos\theta, \quad H_{21} = H_1 \sin\theta, \quad H_{31} = 0 \quad (\nu = 3);$$
$$H_{11} = H_1 \sin\alpha, \quad H_{21} = H_1 \sin\alpha \sin\varphi, \quad H_{31} = H_1 \sin\alpha \cos\varphi \quad (\nu = 2).$$

In this case we have from Eqs. (7.70)

$$H_r = H_1 \cos\theta \left(\frac{r_0}{r}\right)^2, \quad H_\theta = -H_1 \sin\theta \frac{\rho r}{\rho_1 r_0}, \quad H_\varphi = 0 \quad (\nu = 3),$$

$$H_r = H_1 \sin\alpha \sin\varphi \frac{r}{r_0}, \quad H_\varphi = H_1 \sin\alpha \cos\varphi \frac{\rho r}{\rho_1 r_0},$$

$$H_z = H_1 \cos\alpha \frac{\rho}{\rho_1} \quad (\nu = 2), \tag{7.71}$$

$$H_r = H_{r1}, \quad H_y = H_{y1} \frac{\rho}{\rho_1}, \quad H_z = H_{z1} \frac{\rho}{\rho_1} \quad (\nu = 1).$$

As already noted, the particular case of a constant initial magnetic field was first considered in Ref. 15, where the magnetic-field components were obtained by direct integration of the induction equation. For $\nu = 3$ and $H_1 = \text{const}$ we can obtain from Eqs. (7.71) the solution (7.43)–(7.44). The case of a cylindrical blast, when the field has only one component H_z, was considered in Sec. 2. Interactions, not related to blast theory, between a plane shock wave and a weak magnetic field were investigated in Ref. 20. Note also that if $\nu = 3$ it is possible to obtain a field perturbation for a blast in an incompressible fluid. In this case it is necessary to put $\rho = \rho_1$ in Eqs. (7.70). From the continuity equations we have in Lagrange variables

$$\frac{r_0}{r} = \Omega^{1/3}\left(\Omega = 1 - \frac{l^3}{r^3}\right).$$

Here l is the cavity radius with a known dependence on t (see Sec. 7 of Chap. 1). In this case Eqs. (7.70) yield $H_r = H_{r1}\Omega^{2/3}$, $H_\theta = H_{\theta 1}\Omega^{-1/3}$, and $H_\varphi = H_{\varphi 1}\Omega^{-1/3}$. Unfortunately, this solution becomes meaningless near the cavity boundary, since H_θ and H_φ become infinite at $r = l$.

6. Shock wave propagation in a medium of finite electric conductivity

6.1. Deformation of weak field

The assumption that the medium has infinite conductivity is not always tenable. Thus, when air moves behind the front of a strong shock wave (M

= 20), the magnetic Reynolds number at characteristic dimensions of order 1 m are of the order of unity. Let us present some known values of the conductivities for air behind the shock-wave front, for air with cesium admixture, for the ionosphere at an altitude 350 km, for sea water, and for mercury:

Medium	Air M = 20	Air M = 12	Air +0.01%Cs M = 12	Ionosphere (350 km)	Sea water	Mercury
$\sigma\,(s^{-1})$	2×10^{12}	10^{11}	2×10^{13}	10^{11}	2×10^{10}	10^{16}

We give here for air the values of σ behind the shock-wave front, with M the Mach number. At large characteristic scales of the phenomenon, for example for powerful blasts in the ionosphere, the R_m numbers can exceed 10, and the approximation of infinite conductivity is here fully justified.

Assuming σ to have finite values, we consider the case of weak magnetic fields. Let the initial magnetic field be uniform. We consider only the cylindrical blast first investigated in Ref. 9. Similar investigations were later carried out by A. Sakurai.[21]

We choose the equations of motion in the form

$$-\rho\frac{dv}{dt}=\frac{\partial p^*}{\partial r}+\frac{2(\nu-1)h_\varphi}{r},\qquad -\frac{1}{\rho}\frac{d\rho}{dt}=\frac{\partial v}{\partial r}-\frac{\nu-1}{r}v,\tag{7.72}$$

$$-\frac{1}{2}\frac{dh_z}{dt}=h_z\left(\frac{\partial v}{\partial r}+\frac{\nu-1}{r}v\right)-r^{1-\nu}h^{1/2}\frac{\partial}{\partial r}\left(\nu_m r^{\nu-1}\frac{\partial h_z^{1/2}}{\partial r}\right),\tag{7.73}$$

$$-\frac{1}{2}\frac{dh_\varphi}{dt}=h_\varphi\frac{\partial v}{\partial r}-h_\varphi^{1/2}\frac{\partial}{\partial r}\left[\nu_m r^{-1}\frac{\partial}{\partial r}(rh_\varphi^{1/2})\right],\tag{7.74}$$

$$\begin{aligned}-\frac{dp}{dt}&=\gamma p\left(\frac{\partial v}{\partial r}+\frac{(\nu-1)v}{r}\right)\\&\quad-2(\gamma-1)\nu_m\left\{\frac{1}{r^2}\left[\frac{\partial}{\partial r}(rh_\varphi^{1/2})\right]^2+\left(\frac{\partial h_z^{1/2}}{\partial z}\right)^2\right\},\end{aligned}\tag{7.75}$$

$$h=h_z+(\nu-1)h_\varphi,\qquad p^*=p+h,\qquad h_\varphi=H_\varphi^2/8\pi,\qquad h_z=H_z^2/8\pi.$$

Let $h_\varphi=0$ and $h_z\neq0$. We introduce the dimensionless variables

$$\frac{v}{D}=f,\qquad G=\frac{H}{H_\infty},\qquad \frac{\rho}{\rho_\infty}=g,\qquad \lambda=\frac{r}{r_2},\qquad P=\frac{p}{\rho_\infty D^2},$$

$$q=\frac{a_*^2}{D^2},\qquad a_*^2=\frac{\gamma}{\rho_\infty}\left(p_\infty+\frac{H_*^2}{8\pi}\right).$$

If we seek the solution of the system (7.72)–(7.73) in the form

$$f=f_0(\lambda)+qf_1(\lambda)+o(q),\qquad g=g_0(\lambda)+qg_1(\lambda)+o(q),$$

$$P=P_0(\lambda)+qP_1(\lambda)+o(q),\qquad G=G_0(\lambda)+qG_1(\lambda)+o(q),$$

we find, repeating the reasoning of the preceding section, that in first-order approximation the quantities f_0, g_0, and P_0 will satisfy the equations and relations for a strong blast in a gas, and for the function G_0 we get from Eq. (7.73)

$$(f_0-\lambda)\frac{dG_0}{d\lambda}+G_0\left(\frac{df_0}{d\lambda}+\frac{f_0}{\lambda}\right)-\frac{k_0}{\lambda}\frac{d}{d\lambda}\left(\frac{dG_0}{d\lambda}\right)=0, \qquad (7.76)$$

$$k_0=\frac{c^2}{4\pi\sigma}\left(\frac{\alpha\rho_\infty}{E_0}\right)=\text{const.}$$

From the conditions on the shock wave ($\lambda=1$), we get for finite conductivity (see Chap. 1; the gas ahead of the wave is at rest)

$$G_{02}=G_{01}, \qquad f_{02}G_{02}-k_{02}\left(\frac{\partial G}{\partial\lambda}\right)_{02}=-k_{01}\left(\frac{\partial G}{\partial\lambda}\right)_{01}.$$

The gasdynamic functions will satisfy the usual conditions on a strong wave. We assume that $k_{02}\neq k_{01}$ in the general case. Flow sets in by virtue of the continuity of the magnetic field ahead of the shock wave (region 1). To first order, however, the velocity f_0 ahead of the shock wave is zero, so that the induction equation (7.76) takes in region 1 the simple form

$$\frac{k_{01}}{\lambda}\frac{\partial}{\partial\lambda}\left(\lambda\frac{\partial G_0}{\partial\lambda}\right)+\lambda\frac{\partial G_0}{\partial\lambda}=0. \qquad (7.77)$$

In region 2 (behind the shock-wave front), we have Eq. (7.76) with $k_0=k_{02}$. Equation (7.77) can be integrated, and its solution with allowance for the condition $G\to 1$ as $\lambda\to\infty$ is

$$G_{01}=1-A_1\int_\lambda^\infty e^{-x^2/2k_{01}}\frac{dx}{x},$$

where A_1 is an arbitrary constant.

For region 2 we obtain from Eqs. (7.76) the set of equations

$$G_0'=\Omega_0/\lambda, \qquad \Omega_0'=\frac{1}{k_{02}}\left[(f_0-\lambda)\Omega_0+\lambda G_0\left(f_0^1+\frac{f_0}{\lambda}\right)\right], \qquad (7.78)$$

where $f_0(\lambda)$ is a function known from the solution of the gasdynamic problem. An investigation of the system (7.78) has shown that it exhibits near $\lambda=0$ a bounded solution given by the expansion

$$G_0=c_1[1+a_1\lambda^2+O(\lambda^4)],$$

where c_1 is an arbitrary constant and $a_1=1/\gamma k_{02}$.

Introducing a new variable $\tilde{G}_0=G_0/c_1$ we get at the symmetry center the conditions

$$\tilde{G}(0)=1, \qquad \tilde{\Omega}(0)=0. \qquad (7.79)$$

The system of equations for $\tilde{G}$ and $\tilde{\Omega}$ in region 2 was integrated numerically from the center to the shock wave. From the conditions on the shock waves

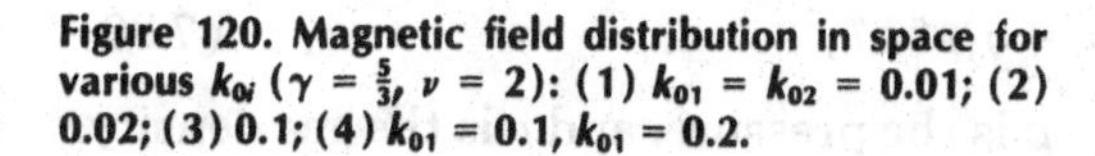

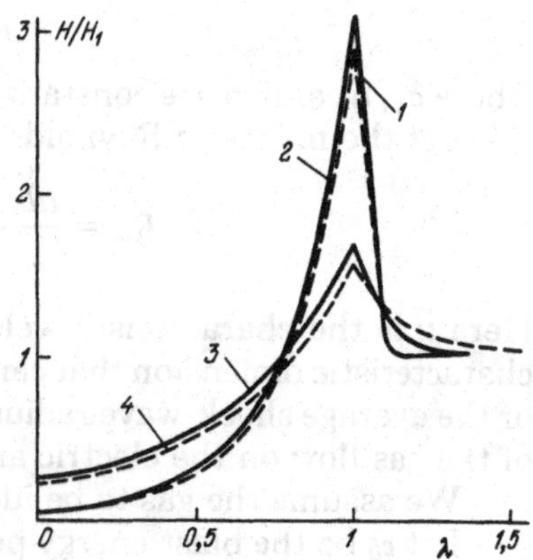

Figure 120. Magnetic field distribution in space for various k_{0i} ($\gamma = \frac{5}{3}$, $\nu = 2$): (1) $k_{01} = k_{02} = 0.01$; (2) 0.02; (3) 0.1; (4) $k_{01} = 0.1$, $k_{01} = 0.2$.

we determined next for the function G_0 the constants c_1 and A_1, and consequently also the function G_0 in the regions 1 and 2. The integral in the equation for G_{01} was taken from the tables.

The solution obtained makes it possible to determine to first order the structure of the magnetic field following passage of the shock wave. The calculations were performed for various values of the parameters k_{0i} and γ. The results of calculations for $\gamma = \frac{5}{3}$ are shown in Fig. 120. The investigations have shown that as the parameters k_{01} decrease ($k_{02} \geq k_{01}$) the maximum of H/H_∞ increases and tends to the ideal value $(\gamma + 1)/(\gamma - 1)$ as $k_{0i} \to 0$, with the H/H_∞ distribution tending to ρ/ρ_0. (This follows from the freezing integral at infinite conductivity.)

We have thus obtained a solution describing the structure of the field variation. By determining the succeeding approximation we can determine the interaction between the motion and the field and describe more precisely the field variation when a shock wave passes through an electrically conducting gas. A similar investigation would be possible for a planar or spherical blast, and in the case of the spherical charge it is necessary to determine two field components that depend on both λ and on the angle θ.

6.2. Flows with small and finite R_m numbers

In the theory of shock tubes and in experiments with discharges along conductors, it may become necessary to study the propagation of planar and cylindrical shock waves at low magnetic Reynolds numbers R_m. These questions arise also in the theory of explosive magnetohydrodynamic generators. Data on cylindrical and planar blasts can be used also to study questions of hypersonic flow around thin blunted bodies in the presence of a magnetic field.

The results that follow were published mainly in 1962.[22] Flows with small R_m were later investigated by P. Lykoudis.[23]

We consider a gas with finite conductivity σ and assume that

$$\sigma = \sigma_1 \rho^n p^m, \tag{7.80}$$

where σ_1, n, and m are constants, p is the pressure, and ρ is the gas density.

Let the magnetic Reynolds numbers be small:

$$R_m = \frac{ul}{\nu_m} < 1 \qquad \left(\nu_m = \frac{c^2}{4\pi\sigma}\right). \tag{7.81}$$

Here u is the characteristic velocity, ν_m the magnetic viscosity, and l is a characteristic dimension that can be taken to be either the shock-tube height or the average shock-wave radius. For small R_m we can neglect the influence of the gas flow on the electric and magnetic field strengths.

We assume the gas to be ideal with constant specific-heat ratio.

Let ε_0 be the blast energy per unit length in the cylindrical case or per unit area in the planar case. Let also the magnetic-field strength vector $\mathbf{H}$ be perpendicular to the flow-velocity vector $\mathbf{v}$. In the cylindrical case, H can have an axial component H_z as well as an azimuthal H_φ. In the planar case the magnetic field is assumed directed along the z axis perpendicular to the gas-motion direction. The electric field vector $\mathbf{E}$ is perpendicular to $\mathbf{H}$.

The solution of this problem reduces to integration of an MHD equation system in the form

$$\rho \frac{d\mathbf{v}}{dt} + \nabla p = \mathbf{f}, \qquad \frac{d\rho}{dt} + \rho \operatorname{div} \mathbf{v} = 0, \qquad \mathbf{f} = \frac{1}{c} \mathbf{j} \times \mathbf{H},$$

$$\frac{dp}{dt} + \gamma p \operatorname{div} \mathbf{v} = (\gamma - 1)\frac{j^2}{\sigma}, \qquad \mathbf{j} = \sigma\left(\mathbf{E} + \frac{1}{c}\mathbf{v} \times \mathbf{H}\right), \tag{7.82}$$

$$\operatorname{rot} \mathbf{H} = \frac{4\pi}{c}\mathbf{j}, \qquad \operatorname{rot} \mathbf{E} = -\frac{1}{c}\frac{\partial \mathbf{H}}{\partial t}, \qquad \operatorname{div} \mathbf{H} = 0. \tag{7.83}$$

It must be recognized here that the vector $\mathbf{v}$ is perpendicular to $\mathbf{E}$ and $\mathbf{H}$ and all the functions sought depend only on the time and on a single coordinate r, the distance from the blast symmetry axis or from the blast plane. From Eqs. (7.82) and (7.83) we have the energy equation

$$\frac{\partial}{\partial t}\left\{r^{\nu-1}\left(\frac{\rho v^2}{2} + \frac{p}{\gamma - 1}\right)\right\} + \frac{\partial}{\partial r}\left\{r^{\nu-1} v\left(\frac{\rho v^2}{2} + \frac{\gamma p}{\gamma - 1}\right)\right\} = r^{\nu-1}\mathbf{jE}. \tag{7.84}$$

The blast will cause a shock wave to propagate through the gas. The conditions on the shock wave are

$$[\rho(v - D)] = 0, \qquad [\rho v(v - D) + p] = 0,$$

$$\left[(v - D)\left(\frac{\rho v^2}{2} + \frac{p}{\gamma - 1}\right) + pv\right] = 0, \tag{7.85}$$

$$[\mathbf{H}] = 0, \qquad [\mathbf{E}_\tau] = 0, \tag{7.86}$$

where the square brackets denote differences between values on opposite sides of the discontinuity surface, D is the shock-wave velocity, and E_τ is the tangential component of the electric field. Assuming a strong shock wave, we can neglect the initial gas pressure compared with the pressure behind the shock-wave front.

Consider now strong cylindrical and planar blasts in a gas under the assumption that R_m are small, the initial gas density ρ_1 is constant, there is no initial electric field, and the coordinate dependence of the initial magnetic field is given by

$$H^2 = \varkappa_1 r^\omega \qquad (\varkappa_1 = \text{const}), \tag{7.87}$$

$$\omega = 0, \quad \text{if} \quad H = H_z, \qquad \omega = -2, \quad \text{if} \quad \nu = 2, \quad H = H_\varphi.$$

The expressions for j^2 and for the component f_r of the force f along the r direction are

$$j^2 = \frac{\sigma^2}{c^2}\{H_z^2 + (\nu - 1)H_\varphi^2\}v^2, \qquad f_r = -\frac{\sigma}{c^2}\{H_z^2 + (\nu - 1)H_\varphi^2\}v. \tag{7.88}$$

In this case the set of equations obtained from Eqs. (7.83) for the gas motion contains the dimensional constant

$$\varkappa = \varkappa_1 \sigma_1 c^{-2}, \qquad [\varkappa] = M^{1-m-n} L^{m+3n-3-\omega} T^{2m-1}.$$

The solution of the problem depends on the defining dimensional parameters r, t, ε_0, ρ_1, and $\varkappa$. It follows from dimensionality theory that the blast problem is self-similar if the condition $\nu(2m - 1) - 2(\omega + 1) = 0$ is met.

We introduce the dimensionless variables V, R, P, and λ defined as

$$v = \frac{r}{t}V(\lambda), \qquad \rho = \rho_1 R(\lambda), \qquad p = \rho_1 \frac{r^2}{t^2}P(\lambda), \qquad \lambda = \frac{r}{r_2}, \tag{7.89}$$

where r_2 is the shock-wave radius.

In view of the self-similarity we have for r_2

$$r_2 = \left(\frac{\varepsilon}{\rho_1}\right)^{1/\nu+2} t^{2/\nu+2}, \qquad \varepsilon_0 = \alpha\varepsilon, \tag{7.90}$$

where ε is a certain constant having the dimension of ε_0 and α is a constant calculated by a method described below.

Taking Eqs. (7.80) and (7.87)–(7.90) into account, we obtain from the system (7.82) the following differential equations for the self-similar functions V, R, and P:

$$\begin{gathered}
\lambda\{(V - \delta)RV' + P\} = R(V - V^2) - 2P - k\lambda^a V R^n P^m, \\
\lambda\{(V - \delta)R' + RV'\} = -\nu RV, \\
\lambda\{(V - \delta)P' + \gamma PV'\} = 2P - (2 + \nu\gamma)PV + (\gamma - 1)k\lambda^a V^2 R^n P^m.
\end{gathered} \tag{7.91}$$

Here

$$\delta = \frac{2}{\nu + 2}, \qquad a = (\nu + 2)(m - 0.5), \qquad k = \varkappa \rho_1^{m+n-1}\left(\frac{\varepsilon}{\rho_1}\right)^{m-0.5}.$$

In this case considered ($\mathbf{E} = 0$) it follows from Eq. (7.84) that the system (7.91) has an energy integral

$$\lambda^{\nu+2}\left\{(V-\delta)\left(\frac{RV^2}{2}+\frac{P}{\gamma-1}+PV\right)\right\}=c_1. \tag{7.92}$$

We label by the subscript 2 quantities behind the shock-wave front. Since $D = dr_2/dt$ and the gas ahead of the shock wave is at rest, we obtain from Eqs. (7.85), after changing to dimensionless variables,

$$\begin{gathered} R_2(V_2-\delta)+\delta=0, \qquad R_2V_2(V_2-\delta)+P_2=0, \\ (V_2-\delta)\left(\frac{R_2V_2^2}{2}+\frac{P_2}{\gamma-1}\right)+P_2V_2=0. \end{gathered} \tag{7.93}$$

The sought functions $V(\lambda)$, $R(\lambda)$, and $P(\lambda)$ should therefore satisfy on the shock wave (at $\lambda = 1$) the conditions (7.93). In addition, it is necessary to meet the condition that the total energy be constant in the region occupied by the moving gas, i.e.,

$$2\{(\nu-1)\pi-(\nu-2)\}\int_0^{r_2}\left(\frac{\rho v^2}{2}+\frac{P}{\gamma-1}\right)r^{\nu-1}\,dr=\varepsilon_0.$$

Changing to dimensionless variables, we obtain

$$\begin{aligned} \alpha(\nu,\gamma,m,n,k) &= 2\{(\nu-1)\pi \\ &\quad -(\nu-2)\}\int_0^1\left(\frac{RV^2}{2}+\frac{P}{\gamma-1}\right)\lambda^{\nu+1}\,d\lambda. \end{aligned} \tag{7.94}$$

From the last equation of Eqs. (7.93) it follows that the constant c_1 in the energy integral (7.92) is zero. We introduce new variables

$$y=\lambda^{\beta}R, \qquad z=\lambda^{\beta}P, \qquad \beta=\frac{(\nu+2)(m-0.5)}{m+n-1} \qquad (m+n-1\neq 0).$$

In these variables, the system (7.91) becomes

$$\begin{gathered} \lambda\{(V-\delta)yV'+z'\}=y(V-V^2)+(\beta-2)z-kVy^nz^m, \\ \lambda\{(V-\delta)y'+yV'\}=\{(\beta-\nu)V-\beta\delta\}y, \\ \lambda\{(V-\delta)z'+\gamma zV'\}=(\beta-2-\nu\gamma)Vz+(2-\beta\delta)z+(\gamma-1)kV^2y^nz^m. \end{gathered} \tag{7.95}$$

The energy integral (7.92) can be rewritten as

$$z=y\frac{(\gamma-1)(\delta-V)V^2}{2(\gamma V-\delta)}. \tag{7.96}$$

Using Eq. (7.96), the solution of the system (7.95) can always be reduced to integration of an equation of first order. Thus, eliminating z' from the first and third equations of Eqs. (7.9) and introducing the new variable $\mu = \ln\lambda$, we obtain

$$\frac{d\mu}{dV} = \chi(y, V) \tag{7.97}$$

$$= \frac{yz - (\delta - V)^2 y}{(2 - 2\delta - \gamma\nu V)z + (1 - V)(\delta - V)Vy + k(\gamma V - \delta)Vy^n z^m}.$$

With allowance for Eq. (7.97), the second equation of Eqs. (7.95) yields

$$\frac{dy}{dV} = \frac{y}{\delta - V}\{1 + [(\nu - \beta)V + \beta\delta]\chi(y, V)\}. \tag{7.98}$$

The function $z(y, V)$ is defined here by Eq. (7.96). Once Eq. (7.98) is integrated, i.e., the function $y(V)$ is determined, we can obtain $z(V)$ from Eq. (7.96) and $\mu(V)$ by quadratures from Eq. (7.97); the complete solution of the problem can thus be found.

It was assumed above that $m + n - 1 \neq 0$. It can be shown that even if $m + n - 1 = 0$ the solution of the system (7.91) can be reduced to integration of one first-order differential equation of the form $d\lambda/dV = F(\lambda, V)$ and one quadrature.

Analysis of the solution of Eq. (7.98) and the system (7.91) has shown that for $n = m = 0$ ($\sigma = \text{const}$) the following asymptotic equations are valid near the center:

$$P = \left[c_1 - \frac{\delta}{\gamma}\ln(k\lambda)\right]\lambda^{-2}, \qquad R = c_2\lambda^{2/\gamma-1},$$

$$y = \lambda^2 R, \qquad V = \frac{\delta}{\gamma} + \frac{Y(\gamma - 1)}{(\gamma - 1)c_1 - 4k\gamma^2\ln y}. \tag{7.99}$$

For $m = 1$, $n = 1$, and $\nu = 2$ the solution has a strong nonintegrable singularity for the function P and becomes meaningless. The case $n = 0$ and $m = 0$ was numerically investigated in detail. The solution was obtained by using the system (7.91) directly. It was also solved for the derivatives and integrated with a computer, alongside a solution of a Cauchy problem with initial data for $\lambda = 1$. Asymptotic equations of type (7.99) were used for the calculation near the blast center. The energy integral was used for a general check on the computational accuracy. In the calculation of the system (7.91) the constant α was determined by integration. The maximum relative error of P determined from Eqs. (7.91) and from the integral (7.96) did not exceed 0.01%. Certain calculation results for $\nu = 1$ and $\nu = 2$ are shown in Figs. 121 and 122. The results in the cylindrical case with $\gamma = 1.4$ were $\alpha = 0.998$ ($k = 0.01$), $\alpha = 1.13$ ($k = 0.1$), $\alpha = 1.56$ ($k = 1$), and $\alpha = \frac{15}{2}$ ($k = 10$), where $k = (\varkappa_1\sigma_1/c^2\rho_1) \times (\varepsilon_0/\alpha\rho_1)^{-1/2}$ is a parameter characterizing the interaction between the gas and the magnetic field.

It follows from Eqs. (7.88) that the electromagnetic force **f** and the velocity **v** are oppositely directed. This slows down the stream compared with the gas velocity for a blast in the absence of a magnetic field. The stream is also subject to Joule dissipation that supplies heat to the particles.

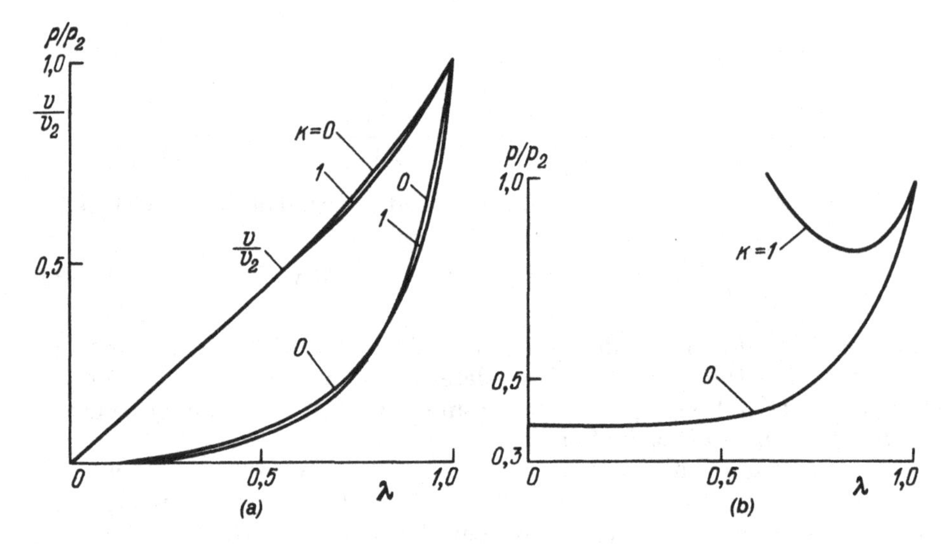

Figure 121. Distribution of the velocities and densities (a) and of the pressures (b) behind a plane wave at low values of R_m ($\gamma = 1.4$, $\nu = 1$, $n = 0$, $m = 0$).

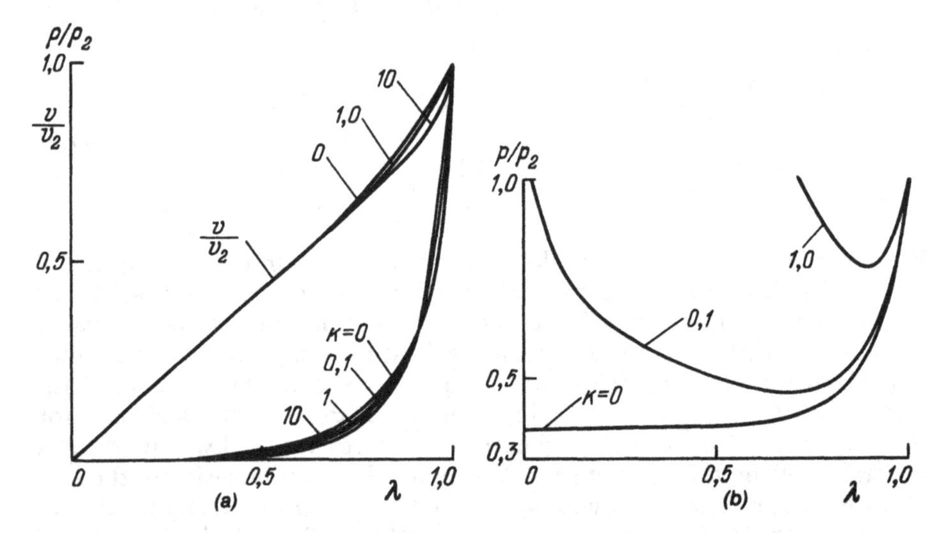

Figure 122. Distribution of velocities and densities (a) and of the pressures (b) behind a cylindrical wave at low values of R_m and various k.

The density of the current flowing through the gas is given by the first relation of Eqs. (7.88). With the current drawn from it, the system can be regarded as a nonstationary magnetohydrodynamic generator. Note that certain systems using the energy of explosives to generate electricity have been considered in Ref. 24. We note also that it has been shown in Ref. 25 that the results of the theory agree well with the experimental results on cylindrical blasts in a magnetic field at small R_m numbers.

The solutions considered can be obtained also for the case of anisotropic conductivity (allowance for the Hall currents).

To solve the strong-blast problem with finite R_m numbers it is necessary to take into account the variation of the magnetic field and use the complete set of MHD equations (7.82) and (7.83).

Let us consider the case of cylindrical symmetry under the conditions $H_z = 0$, $m = 0$, and $n = 0$, i.e., a strong cylindrical point blast with constant conductivity of the gas in the discontinuity region of the hydrodynamic parameters, with the initial magnetic field variable, and with $H^2 = H_\varphi^2 = \varkappa_1 r^{-2}$. It is possible in this case to assume the system of equations (7.72), (7.74), and (7.75) and to take into account the relations

$$j_z = \frac{c}{4\pi r}\frac{\partial}{\partial r}(rH_\varphi), \qquad E_z = \frac{j_z}{\sigma} - \frac{v}{c}H_\varphi.$$

Introducing the dimensionless variables H^0, E^0, and $\mathcal{J}$ defined by

$$h_\varphi = \rho_1 \frac{r^2}{t^2} H^0, \qquad E_z = \frac{\sqrt{8\pi\varepsilon}}{ct} E^0, \qquad j_z = \frac{c}{t}\sqrt{\frac{\rho_1}{2\pi}}\,\mathcal{J}$$

and using Eqs. (7.89), we get

$$\lambda\left\{\left(V - \frac{1}{2}\right)RV' + P' + H^{0\prime}\right\} + RV(V-1) + 2(P + H^0) = 0,$$

$$\lambda\left\{\left(V - \frac{1}{2}\right)R' + RV'\right\} + 2RV = 0; \tag{7.100}$$

$$\lambda\left\{\left(V - \frac{1}{2}\right)P' + \gamma PV'\right\} - 2P + 2(\gamma+1)PV - \frac{2(\gamma-1)}{\lambda^4} A\left\{\frac{d}{d\lambda}(\lambda^2\sqrt{H^0})\right\}^2 = 0,$$

$$\lambda\left\{\left(V - \frac{1}{2}\right)H^{0\prime} + 2H^0V'\right\} - 2H^0 + 4H^0V - \frac{2}{\lambda}\sqrt{H^0}A\frac{d}{d\lambda}\left\{\frac{1}{\lambda}\frac{d}{d\lambda}(\lambda^2\sqrt{H^0})\right\} = 0.$$

For E^0 and $\mathcal{J}$ we have

$$E^0 = \frac{A}{\lambda}\frac{d}{d\lambda}(\lambda^2\sqrt{H^0}) - \lambda^2 V\sqrt{H^0},$$

$$\mathcal{J} = \frac{1}{\lambda}\frac{d}{d\lambda}(\lambda^2\sqrt{H^0}) \qquad \left[A = \nu_m\left(\frac{\rho_1}{\varepsilon}\right)^{1/2}\right]. \tag{7.101}$$

Here A is a dimensionless constant parameter. Note that a system of self-similar equations, equivalent to Eqs. (7.100), was indicated in Ref. 27. The system (7.100) has the following integrals:

$$\frac{RV^2}{2} + \frac{P}{\gamma - 1} + H^0 - 2V\left(\frac{RV^2}{2} + \frac{\gamma}{\gamma - 1}P + 2H^0\right) + \frac{4A}{\lambda^2}\sqrt{H^0}\frac{d}{d\lambda}(\lambda^2\sqrt{H^0}) = c_2, \quad (7.102)$$

$$\lambda^2(V - 0.5)\sqrt{H^0} - A\frac{1}{\lambda}\frac{d}{d\lambda}(\lambda^2\sqrt{H^0}) = c_3.$$

The last integral was pointed out by Greenspan[26] in an analysis of a similar problem for motions with conductivity jumps. The constants c_2 and c_3 are obtained from the boundary conditions. The use of this integral can lower the order of the system (7.100).

We consider now a case where the gas motion is accompanied by the onset of a shock wave. From Eqs. (7.85) and (7.86) we obtain the boundary conditions on the shock-wave front (at $\lambda = 1$) for the dimensionless functions V, R, P, and H^0:

$$[R(V - \delta)] = 0, \qquad [RV(V - \delta) + P] = 0,$$

$$\left[(V - \delta)\left(\frac{RV^2}{2} + \frac{P}{\gamma - 1}\right) + PV\right] = 0, \quad (7.103)$$

$$[A(4H^0 + H^{0\prime}) - 2VH^0] = 0, \qquad [H^0] = 0.$$

In addition, the conditions $V_\infty = 0$, $R_\infty = 1$, $P_\infty = 0$, and $H^0_\infty = 0$ at $\lambda = \infty$ must be met, as well as the condition that the velocity be zero on the symmetry axis. We label, as above, the quantities ahead of (region 1) and behind (region 2) the shock-wave front by the subscripts 1 and 2, respectively. The shock-wave radius is given by Eqs. (7.90). The integral energy-conservation law yields an equation for α.

For a complete solution of the problem it is necessary to integrate the system (7.100) in the regions 1 and 2 with allowance for the energy integral (7.102). It is necessary here to satisfy the boundary conditions (7.103), the conditions at infinity, and the condition at the symmetry center, and to calculate the constant α. The current density and the electric field are obtained from Eqs. (7.101).

We call attention to Ref. 28, where cylindrical gas expansion at finite R_m was considered.

7. Excitation of electromagnetic waves by strong shock waves with conductivity discontinuities

Propagation of strong shock waves alters radically the properties of the gas after the passage of the shock-wave front; namely, the pressure, temperature, density, and the electric conductivity are greatly increased. The presence of interaction between a strong shock wave and the electromagnetic field is due to the abrupt increase of the electric conductivity of the gas behind the discontinuity front. For high-power nuclear blasts, the increase of the electric conductivity can be due to the thermal ionization of the gas and to other factors.[20] An appreciable increase of the electric conductivity of a gas was also noted for blast waves produced by detonation of chemical explosives.[30]

If the shock wave propagates in a space in which magnetic and electric fields are present, the change of the properties of the medium on going through the shock-wave front produces magnetic and electric field perturbations that propagate in the form of electromagnetic waves. Questions connected with emission of electromagnetic waves by shock waves have been the subject of a number of investigations.[1,15,30–32]

We study below the emission of electromagnetic waves by spherical and plane shock waves propagating in weak magnetic or electric fields. We assume that the electromagnetic-wave excitation is due to the onset of nonstationary currents and of a conductivity jump on passage of the shock wave through the gas. Hereafter, we follow our earlier papers.[15,31]

Let a strong shock wave propagate with velocity $D(t)$ (t is the time) in a gas occupying an unbounded large volume. Since the electric and magnetic fields ahead of the shock-wave front are assumed to be weak, we neglect their effect on the gas motion behind the wave front. This means that the parameters of the wave front will be the same as obtained by solving the gasdynamic problem. We assume this problem to be completely solved and hence the velocity $D(t)$ is given.

We examine the initial and boundary conditions. Let the magnetic and electric field vectors $\mathbf{H}$ and $\mathbf{E}$ be constant at the initial instant $t = 0$. We assume that the gas conductivity is zero ahead of the shock-wave front and infinite behind. Since there are no magnetic charges on the shock-wave front, the normal component H_n of the magnetic field is continuous (the magnetic permeability is assumed equal to unity). It follows from the Maxwell equations that the tangential component $\mathbf{E}_\tau$ of the electric field vector is also continuous. Note that the continuity conditions on H_n and $\mathbf{E}$ coincide with analogous conditions on shock waves propagating in a medium of infinite conductivity. We regard[1,32] the magnetic-field tangential component $\mathbf{H}_\tau$ as continuous. In accordance with the results of Refs. 1 and 32, we assume that the magnetic-viscosity coefficient ν_m in the shock layer is larger than the other dissipative coefficients. Denoting by the subscripts 1 and 2, respectively, quantities ahead of and directly behind the shock-wave front, we can write

$$\mathbf{H}_1 = \mathbf{H}_2, \qquad \mathbf{E}_{\tau 1} = \mathbf{E}_{\tau 2}. \tag{7.104}$$

Since the gas behind the shock-wave front has infinite conductivity, the following relation between $\mathbf{H}$ and $\mathbf{E}$ holds in a coordinate frame comoving with the wave:

$$\mathbf{E}_2 = -\frac{1}{c}\,\mathbf{v}_2 \times \mathbf{H}_2 \tag{7.105}$$

(c is the speed of light).

With allowance for Eqs. (7.104) we have in an immobile coordinate frame

$$\mathbf{E}_{\tau 1} = -\frac{1}{c}\,[(\mathbf{v}_2 - \mathbf{D}) \times \mathbf{H}_1]_\tau. \tag{7.106}$$

The boundary conditions (7.104) and (7.106) on the shock-wave front must be taken into account when the above problems dealing with emission of electromagnetic waves are solved. Since the solution of the gasdynamic problem is assumed known, the quantities $\mathbf{v}_2$ and D in Eq. (7.106) are assumed given.

The propagation of electromagnetic waves in a medium with zero conductivity and with $\mu = \varepsilon = 1$ are described by the Maxwell equations

$$\frac{1}{c}\frac{\partial E}{\partial t} = \text{rot}\,\mathbf{H}, \qquad \frac{1}{c}\frac{\partial \mathbf{H}}{\partial t} = -\text{rot}\,\mathbf{E}, \qquad \text{div}\,\mathbf{H} = 0, \qquad \text{div}\,\mathbf{E} = 0. \tag{7.107}$$

The system (7.107) with boundary conditions (7.106) and with specified initial conditions makes it possible to determine the laws governing the electromagnetic-wave propagation.

On the other hand, to determine the electric and magnetic field and the current distribution in the region of motion behind the shock-wave front it is necessary to use the conditions (7.104) and the MHD equations.

7.1. Spherical shock waves

Let the initial magnetic and electric fields be $\mathbf{H}_0$ and $\mathbf{E}_0$. We introduce the spherical coordinate frame (r, θ, φ) with θ measured from the direction of the vector $\mathbf{H}_0$. Recognizing that $\mathbf{v}$ and $\mathbf{D}$ are directed along the radius, the boundary conditions (7.105) and (7.106) expressed in terms of the components become

$$H_{r1} = H_{r2}, \qquad H_{\theta 1} = H_{\theta 2}, \qquad H_{\varphi 1} = H_{\varphi 2}, \tag{7.108}$$

$$E_{\theta 1} = \frac{1}{c}\,(v_2 - D)H_{\varphi 1}, \qquad E_{\varphi 1} = -\frac{1}{c}\,(v_2 - D)H_{\theta 1}. \tag{7.109}$$

Since strong shock waves in an ideal gas with a specific-heat ratio γ satisfy the relation $v_2 = 2D/(\gamma + 1)$, the conditions (7.108) and (7.109) can be written in the form

$$E_{\theta 1} = -\frac{a}{c}\,DH_{\varphi 1}, \qquad E_{\varphi 1} = \frac{a}{c}\,DH_{\theta 1} \qquad \left(a = \frac{\gamma - 1}{\gamma + 1}\right). \tag{7.110}$$

Consider some typical solutions of electromagnetic-wave propagation problems. Let $E_0 = 0$ and $H_0 \neq 0$. Assume that the shock-wave motion is specified by the relations

$$D = \frac{dr_2}{dt} = \frac{c\varphi(\xi_2, r_2)}{a\psi(\xi_2, r_2)}, \qquad \xi_2 = ct - r_2, \qquad r_2(0) = 0,$$

$$\varphi(\xi_2, r_2) = \xi_2 \sum_{k=0}^{m} g_k \xi_2^k [\xi_2 + (k+2) r_2], \qquad 0 \le m < \infty, \quad (7.111)$$

$$\psi(\xi_2 r_2) = r_2^2 - \xi_2 \sum_{k=0}^{m} g_k \xi_2^k [\xi_2 + (k+2) r_2 + \xi_2^2 r_2^{-1} (k+3)^{-1}].$$

Here g_k are constants.

In this case the solution of the problem is given by

$$E_r = E_\theta = 0, \qquad H_\varphi = 0,$$

$$E_\varphi = H_\theta \sin\theta \frac{\xi}{r^2} \sum_{k=0}^{m} g_k \xi^k [\xi + (k+2) r],$$

$$H_r = -\frac{2H_0}{r^3} \xi^2 \cos\theta \sum_{k=0}^{m} \frac{g_k}{k+3} \xi^k [\xi + (k+3) r] + H_0 \cos\theta, \quad (7.112)$$

$$H_\theta = -\frac{H_0}{r} \xi \sin\theta \sum_{k=0}^{m} g_k \xi^k \left[k + 2 + \frac{\xi}{r} + \frac{\xi^2}{(k+3) r^2} \right] - H_0 \sin\theta.$$

Note that the value $\xi = 0$ corresponds to the front of the electromagnetic wave. It follows from the solution (7.112) that at $\xi = 0$ the components of the vectors **E** and **H** take on their initial values.

Let $D = D_0 =$ const, $\mathbf{H}_0 \neq 0$ and $\mathbf{E}_0 \neq 0$, and assume for simplicity that $\mathbf{H}_0$ is perpendicular to $\mathbf{E}_0$. A strong shock wave of constant velocity can be produced, for example, by a spherical piston that expands in the gas from a certain point (taken to be the origin) at constant velocity. The problem is then self-similar and its exact solution can be obtained by introducing the self-similar independent variable $\lambda = r/D_0 t$, separating the variables in the system (7.107), and solving a system of linear ordinary differential equations followed by a choice of the arbitrary constants from the boundary condition. The solution is of the form

$$H_r = H_0 f_1(\lambda) \cos\theta, \qquad H_\theta = H_0 f_2(\lambda) \sin\theta + E_0 f_3(\lambda) \cos\varphi,$$

$$H_\varphi = -E_0 f_3(\lambda) \sin\varphi \cos\varphi, \qquad E_r = E_0 f_4(\lambda) \sin\varphi \sin\theta, \quad (7.113)$$

$$E_\theta = E_0 f_5(\lambda) \sin\varphi \cos\theta, \qquad E_\varphi = H_0 f_3(\lambda) \sin\theta + f_5(\lambda) \cos\varphi,$$

where $f_i(\lambda)$ $(i = 1, \ldots, 5)$ are known functions; they are given in our earlier paper.[31]

Note that the solution (7.113) for $E_0 = 0$ can be obtained from Eqs. (7.112) with $m = 0$.

The solution of the Maxwell equations in the form (7.112) can be used to determine approximately the electromagnetic-wave parameters when the laws governing $D(t)$ differ from Eqs. (7.111) and simulate the dependences for the blast. To this end it is necessary to substitute in Eqs. (7.112) the constants g_k obtained from the condition that the given $D(t)$ or $D(r_2)$ dependence be specified with the aid of Eqs. (7.111). In particular, this solution can be used when tables of $D(t)$ and $r_2(t)$ are available. It must be noted here that if the electromagnetic wave is initially at $r = r_0$, the variable ξ in Eqs. (7.112) must be given by $\xi = r - r_0 - ct$.

7.2. Case of plane waves

We use a Cartesian frame with the Oy and Oz axes parallel to $\mathbf{E}$ and $\mathbf{H}$, respectively, with E and H having in the unperturbed region constant values E_{y0} and H_{z0}. The condition (7.106) then takes the form

$$E_{y1} = -\frac{1}{c}(v_2 - D)H_{z1}. \tag{7.114}$$

From Eqs. (7.107) we have

$$\frac{\partial E_y}{\partial x} = -\frac{1}{c}\frac{\partial H_z}{\partial t}, \qquad -\frac{\partial H_z}{\partial x} = \frac{1}{c}\frac{\partial E_y}{\partial t}. \tag{7.115}$$

It follows from Eqs. (7.115) that E_y and H_z satisfy the wave equations; in addition,

$$E_y + H_z = \Phi(\xi), \qquad E_y - H_z = F(\eta), \tag{7.116}$$

where $\xi = x - ct$, $\eta = x + ct$, and $\Phi(\xi)$ and $F(\eta)$ are arbitrary functions.

Let us consider specific problems.

(A) Propagation of a plane shock wave. Let a strong shock wave begin to propagate in the gas at the initial instant $t = 0$ and let the pressure on its front vary as

$$p_2 = p_0(x_0/x_2)^\beta, \tag{7.117}$$

where p_0 and β are positive constants, x_0 is the initial position of the shock wave, and $x_2(t)$ is the shock-wave coordinate. The case $\beta = 0$ in Eq. (7.117) was considered earlier in Ref. 1. The gasdynamic conditions on the strong waves are

$$v_2 = \frac{2}{\gamma + 1}D, \qquad D^2 = \frac{\gamma + 1}{2}\frac{p_2}{\rho_1} \qquad \left(D = \frac{dx_2}{dt}\right). \tag{7.118}$$

From Eqs. (7.117) and (7.118) we obtain the law of motion of the shock wave

$$x_2 = x_0 \frac{0.5\beta}{1 + 0.5\beta}\left[x_0 + (0.5\beta + 1)\left(\frac{p_0}{2}\frac{\gamma + 1}{\rho}\right)^{1/2} t\right]\frac{1}{1 + 0.5\beta}. \quad (7.119)$$

An electromagnetic wave traveling ahead of the shock wave in the positive Ox direction alters the initial field E_{y0}, H_{z0} so that

$$E_{y1} - H_{z1} = E_{y0} - H_{z0}. \quad (7.120)$$

E_{y1} and H_{z1} on the shock-wave front are related by Eq. (7.106), which takes, when account is taken of Eqs. (7.118), the form

$$E_{y1} = \frac{D}{c}\frac{\gamma - 1}{\gamma + 1} H_{z1}. \quad (7.121)$$

Using Eqs. (7.116) and (7.119)–(7.121) we get

$$H_{z1} = \frac{E_{y0} - H_{z0}}{\left(\frac{p_0}{\rho_1}\frac{\gamma + 1}{2}\right)^{1/2}\left(\frac{\gamma - 1}{\gamma + 1}\right)\left[\frac{x_0}{f(\xi)}\right]^{\beta/2}\frac{1}{c} - 1}, \qquad E_{y1} = H_{z1} + E_{y0} - H_{z0},$$

where $f(\xi)$ is obtained from the relation

$$\xi = f + c\left(\frac{2}{\gamma + 1}\right)^{1/2}\left[\left(\frac{f}{x_0}\right)^{1+0.5\beta} - 1\right]\frac{1}{1 + 0.5\beta}.$$

Assuming the entire solution of the gasdynamic problem known and using for the gas motion behind the shock wave the freezing condition $H_z = \psi(s)\rho$, where $\psi(s)$ is an arbitrary function of the Lagrange coordinate s, we can obtain also the $H_z(x, t)$ or $H_z(s, t)$ dependence in the gas-flow region. The latter dependence is of the form

$$H_z = \frac{\gamma - 1}{\gamma + 1}\frac{E_{y0} - H_{z0}}{\rho_1}\frac{\rho(s, t)}{\left(\frac{p_0}{\rho_1}\frac{\gamma + 1}{2}\right)^{1/2}\left(\frac{\gamma - 1}{\gamma + 1}\right)\left(\frac{s}{x_0}\right)^{\beta/2}\frac{1}{c} - 1}.$$

(B) Strong blast along a plane. A blast of a charge in the form of a plane is produced at $t = 0$. The solution of the gasdynamic problem is known, and the shock-wave motion is given by

$$x_2 = \left(\frac{\tilde{E}}{\rho_1}\right)^{1/3} t^{2/3},$$

where $\tilde{E}$ is a constant connected with the blast energy (see Chap. 2).

We distinguish between two solutions of the problem:

(a) $E_{y0} \equiv 0$; $H_{z0} = 0$ for $0 \le x < x_0$; $H_{z0} \neq 0$ for $x > x_0$, where $x_0 \ge \tilde{E}/\rho_1 c^2$;

(b) a plane electromagnetic wave is incident at the instant $t_0 > \tilde{E}/\rho_1 c^3$ on the blast-induced shock wave.

Case a. Arguments similar to those for case A yield for $x < x_0$

$$H_{z1} = 0, \qquad E_{y1} = 0;$$

and for $x \geq x_0$

$$H_{z1} = 1 - \frac{H_{z0}}{1 - \dfrac{2}{3c}\left(\dfrac{\tilde{E}}{\rho_1}\right)^{1/3}[\tau(\xi)]^{1/3}\left(\dfrac{\gamma-1}{\gamma+1}\right)}, \qquad E_{y1} = H_{z1} - H_{z0},$$

where $\tau(\xi)$ is given by

$$\tau = \frac{1}{c}\left[\left(\frac{\tilde{E}}{\rho_1}\right)^{1/3}\tau^{2/3} - \xi\right] \qquad (\tau \geq t_0).$$

Case b. The problem of reflection of an electromagnetic wave from a shock-wave surface can be solved by using the preceding conclusions. It need only be recognized that E_{y0} and H_{z0} are in this case not arbitrary, since $E_{y0} = H_{z0}$. This follows from the fact that prior to the collision of the shock and electromagnetic waves there was no electric or magnetic field in front of the shock wave.

We note in conclusion that the results obtained in this section can be used to determine the parameters of a gasdynamic shock wave if the parameters of the radiated electromagnetic wave are known.

8. Numerical investigations of cylindrical MHD blast waves in a radiating gas

8.1. Basic equations

We consider the mathematical models and the numerical solutions of problems involving a blast from a cylindrical charge in a gas in the presence of an electromagnetic field.[33] We choose the system of equations describing one-dimensional motion of an inviscid, non-heat-conducting, radiating gas with cylindrical waves in a magnetic field directed along the symmetry axis, in the form

$$\frac{1}{\rho} = v = \frac{1}{2}\frac{\partial r^2}{\partial m}, \qquad u = \frac{\partial r}{\partial t}, \qquad m = \int_0^r \frac{1}{v} r\, dr,$$

$$\frac{\partial u}{\partial t} = -r\frac{\partial}{\partial m}\left(p + \frac{H^2}{8\pi}\right), \qquad \frac{\partial \varepsilon}{\partial t} + p\frac{\partial v}{\partial t} = q,$$

$$\frac{1}{c}\frac{d\varphi}{dt} = 2\pi r E, \qquad \frac{c}{4\pi}\frac{\partial H}{\partial r} = \sigma E, \qquad \varphi = 2\pi\int_0^r H(\xi, t)\xi\, d\xi, \quad (7.122)$$

$$\Omega \nabla I_\omega = k(B_\omega - I_\omega), \qquad q = v\left[\sigma E^2 - \frac{1}{2\pi}\int_{4\pi}\int_0^\infty k(B_\omega - I_\omega)\, d\omega\, d\Omega\right],$$

$$p = p(T, v), \qquad \varepsilon = \varepsilon(T, v), \qquad \sigma = \sigma(T, v), \qquad k = k(T, v, \omega).$$

Here m and r are Lagrange and Euler coordinates, respectively, v the specific volume, u the velocity, ε the internal energy, H and E the magnetic and electric field strengths in the intrinsic coordinate system, c the speed of light, φ the magnetic field flux, σ the conductivity, I_ω the radiation intensity at the frequency ω, Ω the unit vector of the light beam, k the absorption coefficient, $B_\omega(T, \omega)$ the Planck radiation function, T the temperature in Kelvins, and q the specific heat supply.

Note that the system (7.122) presupposes local thermodynamic equilibrium and the absence of radiation scattering.

When actual problems are solved it is necessary to add to the above system the relevant boundary and initial conditions. In addition, it is necessary to provide in some specific form the functions that determine the connection between p, ε, σ, and k with T and v, and for k also the dependence on the frequency ω.

In general, the solution of the system (7.122) entails great difficulties in view of its integro-differential form. We consider therefore the transport equation in the diffusion approximation,[44] and assume the absorption coefficients to be independent of frequency, i.e., we assume a "gray-body" gas. With allowance for these simplifications, the total radiation intensity I takes the form

$$I = \frac{1}{2\pi}\int_0^\infty I_\omega d\omega = I(r,\Omega) = I_0(r) + 3I_1(r)\mu,$$

where μ is the cosine of the angle between the vector Ω and the radial direction, q in the heat-influx equation is given by

$$q = v[\sigma E^2 + 4k(\hat{a}T^4 - I_0)],$$

and the transport equation becomes

$$\frac{1}{r}\frac{d}{dr}(I_1 r) + k\left(I_0 - \frac{\hat{a}T^4}{\pi}\right) = 0, \qquad \frac{dI_0}{dr} + 3kI_1 = 0, \qquad (7.123)$$

where $\hat{a}$ is the Stefan–Boltzmann constant.

Let r_*, v_*, t_*, and T_* be, respectively, the characteristic distance, the specific volume, the time, and the temperature. We introduce the dimensionless independent variables and functions

$$\bar{r} = \frac{r}{r_*}, \quad \bar{v} = \frac{v}{v_*}, \quad \bar{t} = \frac{t}{t_*}, \quad \bar{T} = \frac{T}{T_*}, \quad \bar{u} = u\frac{t_*}{r_*} = \frac{u}{u_*},$$

$$\bar{p} = p\frac{v_*}{u_*^2} = \frac{p}{p_*}, \quad \bar{H} = \frac{H}{\sqrt{8\pi p_*}} = \frac{H}{H_*}, \quad \bar{E} = \frac{Ec}{H_* u_*}, \quad \bar{k} = kr_*,$$

$$\bar{I} = \frac{I\pi}{\hat{a}T^4}, \quad \bar{m} = \frac{mv_*}{r_*^2}, \quad \bar{q} = \frac{qt_*}{p_* v_*}, \quad \bar{\varphi} = \frac{\varphi}{r_*^2 H_*}, \quad \bar{e} = \frac{e}{p_* v_*}.$$

The system (7.122) for the dimensionless quantities takes under the indicated simplifying assumptions the form (the superior bars are omitted):

$$\frac{1}{\rho} = v = \frac{1}{2}\frac{\partial r^2}{\partial m}, \qquad u = \frac{\partial r}{\partial t}, \qquad m = \int_0^r \frac{r}{v}\,dr,$$

$$\frac{\partial u}{\partial t} = -r\frac{\partial}{\partial m}(p + H^2), \qquad \frac{\partial e}{\partial t} + p\frac{\partial v}{\partial t} = q,$$

$$\frac{\partial \varphi}{\partial t} = 2\pi r E, \qquad \frac{\partial H}{\partial r} = R_m E, \qquad \varphi = 2\pi\int_0^r Hr\,dr,$$

$$\frac{1}{2}\frac{d}{dr}(I_1 r) + k(I_0 - T^4) = 0, \qquad \frac{dI_0}{dr} + 3kI_1 = 0,$$

$$q = v\left[2R_m E^2 - \frac{k}{\mathrm{Bo}}(T^4 - I_0)\right].$$

Here R_m and Bo are the magnetic Reynolds number and the Boltzmann number, defined as

$$R_m = \frac{4\pi\sigma}{c^2} r_* u_*, \qquad \mathrm{Bo} = \frac{u_*^3}{v_* 4\hat{a} T_*^4}.$$

8.2. Finite-difference method

The numerical method of solving radiative-magnetohydrodynamics problems is based on the known numerical scheme of solving the one-dimensional nonstationary equations of gasdynamics[33–35] in which artificial viscosity is introduced. The functions $p, v, e, T, \sigma, k, H, I$, and I_0 are defined at integer instants of time and at the centers of the mass intervals. The quantities r, E, and I_1 refer to integer time instants and to the end points of the mass intervals. The values of Q* and q are defined at half-integer time instants and at the centers of the mass intervals, while u is defined at half-integer instants and at the end points of the mass intervals. The subscripts in the difference equations that follow pertain to spatial points, and the superscript to temporal layers.

The finite-difference equations take the following form:

$$u_j^{n+1/2} = u_j^{n-1/2} - \frac{\Delta t^n}{\Delta m_j} r_j^n \left[p_{j+1/2}^n - p_{j-1/2}^n + Q_{j+1/2}^{n-1/2} - Q_{j-1/2}^{n-1/2} + (H_{j+1/2}^n)^2 - (H_{n-1/2}^n)^2\right],$$

where

$$\Delta m_j = (\Delta m_{j+1/2} + \Delta m_{j-1/2})/2, \qquad \Delta t^n = (\Delta t^{n+1/2} + \Delta t^{n-1/2})/2,$$

$$Q_{j-1/2}^{n-1/2} = \frac{C_1(\Delta m_{j-1/2})^2 (v_{j-1/2}^n - v_{j-1/2}^{n-1})^2}{2v_{j-1/2}^{n-1/2}(\Delta t^{n-1/2})^2 (r_{j-1/2}^n)^2} + \frac{C_2 \Delta m_{j-1/2}\,|v_{j-1/2}^n - v_{j-1/2}^{n-1}|}{2v_{j-1/2}^{n+1/2}\Delta t^{n-1/2} r_{j-1/2}^n}$$

for $\quad v_{j-1/2}^n < v_{j-1/2}^{n-1} \quad$ and $\quad Q_{j-1/2}^{n-1/2} = 0 \quad$ if $\quad v_{j-1/2}^n \geq v_{j-1/2}^{n-1}$.

* Q is the "pressure" due to the artificial viscosity.

The equation for the Euler coordinate is

$$r_j^{n+1/2} = r_j^n + u_j^{n+1/2} \Delta t^{n+1/2},$$

the continuity equation

$$v_{j-1/2}^{n+1} = \frac{(r_j^{n+1})^2 - (r_{j-1}^{n+1})^2}{2\Delta m_{j-1/2}},$$

the radiation-transport equations

$$(I_1 r)_{j+1}^n - (I_1 r)_j^n + (r[I_0 - T^4])_{j+1/2}^n \Delta\tau_{j+1/2}^n = 0,$$

$$(I_0)_{j+1/2}^n - (I_0)_{j-1/2}^n + 3(I_1 \Delta\tau)_j^n = 0,$$

where

$$\Delta\tau_j^n = (\Delta\tau_{j+1/2}^n + \Delta\tau_{j-1/2}^n)/2, \qquad \Delta\tau_{j+1/2}^n = k_{j+1/2}^n \Delta r_{j+1/2}^n,$$

$$r_{j+1/2}^n = (r_{j+1}^n + r_j^n)/2,$$

the electromagnetic field equations

$$H_{j+1/2}^{n+1}[(r_{j+1}^{n+1})^2 - (r_j^{n+1})^2] - H_{j+1/2}^n[(r_{j+1}^n)^2 - (r_j^n)^2]$$
$$= \Delta t^{n+1/2}[(rE)_{j+1}^{n+1} - (rE)_j^{n+1} + (rE)_{j+1}^n - (rE)_j^n],$$

$$H_{j+1/2}^{n+1} - H_{j-1/2}^{n+1} = (R_m)_j^n (E)_j^{n+1} \Delta r_j^{n+1},$$

where

$$\Delta r_j^{n+1} = (\Delta r_{j+1/2}^{n+1} + \Delta r_{n-1/2}^{n+1})/2, \qquad \Delta r_{j+1/2}^n = r_{j+1}^n - r_j^n,$$

$$(R_m)_j^n = [(R_m)_{j+1/2}^n + (R_m)_{j-1/2}^n]/2,$$

and the heat-influx equation

$$e_{j+1/2}^{n+1} = e_{j+1/2}^n - [(p_{j+1/2}^{n+1} + p_{j+1/2}^n)/2 + Q_{j+1/2}^{n+1/2}](v_{j+1/2}^n - v_{j+1/2}^{n+1})$$
$$+ q_{j+1/2}^{n+1/2} \Delta t^{n+1/2},$$

where

$$q_{j+1/2}^{n+1/2} = v_{j+1/2}^{n+1/2} \left\{ (2R_m)_{j+1/2}^n (E_{j+1/2}^{n+1/2}) - \frac{k_{j+1/2}^n}{\mathrm{Bo}} [(T_{j+1/2}^n)^4 - (I_0)_{j+1/2}^n] \right\},$$

$$v_{j+1/2}^{n+1/2} = (v_{j+1/2}^{n+1} + v_{j+1/2}^n)/2, \qquad E_{j+1/2}^{n+1/2} = (E_j^{n+1} + E_{j+1}^{n+1} + E_j^n + E_{j+1}^n)/4.$$

The calculation of one step in time is carried out in the following sequence.

The radiation transport equations that determine the radiation energy lost and supplied are solved for the nth time layer, using the gasdynamic parameters known for it, by the flux sweeping method.[36] The velocity, Euler coordinate, and specific volume are determined explicitly from the difference equations for the $(n + 1)$st time layer. This is followed by solution of the implicit equations for the electromagnetic field, with the conductivity σ calculated from the parameters of the nth time layer. The calculations are carried out by the flux sweeping method. The results and the equation for the heat influx are used to calculate the internal energy, the pressure, and the temperature on the $(n+1)$st time layer, using the corresponding equations of state. The permissible time step for this finite-difference scheme was chosen to be the minimal time step meeting the conditions connected with the artificial viscosity[35] and the Courant condition, in which the sound is taken to have the speed $a = \sqrt{a_s^2 + 2vH^2}$ of a fast magnetosonic wave, where a_s is the gasdynamic speed of sound. In addition, it was required that the time step not exceed a duration in which the internal energy changes by $\approx 1\%$ of its value in the nth layer. This condition is imposed because the heat-flux equation contains terms connected with energy supply.

Since the electromagnetic field equations were chosen to be implicit, they impose no limit on the temporal step.

8.3. Examples of solutions

The described numerical method can be used to solve a large group of one-dimensional problems. We consider by way of example the problem of a blast from a condensed explosive in a quiescent cold gas of density ρ_0, pressure p_0, and temperature T_0 in the presence of a uniform magnetic field H_0.

We assume that the cylindrical detonation wave is induced in the explosive by applying an infinitesimal amount of energy along the cylinder axis, and that the wave constitutes a gasdynamic discontinuity on which the conservation laws and the Chapman–Jouguet condition are satisfied, and an energy Q_* is released per unit mass. The detonation products are assumed to be electrically nonconductive and optically transparent, and their intrinsic radiation is disregarded. In addition, we neglect the influence of the leading radiation on the parameters of the unexcited gas ahead of the shock wave produced by expansion of the detonation products.

In accord with the foregoing, the boundary conditions reduce to the following: $u = 0$ on the symmetry axis, p, u, and H are continuous, and $I(\Omega) = I(-\Omega)$ on the surface bounding the detonation product, $T = T_0$, $\rho = \rho_0$, and $H = H_0$, and $I(\Omega) = 0$ for $(\Omega \times r) < 0$ at infinity. The last condition means that there is no radiation flux at infinity.

In the actual calculations we need some specific form of the equation of the state of the medium whose motion is investigated. In our present case we calculate here the thermodynamic functions of the detonation products from the known analytic relations, which are of the form

$$p = \frac{1}{\beta v}(e - e^0) + p^0, \qquad \ln T = \sum_{k=0}^{4} T_k (\ln v)^k, \qquad \frac{1}{\beta} = \sum_{k=0}^{3} B_k (\ln v)^k,$$

$$\ln e^0 = \sum_{k=0}^{4} e_k (\ln p^0)^k, \qquad \ln p^0 = \sum_{k=0}^{4} p_k (\ln v)^k,$$

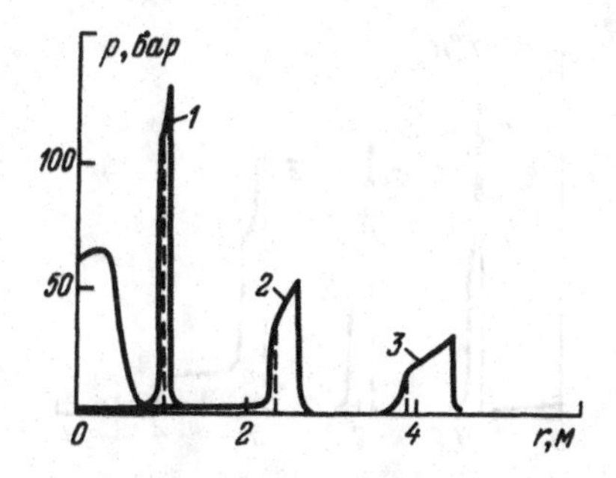

Figure 123. Pressure distributions for a blast produced in air by a cylindrical hexogen charge of initial radius 10 cm. Dashed lines—location of contact surface, T_0 = 300 K, p_0 = 100 Torr, $H_0 = 0$; (1) t_1 = 0.154 ms, (2) t_2 = 0.546 ms, (3) t_3 = 1.194 ms.

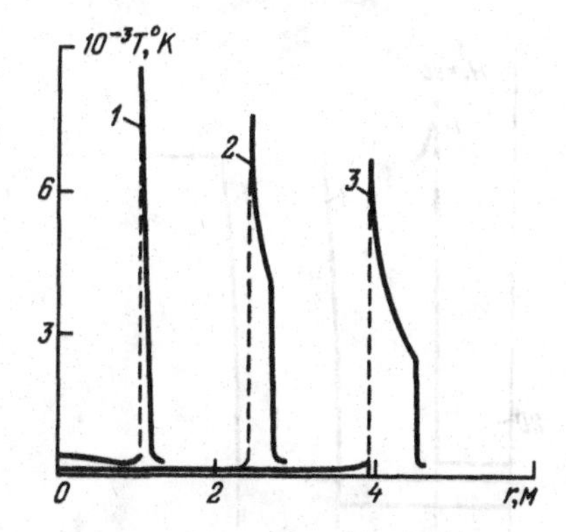

Figure 124. Temperature distributions for a blast produced in air by a cylindrical hexogen charge of initial radius 10 cm. The initial parameters and the times are the same as in Fig. 123.

where p_k, e_k, T_k, and B_k are constants that depend on the physico-chemical properties of the explosive. They are determined by averaging the data obtained from WKB equations of state.[37] The values of the functions $e(T, v)$, $p(T, v)$, $\sigma(T, v)$, and $k(T, v)$ for the ambient gas were calculated by linear approximation of tabulated data that take the real physical processes into account.[38–40]

Determination of the motion of the detonation products prior to the instant of the emergence of the detonation to the charge surface reduces to solving a self-similar problem analogous to that given in Ref. 41, but with a more complicated equation of state for the detonation products, i.e., in final analysis, to integration of a system of ordinary differential equations with corresponding initial conditions on the Chapman–Jouguet detonation wave. The calculations were performed for hexogen (RDX) with $\rho = 1.8$ g/cm^3. They show that the distribution of the gasdynamic parameters behind the wave is similar to that for detonation propagating in a combustible gas.[41]

It is of interest to note that when detonation products expand in a vacuum (this process has already been calculated from finite-difference equations) the radial velocity distribution becomes rapidly linear in the course of motion, and the velocity of the boundary of the detonation products becomes constant at 9 km/s.[33]

Figures 123 and 124 show the distributions of the pressure p and the temperature T for a blast produced in air at T_0 = 300 K and p_0 = 100 Torr

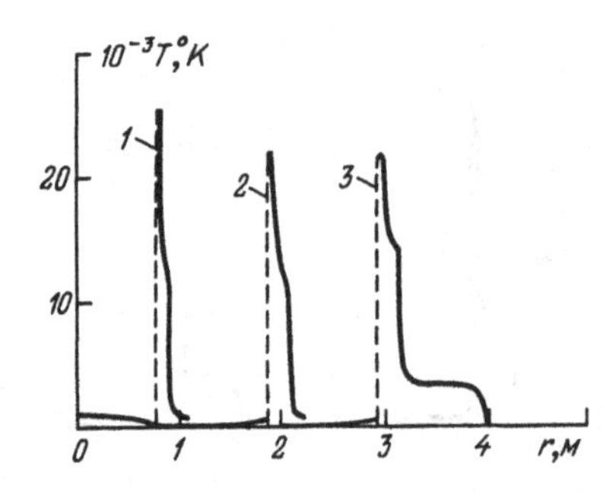

Figure 125. Cylindrical blast in argon. Temperature distribution for three instants of time (p_0 = 100 Torr, T_0 = 300 K, H_0 = 50 kG). Curves 1, 2, and 3 correspond to 0.1, 0.2, and 1.2 ms, respectively.

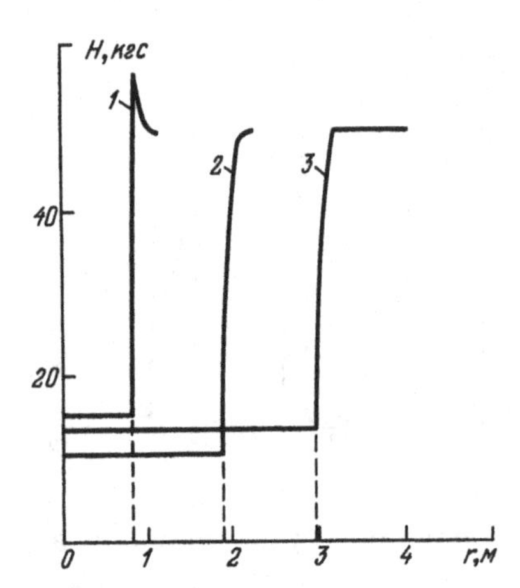

Figure 126. Magnetic field distribution for argon with the parameters of Fig. 125. Dashed lines—locations of contact surface.

by an explosive charge of radius 10 cm in the case $H_0 = 0$ for three instants of time.

Curves 1–3 pertain to t_1, t_2, and t_3 equal to 0.154, 0.546, and 1.194 ms, respectively. The vertical dashed lines mark the detonation-product boundary. One can clearly see a powerful shock wave propagating in air and a compression wave that moves in the detonation products away from their boundary as the flow develops.

Comparison with the results of a calculation with the radiation neglected has shown that, under equal conditions, the radiation, on the whole, has little influence on the character of the flow in this case, but allowance for the radiation leads nevertheless to a lowering of the maximum air temperature near the contact surface, by approximately 2000 K.

According to calculations, the energy lost to radiation by a blast in argon at $H_0 = 0$ is higher, up to 10% of the explosive-charge energy. Typically, the argon temperature behind the shock wave is approximately double that in air. The qualitative character of the flow is on the whole similar in both cases.

A substantial change in the flow pattern occurs in the presence of a strong external magnetic field.

Figures 125–127 show the distributions of the temperature T, the magnetic field strength H, and the radiation flux F in argon at H_0 = 50 kG, p_0

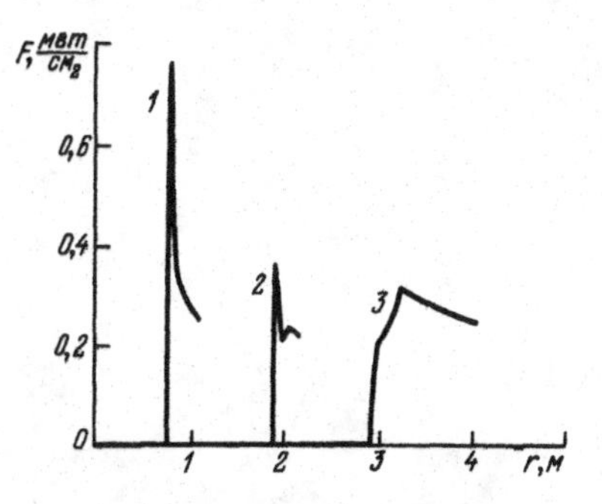

Figure 127. Radiation-flux distribution for a cylindrical blast in argon for three instants of time.

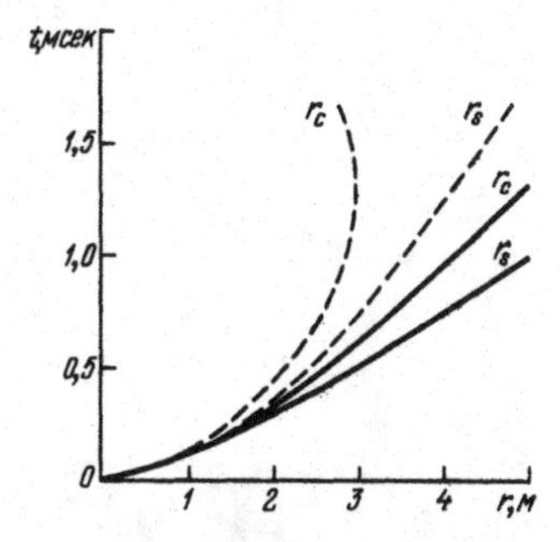

Figure 128. Time variation of the radii r_s of the shock wave and r_c of the contact surface. The solid and dashed curves correspond to H_0 = 0 and 50 kG, respectively.

= 100 Torr, and t_0 = 300 K for a blast produced by a hexogen charge of 10 cm radius. The interaction of the ionized argon with the strong magnetic field decelerates the detonation products in this case. Their boundary is established at an approximate distance of 3 m and begins to move in the opposite direction. Figure 128 shows for comparison the motions $r_s(t)$ of the shock wave and r_c of the contact surface in argon at $H_0 = 0$ (solid curve) and H_0 = 50 kG (dashed).

It is of interest to note that after an appreciable weakening of the shock wave, when the conductivity behind is low, the field interacts not with the entire moving gas mass but only with some part of it near the detonation-products boundary. By the time this boundary stops, the radiated energy is about 5% of the charge energy. The gas, however, is still quite hot and its emission is intense.

The interaction of a magnetic field with plasma streams has been numerically investigated by A. A. Samarskii and co-workers (see, e.g., Ref. 42).

Chapter 8
Propagation of perturbations in solar flares

1. Some data on the main parameters of solar flares and of the interplanetary medium

The onset of chromospheric solar flares and the propagation of the resultant perturbation in interplanetary space have been the subject of many studies. We investigate below only the propagation of flare-induced perturbations through the solar corona and the interplanetary medium.

We know (see, e.g., Refs. 1–4) that solar chromospheric flares are generated and take place in a relatively small volume (the area of a flare is about 0.1% of the area of the solar disk, and the height of the flare layer is of the order of 10^9 cm). The time of evolution of the processes in the flare nucleation site is, as a rule, 3×10^2–4×10^3 s.

A sizable energy is released when a flare develops in the chromosphere. Let E_0^* be the total energy released in the flare. This energy varies with the flare and is of the order of 10^{29}–10^{34} erg. A noticeable fraction of this energy is dissipated in the surrounding space by emission of optical and x rays, and also by fast particles (solar cosmic rays). The bulk of the energy, however, is consumed by the motion of the gas in the solar corona and in the interplanetary medium, i.e., by production of corpuscular streams. We denote this part of the energy by E_0. We shall be interested hereafter only in the energy E_0 and consider the motion of the medium outside the nucleation site of the flare, where the details of the onset and evolution of the flare are no longer of great significance.

Propagation of perturbations from a solar sphere takes place in the solar corona and in the interplanetary medium, the leading front of the perturbation reaching a distance on the order of an astronomical unit (AU) as a rule in a time 20–50 h after the start of the flare. It is known that a magnetic storm sets in on earth during these time intervals. Some specific observational data will be given in Sec. 6.

Let us consider the main properties of a quiescent interplanetary medium, i.e., prior to the start of the evolution of a high-power flare.

The sun emits continuously a stream of particles called the solar wind.[1] In the earth's orbit, the particle density is of order 1–10 cm^{-3}. The velocity v_1 of the solar wind does not decrease with the distance from the sun. Thus, radar measurements of the solar corona[4] have shown that the solar wind velocity at a height 3.5×10^{10} cm above the photosphere at the instant of measurement was approximately 1.6×10^6 cm/s, whereas in the earth's orbit this velocity exceeded 3×10^7 cm/s. Altogether, data of many rocket measurements show that at a distance of 1 AU the velocity v_1 equals 330–700 km/s. The interplanetary-plasma density decreases with the distance from the sun.[1,6]

We introduce a spherical polar coordinate frame (r, θ, φ) with center on the sun and with the polar axis aligned with the sun's rotation axis. In terms of this coordinate frame the gas density ρ_1, the solar-wind velocity $\mathbf{v}_1$, the gas pressure p_1, the magnetic field $\mathbf{H}_1$, and other parameters of the initial interplanetary medium dependent in general not only on the radial coordinate r, but also on the angles θ and φ and on the time t. This circumstance, and also the presence of so many unknown functions and additional parameters, makes it most difficult to determine theoretically the gas motion due to a solar flare. To proceed, we make several simplifying assumptions. First, we assume that the velocity $\mathbf{v}_1$ of the quiet solar wind is directed along the radius. Second, on the basis of the results of Refs. 1 and 4–8, devoted to theoretical and experimental determination of interplanetary-gas parameters, it can be concluded that at an initial gas density ρ_1 and a quiet solar-wind velocity $\mathbf{v}_1$ it is possible to assume, within a certain solar angle, $\varkappa < 2$ as seen from the sun, and in the region between the sun and the earth's orbit we can assume the following approximate relations:

$$\rho_1 = Ar^{-\omega}, \qquad v_1 = Br^{\omega-2}. \tag{8.1}$$

Here A, B, and ω are quantities which we shall regard as constants independent of the angles θ and φ and the time t. The actual values of A, B, and ω can be obtained from rocket-measurement and radar-measurement data (see Refs. 1, 4, 5, 8). To describe the gas motion we can invoke the following theoretical model.

(1) A kinetic description of the processes in a magnetized high-temperature three-component plasma consisting of electrons, ions, and neutral particles.

(2) Use of the magnetohydrodynamic equations of a two- or three-component plasma.

(3) Use of hydrodynamic approximations of a plasma with an anisotropic "pressure"[9,10] (see Chap. 1).

(4) Hydrodynamic model, in which a hydrodynamic approximation is used to describe the motion of the medium, and the equations of ordinary gasdynamics or of single-fluid MHD are employed.

In view of the extreme complexity of the considered phenomenon, we confine ourselves to a study of only the hydrodynamic model.

2. Use of dimensionality analysis and simplest similarity laws

We use the one-fluid hydrodynamic approximation to describe the motion of the medium, and assume the medium to be an ideal gas. Neglecting the energy-release time, the influence of the initial gas pressure p_1, the electromagnetic forces, and the viscosity and heat conduction, the system of the main characteristic gas-motion parameter is

$$r, \quad \theta, \quad \varphi, \quad t, \quad E_0, \quad A, \quad B, \quad \omega, \quad \gamma, \quad g_\odot, \quad R_\odot, \quad l, \quad \Omega, \tag{8.2}$$

where γ is the effective adiabatic exponent of the gas, Ω the average angular velocity of the sun's rotation, $g_\odot$ is the acceleration due to gravity on the sun's surface, $R_\odot$ the sun's radius, and l is the characteristic linear dimension of the flare nucleation site.

The quantities A, B, and ω can take on various values, depending on the time of the year and on the state of the interplanetary medium, while the energy E_0 is different for different flares. It is useful therefore to shed light on those basic dimensionless parameters which characterize the investigated influence, and consider questions arising when the functions describing the gas motion are recalculated when the parameters of the medium and the energy E_0 are varied.[11]

Since gravitation and the sun's own rotation have little influence on the perturbation propagation in flares, the parameters Ω and $g_\odot$ can be disregarded for rough estimates of the moving-gas properties. The dimensional parameters E_0, A, and B can be combined to make up the following quantities with dimension of length: the kinetic characteristic length $r^* = (E_0/AB^2)^{1/(\omega-1)}$ and the kinetic characteristic time $t^* = (r^*)^{3-\omega}/B$. For any dimensionless property of the flow (e.g., the density $g = \rho/\rho_1$) we can write then, on the basis of the π-theorem of dimensionality theory,

$$g = g\left(\frac{r}{r^*}, \frac{t}{t^*}, \theta, \varphi, \gamma, \omega, \alpha_1, \alpha_2\right), \tag{8.3}$$

where $\alpha_1 = R_\odot/r^*$, $\alpha_2 = l/r^*$, and $\alpha_2 \ll \alpha_1$. It follows from Eq. (8.3) that at fixed γ, α_1, and α_2 the dimensionless functions such as Eq. (8.3) will describe a class of flows for different parameters E_0, A, and B. If the characteristic dimensions of the motion region are much larger than $R_\odot$ we can disregard the effect of the finite radius of the sun on the gas motion in a certain vicinity of the leading front of the perturbations, i.e., the influence of the parameter α_1, so that α_2 can also be neglected.

If the solar-wind velocity is assumed within a certain solid angle to be directed along the radius, and the conditions of energy release can be assumed as corresponding to spherical symmetry of the flow, the gas flow within the considered solid angle $\varkappa$ can be regarded as spherically symmetric.

We introduce the notation $x = r/r^*$ and $\tau = t/t^*$. For the spherically symmetric flow model the functions such as Eq. (8.3) depend only on the two variable parameters x and τ, i.e.,

$$g = g(x, \tau, \gamma, \omega). \tag{8.4}$$

For the time of arrival of the leading front of a perturbation such as a shock wave at a point $r = r_a$, we have $t_a = t^*\tau(x_a, \gamma, \omega)$ and $x_a = r_0/r^*$, where the $\tau(x_a, \lambda, \omega)$ dependence is determined when the problem is solved.

Naturally, in a global analysis of perturbation propagation in a flare account must be taken of the dependence of the initial density and of the solar-wind velocity components on the angles θ and φ, and possibly also on the time t. In this case the sought solution of the hydrodynamic problem will depend on a large number of variables, and the functions in this solution will be of the form (8.3).

Consider a spherically symmetric model of the motion of the medium. If the motion of the quiescent solar wind is regarded as isothermal and it is desired to take the initial pressure p_1 into account, we have for the one-dimensional model

$$p_1 = Cr^{-\omega}. \tag{8.5}$$

In this case we have a new dimensional constant C, which makes it possible to introduce a new characteristic length $r^0 = (E_0/C)^{1/(3-\omega)}$. Equations such as Eq. (8.4) will contain one more dimensionless parameter

$$\alpha_3 = r^0/r^* = E_0^{\frac{1}{(3-\omega)} - \frac{1}{\omega-1}} C^{-\frac{1}{(3-\omega)}} (AB^2)^{\frac{1}{\omega-1}}.$$

Our estimates of the total initial solar-wind energy contained in a certain solid angle (see Sec. 4) show that the most important quantity over distances of the order of 1 AU is the kinetic energy of the quiet solar wind, and it exceeds by approximately an order of magnitude the total initial thermal energy of the gas (and is larger than the initial energy of the magnetic field). The parameter α_3 can therefore be neglected in a rough qualitative analysis of the considered phenomenon.

To describe the order of magnitude of the introduced kinetic characteristic length r^*, we refer r^* to a certain radius r_a and exclude AB^2 from the expression for r^* on the basis of the equations for ρ_1 and v_1. We obtain then

$$\frac{r^*}{r_a} = \left(\frac{E_0}{k_*}\right)^{1/(\omega-1)}, \qquad k_* = \rho_1(r_a)[v_1(r_a)]^2 r_a^3.$$

For α_2 we can write

$$\alpha_3 = E_0^{1/(3-\omega)} k^{-1/(3-\omega)} k_*^{1/(\omega-1)}, \qquad k = p_1(r_a) r_a^3.$$

Figure 129 shows for different ω the dependences of r^*/r_a on E_0 for the case when $r_a = 1.5 \times 10^{13}$ cm (the astronomical unit), $k_* = (1.5)^3 \times 10^{31}$ and $5 \times (1.5)^3 \times 10^{32}$ g cm^2/s^2 (dashed and solid lines, respectively), while the flare energy E_0 changes from 10^{28} to 10^{35} erg. The foregoing calculations show that for large E_0 the value of r^* can exceed an astronomical unit. Figure 130 shows a plot of $\alpha_3(k)$ for $k_* = (1.5)^3 \times 10^{32}$ g cm^2/s^2, $E_0 = 10^{33}$ erg, and $\omega = 2.5$ (the range of variation of k in this plot corresponds to a pressure change from 10^{-8} to 10^{-11} dyn/cm^2).

The dependences of the sought functions on the dimensionless parameters can be determined theoretically or experimentally by measurements with outer-space rockets during solar flares. In theoretical and experimental

Figure 129. Characteristic length r^* vs the flare energy E_0 for two values of k_*. (1) $\omega = 2$; (2) $\omega = 2.5$; (3) $\omega = 2.9$ (E_0 is referred to 1 erg and r^* to 1 AU).

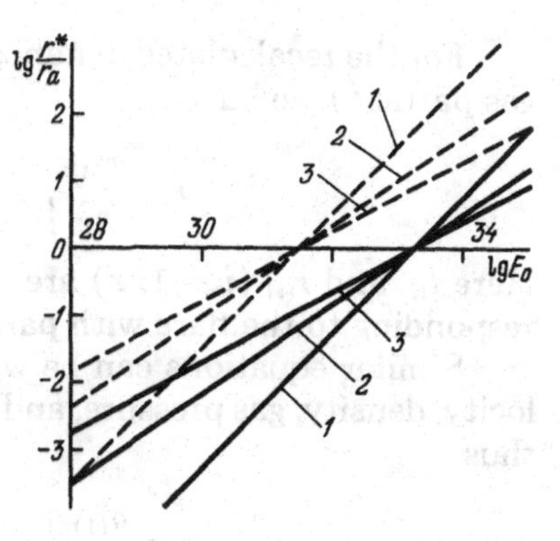

Figure 130. Ratio of the characteristic lengths versus the average-thermal-energy parameter k.

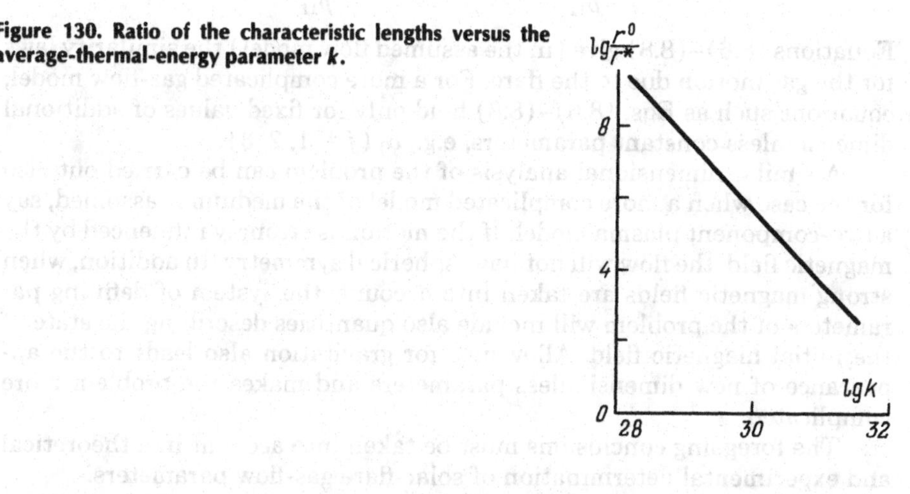

measurements of the gas parameters, the dependences of the density, pressure, and velocity on the coordinates and on the time, as well as the time of arrival of the perturbation at a given point, can be obtained only for certain fixed $E = E_{01}$, $A = A_1$, and $B = B_1$.

We consider therefore the recalculation of the data obtained for other values E_{02}, A_2, and B_2 of these constants. A possible procedure is to obtain the dimensionless time τ and the dimensionless coordinate x for the state E_{01}, A_1, B_1 and use the relations $r = r^* x$ and $t = t^* \tau$ to obtain the coordinates and the time for the state E_{02}, A_2, and B_2.

For the recalculated time we get

$$\tau = \frac{t_{(1)}}{t_1^*}, \quad t_{(2)} = t_2^* \tau = \frac{t_2^*}{t_1^*} t_{(1)}, \tag{8.6}$$

where

$$t_i^* = (r_i^*)^{3-\omega}/B_i, \quad r_i^* = \left(\frac{E_{0i}}{A_i \beta_i^2}\right)^{1/(\omega-1)}, \quad i = 1, 2.$$

For the recalculated distances (the Euler or Lagrange coordinate of the gas particle) we have

$$x = \frac{r_{(1)}}{r_1^*}, \qquad r_{(2)} = r_2^* x = \frac{r_2^*}{r_1^*} r_{(1)}. \tag{8.7}$$

Here $t_{(i)}$ and $r_{(i)}$ $(i = 1, 2)$ are the dimensional time and coordinates corresponding to the flare with parameters E_{0i}, A_i, and B_i.

Similar equations can be written also for the recalculation of the velocity, density, gas pressure, and other quantities. For the density ρ we have thus

$$g = \frac{\rho_{(1)}}{\rho_{11}}, \qquad \rho_{(2)} = \rho_{12} g = \frac{\rho_{12}}{\rho_{11}} \rho_{(1)}. \tag{8.8}$$

Equations (8.6)–(8.8) give (in the assumed flow model) the similarity laws for the gas motion due to the flare. For a more complicated gas-flow model, equations such as Eqs. (8.6)–(8.8) hold only for fixed values of additional dimensionless constant parameters, e.g., α_j $(j = 1, 2, 3)$.

A similar dimensional analysis of the problem can be carried out also for the case when a more complicated model of the medium is assumed, say a two-component plasma model. If the motion is strongly influenced by the magnetic field, the flow will not have spherical symmetry. In addition, when strong magnetic fields are taken into account, the system of defining parameters of the problem will include also quantities describing the state of the initial magnetic field. Allowance for gravitation also leads to the appearance of new dimensionless parameters and makes the problem more complicated.

The foregoing conclusions must be taken into account in a theoretical and experimental determination of solar-flare gas-flow parameters.

3. Point blast in an inhomogeneous moving medium

The foregoing features of the evolution of a flare during its initial stage lead to the conclusion that in a number of cases this phenomenon can be simulated by a point blast in a gas whose initial parameters are governed by the state of the solar corona and the interplanetary plasma. Consider the formulation of the problem in the simplest case of spherical symmetry. Assume that the gas is ideal, with a constant value of γ. We disregard the viscosity and thermal conductivity of the gas. The influence of the magnetic field and of gravitation on the gas motion is neglected. We arrive then at the following problem (see also Sec. 1 of Chap. 3).[12]

Let an energy E_0 be released from the point $r = 0$ at $t = 0$ in a gas with initial states $v_1(r)$, $\rho_1(r)$, and $p_1(r)$. It is required to determine the ensuing gas motion.

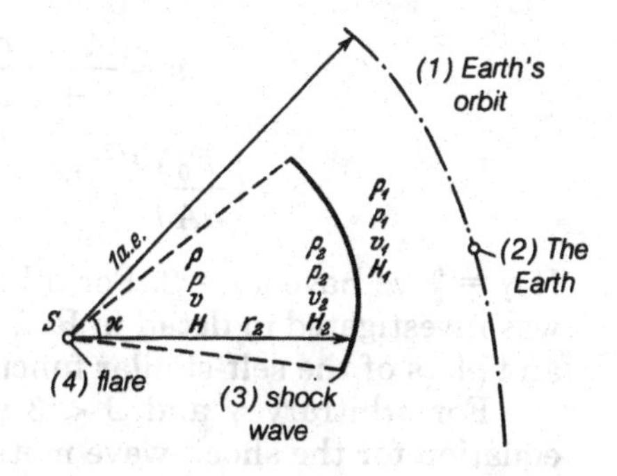

Figure 131. Pattern of shock-wave propagation. (1) Earth's orbit; (2) Earth; (3) shock wave; (4) flare.

Assuming for simplicity that the gas is initially isothermal, we have in accordance with approximate relations (8.1) the following initial data for v_1, ρ_1, and p_1 (at $r > 0$):

$$v_1 = Br^{\omega-2}, \qquad \rho_1 = Ar^{-\omega}, \qquad p_1 = Cr^{-\omega}. \tag{8.9}$$

To determine the ensuing gas flow it is necessary to integrate the system (3.5) of the gasdynamic equations for adiabatic perturbed motion. We denote by $r_2(t)$ the leading front of the gas perturbation, which we assume to be an ordinary shock wave. We have then in addition to the initial conditions also the condition

$$r_2(0) = 0 \tag{8.10}$$

and the condition that a finite energy E_0 be released at the point $r = 0$.

The boundary conditions on the shock wave at $r = r_2$ are of the form (3.6) (see Chap. 3). Besides the conditions on the shock wave, we have the condition at the symmetry center. Thus, if there is no continuously acting mass source at the center, we have $v(0, t) = 0$. It must be noted here once more that, strictly speaking, the initial functions $v_1(r)$, $\rho_1(r)$, and $p_1(r)$ should by themselves comprise a solution of stationary gasdynamic equations with allowance for gravitational and other forces, for otherwise motion will begin ahead of the front of the principal perturbation. For a rough qualitative analysis we assume that the effects of gas motion ahead of the shock-wave front can be neglected if approximate values of v_1, ρ_1, and p_1 are specified. An overall picture of the gas flow is shown in Fig. 131.

For $v_1 = 0$ and $p_1 = 0$ (or $B = C = 0$) we have the case of the well-investigated self-similar problem of a strong blast in a gas with variable initial density. The exact solution of this problem was investigated in Refs. 13–15 and in Chap. 2.

Let us examine some questions encountered in the solution of this problem. If $\omega = \omega_1 = (7 - \gamma)/(\gamma + 1)$ then, as already indicated in Chap. 2, the self-similar strong-blast problem has the analytic solution[13]

$$\rho = \frac{\gamma+1}{\gamma-1}\rho_1\frac{r}{r_2}, \qquad p = \frac{2\delta_1^2}{\gamma+1}\rho_1\frac{r_2^2}{t_2}\left(\frac{r}{r_2}\right)^3, \tag{8.11}$$

$$v = \frac{2\delta_1}{\gamma+1}\frac{r}{t}, \qquad \delta_1 = \frac{2}{5-\omega_1}, \tag{8.12}$$

$$r_2 = \left(\frac{E_0}{\alpha A}\right)^{\delta_1/2} t^{\delta_1}, \qquad \alpha = \frac{2\varkappa(\gamma+1)}{3(\gamma-1)(3\gamma-1)^2}. \tag{8.13}$$

If $\gamma = \frac{5}{3}$ we have $\omega_1 = 2$. For other values of γ and for $\omega = 2$ the solution was investigated in detail in Ref. 15, where the values of α were calculated and plots of the self-similar functions given.

For arbitrary γ and $\omega < 3$ we have in analogy with Eqs. (8.13) an equation for the shock-wave motion:

$$r_2 = \left(\frac{E_0}{\alpha A}\right)^{\delta/2} t^{\delta}, \qquad \delta = \frac{2}{5-\omega}, \tag{8.14}$$

where $\alpha(\gamma, \omega)$ is a constant introduced in Chap. 2 and determined from the integral energy-conservation law.

If $B \neq 0$ or $C \neq 0$ the problem is not self-similar and it is better to solve it by numerical or approximate methods.

The foregoing dimensionality analysis is fully applicable to our problem. We note here only the following. If a heavy point of mass M is assumed to be at the center of the blast and to interact with the gas in accord with Newton's law, it is necessary to add to the right-hand side of the system (3.5) the acceleration due to gravity:

$$F_g = -\frac{Mf}{r^2}. \tag{8.15}$$

In this case the system of defining parameters acquires one more dimensional constant $\zeta = Mf$ and a new characteristic length

$$r_g = \left(\frac{E_0}{\zeta A}\right)^{(2-\omega)/3}. \tag{8.16}$$

If $\omega = 2$, r_g is an abstract parameter, for in this case ζA has the dimension of energy. Note that if $C = B = 0$ the problem is self-similar even when gravitational forces are taken into account. Problems of this type were investigated in detail by M. L. Lidov.[16]

The dimensional analysis has shown that even in the simplest spherically symmetric hydrodynamic model of the flow the solution depends in the general case substantially on the constant parameters γ, ω, and α_3. This raises additional difficulties in the solution of the problem. To clarify the influence of the parameters B and C on the solution one can use the method, treated in detail in Chap. 3, of linearization about a self-similar solution. Two cases can be considered separately, $B \neq 0$, $C = 0$ and $B = 0$, $C \neq 0$.

The first case corresponds to allowance for the velocity of the unperturbed medium, and the second to allowance for variable back pressure. The solution of the problem for these two cases makes it possible to treat (in the linear formulation) both parameters simultaneously, using the principle of superposition of linear solutions.

4. Application of problem solutions to solar-flare propagation perturbations

The applicability of the solutions considered above to solar-flare propagation phenomena are subject to three basic limitations.

The first concerns the validity of a hydrodynamic description of the interplanetary medium. It is known that the plasma-particle mean free paths calculated from Coulomb collisions are much larger than 1 AU, raising the question of applicability of the continuous-medium model and the shock-wave theory. Investigations[1,17,18] have shown, however, that in a magnetized plasma such as the interplanetary medium the Larmor radius r_L as a characteristic physical parameter plays an important role, and the so-called collisionless shock waves can be realized (r_L of the interplanetary plasma is much less than 1 AU). It must be noted, on the other hand, that during the initial stage of the flare evolution the plasma is quite dense (the chromosphere density is 10^{12} cm^{-3}) and the use of a hydrodynamic description is quite acceptable here. Furthermore, starting with 1964, observations have been reported[4,19,20] of magnetohydrodynamic shock waves and other blasts in the interplanetary medium. It should also be noted that the fact that the leading front of the main perturbation reaches the earth's orbit after approximately a day cannot be readily explained without invoking shock-wave theory.[21]

The foregoing arguments in favor of the use of the hydrodynamic model give grounds for expecting a successful application of shock-wave theory to the study of phenomena in solar flares.

The second limitation of the proposed models of flare phenomena is that the gas motion is not one-dimensional, since both the initial properties and the ensuing motion are essentially three-dimensional. The use of one-dimensional spherically symmetric solutions, however, can be useful for the analysis of flows within a certain solid angle measured from the center of the flare or from the sun's center.

The third substantial limitation is connected with the assumption that the energy of the flare is released instantaneously. The actual duration of the energy release is in a number of cases about $\frac{1}{20}$ of the time required by the shock wave to travel a distance of 1 AU. To describe processes in a one-dimensional formulation it would then be necessary to assume that E_0 is a function of time. The simplest cases of such problems were considered in Refs. 1 and 13. An approximate formulation of the problem using instantaneous energy release can be used to estimate the gas-flow parameters in cases when the energy-release time does not exceed $\frac{1}{25}$ of the time for the wave to reach a certain point. Assuming that the flare energy is released within a certain period $E_0 = E_0(t)$, the flare can be simulated by piston expansion in a gas, assuming that the piston velocity depends on the time. It is possible to study also a more general model in which a point blast of energy E_0 is followed by motion, in the gas, of a piston with variable velocity, simulating the expansion of the gas from the flare nucleation center. A similar problem has already been investigated for ordinary gasdynamics.[22]

In the models noted we have neglected mass supply to the gas flow. Actually, however, ejection of a mass from the solar chromosphere into a moving stream can exert a definite influence on the character of the flow. To take this circumstance into account, the flare model must be additionally complicated. The following formulation of the problem can be proposed. Assume that at the initial instant $t = 0$ there is released instantaneously an energy E_0 from a point $r = 0$ into a solid angle $\varkappa$. In addition, an energy $N(t)$ and a mass $Q(t)$ are fed to the same point and play a part in the establishment of the motion inside the solid angle. If N and Q are power-law functions of t, and $E_0 = 0, p_1 = 0$, and $v = 0$, it is possible to distinguish here a class of self-similar solutions. A self-similar problem for a constant initial density was investigated in Ref. 23. An approximate analysis of the self-similar problem shows that the motion will have here a complicated character. A contact surface and a second shock wave are produced in the stream. (This is discussed in Sec. 7 of the present chapter.)

Note also that the dependences of the initial parameters ρ_1, p_1, and v_1 on the angles θ and φ could also be taken into account in a sector approximation, using a rough scheme similar to that used in Ref. 24.

It is of interest to estimate the initial energy of a quiet solar wind in a unit solid angle ($\varkappa = 1$) at a distance of 1 AU.

Using the approximation (8.9), we put

$$A = a \times 10^{-22} r_a^{\omega}, \qquad 0.01 \le a \le 1,$$

$$B = v_1(r_a) r_a^{2-\omega}, \qquad v_1(r_a) = 1.5k \times 10^7 \text{ cm/s},$$

$$1.5 \le k \le 4, \qquad C = \beta A \times 10^{12}, \qquad 0.01 \le \beta \le 15,$$

$r_a = 1$ AU, and the constants A, B, and C are given in grams, centimeters, and seconds. For the initial energy per unit solid angle, neglecting the sun's radius, we have

$$E_1 = \int_0^{r_a} r^2 \left(\frac{\rho_1 v_1^2}{2} + \frac{p_1}{\gamma - 1} \right) dr. \tag{8.17}$$

Taking Eqs. (8.9) into account, we get

$$E_1 \sim A \left[\frac{r_a^{\omega-1}}{\omega - 1} \frac{B^2}{2} + \frac{r_a^{3-\omega}}{(\gamma - 1)(3 - \omega)} \frac{C}{A} \right]. \tag{8.18}$$

The first term in Eqs. (8.17) and (8.18) corresponds to the initial kinetic energy E_{1k}, the second to the thermal energy E_{1T}, and other forms of energy are disregarded. For $\omega = 2$, $a = 0.1$, $\beta = 5$, and $k = 1.5$ we get $E_{1k} \sim 5 \times 10^{31}$ and $E_{1T} \sim 10^{30}$.

The bulk of the energy is thus contained in E_{1k}. The thermal and electromagnetic energies are substantially lower than the initial kinetic energy. On the other hand, the energy E_{1k} at a distance on the order of 1 AU and larger is comparable with the total flare energy, and it is desirable to take it into account in the theoretical models of the gas motion. An example of such an account was given in Sec. 3 of the present chapter and in Chap. 3.

For powerful, strongly localized in space, and rapidly developing flares we can use the results of the theory of a strong point blast. The simplest solution (8.11)–(8.13) in the particular case $\gamma = \frac{5}{3}$ and $\omega_1 = 2$ was used by Parker[1] to describe gas motion in flares (he disregarded completely the influence of the solar wind).

Using the results obtained for these very simple problems, we can attempt to investigate in greater detail questions connected with application of blast theory to solar flares. With respect to these flares, we note the following conclusions of our earlier investigation.[25]

(1) Using strong-blast theory we can determine the time of arrival of a shock wave at the earth's orbit or at some other point, if the flare energy is known. In fact, from Eqs. (8.14) we have

$$t_a = \left(\frac{\alpha A}{E_0}\right)^{1/2} r_a^{(5-\omega)/2}. \tag{8.19}$$

From Eq. (8.19), at equal values of ω we get for different flares

$$\frac{t_2}{t_1} = \left(\frac{A_2}{A_1}\frac{E_{01}}{E_{02}}\right)^{1/2}.$$

If $A_2 = A_1$ we have

$$\frac{t_2}{t_1} \sim \left(\frac{E_{01}}{E_{02}}\right)^{1/2}. \tag{8.20}$$

Since α and A can be regarded as known for the various γ and ω, we can approximately determine from Eq. (8.19) the time when the leading front of the perturbation reaches the earth. Equation (8.20) shows that the ratio of the squares of the times of perturbation arrival at a certain point is inversely proportional to the ratio of the flare energies.

(2) The flare energy can be determined if the arrival time t_a of the perturbation at a specified point is known:

$$E_0 = \alpha \frac{r_a^{5-\omega}}{t_a^2} A. \tag{8.21}$$

If $t_a = 10^5$ s and $r_a = 1$ AU, then at $\omega = 2$ we have $E_0 \sim a\alpha(1.5)^5 \times 10^{33}$ erg. For $\alpha \sim 1$ and $a \sim 0.1$ we have $E_0 \sim 10^{33}$ erg. More accurate estimates for specific flares will be given in Sec. 6.

(3) Similar conclusions and estimates can be obtained also when account is taken of the initial velocity v_1 (see Sec. 6 below) and of the initial pressure p_1. We note here that an analysis of the solutions obtained with allowance for back pressure in Chap. 3 shows that in a certain region behind the shock-wave front the density increases as r decreases, i.e., a negative density gradient is present. This effect does not occur if p_1 and ρ_1 are constant.

(4) To process the measured shock-wave parameters it is necessary to use the conclusions of dimensionality theory and the similarity laws.

(5) The effect of the flow on the magnetic fields and the influence of the gravitation on the gas flow can be taken into account.

The deformation of the magnetic field by the gas flow in the flare will be treated in greater detail in the next section.

5. Change of magnetic field in the moving gas

We describe the magnetic field in the spherical coordinate frame introduced in Sec. 1. Let $\mathbf{\Omega}$ be a vector directed along the sun's rotation axis, taken to be the polar axis. The components of the magnetic field $\mathbf{H}$ at the point (r, θ, φ) are H_r, H_θ, and H_φ, with H_r directed radially from the sun, H_φ along the vector $\mathbf{\Omega} \times \mathbf{r}$, and H_θ in the southern direction (we assume here the usual right-hand coordinate frame). Parker[1] has proposed an unperturbed magnetic field model in which the solar wind is radially directed outside a certain distance $r = r_0$ and has a constant velocity v_1 [the case $\omega = 2$ in Eqs. (8.1)]. The gas was assumed to be perfectly conducting. We consider a similar model for a variable velocity $v_1 = Br^{\omega-2}$.

The equation of the magnetic force lines in the (r, φ) plane is

$$\frac{dr}{r\,d\varphi} = \frac{H_r}{H\varphi} \sin\theta. \tag{8.22}$$

By analogy with Refs. 1 and 26, we can write for the components of the initial magnetic field

$$H_{r1} = H_{r0}\left(\frac{r_0}{r}\right)^2, \qquad H_{\theta 1} = H_{\theta 0} = 0, \tag{8.23}$$

$$H_{\varphi 1} = -H_{\varphi 0}\left(\frac{r_0}{r}\right)^2\left[\frac{\Omega}{v_1}(r - r_0)\sin\theta\right].$$

Here H_{r0}, $H_{\theta 0}$, and $H_{\varphi 0}$ denote the components of the field H at a certain point $(r_0, \theta_0, \varphi_0)$ at which the force line passing through the point (r, θ, φ) is formed.

Integrating Eq. (8.22) with allowance for Eqs. (8.23), we obtain the equation for the force lines

$$\varphi = \varphi_0 + \frac{r_0}{2-\omega}\left[1 - \left(\frac{r_0}{r}\right)^{2-\omega}\right]\frac{\Omega}{v_1} - \frac{r - r_0}{3-\omega}\frac{\Omega}{v_1}, \qquad \omega \neq 2,$$

$$\varphi = \varphi_0 + \left[1 - \frac{r}{r_0} + \ln\frac{r}{r_0}\right]\frac{r_0\Omega}{v_1}, \qquad \omega = 2. \tag{8.24}$$

Here $\Omega = 2.9 \times 10^{-6}$ rad/s is the average angular velocity of the sun's rotation.

The plots of Eqs. (8.24) in the (r, φ) plane are a family of spirals that depend on the parameters φ_0, r_0, ω, Ω, and B.

Relations (8.23) and (8.24) provide a more general model (compared with the Parker model[1]) of a quiet interplanetary field. Since the total magnetic field energy is substantially lower than the flare energy E_c, and $\rho_2 v_2$ is much larger than the initial pressure $H_1^2/8\pi$, we can disregard in first-order approximation the influence of the magnetic field on the motion of the interplanetary plasma.[27] We can calculate then the perturbed magnetic field parameters for spherically symmetric gas flows due to a strong blast in a medium of variable density $\rho_1 = Ar^{-\omega}$. In the one-fluid MHD approximation, we have on the basis of the equations given in Chaps. 2 and 7 (see also Refs. 13, 14, 27, and 28):

$$H_r = H_{r1}\left(\frac{\xi}{r}\right)^2, \qquad H_\theta = H_{\theta 1}\frac{\rho}{\rho_1}\frac{r}{\xi}, \qquad H_\varphi = \frac{\rho}{\rho_1}\frac{r}{\xi}H_{\varphi 1},$$
$$\frac{r}{r_2} = \Phi_1(\omega, \gamma, \mu), \qquad \frac{\xi}{r_2} = \Phi_2(\omega, \gamma, \mu), \qquad \frac{\gamma+1}{2\gamma} \le \mu \le 1. \tag{8.25}$$

Here ξ is the Lagrange coordinate of the particle, μ is a parameter, and the functions $\Phi_1(\omega, \gamma, \mu)$ and $\Phi_2(\omega, \gamma, \mu)$ are given by Eqs. (2.5) and (2.17).

Relations (8.22)–(8.25) allow us to write a differential equation for the force lines in the region of the perturbed motion of the gas

$$\frac{d\varphi}{dr} = \frac{(r_0 - r)\Omega}{v_1(\xi)}\frac{r^2}{\xi^2}\frac{\rho}{\rho_1}.$$

We denote by u the angle between the direction of the radius vector and the magnetic field; this angle is given by

$$\tan u = r\frac{d\varphi}{dr}. \tag{8.26}$$

In the general case of a strong shock wave we have for the difference between $\tan u$ at an arbitrary point of the stream and the initial value $\tan u_1$ (Ref. 29)

$$\tan u - \tan u_1 = \frac{H_{\varphi 1}}{H_{r1}\sin\theta}\left[\left(\frac{\xi}{r}\right)^2\frac{\rho_1}{\rho} - 1\right]. \tag{8.27}$$

From Eq. (8.27) we easily obtain an expression for the discontinuity of tan u on going through the shock wave:

$$\tan u_2 - \tan u_1 = +\frac{2}{\gamma - 1}\frac{H_{\varphi 1}}{H_{r1}\sin\theta}. \tag{8.28}$$

The magnetic field configurations obtained in this section can be useful for the reduction of results of interplanetary-medium observations. The calculated magnetic field configurations can be used to determine the motion of high-energy particles as they cross a magnetic field discontinuity. The electric induction field $\mathbf{E} = -(1/c)\mathbf{v} \times \mathbf{H}$ can accelerate these particles. Calculations of the particle motion (see Sec. 3 of Chap. 1) have shown that the electrons can be accelerated twofold on crossing a shock-wave front (if the initial magnetic field is uniform). We shall not consider this question here in detail.

6. Comparison of the conclusions of the theory with some observation data

Many results of observation of chromospheric solar flares and measurement of the parameters of shock waves produced in these flares and propagating in interplanetary space have by now been accumulated. Analysis of the optical and radio-astronomy measurements, observation data on the onset of magnetic storms on the earth, and measurements of shock-wave parameters and cosmic-ray intensities with satellites, space rockets, and automatic interplanetary stations[1,4,8,19,20,26,30–36] lead to the following general conclusions, some of which have already been mentioned above:

(1) the time of arrival of a perturbation at the earth's orbit ranges from 16 to 70 h;

(2) passage of a shock wave near the earth's orbit increases the plasma density by 2–3 times;

(3) optical measurements yield an estimate of 10^{30}–10^{32} erg for the flare energy;

(4) the averaged shock-wave velocity on the earth's orbit is of the order of 500–800 km/s;

(5) the average shock-wave velocities on the earth's orbit are as a rule not more than 2–3 times larger than the velocity of the quiet solar wind.

The foregoing experimental data were reduced by us[37–39] and the information we shall need is summarized in Table 6. This table lists the dates and time of the flare (universal time, UT), the coordinates on the sun, and the strength (B). We use the blast-wave model considered above to describe the perturbation propagation. In accordance with the indicated method of determining the flare energy from the known values of α, A, and the time t_a of arrival of the perturbation at a specified r_a, we have from the strong-blast equation (8.21)

$$E_0 = \alpha(\gamma) \frac{r_2^3}{t^2} m_p n_{1a} r_a^2 \qquad (\omega = 2), \tag{8.29}$$

where n_{1a} is the density of the particles on the earth's orbit or at some other measurement point, and m_p is the proton mass.

We assume that the flare energy went to move the gas inside a solid angle $\varkappa = 2\pi$. The effective adiabatic exponent γ can vary with the state of the interplanetary medium. For a monatomic gas and sufficiently frequent collisions we have $\gamma = \frac{5}{3}$. If ionization and radiation effects are taken into account, this exponent is close to 1.3. For a tenuous plasma with very infrequent collisions, $\gamma = 3$ is more reasonable. Indeed, as noted in Chap. 1 for the MHD equations for a tenuous plasma, if $p_{\parallel} > p_{\perp}$ and for a weak magnetic field it will follow from the hydrodynamic equations and from the conditions on the shock wave that $\gamma = 3$. Note that in the case $\varkappa = 2\pi$ we have $\alpha = 1.047$ ($\gamma = \frac{5}{3}$) and $\alpha \sim 0.2$ ($\gamma = 3$). The energies estimated from Eq. (8.29) are listed in Table 6. Changing Eq. (8.16) to dimensionless form, we can verify the similarity law[11] for the flares by means of the $r_2(t)$ dependence. We have taken here for E_0 the values calculated at $\gamma = \frac{5}{3}$ for the coordinate closest to the sun. When the arrival time was measured with detectors at two points of space, the dimensionless parameters $x_2 = r_2/r^*$ and $\tau = t/t^*$ were calculated for both points. A check on the locations, on the plot, of the points corresponding to $r_2^{(2)}$ and t_2 indicates good satisfaction of the similarity law (Table 6).

For the case of measurement at two points of space, we determined also the average propagation velocities D_{av}. The theoretical equation for the shock-wave velocity with the wind disregarded is

$$\frac{D}{v_1} = \frac{dx_2}{dt} = y^{-1} = \frac{\sqrt{\alpha}}{\delta}\, x_2^{(\omega-1)/2}. \tag{8.30}$$

When drift v_1 by the solar wind is taken into account we have, in accordance with the results of Sec. 4 of Chap. 3,

$$x_2 = \left(\frac{\delta}{\sqrt{a}}\right)^{2/(\omega-1)} y^{2/(\omega-1)} \exp\frac{2A_{11}y}{\omega-1}, \tag{8.31}$$

$$\tau = \tau_0 + \int_{x_{20}}^{x_2} y(x)\,dx, \qquad x_{20} = \alpha^{\delta/2}\tau_0^{\delta}. \tag{8.32}$$

Let $\omega = 2$ and $\gamma = \frac{5}{3}$, and then $A_{11} = 0.563$. The flare energies were estimated from the $x_2(\tau)$ dependence that follows from Eqs. (8.31) and (8.32). The corresponding values of E_0 turned out to be lower than those calculated by the strong-blast theory without allowance for the wind (see Table 6).

A theoretical plot of $x_2(\tau)$ with allowance for the wind is shown in Fig. 132 (solid lines), where the dash-dot line corresponds to the self-similar $x_2(\tau)$ dependence. The theoretical dependences of $D/v_1 = D$ on x_2, obtained from Eqs. (8.30) and (8.31), are shown in Fig. 133. The same figure shows the data obtained by calculating the averaged velocities $D_{av} = (r_2^{(2)} - r_2^{(1)})/(t_2 - t_1)$. The coordinate of the point for D_{av} was taken to be $(r_2 + r_2^{(2)})/2$. In addition, from the values of D and v in Eq. (8.31) we determined the values of x_2 and the flare energy E_0, and the results were then used to find, for the time $t_{av} = (t_1 + t_2)/2$, the corresponding values of τ ($\tau_{av} = t_{av}/t^*$). These data on the connection between x_2 and τ are marked by circles in Fig. 132. The vertical lines in Fig. 133 correspond to the expected errors in the determined values of D_{av}.[33,35] The numbers of the points correspond to the flare numbers indicated in Table 6.

Another possible method of determining E_0 is to use the measured density ratios ρ_2/ρ_1, using them to calculate y from the conditions on the shock wave, and determining E_0 from the $x(y_0)$ dependence using Eq. (8.31). Naturally, it is possible also to take the back pressure into account. The

Table 6. Flare observation data. Key: (1) flare description; (2) propagation time and distance; (3) parameters of medium; (4) estimate of energy E_0, 10^{31} erg; (5) strong-blast model; (6) allowance for wind v^1.

		(2)				(3)			(4)			
		t, h		r_2, 10^6 km					(5)		(6)	
N	(1)	t_1	t_2	$r_2^{(1)}$	$r_2^{(2)}$	$n_1, \frac{1}{cm^3}$	$v_1, \frac{km}{s}$	D mean	$\gamma = \frac{5}{3}$	$\gamma = 3$	E_0 from $x_2(t)$	E_0 from $x_2(\gamma)$
1	4 X 1965 0937 *UT* S24W31, 2+	17,1	26	150	170	6	350	620	201	38,5	165	17
2	18 I 1966 2254 *UT* N20E07, 2*B*	44	55	128	147	9	380	480	28,0	5,42	14,0	9,9
3	7 VII 1966 0027 *UT* N34W48, 2*B*	44,5	44,52	152	152,02	4	400	710	21,0	4,00	11,6	14
4	18 IX 1967 1918 *UT* N18W84, 2*B*	16	27	116	150	20	450	860	210	40,5	168	55
5	17 IX 1967 2314 *UT* N16W61, 2*B*	28	46	117	150	20	450	510	71,8	14,0	41,0	13,2

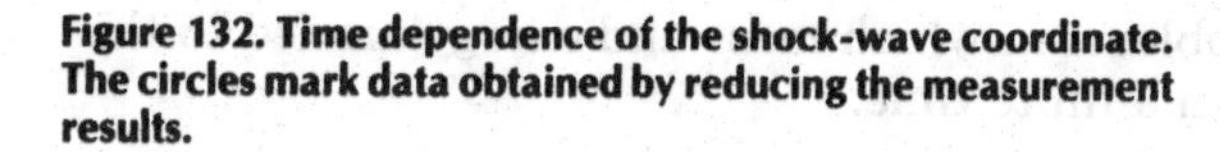

Figure 132. Time dependence of the shock-wave coordinate. The circles mark data obtained by reducing the measurement results.

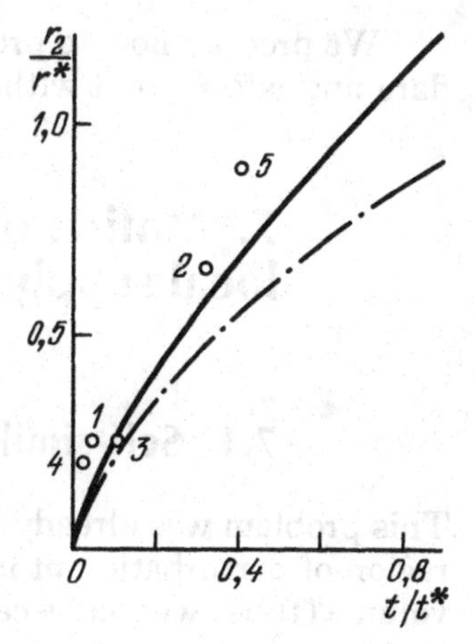

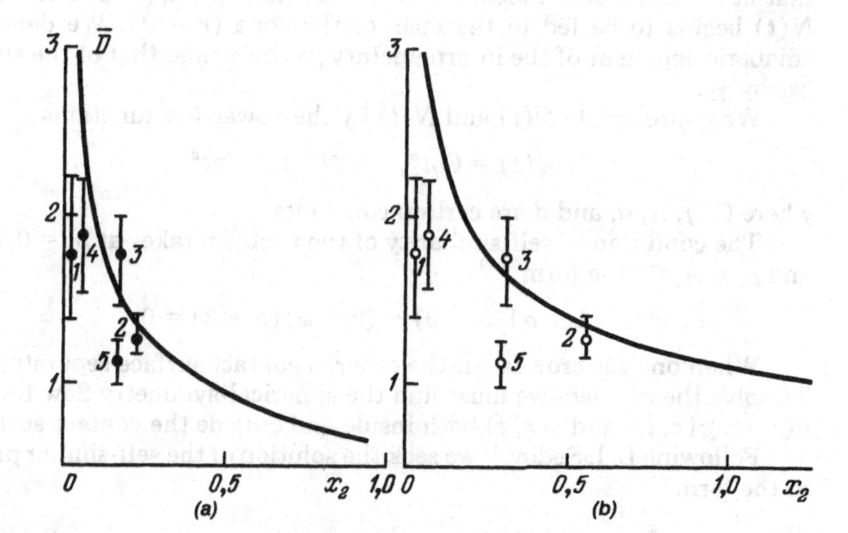

Figure 133. Plot of D/v_1 against x_2 for a strong shock wave without (a) and with (b) allowance for the wind. The circles and points mark data obtained by reducing the measurement results.

foregoing results indicate a satisfactory agreement between the observed and theoretical values.

Note also that estimates of the effective γ from the conditions on the shock wave, from the values of ρ_2/ρ_1 given in Ref. 35, and from the values of D_{av} have shown[37] that γ ranges from 2 to 3, and the smaller n_1 and the larger D_{av} the closer γ is to 3.

The presented comparison of the point-blast-theory data with the observations allows us to conclude that it is possible to use this theory for a qualitative and rough quantitative analysis of the propagation of perturbations from strong solar flares.

We proceed now to problems in which mass and energy are fed to the flare nucleation point within a finite time.

7. Motion of interplanetary gas following local supply of mass and energy

7.1. Self-similar problem

This problem was already briefly formulated above. We shall simulate the region of perturbation of interplanetary gas by a flare by a certain conical volume (tube) with cone center near the flare nucleation point. We assume that at $t = 0$ a gaseous medium with mass flow $Q = Q(t)$ and energy flux $N(t)$ begins to be fed to the apex of the cone ($r = 0$). We denote the adiabatic exponent of the interplanetary gas by γ_1 and that of the supplied gas by γ_2.

We approximate $Q(t)$ and $N(t)$ by the power-law functions

$$Q(t) = Cqt^{\alpha}, \qquad N(t) = Cnt^{\beta},$$

where C, q, n, α, and β are certain constants.

The condition of self-similarity of the problem takes at $p_1 = 0$, $v_1 = 0$, and $\rho_1 = A_1 r^{-\omega}$ the form[38,39]

$$(1+\alpha)(5-\omega) - (3-\omega)(\beta+3) = 0. \tag{8.33}$$

When one gas crowds out the other, a contact surface separates them. To solve the problem we must find the spherical-symmetry flow functions $v(r,t)$, $p(r,t)$, and $\rho(r,t)$ both inside and outside the contact surface.

Following L. I. Sedov,[11] we seek the solution of the self-similar problem in the form

$$v = \frac{r}{t}V(\lambda), \qquad \rho = A_1 r^{-\omega} R(\lambda), \qquad p = A_1 r^{2-\omega} t^{-2} P(\lambda),$$

$$\lambda = \lambda_0 \sqrt{\frac{q}{n}}\, r t^{-\delta}, \qquad \lambda_0 = \text{const}, \qquad \delta = \frac{\beta+3}{5-\omega}.$$

The functions $V(\lambda)$, $R(\lambda)$, and $P(\lambda)$ are the solution of the system of Sedov's equations[11] in the spherical case ($\nu = 3$)

$$\frac{dz}{dV} = \frac{z}{(V-\delta)W(V,z)}\{[2(V-1)+\nu(\gamma_i-1)V](V-\delta)^2$$
$$-(\gamma_i-1)V(V-1)(V-\delta) - [2(V-1)+\varkappa_i(\gamma_i-1)z]\}, \tag{8.34}$$
$$\frac{d\ln\lambda}{dV} = \frac{z-(V-\delta)^2}{W(V,z)},$$

$$(V-\delta)\frac{d\ln R}{d\ln\lambda}=(\omega-\nu)V-\frac{W(V,z)}{z-(V-\delta)^2},$$

$$W=[V(V-1)(V-\delta)-(\varkappa_i-\nu V)z],$$

$$z=\frac{\gamma_i P}{R},\qquad \varkappa_i=\frac{2+\delta(\omega-2)}{\gamma_i}.$$

The self-similar problem was investigated by L. V. Shidlovskaya,[30] and the case $\omega=0$, $V_1=0$ was investigated earlier in Ref. 23. By analogy with Ref. 23, a qualitative analysis of the field of the integral curves was carried out and it was shown that the solution contains, in addition to the contact discontinuity and the shock wave due to the "piston," one more shock wave in the vicinity of the added gas. Near the symmetry center, the following asymptotic relation holds for the considered class of the solutions

$$\lambda=C_1(V-1)^{\delta-1}V^{\delta},\qquad R=C_2(V-1)^{3-\omega}(V-1)^{-1},\qquad P=0.$$

Under the condition (8.33), the problem becomes self-similar if

$$v_1=Br^{(\delta-1)/\delta},$$

i.e., the initial velocity differs from zero. In this case we assume for the unperturbed motion of the gas

$$P_1=0,\qquad R_1=1,\qquad z_1=0,\qquad 0\le V\le\delta,\qquad B=\delta(\lambda_0\sqrt{q/n})^{-1/\delta},$$

and from the condition on the strong shock wave we have the initial data on the forward discontinuity ($\lambda=1$):

$$V_{12}=\delta+\frac{\gamma_1-1}{\gamma_1+1}(V_{11}-\delta),\qquad R_{12}=\frac{\gamma_1+1}{\gamma_1-1},$$
$$z_{12}=\frac{2\gamma_1(\gamma_1-1)}{(\gamma_1+1)}(V_{11}-\delta)^2,\tag{8.35}$$

where the subscript 11 corresponds to the unperturbed interplanetary gas, and the subscript 12 to the state on the leading shock wave.

The system (8.34) with the initial conditions (8.35) was integrated numerically by the Runge–Kutta method.[40] The values of V_1, γ_1, and γ_2, and the coordinate $\lambda=1$ of the forward shock wave, were specified, while the values of q and n were obtained in the course of the calculation.

The results of the solution for the case $\gamma_1=\gamma_2=1.3$, $\omega=2.2$, $V_1=0.25$, and $\delta=0.75$ are shown in Fig. 134. It shows plots of the dimensionless characteristics of the gas motions as functions of λ. The dashed line marks the contact discontinuity. A second shock wave introduced at $\lambda=0.52$ follows from an analysis of the field of the integral curves in the region $V_1<V<\infty$ and $0\le\lambda\le1$.

Note that secondary shock waves are frequently produced in the expansion flows behind the contact surface (for example, when a diaphragm is uncovered in a shock tube, or when explosive-detonation products are

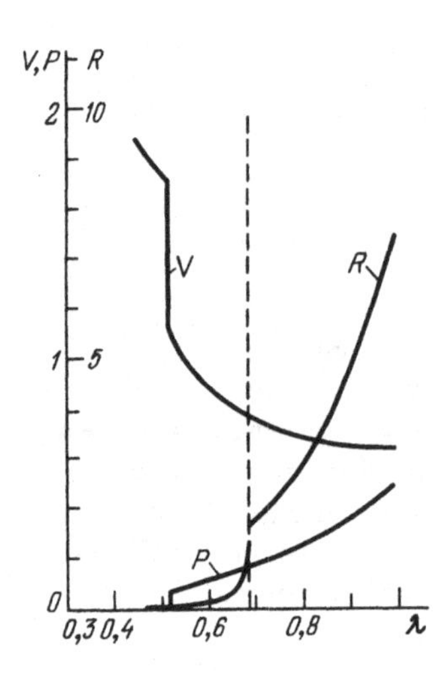

Figure 134. Distributions of the dimensionless density, pressure, and velocity with respect to λ: $\gamma_1 = \gamma_2 = 1.3$, $\delta = 0.75$, $\omega = 2.2$. The dashed line represents the contact surface.

expanded). The physical meaning of their formation is that when the stream is expanded again near the center a compression wave is produced behind the forward shock wave and moves through the particles towards the center and soon is transformed into a second shock wave. This wave is strongly carried away by the opposing stream. This configuration is called a wave pair in the theory of interplanetary disturbances.[41] Since we can write for the wave velocities the equations

$$D_1 = \delta r_1/t, \qquad D_2 = \delta r_2/t,$$

it can be seen that $D_1 - D_2 = \delta(r_1 - r_2)/t > 0$, i.e., $D_1 > D_2$. Similar reasoning shows also that the velocity D_c of the contact discontinuity satisfies the inequalities

$$D_2 < D_c < D_1.$$

If a coordinate frame tied to D_c is mentally introduced, the waves D_1 and D_2 will move in this frame in opposite directions ($D_2 - D_c < 0$, $D_1 - D_c > 0$). This situation occurs also for a pair of waves. As applied to our problem it can be stated that in a coordinate frame connected to the contact surface the solar-plasma particles "impinge" on the contact surface and are stopped by the interplanetary gas. This produces a second shock wave that equalizes the pressures on the contact discontinuity. Analysis of the above self-similar problem thus explains theoretically the "wave pair" or the secondary (suspended) discontinuities in the expansion flows.

Figure 135 shows the distributions of $v(t)$ and $H(t)$ for $r = r_0$; for the example considered, $H(t)$ was obtained from Eqs. (8.25) above. It was assumed that on the earth's orbit $n = 10\ \mathrm{cm}^{-3}$ and $v_1 = Br_a^{(\delta-1)/\delta} = 225$ km/s. The field at the initial point $r_0 = 0.001 r_a$ was taken to be $H_{r0} = 10^{-3}$ G. Comparison with the measurement data for the flare of February 1967 shows good agreement of the plots of $v(t)$ and $H(t)$.[40,41]

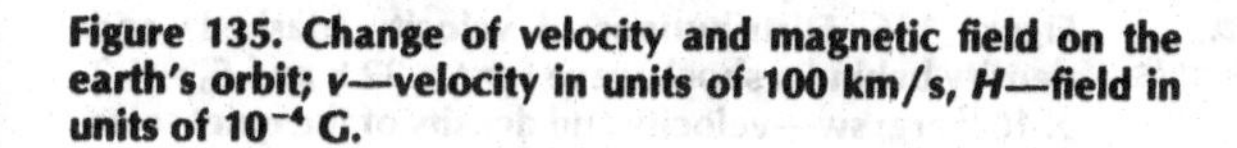

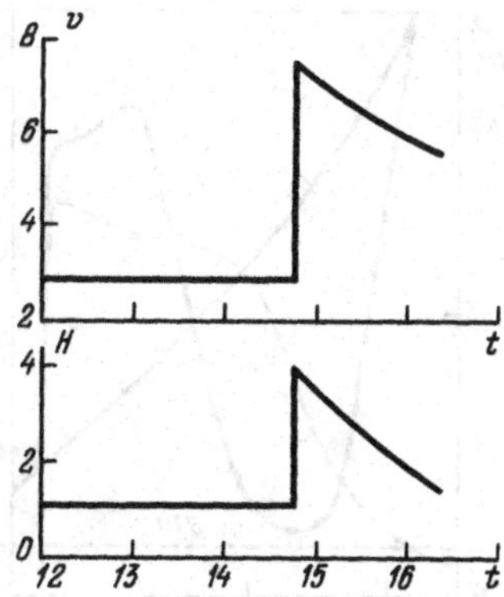

Figure 135. Change of velocity and magnetic field on the earth's orbit; v—velocity in units of 100 km/s, H—field in units of 10^{-4} G.

7.2. Numerical solution of one-dimensional problems

Finite-difference problems are being extensively used of late to investigate the hydrodynamics of interplanetary-gas disturbances. It is shown in the very first paper on this question[41] that many aspects of solar-flare plasma dynamics can be described even in the framework of one-dimensional flows.

We shall follow hereafter Refs. 43 and 44. We consider first one-dimensional spherically symmetric flows. The gas motion is adiabatic with an exponent $\gamma = \frac{5}{3}$. The solar gravitational forces are included in the equations of motion. The effect of the magnetic field on the plasma motion is disregarded.

The gas flow described by the system of the hydrodynamic equations is considered inside a region between an inner spherical surface of radius r_1 and an outer surface of radius r_2. It is assumed that a stationary flow (v_1, ρ_1, p_1) is set in motion in this region at the instant $t = 0$, and is calculated using a special program.

It is assumed that an additional influx of mass and energy through the inner boundary $r = r_1$ is initiated at $t = 0$, i.e., the parameters v_1, ρ_1, and p_1 increase abruptly to v_2, ρ_2, and p_2, and that the leading front of the disturbance is a shock wave with velocity D.

It is required to calculate the one-dimensional plasma motion that develops in the succeeding instants under the condition that the increased parameters ρ_2 and v_2 are maintained on the surface $r = r_1$ for a time τ. After the lapse of the time τ these parameters assume anew and maintain thereafter the values v_1, ρ_1, and p_1.

This problem was solved by the numerical "large-particle end-to-end counting" method,[45] modified somewhat to fit the peculiarities of the problem. Secondary discontinuities were not distinguished.

Given the time τ and the values of p, ρ, and v, the equation of state, and the value of the gravitational energy, it is possible to calculate the total energy supply E_0 and the supplied mass M in the considered flare model. It can be approximately assumed that $E_0 = 4\pi r_1^2 \tau(\rho_2 v_2^3 - \rho_1 v_1^3)/2$ and

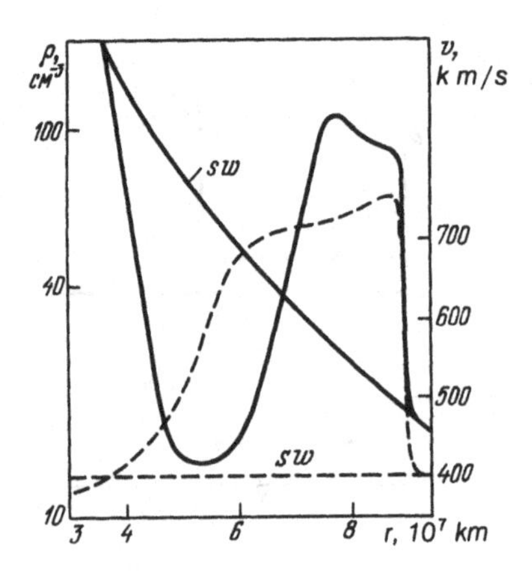

Figure 136. Distributions of velocity (dashed) and density behind a shock wave for $t = 32$ h and $E_0 = 4.2 \times 10^{32}$ erg; sw—velocity and density of the quiet solar wind.

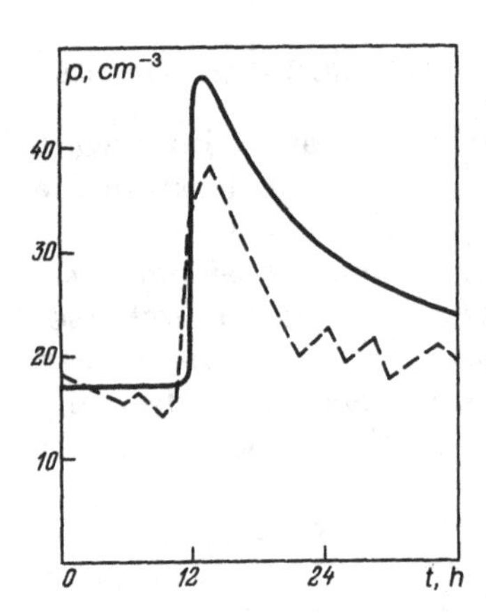

Figure 137. Time dependence of the density in the vicinity of the shock-wave front; $E_0 = 1.9 \times 10^{32}$ erg, $\tau = 0.3$ h. Dashed line—observation data of 26 April 1967.

$M = 4\pi r_1^2 \tau(\rho_2 v_2 - \rho_1 v_1)$. It was also assumed that $r_1 = 0.1$ AU and $r_2 = 1.1$ AU.

Figure 136 shows the velocity and density distributions for the case $E_0 = 4.2 \times 10^{32}$ erg, $v_1 = 314$ km/s, $\rho_1 = 2.2 \times 10^3$ cm^{-3} (number density), $\tau = 1.33$ h, and initial disturbance shock-wave velocity $D = 1250$ km/s at the instant of arrival of the leading front at the point $r = 0.94$ AU ($t = 32$ h). At this time the gas velocity increases sharply to 700 km/s, and the density ρ increases from 17 cm^{-3} to 80 cm^{-3}.

The same figure shows the velocities and densities of the quiet solar wind (curve labeled *sw*). Figure 137 shows the time variation of the density in the vicinity of the shock-wave leading front (case $E_0 = 1.9 \times 10^{32}$ erg, $D = 10^3$ km/s, $\tau = 0.83$ h; solar-wind parameters: $u_1 = 200$ km/s, $\rho_1 = 2 \times 10^3$ cm^{-3}, $T_1 = 1.06 \times 10^6$ K). The dashed lines correspond here to observed data for the flare of 26 April 1967.[41,46] The density plots show good qualitative and satisfactory quantitative agreement of the theoretical and experimental data. The changes of the magnetic fields can be obtained from Eqs. (8.25) and are given in Ref. 43.

In addition to the foregoing calculations, a numerical investigation was made of the propagation of a disturbance from two successive flares ("blast in a blast"). It was found,[44] in particular, that at the instant of coalescence the parameters of the leading fronts of the dual flare are higher, but the propagation is somewhat slower than that of a single flare having the total energy of the double flash, the total disturbance times τ being equal.

After coalescence of the leading fronts of the disturbances of a dual flare, the resultant disturbance moves more rapidly than the analogous one for a single flare.

As a result, a disturbance from a dual flare arrives earlier at a distance of order 1 AU, and the shock-wave velocity and the densities and velocities behind it are higher for a dual flare. Since dual flares are observable, these calculations are of interest for estimates of their energy.

Flares in a model of a tenuous magnetized two-component plasma were investigated in Ref. 47 in the one-dimensional approximation.

The numerical-modeling results have shown that, depending on the disturbance parameters (amplitude, duration), there is produced in a tenuous inhomogeneous plasma either a solitary wave or a collisionless shock wave; solitary waves can become transformed into other types of disturbance. These questions are of considerable interest for the theory of solar flares and require further study.

7.3. Two-dimensional calculations

The finite size of the energy-release region, and also the two-dimensional structure of the magnetic fields, requires a more exact account of the spatial configurations of the gas flow. It is necessary here to use two- and three-dimensional numerical schemes of solving the magnetohydrodynamic equations.

To this end, a numerical method was developed,[44,49] based on the idea of the large-particles method.[50] The basis here is a MHD equation system (7.34) in spherical coordinates, to which are added the mass gravitation force in the equation of motion $F_r = \rho G M_s / r^2$, $F_\theta = 0$, $F_\varphi = 0$, and the equation for the gravitational potential in the energy equation $U = -GM_s / r$ (here M_s is the mass of the sun and G is the gravitational constant).

We choose a spherical coordinate frame with center on the sun and assume that the flow is symmetric about a certain axis passing through the center. The functions sought will depend only on the time t, the radius r, and the polar angle θ measured from the symmetry axis (polar axis). The polar axis is chosen such that it passes through the center of the region occupied by the flare and constitutes, by assumption, a truncated cone, as shown in Fig. 138.

We assume that the gas motion of interest to us takes place in a region bounded by the spherical surfaces $r = r_0$ and $r = r_N$ as the angle changes from 0 to π. To determine the initial stationary flow of the solar wind, we assume that $v_\theta = H_\theta = 0$ and that the flow is one-dimensional. This ap-

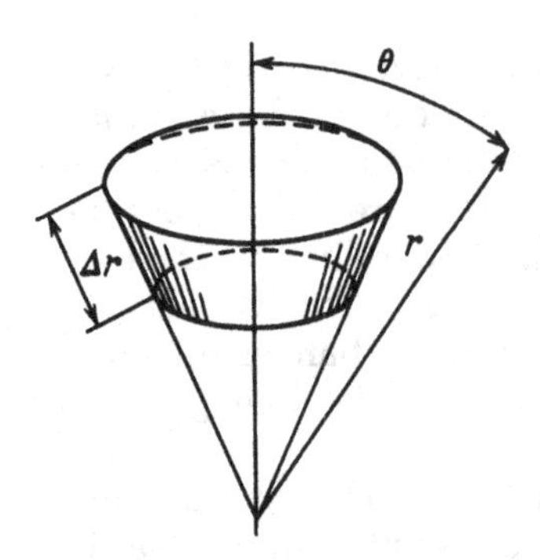

Figure 138. Initial configuration of disturbed motion and coordinate frame.

proximation is admissible, since we shall consider a solution in the indicated region under conditions of rather strong disturbances, and the components v_θ, v_φ, H_θ, and H_φ are small compared with the radial components of the same quantities. (This is known for numerous observations.)

Under these assumptions, the stationary flow is described by the approximate system

$$\frac{\partial}{\partial r}(\rho v_r r^2) = 0, \qquad \frac{1}{r^2}\frac{\partial}{\partial r}(\rho v_r^2 r^2) = -\frac{\partial p}{\partial r} - \rho\frac{GM_s}{r^2},$$

$$\frac{\partial p}{\partial \theta} = -\frac{\partial}{\partial \theta}\left(\frac{H_r^2}{8\pi}\right), \qquad \frac{\partial}{\partial r}[(\rho\varepsilon + p)v_r r^2] = 0, \qquad \frac{\partial}{\partial r}(r^2 H_r) = 0, \qquad (8.36)$$

which was used to determine the undisturbed state under the boundary conditions $v_r = v_{r0}$, $\rho = \rho_0$, $p = p_0$, and $H_r = H_{r0}$ for $r = r_0$.

The system (8.36) was integrated numerically in the course of the solution. We now formulate the problem. It is required to determine the functions $v_r(t, r, \theta)$, $v_\theta(t, r, \theta)$, $p(t, r, \theta)$, $\rho(t, r, \theta)$, $H_r(t, r, \theta)$, and $H_\theta(t, r, \theta)$ in the indicated region under the following initial and boundary conditions (the influence of the azimuthal-field component H_φ is disregarded in this formulation). The quantities specified at $t = 0$ are the distribution of the parameters of the quiet solar wind and the disturbance in the truncated cone $r_0 \le r \le r_0 + \Delta r$, $0 \le \theta \le \theta_*$ with the parameters v_r, p_1, ρ_1, and H_{r1} ($v_{\theta 1} = 0$, $H_{\theta 1} = 0$), which were assumed in the actual calculations to be equal to the parameters behind a shock wave moving with velocity D at a point r_0 of a quiet solar wind.[48]

The boundary conditions assumed are the following:

$$v_r, v_\theta, \rho, p, H_r, H_\theta = \begin{cases} v_{r0}, \ 0, \ \rho_0, \ p_0, \ H_{r0}, \ 0 \\ \text{for} \quad r = r_0, \quad \theta_* < \theta \le \pi, \quad t \ge 0; \\ v_{r1}, \ 0, \ \rho_1, \ p_1, \ H_{r1}, \ 0 \\ \text{for} \quad r_0 \le r \le r_0 + \Delta r, \quad 0 \le \theta \le \theta_*, \quad t \le \tau; \\ v_{r0}, \ 0, \ \rho_0, \ p_0, \ H_{r0}, \ 0 \\ \text{for} \quad r_0 \le r \le r_0 + \Delta r, \quad 0 \le \theta \le \theta_*, \quad t > \tau. \end{cases}$$

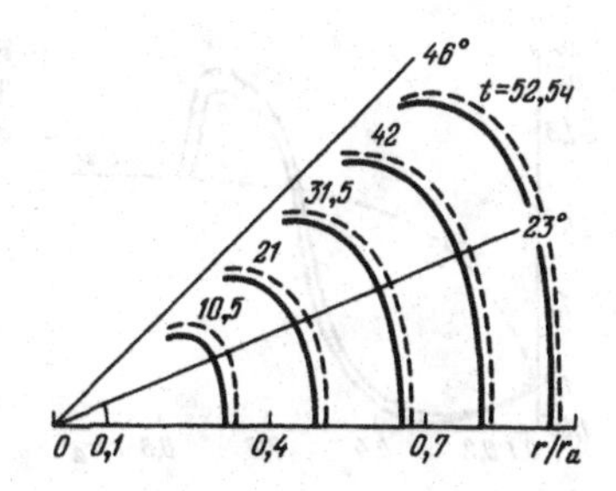

Figure 139. Form of shock wave in polar coordinates (r/r_a, θ) for a number of instants of time: t_1 = 10.5 h, t_2 = 21 h, t_3 = 31.5 h, t_4 = 42 h, t_5 = 52.5 h, $E_0 = 6 \times 10^{29}$ erg.

In addition, we have from the conditions of the flow symmetry

$$v_\theta = H_\theta = 0 \quad \text{for} \quad r_0 \leq r \leq r_N, \quad \theta = 0.$$

The solution of the above problem was obtained numerically up to the instant at which the shock wave reaches the earth's orbit. The general computation procedure is given in Appendix III.

It was assumed in the calculations that the thickness Δr of the initial disturbance layer is equal to the mesh of a grid in the radial direction, while the energy and mass in the flare ejections were defined as

$$E_0 = 4\pi r_0^2 \frac{(1 - \cos\theta_*)}{2} \tau \left[\frac{1}{2} \rho_1 v_1^3 - \frac{1}{2} \rho_0 v_{r0}^3 \right],$$

$$M = 4\pi r_0^2 \frac{(1 - \cos\theta_*)}{2} \tau [\rho_1 v_{r1} - \rho_0 v_{r0}].$$

Since the calculations have shown that the influence of the magnetic field on the gas flow is small, the field configurations were subsequently calculated as follows: the component H_θ was neglected as small, the component H_r was taken from the calculation, and the azimuthal-field component H_φ was calculated from the freezing for the initial helical field in the aforementioned Parker model (see Sec. 4).

Let us consider certain calculation results for the following initial conditions: r_0 = 0.1 AU, r_N = 1.1 AU, and for $r = r_0$ we choose v_{r0} = 339 km/s, $T_0 = 7.9 \times 10^5$ K, $\rho_0 = 1.26 \times 10^3$ cm^{-3}, and $H_{r0} = 4 \times 10^{-3}$ G. This corresponded to the following parameters of the quiet solar wind: v_{r0} = 417 km/s, ρ_0 = 10 cm^{-3}, and $H_{r0} = 5.54 \times 10^{-5}$ G ($E_0 = 6 \times 10^{29}$ erg, initial disturbance shock-wave velocity D = 1500 km/s, Δr = 0.01 AU, θ_* = 23°, and τ = 0.1 h).

Figure 139 shows the forms of the shock wave for the following instants of time:

$$t_1 = 10.5 \text{ h}, \quad t_2 = 21 \text{ h}, \quad t_3 = 31.5 \text{ h}, \quad t_4 = 42 \text{ h}, \quad t_5 = 52.5 \text{ h}.$$

It can be seen that for this time the disturbance is localized in a cone with half-apex angle θ = 46°.

The dashed lines correspond to waves without allowance for the influence of the magnetic field. The front of the MHD shock wave reaches a distance of 1 AU at θ = 0° after 60.4 h. The magnetic field "brakes" the

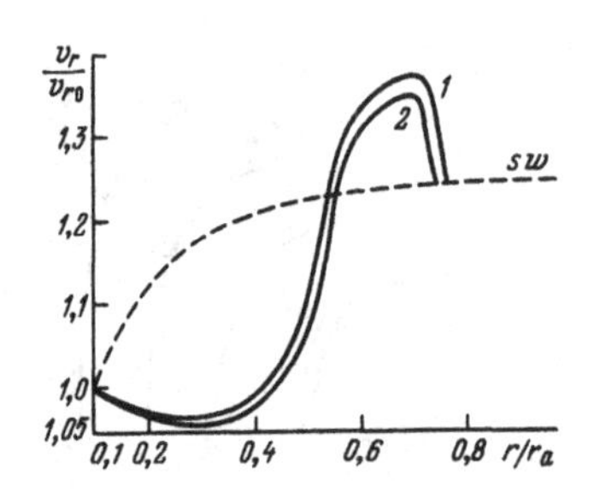

Figure 140. Change of velocity behind the shock wave, t = 36 h, for the angles $\theta = 0°$ (1) and $\theta - 23°$ (2). $E = 6 \times 10^{29}$ erg.

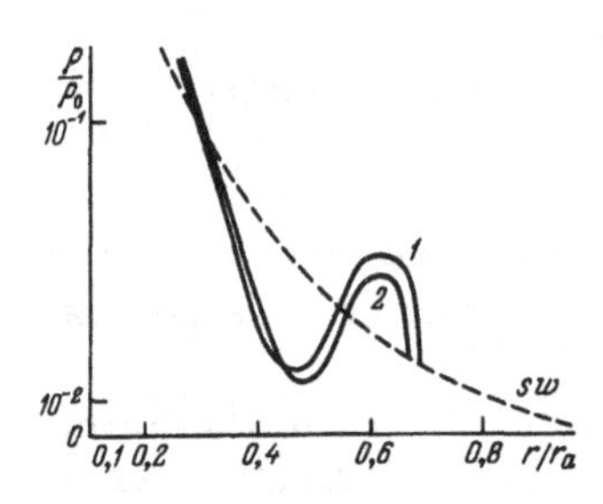

Figure 141. Density distribution behind the shock wave at the instant t = 36 h and for $\theta = 0°$ (1) and 23° (2).

shock waves and its arrival time at $\theta = 0°$ increases by approximately 5 hours compared with analogous disturbance without a field.

Note that for observation points located at equal distances from the sun, but at different angles to the symmetry axis, the time of shock-wave front arrival will be different, as will also the values of the flow velocities and densities.

Figure 140 shows plots of v_r/v_{r0} for $\theta = 0°$ and $\theta = 23°$ at the instant $t = 36$ h. The dashed line shows the parameters of the quiet wind.

A strong deceleration of the gas in the tail part of the disturbed flow is noted. Figure 141 shows plots of the density ρ/ρ_0 for the same angles. The transverse velocities v_θ are substantially lower in this calculation than the radial ones and reach 70 km/s at distances 0.1–0.5 AU from the initial disturbance. However, they make no substantial contribution to the total velocity at large distances (0.7–1 AU).

Calculations have shown that in the radial region 0.2–0.4 AU a noticeable rarefraction takes place (the particles are swept out), and the tangential velocities v_θ here are negative, i.e., particles are "drawn" into the central region from the sides.

Figure 142 shows stream lines determined by numerical integration of the equation $(1/r)(dr/d\theta) = v_r/v_\theta$. The figure shows also the positions of the shock wave and contact surface for the instant $t = 36$ h.

Figure 143 shows the variation of the magnetic field $H = \sqrt{H_r^2 + H_\varphi^2}$ with time at a distance $r = r_a = 1$ AU. The dependence obtained agrees well qualitatively with spacecraft measurements. On passage of a shock wave the magnetometers record an abrupt jump of the magnetic field intensity, corresponding to a sudden start of a geomagnetic storm on earth, followed by an abrupt decrease of the intensity compared with the field of

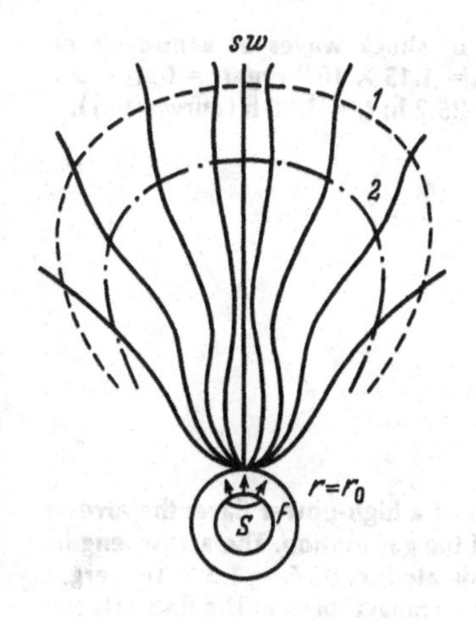

Figure 142. Velocity stream lines. 1—shock wave; 2—contact surface for t = 36 h. S—sun, F—flare region, r_0—radius of inner boundary of the disturbed region.

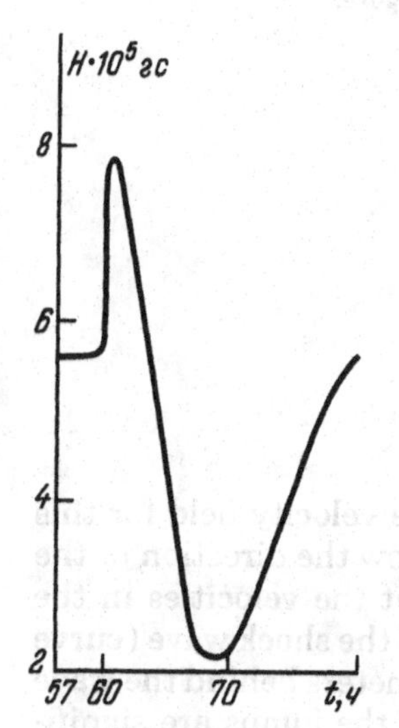

Figure 143. Time dependence of the magnetic field at r = 1 AU, = 0, $E_0 = 6 \times 10^{29}$ erg.

the quiet solar wind, corresponding to the succeeding (principal) phase of the geomagnetic storm.

The foregoing calculation is illustrative and corresponds to a weak flare. Besides this calculation, numerical solutions were obtained[48,49] for a number of variants with more powerful flares, $E_0 = 10^{31}$–10^{32} erg. Thus, if $E_0 = 1.2 \times 10^{31}$ erg, τ = 0.6 h, and D = 1600 km/s, the shock wave reaches the earth's orbit in 38.3 h. The form of the shock waves agrees qualitatively with that of the preceding case, but the disturbance subtends an angle θ = 57° (Fig. 144). The times chosen here were $t_1 = 0$, $t_2 = 8.4$ h, $t_3 = 16.8$ h, $t_4 = 25.2$ h, and $t_5 = 37.8$ h.

It can be seen that within an angle $\theta = 23°$ the flow becomes close to spherically symmetric after a long time. Calculations have shown that in this case v_θ on the earth's orbit is approximately 35 km/s (at $\theta = 0$); this is of the order of the tangential velocities observed for high-power flares.

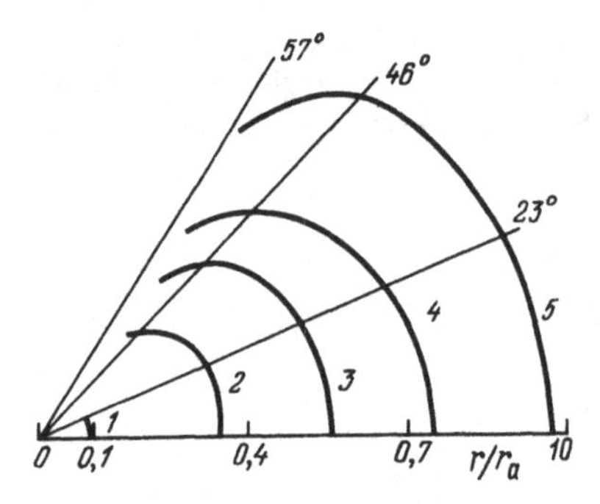

Figure 144. Form of shock waves at a number of instants of time: $E_0 = 1.15 \times 10^{31}$ erg, $t_1 = 0$; $t_2 = 3.4$ h; $t_3 = 16.8$ h; $t_4 = 25.2$ h; $t_5 = 37.8$ h (curves 1–5).

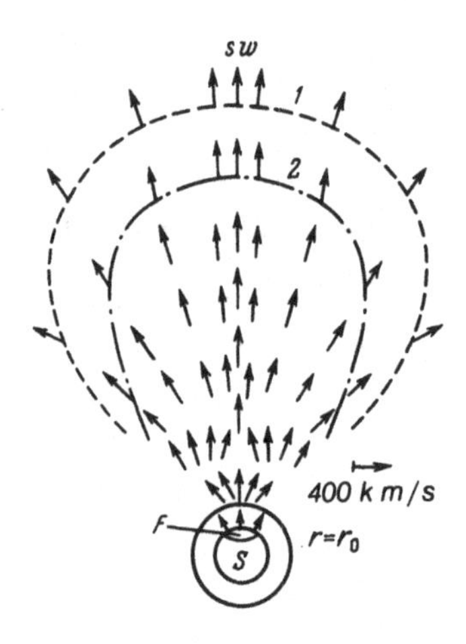

Figure 145. Velocity field of a high-power flare; the arrows indicate the directions of the gas motion. The arrow lengths give the moduli in the indicated scale; $E_0 = 1.2 \times 10^{31}$ erg, $t = 38$ h; 1—shock wave, 2—contact surface. The flare scheme is shown at the bottom of the figure.

Figure 145 shows an approximate picture of the velocity field for this flare ($E_0 = 1.2 \times 10^{31}$ erg, $t \approx 38$ h). The arrows show the direction of the gas motion, and the lengths of the arrows represent the velocities in the scale indicated on the figure. The contact surface and the shock wave (curve 1) are also marked. The distribution of the gas parameters behind the wave front corresponds qualitatively to a weak flare, but the jumps are significantly larger: for $t = 36$ h ($r \approx r_a$) we have $\Delta v_r / v_{r0} \approx 0.5$ on the leading front (as opposed to $\Delta v_r / v_{r0} = 0.15$ for $E_0 = 6 \times 10^{29}$ erg for the same time ($r \approx 0.8 r_a$). It should be noted that in view of the coarse spatial grid the onset of secondary waves could not be observed in these examples. The foregoing results illustrate the importance of two-dimensional effects, and the qualitative and quantitative correspondence of the flow patterns obtained, to many aspects of the dynamics of solar-flare evolution.[48,49]

Numerical simulation of perturbations from a number of specific flares (including dual ones) was carried out in Ref. 51. Comparison with the observation data shows good agreement between experiment and theory.

Further development of two-dimensional flare models, aimed at a more accurate account of the physical properties of the interplanetary plasma and the flare jets, will provide a more accurate picture of the flow evolution and will permit the performance of computer experiments on disturbance propagations accompanying solar flares.

Appendix I. Finite-difference method of calculation with Lagrange coordinates

We choose the gas equations of motion in the form

$$\frac{1}{\rho} = V = \frac{1}{\nu}\frac{\partial r^{\nu}}{\partial m}, \qquad \nu = 1, 2, 3,$$

$$u = \frac{\partial r}{\partial t}, \qquad m = \int_0^r \rho r^{\nu-1}\,dr, \qquad \frac{\partial u}{\partial t} = -r^{\nu-1}\frac{\partial}{\partial m}(p+Q), \tag{I.1}$$

$$\frac{\partial \varepsilon}{\partial t} + (p+Q)\frac{\partial V}{\partial t} = q, \qquad p = p(T, V), \qquad \varepsilon = \varepsilon(T, V),$$

where q is the heat influx and Q is an artificial viscosity:

$$Q = Q_k = -\frac{(c_k\Delta m)^k}{2Vr^{k(\nu-1)}}\left(\frac{\partial V}{\partial t}\right)^{k-1}\left[\frac{\partial V}{\partial t} - \left|\frac{\partial V}{\partial t}\right|\right], \qquad c_k = \text{const}, \qquad k = 1, 2. \tag{I.2}$$

Let the subscripts n and j indicate steps in time and space (in mass), respectively.

Equations (I.1) and (I.2) in finite difference form are written as

$$V_{j-1/2}^{n+1} = \frac{1}{\rho_{j-1/2}^{n+1}} = \frac{(r_j^{n+1})^{\nu} - (r_{j-1}^{n+1})^{\nu}}{\nu\Delta m_{j-1/2}},$$

$$\Delta m_j = \tfrac{1}{2}\Delta m_{j+1/2} + \tfrac{1}{2}\Delta m_{j-1/2},$$

$$r_j^{n+1} = r_j^n + u_j^{n+1/2}\Delta t^{n+1/2}, \tag{I.3}$$

$$u_j^{n+1/2} = u_j^{n-1/2} - \frac{\Delta t^n}{\Delta m_j}(r_j^n)^{\nu-1}[p_{j+1/2}^n - p_{j-1/2}^n + Q_{j+1/2}^{n-1/2} - Q_{j-1/2}^{n-1/2}],$$

$$\Delta t^n = \tfrac{1}{2}\Delta t^{n+1/2} + \tfrac{1}{2}\Delta t^{n-1/2};$$

$$Q_{j-1/2}^{n+1/2} = \frac{c_2(\Delta m_{j-1/2})^2(V_{j-1/2}^{n+1} - V_{j-1/2}^n)^2}{(V_{j-1/2}^{n+1} + V_{j-1/2}^n)[(r_j^{n+1} + r_{j-1}^{n+1})/2]^{2(\nu-1)}(\Delta t^{n+1/2})^2} + \frac{c_1\Delta m_{j-1/2}|V_{j-1/2}^{n+1} - V_{j-1/2}^n|}{(V_{j-1/2}^{n+1} + V_{j-1/2}^n)(\Delta t^{n+1/2})[(r_j^{n+1} + r_{j-1}^{n+1})/2]^{\nu-1}} \tag{I.4}$$

for $\quad V^{n+1} < V^n; \qquad Q_{j-1/2}^{n+1/2} = 0 \quad$ for $\quad V^{n+1} \geq V^n;$

$$\varepsilon_{j-1/2}^{n+1} = \varepsilon_{j-1/2}^n + (p_{j-1/2}^{n+1/2} + Q_{j-1/2}^{n+1/2})(V_{j-1/2}^n - V_{j-1/2}^{n+1}) + q_{j-1/2}^{n+1/2}\Delta t^{n+1/2},$$

$$\varepsilon^{n+1} = \varepsilon(T^{n+1}, V^{n+1}), \qquad p^{n+1} = p(T^{n+1}, V^{n+1}), \tag{I.5}$$

$$p_{j-1/2}^{n+1/2} = \tfrac{1}{2}p_{j-1/2}^{n+1} + \tfrac{1}{2}p_{j-1/2}^n.$$

The system of difference equations (I.3) and (I.4) makes it possible to determine the sought quantities at the instant t^{n+1} given their values at t^n.

The nonlinear equation (I.5) is solved in each step by iterations and is an implicit element of the scheme. The scheme itself is explicit.

It is expedient in the calculations to monitor the stability in accordance with the Courant criterion and in accordance with the criterion of the explicit schemes of the heat-conduction equations for the choice of the viscosity step Q. Details of the scheme are given in H. L. Brode's papers (see Chap. 2, Ref. 30 and Chap. 4, Ref. 10). This scheme was used to calculate the gasdynamic functions for problems with radiation (see Chap. 2).

Appendix II. Calculation procedure of the integral-equations method

We start with Eqs. (4.39)–(4.42) of Chap. 4. We designate in general form the sought functions by Ω_S and in the examples by $\Omega_1 = m$, $\Omega_2 = g$, $\Omega_3 = e$, and $\Omega_4 = \psi$.

In addition, the values of all the sought functions on each boundary of the considered bands will be marked by a suitable index. The system of integral equations (4.39)–(4.42), with allowance for expressions (4.33) for the functions on the shock wave and the asymptotic relations (4.37), can be reduced to the form

$$\mathcal{J}'_{S0} + k_S \Omega_{S0} \xi'_0 + b_S \xi_0 \Omega'_{S0} + \varkappa_S \Omega'_{40} + \delta^{-1}\delta' X = q^{-1} Y_S,$$
$$\mathcal{J}'_{Si} + \Omega_{Si} \xi'_i - \delta^{-1}\delta' W_{Si} = (W_{Si} + \beta_S \mathcal{J}_{Si}) q^{-1}, \tag{II.1}$$

where

$$\mathcal{J}_{Si} = \int_{\xi_i}^{1} \Omega_S d\xi,$$

$$X_1 = \frac{1}{\nu}\mathcal{J}_{10} - \frac{\psi_0^\gamma}{\gamma} + \frac{q}{\gamma} + \frac{\nu-1}{\nu}\mathcal{J}_{00} + \frac{k_1 m_0 \xi_0}{\nu}[1 + (\nu - 1)\varphi_0],$$

$$X_2 = g_0\xi_0 - 1 + \mathcal{J}_{20}, \qquad X_3 = e_0\xi_0 - Y_3 + \mathcal{J}_{30},$$

$$X_4 = \psi_0\xi_0 - \psi_n g_n^{-1} + \mathcal{J}_{40}, \qquad Y_1 = \tfrac{1}{2}(\mathcal{J}_{10} + k_1 m_0 \xi_0) - X_1,$$

$$Y_2 = \beta_4 g_0 \xi_0 - X_2, \qquad Y_3 = \frac{q}{\gamma - 1}, \qquad Y_n = \beta_4(\psi_0\xi_0 + \mathcal{J}_{40}) - X_4,$$

$$W_{1i} = g_i f_i^2 + \frac{\psi_i^\gamma}{\gamma} - \frac{q}{\gamma} - \frac{\nu-1}{\nu}\mathcal{J}_{0i} - \frac{\mathcal{J}_{1i}}{\nu} - \xi_i m_i,$$

$$W_{2i} = g_i(\varphi_i - 1)\xi_i + 1 - \mathcal{J}_{2i},$$

$$W_{3i} = e_i(\varphi_i - 1)\xi_i + Y_3 - \mathcal{J}_{31} + \psi_i^\gamma \varphi_i \xi_i,$$

$$W_{4i} = \psi_i(\varphi_i - 1)\xi_i + \psi g_n^{-1} - \mathcal{J}_{4i},$$

$$k_1 = \frac{\nu(\gamma-1)}{\nu + 2(\gamma-1)(\nu-1)}, \qquad k_2 = k_3 = k_4 = 1,$$

$$b_1 = k_1, \qquad b_2 = b_3 = b_4 = 0,$$

$$\varkappa_1 = 0, \qquad \varkappa_2 = g_0\xi_0/\psi_0, \qquad \varkappa_3 = \gamma\xi_0 e_0/\psi_0, \qquad \varkappa_4 = \xi_0,$$

$$\beta_1 = \tfrac{1}{2}, \qquad \beta_2 = 0, \qquad \beta_3 = 1, \qquad \beta_4 = 1/\gamma.$$

Equation (II.1) contains the quantities m_i, g_i, φ_i, and e_i, the relations between which are

$$\varphi_l = \frac{m_l}{g_l}\xi_l^{(\nu-2)/\nu}. \tag{II.2}$$

$$e_0 = \frac{\psi_0^\gamma}{\gamma - 1}, \qquad e_i = \frac{\gamma}{2}\varphi_i^2 g_i \xi_i^{2/\nu} + \frac{\psi_i^\gamma}{\gamma - 1}. \tag{II.3}$$

Equations (II.1) are basic for the further development of the method of calculating the problem. It follows from the foregoing that the system (II.1) contains only $3n$ independent integral equations. To obtain from Eqs. (II.1) an approximate system of ordinary differential equations, we shall approximate the integrands by interpolation Lagrange polynomials with interpolation nodes on the lines $\xi = \xi_k$ $(k = 0, 1, \ldots, n)$. One can use both "end-to-end" approximations, in which the degree of polynomials is equal to the number of bands n, and "piecewise" approximations (for $n > 1$), in which the degree of the polynomials is lower than n.

If the degree of the interpolation polynomials is equal to n, we have

$$\Omega_S = \sum_{k=0}^{n} \frac{(-1)^{n-k} z(z-1)(z-2)\ldots(z-n)}{k!(n-k)(z-k)} \Omega_{Sk},$$

where $z = (\xi - \xi_0)n/(1-\xi_0)$. For the integrals $\mathcal{J}_{Sl}$ we obtain

$$\mathcal{J}_{Sl} = (1-\xi_0) \sum_{i=0}^{n} A_{lk} \Omega_{Sk}, \tag{II.4}$$

where the coefficients A_{lk} for all n are called the generalized Cotes polynomials:

$$A_{kl} = \frac{(-1)^{n-k}}{k!(n-k)!n} \int_l^n \frac{z(z-1)\ldots(z-n)}{z-k} dz. \tag{II.5}$$

If the approximation is carried out by a set of interpolation polynomials having a degree lower than n, the expressions for the approximate representation of the integrals can be written as before in the form (II.4). Note that the use of interpolation polynomials of this type (e.g., second-order polynomials) is advantageous in computation schemes with higher-order approximations ($n > 4$), when the required accuracy can be obtained on account of the large number of bands. Such "piecewise" approximations simplify the structure of the approximating system and can turn out to be more reliable than the "end-to-end" approximations in regions where the represented functions vary strongly.

Numerical values of the generalized Cotes coefficients are given in Ref. 33 (Chap. 4).

Substituting the quadrature equations (II.4) in the system of integral equations (II.1), we arrive at an approximating system of $4n$ ordinary differential equations with respect to q for the $4n$ sought functions. We shall obtain g_0, however, not from this system but from the corresponding asymptotic equation (4.37), discarding in this case one of the equations of the approximating system.

We obtain then for the $4n - 1$ unknown functions m_0, φ_0, δ, m_i, g_i, e_i, and ψ_i a system of differential equations that can be written in the form

$$\begin{aligned} &[b_\lambda \xi_0 + (1-\xi_0)(A_{00} - \sum_{i=1}^{n-1} \omega_i A_{i0})]\Omega'_{\lambda 0} + K_\lambda \xi'_0 + \varkappa_\lambda \Omega'_{40} + (X_\lambda + \sum_{i=1}^{n-1} \omega_i W_{\lambda i}) \frac{\delta'}{\delta} \\ &= \frac{1}{q}[Y_\lambda - \sum_{i=1}^{n-1} \omega_i (W_{\lambda i} + \beta_\lambda \mathcal{J}_{\lambda i})] - (1-\xi_0)(A_{0n} - \sum_{i=1}^{n-1} \omega_i A_{in})\Omega'_{\lambda n}, \end{aligned} \tag{II.6}$$

$$\Omega'_{Sj} = \frac{1}{1-\xi_0}\left[\Omega_{Sj}\xi'_0 + \sum_{i=1}^{n-1} \frac{\Delta_{ij}}{\Delta_n} F_{Si}\right] \tag{II.7}$$

$$(\lambda = 1, 3, 4), \qquad (S = 1, 2, 3, 4), \qquad (j = 1, 2, \ldots, n-1).$$

We have introduced here the notation

$$K_\lambda = (k_\lambda - A_{00})\Omega_{\lambda 0} - A_{0n}\Omega_{\lambda n} + \sum_{i=1}^{n-1} \omega_i z_{\lambda i},$$

$$z_{Si} = A_{i0}\Omega_{S0} + A_{in}\Omega_{Sn} - \frac{n-i}{n}\Omega_{Si},$$

$$F_{Si} = z_{Si}\xi_0' - (1-\xi_0)(A_{i0}\Omega_{S0}' + A_{in}\Omega_{Sn}') + \frac{\beta_s}{q}\mathcal{I}_{Si} + \mu W_{Si},$$

$$\omega_i = \frac{1}{\Delta n}\sum_{j=1}^{n-1} A_{0j}\Delta_{ij}, \qquad \Delta_n = |A_{ij}|.$$

Δ_{ij} are the cofactors of the elements A_{ij} of the determinant Δ_n. The derivative ξ_0' in Eqs. (II.6) and (II.7) is eliminated with the aid of Eq. (4.38).

We add to the system (II.6)–(II.7) one more equation that follows from the asymptotic relations (4.37):

$$\gamma\varphi_0(1 + q\delta^{-1}\delta') - 1 + \gamma q\psi_0^{-1}\psi_0' = 0. \tag{II.8}$$

This equation can be used instead of Eq. (II.6) or Eq. (II.7). The system (II.6) and (II.7) then makes it possible to devise different variants of the computation schemes. In all these schemes the $\tau(q)$ dependence is determined from the differential equation (4.26).

A general check of the solution's accuracy is the satisfaction of the adiabaticity condition in the particle with coordinate η_0 and by satisfaction of the integral mass and energy conservation laws (4.23) and (4.24), which can be written in the present notation in the form

$$Y_2 = 0, \qquad X_3 - 1/\delta = 0.$$

These equations are subject to respective errors ε_{1n} and ε_{2n} given by

$$\varepsilon_{1n} = Y_2, \qquad \varepsilon_{2n} = \left(X_3 - \frac{1}{\delta}\right)\left(\frac{1}{\delta} + \frac{q}{\gamma - 1}\right)^{-1}.$$

Note that some computation schemes exhibit other methods of checking the accuracy.

We have analyzed the following computation variants.

Variant I. We choose a system comprising Eq. (II.8), the first and second equation of Eqs. (II.6), i.e., equations with subscripts $\lambda = 1$ and $\lambda = 3$, and the system (II.7). Integration yields the functions τ, δ, m_l, ψ_l, g_i, and e_i. The values of φ_i are obtained from Eq. (II.2). Equations (II.3) serve as an additional control of the computational accuracy.

Variant II. We choose a system comprising Eqs. (II.6) with subscripts $\lambda = 1$ and $\lambda = 4$, and the system (II.7). Integration yields the functions τ, δ, m_l, ψ_l, g_l, and e_l. The values of φ_l are obtained from Eq. (II.2). The values of e_i are also determined from Eq. (II.3), providing an additional accuracy check.

Variant III. The chosen system comprises the second and third equations of (II > 6), i.e., equations with subscripts $\lambda = 3$ and $\lambda = 4$, and the equations of (II.7) with subscripts $S = 2, \ldots, 4$. Integration yields the functions τ, δ, ψ_l, g_i, and e_i. The relative dimensionless velocities φ_i are obtained from the following algebraic equation that follows from Eqs. (II.3):

Table 7. Point-blast gasdynamic functions.

$\nu = 1$ $\gamma = 1.4$

q	τ	R_n	ξ_0	$\frac{p_n}{p_\infty} q$	$\frac{\rho_n}{\rho_\infty}$	$\frac{v_n}{D}$	$\frac{T_n}{T_\infty} q$	$\frac{E_n}{p_\infty} q$
0.05	0.0021	0.0163	0.600	1.158	4.800	0.792	0.241	5.002
10	0066	0358	476	1.150	4.000	750	288	4.450
15	0151	0641	379	1.142	3.429	708	333	4.058
20	0277	0999	304	1.133	3.000	667	378	3.767
0.25	0.0446	0.142	0.249	1.125	2.667	0.625	0.422	3.542
30	0687	196	202	1.117	2.400	583	465	3.363
35	102	266	163	1.108	2.182	542	508	3.219
40	150	358	129	1.100	2.000	500	550	3.100
45	219	483	0995	1.092	1.846	458	592	3.001
0.50	0.318	0.653	0.0754	1.083	1.714	0.417	0.632	2.917
55	465	0.892	0559	1.075	1.600	375	672	2.845
60	0.687	1.24	0404	1.067	1.500	333	711	2.783
65	1.04	1.76	0285	1.058	1.412	292	749	2.730
70	1.61	2.59	0194	1.050	1.333	250	788	2.683
0.75	2.67	4.06	0.0123	1.042	1.263	0.208	0.825	2.643
80	5.17	7.40	0068	1.033	1.200	167	861	2.607
85	1.8	16.0	0031	1.025	1.143	125	897	2.575
90	25.4	33.2	0015	1.017	1.091	083	932	2.547

$\nu = 1$ $\gamma = 1.4$

q	$\frac{p_0}{p_n}$	$\frac{p_1}{p_n}$	$\frac{p_2}{p_n}$	$\frac{p_3}{p_n}$	$\frac{p_4}{p_n}$	$\frac{p_5}{p_n}$	$\frac{p_6}{p_n}$	$\frac{p_7}{p_n}$
0.05	0.423	0.442	0.470	0.506	0.557	0.625	0.718	0.841
10	388	402	423	456	506	578	683	827
15	364	384	402	432	480	555	662	809
20	354	380	396	426	473	648	654	802
0.25	0.360	0.382	0.399	0.430	0.479	0.552	0.658	0.805
30	370	390	406	436	485	558	663	808
35	385	402	417	447	494	566	670	812
40	407	421	434	460	505	575	677	818
45	435	446	457	479	519	585	684	822
0.50	0.470	0.479	0.487	0.504	0.536	0.597	0.692	0.826
55	513	519	524	534	559	611	701	831
60	561	565	567	572	587	628	710	836
65	612	614	615	617	624	650	720	841
70	664	666	666	667	669	680	732	846
0.75	0.717	0.720	0.720	0.720	0.720	0.720	0.749	0.848
80	771	774	774	774	774	773	778	844
85	826	830	829	830	829	830	828	851
90	881	886	886	886	885	886	885	890

Table 7. (continued)

$\nu = 1$ $\gamma = 1.4$

q	$\frac{\rho_0}{\rho_n}$	$\frac{\rho_1}{\rho_n}$	$\frac{\rho_2}{\rho_n}$	$\frac{\rho_3}{\rho_n}$	$\frac{\rho_4}{\rho_n}$	$\frac{\rho_5}{\rho_n}$	$\frac{\rho_6}{\rho_n}$	$\frac{\rho_7}{\rho_n}$
0.05	0.198	0.203	0.254	0.318	0.399	0.502	0.631	0.790
10	133	167	227	302	399	523	674	815
15	108	158	230	311	410	532	674	815
20	102	168	252	337	433	543	666	816
0.25	0.0984	0.185	0.281	0.377	0.473	0.570	0.686	0.828
30	0972	208	322	420	505	596	706	840
35	0981	242	370	460	535	621	725	849
40	101	292	421	498	566	644	739	858
45	104	353	475	537	594	662	751	864
0.50	0.110	0.423	0.530	0.575	0.620	0.678	0.761	0.869
55	116	497	584	614	646	695	771	874
60	124	571	638	653	675	712	780	879
65	131	641	688	696	709	732	789	882
70	139	704	737	740	748	758	800	888
0.75	0.147	0.761	0.784	0.784	0.789	0.790	0.813	0.889
80	154	812	830	828	832	832	835	886
85	162	859	873	870	874	876	873	891
90	170	902	916	912	916	918	917	919

$\nu = 1$ $\gamma = 1.4$

q	$\frac{v_0}{D}$	$\frac{v_1}{D}$	$\frac{v_2}{D}$	$\frac{v_3}{D}$	$\frac{v_4}{D}$	$\frac{v_5}{D}$	$\frac{v_6}{D}$	$\frac{v_7}{D}$
0.05	0.406	0.445	0.485	0.529	0.575	0.624	0.677	0.733
10	292	335	381	430	483	542	606	676
15	205	249	296	346	404	469	542	622
20	142	184	229	279	338	407	485	572
0.25	0.099	0.138	0.181	0.230	0.287	0.355	0.436	0.527
30	066	100	138	183	238	307	388	482
35	040	067	100	141	193	261	344	438
40	022	043	068	102	151	217	299	395
45	011	024	043	070	112	174	256	352
0.50	0.004	0.012	0.024	0.043	0.077	0.134	0.214	0.310
55	001	005	011	023	048	096	172	268
60	000	002	004	010	024	062	132	227
65	000	000	001	003	010	033	093	186
70	000	000	000	001	002	012	057	146
0.75	0.000	0.000	0.000	0.000	0.000	0.001	0.025	0.103
80	000	000	000	000	000	−0.001	0.003	056
85	000	000	000	000	000	0.001	−0.001	017
90	000	000	000	000	000	001	0.000	003

Table 7. (continued)

$\nu = 2$ $\gamma = 1.4$

q	τ	R_n	ξ_0	$\frac{p_n}{p_\infty}q$	$\frac{\rho_n}{\rho_\infty}$	$\frac{v_n}{D}$	$\frac{T_n}{T_\infty}q$	$\frac{E_n}{p_\infty}q$
0.05	0.0097	0.0999	0.380	1.158	4.800	0.792	0.241	5.002
10	0207	148	292	1.150	4.000	750	288	4.450
15	0339	193	234	1.142	3.429	708	333	4.058
20	0498	238	194	1.133	3.000	667	378	3.767
0.25	0.0687	0.285	0.160	1.125	2.667	0.625	0.422	3.542
30	0901	333	132	1.117	2.400	583	465	3.363
35	116	387	110	1.108	2.182	542	508	3.219
40	148	448	0877	1.100	2.000	500	550	3.100
45	186	518	0698	1.092	1.846	458	592	3.001
0.50	0.234	0.599	0.0543	1.083	1.714	0.417	0.632	2.917
55	295	700	0409	1.075	1.600	375	672	2.845
60	376	826	0295	1.067	1.500	333	711	2.783
65	485	0.988	0202	1.058	1.412	292	749	2.730
70	637	1.21	0131	1.050	1.333	250	788	2.683
0.75	0.861	1.52	0.0081	1.042	1.263	0.208	0.825	2.643
80	1.20	1.98	0047	1.033	1.200	167	861	2.607
85	1.86	2.83	0022	1.025	1.143	125	897	2.575
90	3.16	4.48	0009	1.017	1.091	083	932	2.547

$\nu = 2$ $\gamma = 1.4$

q	$\frac{p_0}{p_n}$	$\frac{p_1}{p_n}$	$\frac{p_2}{p_n}$	$\frac{p_3}{p_n}$	$\frac{p_4}{p_n}$	$\frac{p_5}{p_n}$	$\frac{p_6}{p_n}$	$\frac{p_7}{p_n}$
0.05	0.378	0.390	0.409	0.439	0.483	0.550	0.650	0.796
10	362	372	388	416	462	533	643	804
15	353	366	384	411	462	539	654	813
20	348	364	382	415	471	554	670	821
0.25	0.347	0.358	0.380	0.421	0.478	0.563	0.678	0.823
30	352	364	385	429	497	582	695	835
35	359	372	397	442	512	603	711	844
40	370	382	412	461	528	622	728	853
45	386	399	431	484	551	640	746	864
0.50	0.407	0.422	0.455	0.508	0.576	0.660	0.762	0.875
55	437	452	484	534	601	680	776	884
60	479	491	519	563	626	700	789	891
65	535	540	559	595	652	722	804	898
70	605	599	606	632	679	744	820	905
0.75	0.680	0.669	0.662	0.674	0.710	0.767	0.836	0.913
80	752	743	729	724	745	791	853	923
85	820	817	809	796	789	819	870	931
90	882	881	879	875	863	857	890	944

Table 7. (continued)

$\nu = 2$ $\gamma = 1.4$

q	$\frac{\rho_0}{\rho_n}$	$\frac{\rho_1}{\rho_n}$	$\frac{\rho_2}{\rho_n}$	$\frac{\rho_3}{\rho_n}$	$\frac{\rho_4}{\rho_n}$	$\frac{\rho_5}{\rho_n}$	$\frac{\rho_6}{\rho_n}$	$\frac{\rho_7}{\rho_n}$
0.05	0.0466	0.0719	0.111	0.165	0.240	0.348	0.502	0.713
10	0351	0608	105	170	263	396	580	791
15	0287	0617	122	202	307	446	621	810
20	0258	0706	150	249	363	497	649	820
0.25	0.0245	0.0837	0.182	0.299	0.420	0.550	0.689	0.838
30	0240	103	224	352	478	598	723	857
35	0239	130	276	409	528	639	750	871
40	0241	169	338	446	570	673	776	883
45	0246	222	404	522	609	702	797	894
0.50	0.0254	0.294	0.470	0.568	0.646	0.726	0.814	0.906
55	0266	380	533	607	677	748	828	913
60	0283	473	589	641	705	769	841	919
65	0305	561	639	675	731	789	853	925
70	0333	643	687	710	755	808	866	931
0.75	0.0362	0.718	0.738	0.747	0.782	0.827	0.880	0.938
80	0388	785	795	788	810	845	893	944
85	0413	846	858	844	844	866	905	950
90	0435	896	911	903	900	896	920	960

$\nu = 2$ $\gamma = 1.4$

q	$\frac{v_0}{D}$	$\frac{v_1}{D}$	$\frac{v_2}{D}$	$\frac{v_3}{D}$	$\frac{v_4}{D}$	$\frac{v_5}{D}$	$\frac{v_6}{D}$	$\frac{v_7}{D}$
0.05	0.307	0.375	0.435	0.492	0.548	0.606	0.666	0.728
10	308	387	433	478	525	576	631	690
15	274	343	386	427	477	530	588	649
20	222	276	322	367	420	477	540	605
0.25	0.170	0.219	0.270	0.322	0.371	0.429	0.493	0.559
30	134	185	226	276	334	389	452	518
35	101	149	189	233	289	350	410	476
40	071	112	150	194	246	309	371	434
45	046	078	113	158	208	270	333	395
0.50	0.026	0.048	0.078	0.121	0.173	0.231	0.294	0.356
55	0.008	0.021	046	086	138	193	255	316
60	−0.004	−0.001	0.018	054	104	157	216	276
65	−0.009	−0.018	−0.004	025	070	123	178	236
70	−0.008	−0.025	−0.021	0.001	039	089	141	196
0.75	−0.005	−0.022	−0.029	−0.018	0.012	0.056	0.106	0.156
80	−0.002	−0.013	−0.024	−0.028	−0.009	0.027	073	120
85	0.000	−0.005	−0.011	−0.021	−0.025	−0.001	038	081
90	000	−0.001	−0.003	−0.006	−0.015	−0.018	007	046

Table 7. (continued)

$\nu = 3$ $\gamma = 1.4$

q	τ	R_n	ξ_1	$\frac{p_n}{p_\infty}q$	$\frac{\rho_n}{\rho_\infty}$	$\frac{v_n}{D}$	$\frac{T_n}{T_\infty}q$	$\frac{E_n}{p_\infty}q$
0.05	0.0151	0.194	0.350	1.158	4.800	0.792	0.241	5.002
10	0284	252	270	1.150	4.000	750	288	4.450
15	0425	300	218	1.142	3.429	708	333	4.058
20	0580	344	181	1.133	3.000	667	378	3.767
0.25	0.0755	0.388	0.151	1.125	2.667	0.625	0.422	3.542
30	0950	432	125	1.117	2.400	583	465	3.363
35	116	477	104	1.108	2.182	542	508	3.219
40	141	524	0848	1.100	2.000	500	550	3.100
45	170	576	0682	1.092	1.846	458	592	3.001
0.50	0.204	0.635	0.0538	1.083	1.714	0.417	0.632	2.917
55	245	702	0412	1.075	1.600	375	672	2.845
60	296	781	0305	1.067	1.500	333	711	2.783
65	359	876	0210	1.058	1.412	292	749	2.730
70	441	0.994	0138	1.050	1.333	250	788	2.683
0.75	0.557	1.16	0.0084	1.042	1.263	0.208	0.825	2.643
80	0.735	1.39	0048	1.033	1.200	167	861	2.607
85	1.02	1.76	0024	1.025	1.143	125	897	2.575
90	1.47	2.33	0008	1.017	1.091	083	932	2.547

$\nu = 3$ $\gamma = 1.4$

q	$\frac{p_0}{p_n}$	$\frac{p_1}{p_n}$	$\frac{p_2}{p_n}$	$\frac{p_3}{p_n}$	$\frac{p_4}{p_n}$	$\frac{p_5}{p_n}$	$\frac{p_6}{p_n}$	$\frac{p_7}{p_n}$
0.05	0.371	0.384	0.404	0.435	0.481	0.550	0.650	0.797
10	357	367	385	415	463	537	649	809
15	348	362	380	413	467	549	668	822
20	343	360	381	419	480	568	690	833
0.25	0.339	0.355	0.331	0.423	0.488	0.579	0.696	0.840
30	340	353	382	434	503	596	710	846
35	346	360	391	448	526	618	730	857
40	355	370	406	466	549	642	748	869
45	366	383	426	489	571	665	766	878
0.50	0.381	0.402	0.450	0.516	0.594	0.687	0.783	0.887
55	402	428	479	546	621	706	801	896
60	433	462	514	577	652	730	818	906
65	480	503	553	609	685	754	835	917
70	547	553	596	645	716	779	850	927
0.75	0.632	0.616	0.644	0.686	0.740	0.805	0.856	0.934
80	722	698	692	733	775	822	887	939
85	814	786	766	781	812	852	901	946
90	884	873	850	839	860	886	923	964

Table 7. (continued)

$\nu = 3$ $\gamma = 1.4$

q	$\frac{\rho_0}{\rho_n}$	$\frac{\rho_1}{\rho_n}$	$\frac{\rho_2}{\rho_n}$	$\frac{\rho_3}{\rho_n}$	$\frac{\rho_4}{\rho_n}$	$\frac{\rho_5}{\rho_n}$	$\frac{\rho_6}{\rho_n}$	$\frac{\rho_7}{\rho_n}$
0.05	0.0371	0.0593	0.0957	0.148	0.222	0.330	0.488	0.706
10	0249	0509	0928	155	248	382	572	792
15	0208	0526	107	188	294	438	622	819
20	0200	0602	133	234	353	496	657	825
0.25	0.0189	0.0696	0.166	0.283	0.414	0.552	0.698	0.849
30	0184	0866	205	340	473	603	733	865
35	0182	110	254	400	531	648	764	880
40	0183	143	314	460	580	686	789	894
45	0185	191	381	518	622	718	811	905
0.50	0.0190	0.256	0.452	0.570	0.658	0.747	0.830	0.914
55	0196	338	518	613	692	771	848	922
60	0206	428	578	651	725	792	863	931
65	0222	516	630	685	756	813	877	939
70	0243	596	678	719	784	834	890	946
0.75	0.0269	0.671	0.723	0.756	0.804	0.855	0.901	0.952
80	0296	742	773	794	834	874	923	942
85	0322	823	824	832	861	892	928	961
90	0341	890	889	876	898	917	945	974

$\nu = 3$ $\gamma = 1.4$

q	$\frac{v_0}{D}$	$\frac{v_1}{D}$	$\frac{v_2}{D}$	$\frac{v_3}{D}$	$\frac{v_4}{D}$	$\frac{v_5}{D}$	$\frac{v_6}{D}$	$\frac{v_7}{D}$
0.05	0.473	0.510	0.547	0.583	0.620	0.660	0.702	0.746
10	401	444	484	523	563	606	652	701
15	345	389	429	470	514	560	611	660
20	294	340	380	422	467	517	571	620
0.25	0.237	0.287	0.333	0.374	0.419	0.468	0.522	0.576
30	190	237	282	329	374	425	478	531
35	152	200	238	286	336	385	438	490
40	117	163	199	244	297	347	399	451
45	085	124	160	204	256	309	359	410
0.50	0.055	0.086	0.122	0.167	0.216	0.271	0.321	0.369
55	029	051	086	132	180	233	283	329
60	0.002	0.019	053	098	149	196	246	290
65	−0.018	−0.010	025	066	119	161	208	252
70	−0.026	−0.032	0.000	037	088	128	171	213
0.75	−0.021	−0.045	−0.021	0.013	0.052	0.097	0.134	0.174
80	−0.012	−0.036	−0.029	−0.003	028	062	100	136
85	−0.004	−0.027	−0.039	−0.024	0.001	031	065	092
90	−0.001	−0.009	−0.024	−0.030	−0.011	008	035	062

$$\varphi_i = \pm\left[\left(e_i - \frac{\psi_i^\gamma}{\gamma - 1}\right)\frac{2}{\gamma g_i}\right]^{1/2} \xi^{-1/\nu}. \qquad \text{(II.9)}$$

The velocity φ_0 is calculated using Eq. (II.8).

For all these variants, the approximating system of ordinary differential equations is integrated by some arbitrary method with a computer. There exists for this system a Cauchy problem, and the initial data for some sufficiently small $q = q_0$ are taken from the linearized blast-problem solution given in Chap. 3.

No special analysis was made of the stability of the computation scheme. Trial checks of the Courant stability condition have shown that the condition is met, albeit without much margin. It must be noted at the same time that for values of q close to the initial q_0 there was observed in certain cases a "buildup" of oscillations of the solution of the approximating system about a known linearized solution of the problem. This became more pronounced for approximating systems with higher numbers n. In all the computed cases, however, these oscillations attenuated as q increased. It turns out that by suitable choice of q_0 and $\xi_0(q_0)$ and the approximation of the integral it is possible to decrease substantially the initial buildup of the solution, which is probably due to insufficiently accurate matching of the initial data.

In the initial stage, variant III was used for the computations. In this variant, however, the calculated φ_i became less accurate in the vicinity of the zero of the radicand of Eq. (II.9). Variant I was chosen subsequently to be the principal one. Isolated computations were performed also in variant II. The agreement of the results of the different variants is better with larger n.

The initial data were specified in the following manner. The quantities q_0 and $\xi_0(q_0)$ were chosen to satisfy, in the region where the linearized solution is valid, i.e., at $q_0 < (\gamma - 1)/2$ (see Sec. 5 of Chap. 3), the conditions

$$\varepsilon_{1n}(q_0) < \varepsilon_{1n}^0, \qquad \varepsilon_{2n}(q_0) < \varepsilon_{2n}^0,$$

$$|\psi(\xi_0, q_0) - \psi(0, q_0)| < \varepsilon_{3n}^0 \psi(0, q_0),$$

where ε_{1n}^0, ε_{2n}^0, and ε_{3n}^0 are specified numbers chosen to fit the required accuracy and the number n of the approximation. In addition, the derivative $\varphi_0'(q_0)$ calculated from the approximating system had to be close to its exact value, known from the solution of the linearized problem, at $q = 0$.

Assuming that the initial values of the central interval $\xi_0(q_0) = \xi_{00}$ in the self-similar problem and in the linearized problem are equal, we have for the calculation of the functions sought at $q = q_0$, with allowance for Eqs. (3.72),

$$g_k = g_{k0} + q_0 g_{k1} \quad (k \neq 0), \qquad \psi_k = \psi_{k0} + q_0 \psi_{k1},$$

$$e_k = e_{k0} + q_0 e_{k1} \quad (k \neq 0), \qquad \varphi_k = \frac{f_k}{\xi_k^{1/\nu}}, \qquad m_k = \frac{\varphi_k g_k}{\xi_k^{(\nu-2)/\nu}}.$$

The functions Ω_{*k0} and Ω_{*k1} are taken, using interpolation, at the points $\xi_{k0} = (n-k)/n\xi_{00} + k/n$. The values of g_0 and e_0 are calculated from the asymptotic relations (4.37). The initial data for τ, R_n, and δ are obtained with the aid of Eqs. (3.47) of Chap. 3.

Some calculation results for $\gamma = 1.4$ and $n = 8$ are given in Table 7.

Appendix III. Two-dimensional variant of the large-particle method for MHD equations

We consider briefly the method of numerical solution of the two-dimensional problem of shock-wave propagation in outer space (Sec. 7, Chap. 8). We follow here the results of L. V. Shidlovskaya (see Chap. 8 and Refs. 44, 48, and 49).

The mass-, momentum-, and energy-conservation equations in Euler variables are

$$\frac{\partial \rho}{\partial t} + \frac{1}{r^2}\frac{\partial}{\partial r}(\rho u r^2) + \frac{1}{r \sin\theta}\frac{\partial}{\partial \theta}(\rho v \sin\theta) = 0, \tag{III.1}$$

$$\frac{\partial}{\partial t}(\rho u) + \frac{1}{r^2}\frac{\partial}{\partial r}(\rho u^2 r^2) + \frac{1}{r\sin\theta}\frac{\partial}{\partial\theta}(\rho u v \sin\theta) = -\frac{\partial p}{\partial r} + \rho\frac{v^2}{r} - \rho\frac{GM_s}{r^2} + \frac{H_\theta}{4\pi}\left[\frac{\partial H_r}{\partial\theta} - \frac{\partial(rH_\theta)}{\partial r}\right], \tag{III.2}$$

$$\frac{\partial}{\partial t}(\rho v) + \frac{1}{r^2}\frac{\partial}{\partial r}(\rho u v r^2) + \frac{1}{r\sin\theta}\frac{\partial}{\partial\theta}(\rho v^2 \sin\theta) = -\frac{1}{r}\frac{\partial p}{\partial\theta} - \rho\frac{uv}{r} - \frac{H_r}{4\pi r}\left[\frac{\partial H_r}{\partial\theta} - \frac{\partial(rH_\theta)}{\partial r}\right], \tag{III.3}$$

$$\frac{\partial}{\partial t}(\rho E) + \frac{1}{r^2}\frac{\partial}{\partial r}[(\rho E + p)ur^2] + \frac{1}{r\sin\theta}\frac{\partial}{\partial\theta}[(\rho E + p)v\sin\theta] = -\frac{(H_r v - H_\theta u)}{4\pi r}\left[\frac{\partial H_r}{\partial\theta} - \frac{\partial(rH_\theta)}{\partial r}\right]. \tag{III.4}$$

Here G is the gravitational constant, M_s is the mass of the sum, E is the specific total energy (disregarding the magnetic-field energy)

$$E = \varepsilon + (u^2 + v^2)/2 - GM_s/r, \tag{III.5}$$

where ε is the specific internal energy, $u = v_r$, and $v = v_\theta$.

In the actual calculations we shall assume hereafter that the medium considered is an ideal gas, i.e., the internal energy and the temperatures are given by

$$\varepsilon = p/\rho(\gamma - 1), \qquad T = m\varepsilon/3k, \tag{III.6}$$

where m is the hydrogen-atom mass, k Boltzmann's constant, and γ the adiabatic exponent.

For the magnetic field $\mathbf{H}$ in a medium of infinite conductivity we have

$$\frac{\partial H_r}{\partial t} + \frac{1}{r\sin\theta}\frac{\partial}{\partial\theta}[\sin\theta(H_r v - H_\theta u)] = 0. \tag{III.7}$$

$$\frac{\partial H_\theta}{\partial t} - \frac{1}{r}\frac{\partial}{\partial r}[r(H_r v - H_\theta u)] = 0. \tag{III.8}$$

$$\frac{1}{r^2}\frac{\partial}{\partial r}(r^2 H_r) + \frac{1}{r\sin\theta}\frac{\partial}{\partial\theta}(\sin\theta\, H_\theta) = 0. \tag{III.9}$$

For the solution to satisfy Eq. (III.9) it suffices to require that the initial data satisfy this equation. We shall therefore use Eq. (III.9) only to determine the initial distribution of the gas-flow parameters. Equations (III.1–III.8) form a closed system.

Assume that the gas moves in a region bounded by the spherical surfaces $r = r_0$ and $r = r_N$, within an angle θ ranging from 0 to π. Since the general form of Eqs. (III.1)–(III.8) is not changed by going over to dimensionless quantities, we can assume that these equations have already been reduced to dimensionless form by choosing the characteristic values of the plasma parameters at $r = r_0$.

To determine the initial stationary flow of the solar wind we assume that the quiet solar wind has only radial components of the velocity and the magnetic field intensity, i.e., $v = H_\theta = 0$ at the initial instant of time.

Using these assumptions and equating to zero the time derivatives in Eqs. (III.1)–(III.4), we obtain a system of ordinary differential equations with boundary conditions at the point $r = r_0$. To find the distributions of the hydrodynamic parameters and the magnetic field intensity in a quiet solar wind, this system of equations was integrated by the Runge–Kutta method.

We specify at the initial instant of time the distribution of the parameters u, v, ρ, p, H_r, and H_θ of a quiet solar wind and a disturbance, contained in a truncated cone, with parameters u_1, $v = 0$, ρ_1, p_1, H_{r1}, and $H_\theta = 0$. It is required to determine the gas flow in the succeeding instant of time, until the shock-wave front reaches the earth's orbit. The boundary conditions are described in Sec. 7 of Chap. 8.

We proceed to describe the numerical method (Chap. 8 and Ref. 50). We subdivide the integration region into an immobile Euler grid into cells whose centers correspond to the points with subscripts i and j. The volume of each cell is $V_{i,j} = 2\pi(r_i)^2 \sin\theta_j \Delta r \Delta\theta$, where $\Delta r = r_{i+1} - r_j$ and $\Delta\theta = \theta_{j+1} - \theta_j$ are the radial and angular mesh dimensions.

If definite values of the magnetohydrodynamic flow parameters are known for each cell at the instant of time $t = n\Delta t$, the problem reduces to finding their values at times $t = (n+1)\Delta t$.

The calculation of each time step breaks up into three stages. In the *first* (Euler) step we neglect the mass flows through the cell boundaries and determine the intermediate values of the parameters with the aid of the equations

$$\rho\frac{\partial u}{\partial t} = \rho\frac{v^2}{r} - \frac{\partial p}{\partial r} - \rho\frac{GM_s}{r^2} + \frac{H_\theta}{4\pi r}\left[\frac{\partial H_r}{\partial\theta} - \frac{\partial(rH_\theta)}{\partial r}\right], \tag{III.10}$$

$$\rho\frac{\partial v}{\partial t} = -\rho\frac{uv}{r} - \frac{1}{r}\frac{\partial p}{\partial\theta} - \frac{H_r}{4\pi r}\left[\frac{\partial H_r}{\partial\theta} - \frac{\partial(rH_\theta)}{\partial r}\right], \tag{III.11}$$

$$\rho\frac{\partial E}{\partial t} = -\frac{1}{r^2}\frac{\partial}{\partial r}(pur^2) - \frac{1}{r\sin\theta}\frac{\partial}{\partial\theta}(pv\sin\theta) - \frac{(H_rv - H_\theta u)}{4\pi r}\left[\frac{\partial H_r}{\partial\theta} - \frac{\partial(rH_\theta)}{\partial r}\right]. \tag{III.12}$$

The finite-difference equations, of first order of accuracy, with respect to time and space are of the form

$$\begin{aligned}\tilde{u}^n_{i,j} = u^n_{i,j} &- \frac{\Delta t}{2\rho^n_{i,j} r_i^2 \Delta r}\left[r^2_{i+1/2}(p^n_{i+1,j} - p^n_{i,j}) - r^2_{i-1/2}(p^n_{i-1,j} - p^n_{i,j})\right] \\ &- \frac{\Delta t}{r_i^2}\left[GM_s - r_i(v^n_{i,j})^2\right] + \frac{(H_\theta)^n_{i,j}\Delta t}{4\pi r_i}\left[\frac{(H_r)^n_{i,j+1/2} - (H_r)^n_{i,j-1/2}}{\Delta\theta}\right. \\ &\left. - \frac{r_{i+1/2}(H_\theta)^n_{i+1/2,j} - r_{i-1/2}(H_\theta)^n_{i-1/2,j}}{\Delta r}\right],\end{aligned} \tag{III.13}$$

$$\tilde{v}^n_{i,j} = v^n_{i,j} - \frac{\Delta t}{2\rho^n_{i,j} r_i \sin\theta_j \Delta\theta}[\sin\theta_{j+1/2}(p^n_{i,j+1} - p^n_{i,j}) - \sin\theta_{j-1/2}(p^n_{i,j-1} - p^n_{i,j})]$$
$$- \frac{\Delta t v^n_{i,j} u^n_{i,j}}{r_i} - \frac{(H_r)^n_{i,j}\Delta t}{4\pi r_i}\left[\frac{(H_r)^n_{i,j+1/2} - (H)^n_{i,j-1/2}}{\Delta\theta} - \frac{r_{i+1/2}(H_\theta)^n_{i+1/2,j} - r_{i-1/2}(H_\theta)^n_{i-1/2,j}}{\Delta r}\right], \tag{III.14}$$

$$\tilde{E}^n_{i,j} = E^n_{i,j} - \frac{\Delta t}{2\rho^n_{i,j} r_i^2 \sin\theta_j \Delta r \Delta\theta}\{\sin\theta_j \Delta\theta [p^n_{i+1/2} r^2_{i+1/2}(u^n_{i+1/2,j} + \tilde{u}^n_{i+1/2,j})$$
$$- p^n_{i-1/2,j} r^2_{i-1/2,j}(u^n_{i-1/2,j} + \tilde{u}^n_{i-1/2,j})]$$
$$+ r_i \Delta r[\sin\theta_{j+1/2} p^n_{i,j+1/2}(v^n_{i,j+1/2} + \tilde{v}^n_{i,j+1/2})$$
$$- \sin\theta_{j-1/2} p^n_{i,j-1/2}(v^n_{i,j-1/2} + \tilde{v}^n_{i,j-1/2})]\} - \frac{\Delta t[(H_r)^n_{i,j} v^n_{i,j} - (H_\theta)^n_{i,j} u^n_{i,j}]}{4\pi r_i}$$
$$\times \left[\frac{(H_r)^n_{i,j+1/2} - (H_r)^n_{i,j-1/2}}{\Delta\theta} - \frac{r_{i+1/2}(H_\theta)^n_{i+1/2,j} - r_{i-1/2}(H_\theta)^n_{i-1/2,j}}{\Delta r}\right]. \tag{III.15}$$

Note that another approach to construction of a difference scheme during the *first stage* is possible, namely, in place of the equation for the total energy E we can use the equation for the internal energy ε. In this case, however, the difference scheme will be nonconservative, and to improve the stability of the computation during the first stage it is necessary to replace the pressure p in Eqs. (III.10)–(III.17) by $p + Q$, where Q is an artificial viscous pressure. If a is the local speed of sound, Q can be expressed in the form (Chap. 8 and Ref. 44)

$$Q^n_{i+1/2,j} = \rho^n_{i+1/2,j}(u^n_{i,j} - u^n_{i+1,j})[Bc^n_{i+1/2,j} + A^2(u^n_{i,j} - u^n_{i+1,j})]$$
$$\text{for} \quad u^n_{i,j} > u^n_{i+1,j},$$
$$Q^n_{i+1/2,j} = 0 \quad \text{for} \quad u^n_{i,j} \le u_{i+1,j}.$$

Introduction of a linear term in the expression for Q in the case of weak shock waves attenuates the oscillations produced behind the shock front. The quadratic term plays a similar role when strong shock waves are considered. In the *second (Lagrangian) stage* we calculate the mass-flux density for gas motion through a cell boundary. We denote by $\Delta M^n_{1=1/2,j}$ the mass flux through the boundary $(j + 1/2)$ of the i, j cell in a time Δt, and by $M^n_{i,j+1/2}$ the analogous mass flux through the boundary $(j + 1/2)$. We can then write

$$\Delta M^n_{i+1/2,j} = \begin{cases} 2\pi r^2_{i+1/2}\sin\theta_j \rho^n_{i,j}\tilde{u}^n_{i+1/2,j}\Delta\theta\Delta t & \text{for} \quad \tilde{u}^n_{i+1/2,j} \ge 0, \\ 2\pi r^2_{i+1/2}\sin\theta_j \rho^n_{i+1,j}\tilde{u}^n_{i+1/2,j}\Delta\theta\Delta t & \text{for} \quad \tilde{u}^n_{i+1/2,j} < 0, \end{cases}$$
$$\Delta M^n_{i,j+1/2} = \begin{cases} 2\pi r_i \sin\theta_{j+1/2}\rho^n_{i,j}\tilde{v}^n_{i,j+1/2}\Delta r\Delta t & \text{for} \quad \tilde{v}^n_{i,j+1/2} \ge 0, \\ 2\pi r_i \sin\theta_{j+1/2}\rho^n_{i,j+1}\tilde{v}^n_{i,j}\Delta r\Delta t & \text{for} \quad \tilde{v}^n_{i,j+1/2} < 0. \end{cases} \tag{III.16}$$

The *third stage* consists of determining the final stream parameters ρ and $X = (u, v, E)$ corresponding to the instant of time $t = (n + 1)\Delta t$ on the basis of the mass-, momentum-, and energy-conservation laws for each cell:

$$\rho^{n+1}_{i,j} = \rho^n_{i,j} + (\Delta M^n_{i-1/2,j} + \Delta M^n_{i,j-1/2} - \Delta M^n_{i+1/2,j} - \Delta M^n_{i,j+1/2})/V_{i,j}, \tag{III.17}$$

$$X_{i,j}^{n+1} = \rho_{i,j}^{n}\tilde{X}_{i,j}^{n}/\rho_{i,j}^{n+1} + [\tilde{X}_{i-1,j}^{n}\Delta M_{i-1/2,j}^{n} + \tilde{X}_{i,j-1}^{n}\Delta M_{i,j-1/2}^{n} - \tilde{X}_{i,j}^{n}(\Delta M_{i+1/2,j}^{n} + \Delta M_{i,j+1/2}^{n})]/\rho_{i,j}^{n+1}V_{i,j}. \qquad \text{(III.18)}$$

The specific internal energy and the temperature can be determined from relations that hold for any instant of time:

$$\varepsilon_{i,j}^{n+1} = E_{i,j}^{n+1} - [(u_{i,j}^{n+1})^2 + (v_{i,j}^{n+1})^2]/2 + GM_s/r_i,$$

$$T_{i,j}^{n+1} = \varepsilon_{i,j}^{n+1} m/3k.$$

In the third stage we calculate also the values of H_r and H_θ for the instant of time $t = (n+1)\Delta t$, which we propose to calculate by using the following finite-difference equations:

$$(H_r)_{i,j}^{n+1} = (H_r)_{i,j}^{n} - \frac{\Delta t}{r_i \sin\theta_j \Delta\theta}\{\sin\theta_{j+1/2}[(H_r)_{i,j}^{n}\tilde{v}_{i,j+1/2}^{n} - (H_\theta)_{i,j}^{n}\tilde{u}_{i,j+1/2}^{n}] - \sin\theta_{j-1/2}[(H_r)_{i,j-1}^{n}\tilde{v}_{i,j-1/2}^{n} - (H_\theta)_{i,j-1}^{n}\tilde{u}_{i,j-1/2}^{n}]\},$$

$$(H_\theta)_{i,j}^{n+1} = (H_\theta^n)_{i,j} + \frac{\Delta t}{r_i \Delta r}\{r_{i+1/2}[(H_r)_{i,j}^{n}\tilde{v}_{i+1/2,j}^{n} - (H_\theta)_{i,j}^{n}\tilde{u}_{i+1/2,j}^{n}] - r_{i-1/2}[(H_r)_{i-1,j}^{n}\tilde{v}_{i-1/2,j}^{n} - (H_\theta)_{i-1,j}^{n}\tilde{u}_{i-1/2,j}^{n}]\}.$$

The difference equations of the third stage are valid only for internal cells of the considered region. If, however, the outer boundary is chosen such that it coincides with the center of the Nth cell, it is necessary only to determine the flow parameters in the imagined outer cell. On the inner boundary r_0 the values of all the hydrodynamic parameters are specified for all instants of time and require no evaluation.

The usual stability condition for Eqs. (III.13)–(III.18) is the Courant condition (the time steps must be shorter than the time interval in which a sound signal reaches the boundaries of adjacent cells). Since the disturbance propagates along a moving solar wind and the disturbed motion follows two directions in space, and also in view of the presence of a two-dimensional disturbed magnetic field (magnetohydrodynamic waves can exist), we use the stability condition in the form where $\beta < 1$ is a certain coefficient, and $a = \sqrt{\gamma p/\rho}$ is the speed of the sound waves in gases.

The accuracy of the calculations must be monitored against the results of a comparison with other numerical and analytic solutions of test problems, and also by performing the calculations at double the steps in radius and angle. For the two-dimensional calculations described in Chap. 8 it was assumed that radial and angular steps $\Delta r = 0.005$ AU and $\Delta\theta = 5°45'$ suffice for the calculation of the investigated problem (doubling the steps leads to basic-parameter differences smaller than 14%).

References

Chapter 1

1. L. I. Sedov, *Lectures on the Mechanics of a Continuous Medium* [in Russian] (Moscow University, Moscow, 1966).

2. L. V. Ovsyannikov, *Lectures on the Principles of Gasdynamics* [in Russian] (Nauka, Novosibirsk, 1967).

3. J. Serrin, *Mathematical Principles of the Classical Mechanics of Liquids,*

4. V. P. Korobeinikov, N. S. Mel'nikova, and E. V. Ryazanov, *Point-Blast Theory* [in Russian] (Fizmatgiz, Moscow, 1961).

5. L. D. Landau and E. M. Lifshitz, *Fluid Mechanics* (Pergamon, New York, 1959).

6. L. I. Sedov, *Mechanics of Continuous Media,* Vols. 1 and 2 [in Russian], 2nd ed. (Nauka, Moscow, 1973).

7. N. E. Kochin, *Vector Analysis and Principles of Tensor Analysis* [in Russian] (USSR Acad. Sci., Moscow, 1961).

8. S. L. Sobolev, *Equations of Mathematical Physics* [in Russian] (Gostekhizdat, Moscow, 1950).

9. A. G. Kulikovskii and G. A. Lyubimov, *Magnetohydrodynamics* [in Russian] (Fizmatgiz, Moscow, 1962).

10. L. D. Landau and E. M. Lifshitz, *Electrodynamics of Continuous Media* (Pergamon, New York, 1959).

11. Shi-I Pai, *Magnetogasdynamics and Plasma Dynamics* (Springer, New York, 1962).

12. S. I. Braginskii, "Contribution to the magnetohydrodynamics of weakly conducting liquids," Sov. Phys.-JETP **10**, No. 5, 1005 (1969).

13. T. G. Cowling, *Magnetohydrodynamics* (Wiley, New York, 1957).

14. G. F. Chew, M. L. Goldberger, and F. E. Low, "The Boltzmann equation and the one-fluid hydromagnetic equations in the absence of particle collisions," Proc. R. Soc. London Ser. A **236**, 112 (1956).

15. L. D. Pichakhchi, "Discontinuities in a tenuous plasma in the approximation of Chew, Goldberger, and Low," Ukr. Fiz. Zh. **5**, No. 4, 450 (1960).

16. T. F. Volkov, "Hydrodynamic description of strongly rarefied plasma," in *Reviews of Plasma Physics,* Vol. 4 (Consultants Bureau, New York, 1970).

17. S. Chapman and T. G. Cowling, *Mathematical Theory of Non-Uniform Gases* (Cambridge University, Cambridge, 1952).

18. S. I. Braginskii, "Transport phenomena in a plasma," in *Reviews of Plasma Physics,* Vol. 1 (Consultants Bureau, New York, 1963).

19. R. K. Jaggi, "Wave motion in a plasma with anisotropic pressure," Phys. Fluids **5**, No. 8, 949 (1962).

20. L. Spitzer, *Physics of Fully Ionized Gases* (Wiley, New York, 1962).

21. C. L. Longmire, *Elementary Plasma Physics* (Wiley, New York, 1963).

22. D. A. Frank-Kamenetskii, *Lectures on Plasma Physics* [in Russian] (Atomizdat, Moscow, 1966).

23. *Magnetohydrodynamics* (Symposium Materials) [in Russian] (Atomizdat, Moscow, 1964).

24. H. Alfvén and K. G. Felthammer, *Cosmical Electrodynamics,* 2nd ed. (Oxford, London, 1963).

25. D. V. Sivukhin, "Drift theory of charged-particle motion in an electromagnetic wave," in *Reviews of Plasma Physics,* Vol. 1 (Consultants Bureau, New York, 1963).

26. I. E. Tamm, *Principles of the Theory of Electricity* [in Russian] (Gostekhizdat, Moscow, 1956).

27. M. N. Kogan, "Shock waves in magnetogasdynamics," Prikl. Mat. Mekh. **72,** 557–563 (1959).

28. R. V. Polovin, "Shock waves in magnetohydrodynamics," Sov. Phys.-Usp. **3,** No. 2 (1960).

29. S. V. Iordanskii, "Zemplen's Theorem in magnetohydrodynamics," Dokl. Akad. Nauk SSSR **121,** No. 4 (1958).

30. R. V. Polovin and G. Ya. Lyubarskii, "Impossibility of rarefaction shock waves in magnetohydrodynamics," Sov. Phys.-JETP **8,** No. 2, 351 (1959).

31. F. A. Baum, S. A. Kaplan, and K. P. Stanyukovich, *Introduction to Cosmic Gasdynamics* [in Russian] (Fizmatgiz, Moscow, 1958).

32. A. G. Kulikovskii and G. A. Lyubimov, "Magnetohydrodynamic gas-ionizing shock waves," Dokl. Akad. Nauk SSSR **129,** 52 (1959).

33. A. A. Barmin and A. G. Kulikovskii, "Gas-ionizing shock waves in the presence of an arbitrarily oriented magnetic field," in *Problems of Hydrodynamics and Mechanics of a Continuous Medium* [in Russian], pp. 33–35 (Nauka, Moscow, 1969).

34. R. Z. Sagdeev, "Collective processes and shock waves in a rarefied plasma," in *Reviews of Plasma Physics,* Vol. 4 (Consultants Bureau, New York, 1970).

35. V. N. Karpman, in *Problems of Hydrodynamics and Mechanics of a Continuous Medium* [in Russian], pp. 36–57 (Nauka, Moscow, 1969).

36. B. L. Rozhdestvenskii and N. N. Yanenko, *Systems of Quasilinear Equations* [in Russian] (Nauka, Moscow, 1968).

37. A. Jeffrey and T. Taniuti, *Nonlinear Wave Propagation* (Academic, New York, 1964).

38. R. V. Polovin, Comment at 2nd Conference on theoretical and applied magnetohydrodynamics, in *Problems of Magnetohydrodynamics and Plasma Dynamics,* (Academy of Sciences of Latvian SSR, Riga, 1962), p. 37.

39. V. P. Korobeinikov and S. P. Lomnev, "Motion of charged particles in a plasma in the presence of a magnetohydrodynamic shock wave," Prikl. Mekh. Tekh. Fiz. No. 6, 89–92 (1964).

40. Y. M. Lynn, "Discontinuities in an anisotropic plasma," Phys. Fluids **10,** No. 10 (1967).

41. V. V. Sobolev, *Course of Theoretical Astrophysics* [in Russian], 2nd ed. (Nauka, Moscow, 1975).

42. J. W. Bond, K. Watson, and J. Welch, *Physical Theory of Gasdynamics* (Addison-Wesley, Reading, MA, 1965).

43. W. G. Vincenti and S. C. Traugott, "The coupling of radiative transfer and gas motion," Annu. Rev. Fluid Mech. **3** (1971).

44. L. I. Sedov, *Similarity and Dimensionality Methods in Mechanics* [in Russian], 9th ed. (1981) [English translation published by Mir, Moscow, 1985)].

45. K. P. Stanyukovich, *Unsteady Motion of Continuous Media* (Pergamon, New York, 1960).

46. Ya. B. Zel'dovich and A. S. Kompaneets, *Detonation Theory* [in Russian] (Gostekhizdat, Moscow, 1955).

47. S. S. Grigoryan, "Cauchy piston problem for one-dimensional unsteady motions," Prikl. Mat. Tekh. Fiz. No. 2, 179–187 (1955).

48. V. P. Korobeinikov, "One-dimensional gas motions accompanied by shock waves in a magnetic field," Prikl. Mat. Tekh. Fiz. No. 2, 47–53 (1960).

49. L. S. Pontryagin, *Continuous Groups* [in Russian] (Nauka, Moscow, 1973).

50. N. G. Chebotarev, *Theory of Lie Groups* [in Russian] (Gostekhizdat, Moscow, 1940).

51. L. V. Ovsyannikov, *Group Properties of Differential Equations* [in Russian] (USSR Academy of Sciences, Novosibirsk, 1962).

52. G. Birkhoff, "Dimensional analysis and partial differential equations," Electr. Eng. (NY) **67**, 1185–1188 (1948).

53. L. I. Sedov and V. V. Lokhin, "Nonlinear tensor functions of several tensor arguments," Prik. Mat. Mekh. **27**, No. 3, 393–417 (1963).

54. G. I. Barenblatt, "Limiting self-similar motions in the theory of nonstationary gas filtration in a porous medium, and boundary-layer theories," Prik. Mat. Mekh. **18**, No. 4, 409–414 (1954).

55. V. P. Korobeinikov, "Invariant solutions of magnetohydrodynamics equations," Magn. Gidrodin. No. 3, 55–58 (1967).

56. V. P. Korobeinikov, "Integral equations of unsteady adiabatic motions of a gas," Dokl. Akad. Nauk SSSR **104**, 509–512 (1955).

57. M. L. Lidov, "Finite integral of equations of one-dimensional self-similar adiabatic gas motions," Dokl. Akad. Nauk SSSR **103**, 35–36 (1955).

58. V. P. Korobeinikov, "One-dimensional self-similar motions of a conducting gas in a magnetic field," Sov. Phys.-Dokl. **3**, 739 (1959).

59. G. E. Shilov, *Mathematical Analysis. Second Special Course* [in Russian] (Nauka, Moscow, 1965).

60. V. S. Vladimirov, *Equations of Mathematical Physics* [in Russian] (Nauka, Moscow, 1967).

61. G. B. Whitham, *Linear and Nonlinear Waves* (Wiley, New York, 1974).

62. E. R. Benton and G. W. Platzman, "A table of solutions of the one-dimensional Burgers equation," Q. Appl. Math. **30**, No. 2 (1972).

63. O. V. Rudenko and S. M. Soluyan, *Theoretical Principles of Nonlinear Acoustics* [in Russian] (Nauka, Moscow, 1975).

64. V. P. Korobeinikov, "Problems of point-blast theory in gases," Tr. Mat. Inst., Akad. Nauk SSSR **119**, 278 (1973).

65. L. I. Sedov, "Propagation of strong blast waves," Prikl. Mat. Mekh. **10**, No. 2, 241–250 (1946).

66. L. I. Sedov, "Motion of air in a strong blast," Dokl. Akad. Nauk SSSR **52**, No. 1, 17–20 (1946).

67. G. I. Taylor, "The formation of a blast by a very intense explosion," Proc. R. Soc. London Ser. A **201**, No. 1065, 159–186 (1950).

68. K. P. Stanyukovich, "Use of particular solutions of gasdynamics equations to study detonation and shock waves," Dokl. Akad. Nauk SSSR **52**, No. 7 (1946).

69. H. Goldstine and J. von Neumann, "Blast wave calculation," Commun. Pure Appl. Math. **8**, No. 2, 327–354 (1955).

Chapter 2

1. L. I. Sedov, "Motion of air in a strong blast," Dokl. Akad. Nauk SSSR **52**, No. 1, 17–20 (1946).

2. L. I. Sedov, *Similarity and Dimensionality Methods in Mechanics*, 9th ed., 1981 [English translation] (Mir, Moscow, 1985).

3. V. P. Korobeinikov, N. S. Mel'nikova, and E. V. Ryazanov, *Point-Blast Theory* [in Russian] (Fizmatgiz, Moscow, 1961).

4. N. S. Burnova (N. S. Mel'nikova), "Investigation of the point-blast problem" (Author's abstract of Candidate's dissertation) (Moscow State University, Moscow, 1953).

5. V. P. Korobeinikov and E. V. Ryazanov, "Presentation of the solution of the blast-point problem in a gas in special cases," Prikl. Mat. Mekh. **23**, No. 2, 384–387 (1959).

6. V. P. Korobeinikov, P. I. Chushkin, and K. V. Sharovatova, *Tables of Gasdynamic Functions of the Initial State of a Point Blast* [in Russian] (Computation Center, USSR Academy of Sciences, Moscow, 1963).

7. K. A. Semendyaev, *Empirical Formulas* [in Russian] (GTTI, 1933).

8. G. M. Fikhtengol'ts, *Course of Differential and Integral Calculus* [in Russian], Vol. 2 (Nauka, Moscow, 1966).

9. N. N. Kochina, "Some exact solutions of the equations of one-dimensional unsteady motion of a perfect gas," Prikl. Mat. Mekh. **21**, 449 (1957).

10. Ya. B. Zel'dovich and Yu. P. Raizer, *Elements of Gas Dynamics and the Classical Theory of Shock Waves* (Academic, New York, 1968).

11. V. P. Korobeinikov, "Problem of strong point blast at zero temperature gradient," Dokl. Akad. Nauk SSSR **109**, 271–273 (1956).

12. O. S. Ryzhov and G. I. Taganov, "Second limiting case of the strong blast problem," Prikl. Mat. Mekh. **20**, No. 4, 545–548 (1956).

13. M. Frommer, "Integral curves of an ordinary first-order differential equation in the vicinity of a rational singularity," Usp. Mat. Nauk **9**, 212–253 (1941).

14. N. S. Mel'nikova, "Unsteady motion of a piston-propelled gas at zero temperature gradient," Bul. Instit. Politeh. Iasi **19**, No. 10, 3–4 (1964).

15. Yu. K. Bobrov and M. G. Shats, "Homothermal motion of gas in the channel of a high-current pulsed arc," Vestnik Politekh. In-ta, Electroenergetika, No. 17 (Wishcha Shkola, Kiev, 1980), pp. 38–42.

16. Yu. K. Bobrov and M. G. Shats, "Detonation model of a pulsed arc in self-similar approximation," Tekhn. Electrodin. No. 3, 67–75 (1981).

17. G. Ludwig and M. Hale, "Theory of a boundary layer with dissociation and ionization," in *Advances in Applied Mechanics*, Vol. 4, edited by H. L. Dryden and T. von Karman (Academic, New York, 1953).

18. F. A. Baum, S. A. Kaplan, and K. P. Stanyukovich, *Introduction to Cosmic Gasdynamics* [in Russian] (Fizmatgiz, Moscow, 1958).

19. A. I. Soloukhin, A. Ya. Yakobi, and A. V. Komin, *Optical Characteristics of a Hydrogen Plasma* [in Russian] (Nauka, Novosibirsk, 1973).

20. C. M. Harris, "Equilibrium properties and equation of state of a hydrogen plasma," Phys. Rev. A **2**, 133 (1964).

21. N. M. Kuznetsov, *Thermodynamic Functions and Shock Adiabats of Air at High Temperatures* [in Russian] (Mashinostroenie, 1965).

22. R. A. Gross, "Continuum radiation behind a blast wave," Phys. Fluids **7**, No. 7 (1964).

23. H. L. Brode, "Blast wave from a spherical charge," Phys. Fluids **2**, No. 2, 217–229 (1959).

24. H. L. Brode, "Blast calculations with a computer. Blast gasdynamics," [Russian translation in an anthology] (Mir, Moscow, 1977).

25. M. I. Volchinskaya, V. Ya. Gol'din, and N. N. Kalitin, "Use of the equation of state in gasdynamic calculations," Preprint No. 53, Inst. Appl. Mech. USSR Acad. Sci., 1975.

26. F. Higashino, "Cylindrical blast waves in ideal dissociating gases," J. Engineering, Tokyo Univ. B **32**, No. 1, 105–116 (1973).

27. W. G. Chase and H. K. Moore, editors, *Exploding Wires* (Plenum, New York, 1962).

28. M. N. Plooster, "Shock waves from line sources," Phys. Fluids **13**, No. 11, 2665–2670 (1970).

29. E. Larish and I. Shekhtman, "Introduction of radiation in the problem of gas-dynamics," Sov. Phys.-Dokl. **2**, 170 (1958).

30. H. L. Brode, "Action of nuclear explosion," [Russian translation in anthology "Action of Nuclear Explosion"] (Mir, Moscow, 1971).

31. G. S. Romanov, L. K. Stanchii, and K. L. Stepanov, "Calculation of averaged radiation paths in a multicomponent multiply ionized plasma," Zh. Prikl. Spektrosk. **30**, 37 (1979).

32. J. W. Bond, K. Watson, and J. Welch, *Physical Theory of Gasdynamics* (Addison-Wesley, Reading, MA, 1965).

33. I. V. Avilova, L. M. Biberman, V. S. Vorob'ev, V. M. Zmanin, G. A. Kobzev, A. N. Lagar'kov, A. Kh. Mnatsakanyan, and G. E. Norman, *Optical Properties of Hot Air* [in Russian] (Nauka, Moscow, 1979).

34. G. M. Bam-Zelikovich, "Propagation of strong blast waves," in *Theoretical Hydrodynamics* [in Russian] No. 4 (Oborongiz, Moscow, 1949).

35. V. P. Korobeinikov, "Investigation of certain problems of unsteady one-dimensional motions of gas" (author's abstract of candidate's dissertation), Math. Inst. Acad. Sci. USSR, 1956.

36. V. P. Korobeinikov, "Propagation of a strong spherical blast wave in a heat-conducting gas," Dokl. Akad. Nauk SSSR **113**, 1006–1009 (1957).

37. V. E. Neuvazhaev, "Propagation of a spherical blast wave in a heat-conducting gas," Prikl. Mat. Mekh. **26**, No. 6, 1094–1099 (1962).

38. V. V. Sychev, "Contribution to the theory of a strong blast in a heat-conducting gas," Prikl. Mat. Mekh. **29**, No. 6, 997–1003 (1965).

39. K. B. Kim, S. A. Berger, M. M. Kamel, V. P. Korobeinikov, and A. K. Oppenheim, "Boundary-layer theory for blast waves," J. Fluid Mech. **71**, part 1, 65–85 (1975).

40. I. O. Bezhaev, "Effect of viscosity and thermal conductivity of a gas on blast propagation," in the anthology *Theoretical Hydrodynamics* [in Russian], No. 11 (Oborongiz, Moscow, 1953).

41. V. P. Shidlovsky, "Effect of dissipative phenomena on the evolution of shock waves," AIAA J. **15**, No. 1, 33–38 (1977).

42. D. N. Brushlinskii and V. P. Korobeinikov, "Self-similar problem of strong blast with allowance for heat transfer by the radiation," Dokl. Akad. Nauk SSSR **259**, No. 5, 1060–1063 (1981).

43. S. C. Traugott, "Non-diffusive effects in the radiation propagation of a thermal pulse," Phys. Fluids **13**, 2242–2252 (1970).

44. B. V. Putyatin, "Initial stage of a point blast in a radiating gas," Izv. Akad. Nauk SSSR, Mekh. Zhidk. Gaza No. 3 (1980).

45. V. P. Korobeinikov and B. V. Putyatin, "Analytic methods of solving problems of high-temperature gas flows," in *High-Temperature Gasdynamics* [in Russian], edited by R. I. Soloukhin (Akad. Nauk BSSR, Minsk, 1983), p. 140.

46. V. P. Ageev, A. I. Barchukov, F. V. Bunkin, V. I. Konov, V. P. Korobeinikov, B. V. Putjatin, and V. M. Hudjakov, "Experimental and theoretical modeling of laser propulsion," Astronaut. Acta **7**, 79–90 (1980).

47. V. P. Ageev, A. I. Barchukov, F. V. Bunkin, A. A. Gorbunov, V. M. Hudjakov, V. I. Konov, V. P. Korobeinikov, and B. V. Putyatin, "Some characteristics of the laser multi-pulse explosive type jet thruster," Astronaut. Acta **8**, No. 5, 625–641 (1981).

48. R. J. Latko and R. Viskanta, "Transient energy transfer in a radiating gas during shock expansion," Phys. Fluids **12,** No. 10, 2036–2045 (1969).

49. J. A. Zinn, "A finite-difference scheme for time-dependent spherical radiation hydrodynamics problems," J. Comput. Phys. **13,** 569–590 (1973).

50. R. D. Richtmeyer and K. W. Morton, *Difference Methods for Initial-Value Problems* (Interscience, New York, 1967).

51. V. P. Korobeinikov, B. V. Putyatin, P. I. Chushkin, and L. V. Shurvalov, "Numerical modeling of the Tunguska catastrophe," in *Proceedings of the International Conference on Numerical Methods in Hydrodynamics* [in Russian], Vol. 2 (Institute of Applied Mechanics, USSR Academy of Sciences, Moscow, 1978), pp. 126–131.

52. I. V. Nemchinov, I. A. Polozova, V. V. Svettsov, and V. V. Shuvalov, "Numerical calculation of a one-dimensional blast with radiation," in *Dynamics of a Radiating Gas* [in Russian], No. 3 (Computation Center, USSR Academy of Sciences, Moscow, 1980), pp. 33–35.

53. V. V. Svettsov, "Calculation of a spherically symmetric blast problem by the method of averaging the transport equation," (Computation Center, USSR Academy of Sciences, Moscow, 1980), pp. 46–47.

54. G. I. Taylor, "The formation of a blast wave by a very intense explosion," Proc. R. Soc. London A **201,** No. 1065, 159–186 (1950).

55. *Effects of Atomic Weapons* (Los Alamos Scientific Laboratory, McGraw, NM, 1950).

56. *Effect of Atomic Weapons* [Russian translation] (Voenizdat, Moscow, 1960).

57. "Effective-energy method in the problem of a strong blast in a real gas," Prikl. Mech. Tekh. Fiz. No. 5 (1968).

58. N. N. Sobolev, "Investigation of electric explosion of thin wires," Zh. Eksp. Teor. Fiz. **17,** 11 (1947).

59. S. V. Lebedev, "Explosion of a wire by an electric current," Sov. Phys.-JETP **5,** No. 2, 243 (1957).

60. W. Chase and H. M. Moore, editors, *Exploding Wires* (Plenum, New York, 1962).

61. W. Chase and H. M. Moore, editors, *Exploding Wires,* Vol. III (Plenum, New York, 1964).

62. G. G. Dolnev and S. L. Mandel'stam, "Density and temperature of a gas in a spark discharge," Zh. Eksp. Teor. Fiz. **24** (1963).

63. A. E. Voitenko and I. Sh. Model', "Obtaining strong shock waves in electric discharges in gaps," Sov. Phys.-JETP **17,** No. 6 (1964).

64. *Magnetohydrodynamics,* a Symposium, edited by R. K. Landshoff (Stanford University, Stanford, CA, 1957).

65. A. C. Kolb, "Production of high-energy plasmas by magnetically driven shock waves," Phys. Rev. **107,** No. 2, 345 (1957).

66. V. V. Korobkin, S. L. Mandel'shtam, P. P. Pashinin, *et al.* "Investigation of a spark produced in air by focused laser radiation," Sov. Phys.-JETP **26,** No. 1 (1968).

67. E. Panarella and P. Savic, "Blast waves from a laser-induced spark in air," Can. J. Phys. **46,** No. 3 (1967).

68. N. G. Basov, V. A. Boiko, O. N. Krokhin, *et al.* "Formation of a long spark in air by weakly focused laser radiation," Dokl. Akad. Nauk SSSR **137,** No. 3 (1967).

69. N. G. Basov, "Lasers in physical research," Priroda No. 10 (1967).

70. *Effects of Laser Emission* [coll. Russian translation] (Mir, Moscow, 1968).

71. J. W. Daiber and H. M. Thompson, "Laser-driven detonation waves in gases," Phys. Fluids **10,** No. 6, 1162–1169 (1967).

72. A. K. Oppenheim and R. I. Soloukhin, "Experiments in gas dynamics of explosions," Annu. Rev. Fluid Mech. **5,** 31–35 (1973).

Chapter 3

1. N. S. Burnova (N. S. Mel'nikova), "Investigation of the point-blast problem" (Author's abstract of candidate's dissertation) (Moscow State University, Moscow, 1953).

2. L. I. Sedov, *Similarity and Dimensionality Methods in Mechanics,* 3rd Russian edition (Gostekhizdat, Moscow, 1954) [Translation of 1st ed., Academic, New York, 1967].

3. A. Sakurai, "On propagation and structure of the blast wave," J. Phys. Soc. Jpn. **8,** No. 5 (1953); **9,** No. 2, 256–266 (1954).

4. V. P. Korobeinikov and E. V. Ryazanov, "On the theory of linearized blast problems," Prikl. Mat. Mekh. **23,** No. 4, 749–759 (1959).

5. V. P. Korobeinikov, "Application of blast theory to the propagation of shock waves in solar flares," in Abstracts of the 3rd All-Union Conference on Theoretical and Applied Mechanics [in Russian] (Nauka, Moscow, 1968), p. 171.

6. V. P. Korobeinikov, "On the gas flow due to solar flares," Solar Phys. **7,** 463–470 (1969).

7. M. L. Lidov, "On the theory of linearized solution near one-dimensional self-similar gas motions, Dokl. Akad. Nauk SSSR **102,** No. 6 (1955).

8. V. P. Korobeinikov, N. S. Mel'nikova, and E. B. Ryazanov, *Point Blast Theory* [in Russian] (Fizmatgiz, Moscow, 1961).

9. H. Mirels and J. F. Mullen, "Aerodynamic blast simulation in hypersonic tunnels," AIAA J. **3,** No. 11 (1965).

10. D. N. Brushlinskii and G. S. Solomakhova, "Investigation of the strong-blast problem with allowance for back pressure," in the colloquium *Theoretical Hydromechanics* [in Russian], Vol. **19,** No. 7 (Oborongiz, Moscow, 1956), pp. 81–100.

11. V. P. Korobeinikov, P. I. Chushkin, and K. V. Sherovatova, *Tables of Gasdynamic Functions of the Initial State of a Point Blast* (Computation Center, USSR Academy of Sciences, Moscow, 1963).

12. V. P. Korobeinikov and P. I. Chushkin, "Calculation of the initial stage of a point blast in various gases," Prikl. Mekh. Tekh. Fiz. No. 4, 48–57 (1963).

13. R. J. Swigart, "Third-order blast theory and its application to hypersonic flow past blunt-nosed cylinders," J. Fluid. Mech. **9** (1960).

14. A. Sakurai, "Blast wave theory," in *Basic Developments in Fluid Dynamics,* Vol. 1 (Academic, New York, 1965), pp. 309–375.

15. A. Sakurai, "Solution of a point source blast wave equation," J. Phys. Soc. Jpn. **51,** 1355–1356 (1982).

16. G. G. Bach and J. H. Lee, "Higher-order perturbation solutions for blast waves," AIAA J. **7,** 742–744 (1969).

17. I. S. Berezin and N. P. Zhidkov, *Calculation Methods* [in Russian], Vol. 2 (Fizmatgiz, Moscow, 1960).

18. G. Guderley, "Starke kugelige and zylindrische Verdichtungsstösse in der Nähe des Kugelmittelpuktes bzw. der Zylinderachse," Luftfahrtforschung **19,** No. 9, 302–312 (1942).

19. K. P. Stanyukovich, *Unsteady Motion of Continuous Media* (Pergamon, New York, 1960).

20. L. D. Landau and E. M. Lifshitz, *Fluid Mechanics* (Pergamon, New York, 1959).

21. G. G. Bach and J. H. Lee, "Initial propagation of impulsively generated converging cylindrical and spherical shock waves," J. Fluid Mechanics **37,** part 3, 513–528 (1969).

22. J. D. Cole, *Perturbation Methods in Applied Mechanics* (Xerox Corporation, Webster, NY, 1968).

23. M. Van Dyke, *Perturbation Methods in Liquid Mechanics* (Academic, New York, 1964).

24. V. P. Korobeinikov, Sov. Phys.-Dokl. **28**, 120 (1983).

Chapter 4

1. L. I. Sedov, Dokl. Akad. Nauk SSSR **85**, No. 4, 723–726 (1952).

2. L. I. Sedov, *Similarity and Dimensionality Methods in Mechanics*, 9th ed. (Nauka, Moscow, 1981) [English translation Mir, Moscow, 1985].

3. D. E. Okhotsimskii, I. L. Kondrashova, Z. P. Vlasova, *et al.*, "Point-blast calculation with allowance for backpressure," Tr. Mineral. Inst., Akad. Nauk SSSR **50** (1957).

4. H. Goldstine and J. von Neumann, "Blast wave calculations," Commun. Pure Appl. Math. **8**, 327–354 (1955).

5. V. P. Korobeinikov, N. S. Mel'nikova, and E. V. Ryazanov, *Point-Blast Theory* [in Russian] (Fizmatgiz, Moscow, 1961).

6. Ya. B. Zel'dovich, *Theory of Shock Waves and Introduction to Gasdynamics* [in Russian], Akad. Nauk SSSR (1946) (Chap. 1 translation published by Academic, New York, 1968).

7. L. D. Landau, "Shock waves at large distances from the blast point," Prikl. Mat. Mekh. **9**, No. 4, 286–292 (1945).

8. Yu. L. Yakimov, "Asymptotic solutions of the equations of one-dimensional unsteady motion of an ideal gas, and asymptotic laws of shock-wave damping," Prikl. Mat. Mekh. **19**, No. 6, 681–692 (1955).

9. G. M. Shefter, "Asymptotic solution of the equations of one-dimensional steady motion of an ideal gas with cylindrical symmetry," Dokl. Akad. Nauk SSSR **116**, 572–575 (1954).

10. H. L. Brode, "Numerical solutions of spherical blast waves," J. Appl. Phys. **26**, No. 6, 766 (1955).

11. D. E. Okhotsimskii and Z. P. Vlasova, "Behavior of shock waves at large distances from the blast point," Vychisl. Mat. Mat. Fiz. **2**, No. 1, 107–124 (1962).

12. P.-C. Chou, Karpp, and H.-L. Huan, "Numerical calculation of shock waves by the method of characteristics" [Russian translation in Raketentech. Tekh. Kosmonavtika **5**, No. 4 (1967)].

13. A. I. Zhukov, "Use of the method of characteristics for numerical solution of one-dimensional gasdynamics problems," Tr. Mineral. Inst., Akad. Nauk SSSR **58** (1960).

14. P. I. Chushkin, "Method of characteristics for 3D supersonic flows" [in Russian], Tr. VTs Akad. Nauk SSSR 1968.

15. V. V. Rusanov, "Characteristics of general gasdynamics equations." Mat. Mat. Fiz. **3**, No. 3, 508–524 (1963).

16. G. G. Chernyi, "The point-blast problem," Dokl. Akad. Nauk SSSR **112**, 213–216 (1957).

17. G. G. Chernyi, "Use of integral relations for the propagation of strong shock waves," Prikl. Mat. Mekh. **24**, No. 1 (1960).

18. N. S. Burnova (N. S. Mel'nikova), "Investigations of the Point-Blast Problem" [in Russian] (Author's abstract of candidate's dissertation, Moscow State University, Moscow, 1953).

19. V. P. Korobeinikov, "Analogy between a cylindrical blast and hypersonic gas flow around bodies," Zh. Prikl. Mekh. Tekh. Fiz. No. 6, 45–49 (1962).

20. N. S. Mel'nikova and T. M. Salamakhin, "Point-blast calculations in various media," Zh. Prikl. Mekh. Tekh. Fiz. No. 4, 155–160 (1964).

21. N. S. Mel'nikova, "Point blast in a medium with variable initial density," Tr. Mineral. Inst., Akad. Nauk SSSR **87**, 66–85 (1955).

22. J. H. Lee, R. Knystautas, and G. G. Bach, "Theory of Explosions," MERL Report 69-10, McGill University, 1969.

23. O. A. Sakurai, "On the propagation of cylindrical shock waves," in *Exploding Wires*, edited by W. C. Chase and H. K. Moore (Plenum, New York, 1959), pp. 264–270.

24. V. V. Adushkin and I. V. Nemchinov, "Approximate deduction of gas parameters behind a shock-wave front from the law of front motion," Prikl. Mat. Tekh. Fiz. No. 4, 58–67 (1963).

25. V. P. Andreev, "Approximate method of calculating one-dimensional flows with shock waves," Izv. Akad. Nauk SSSR Mekh. Zhidk. Gaza, No. 6 (1967).

26. A. A. Dorodnitsyn, "One method of numerically solving certain nonlinear aerodynamics problems," in Proceedings of the 3rd All-Union Mathematical Congress [in Russian], Akad. Nauk SSSR **3** (1958).

27. O. M. Belotserkovskii and P. I. Chushkin, "Numerical integral-equations method," Zh. Vychisl. Mat. Mat. Fiz. **2**, 731–759 (1962).

28. O. M. Belozerkovsky and P. I. Chushkin, "The numerical solution of problems in gasdynamics," in *Basic Developments in Fluid Dynamics*, Vol. 1 (Academic, New York, 1965).

29. V. P. Karlikov, V. P. Korobeinikov, and E. V. Ryazanov, "Approximate method of solving the blast problem in certain ideal compressible media," Prikl. Mekh. Tekh. Fiz. No. 2, 132–134 (1963).

30. V. P. Korobeinikov and A. P. Chushkin, "Calculation method of a point blast in a gas," Dokl. Akad. Nauk SSSR **154**, 549–552 (1964).

31. V. P. Korobeinikov and P. I. Chushkin, "Planar, cylindrical, and spherical blast in a gas with backpressure," Tr. Mineral. Inst., Akad. Nauk SSSR **87**, 4–34 (1966).

32. V. P. Korobeinikov, P. I. Chushkin, and K. V. Sharovatova, (a) *Tables of gasdynamic functions* [in Russian]; Soobshch. Vychisl. Matem. VTs Akad. Nauk SSSR, No. 2 (1963); (b) *Point-blast gasdynamic functions* [in Russian], Proc. Comput. Center USSR Akad. Sci. (1969).

33. V. P. Korobeinikov, "Problems in the theory of point blasts in gases," Tr. Mineral. Inst., Akad. Nauk SSSR **119** (1973).

34. Kh. S. Kestenboim, G. S. Roslyakov, and L. A. Chudov, *Point Blast Calculation Methods. Tables* [in Russian] (Nauka, Moscow, 1974).

35. N. A. Arkhangel'skii, "Algorithms for numerical cylindrical-blast calculation by the grid method, with allowance for backpressure," Vychisl. Mat. Mat. Fiz. **22**, 222–236 (1971).

36. V. P. Korobeinikov, "Motion induced by an external energy source," Inzh. Fiz. Zh. **25**, 1121–1126 (1973).

37. L. H. Back and G. Varis, "Detonation propulsion for high-pressure environments," AIAA J. **12**, 1123–1130 (1974).

38. V. P. Ageev, A. I. Barchukov, F. V. Bunkin, V. I. Konov, V. P. Korobeinikov, B. V. Putyatin, and V. M. Hudjakov, "Experimental and theoretical modeling of laser propulsion," Astronaut. Acta **7**, 79–90 (1980).

39. O. M. Belotserkovskii and Yu. M. Davydov, "Nonstationarily large-particles method for gasdynamic calculations," Vychisl. Mat. Mat. Fiz. **2**, No. 1, 182–207 (1971).

40. B. A. Rozhdestvenskii and N. N. Yanenko, *Systems of Quasilinear Equations* [in Russian] (Nauka, Moscow, 1968).

41. M. A. Sadovskii, "Mechanical action of blast waves in air from experimental data" [in Russian], in Fiz. Vzryva, Akad. Nauk SSSR, No. 1952.

42. M. A. Tsikulin, *Shock Waves Accompanying Large Meteor Trails in the Atmosphere* [in Russian] (Nauka, Moscow, 1969).

43. H. Shardin, "Measurement of spherical shock waves," Commun. Pure Appl. Math. **7,** No. 1 (1954).

44. *Effect of Atomic Weapons* (Los Alamos Scientific Laboratory, McGraw, NM, 1950).

45. Yu. S. Yakovlev, *Blast Hydrodynamics* [in Russian] (Sudpromgiz, 1961).

46. S. A. Lovlya, *et al., Explosions* [in Russian] (Nedra, 1966).

47. A. S. Fonarev and S. Yu. Chernyavskii, "Calculation of shock waves of spherical exploding charges in air," Izv. Akad. Nauk SSSR, Mekh. Zhidk. Gaza, No. 5 (1968).

48. J. W. Miles, "Decay of spherical blast waves," Phys. Fluids **10,** 2706 (1967).

49. V. P. Korobeinikov, "Approximate equations for the characteristics of the shock-wave front of a point blast in a gas," Dokl. Akad. Nauk SSSR **111,** 557–559 (1956).

50. R. F. Fletcher, D. Cerneth, and C. Goodman, "Explosion of propellants," AIAA J. **4,** No. 4 (1966).

51. A. V. Zolotov, (a) "On the possibility of a "thermal" blast and on the structure of the Tunguska meteorite," Dokl. Akad. Nauk SSSR **172,** No. 4, 805–808 (1967); (b) Estimate of the parameters of the Tunguska meteorite from new data," *ibid.* **172,** No. 5, 1049–1052 (1967).

52. H. Brode, "Blast waves from a spherical charge," Phys. Fluids **2,** No. 2, 217–229 (1959).

53. V. P. Korobeinikov, "Similarity of energy transport processes in nonstationary high-temperature gas flows," in *Certain Problems in the Mechanics of a Continuous Medium* [in Russian], edited by S. S. Grigoryan (Moscow University, Moscow, 1978), p. 188.

54. G. G. Chernyi, *Gas Flow at High Supersonic Velocity* [in Russian] (Fizmatgiz, Moscow, 1959).

55. W. D. Hayes and R. F. Probstein, editors, *Hypersonic Flow Theory,* 1st ed. (Academic, New York, 1957).

56. P. I. Chushkin and N. P. Shulishina, *Tables of Supersonic Flows Around Blunted Bodies* [in Russian] (Computation Center, USSR Academy of Sciences, Moscow, 1961).

57. V. V. Lunev, K. M. Magomedov, and V. G. Pavlov, *Hypersonic Flow Around Blunted Cones with Allowance for Equilibrium Physico-Chemical Transformations* [in Russian] (Computation Center, USSR Academy of Sciences, Moscow, 1968).

58. E. I. Andriankin, "Perturbation method for the strong-blast problem," Izv. Akad. Nauk SSSR Mekh. No. 12 (1958).

59. T. S. Chang and O. Laporte, "Reflection of strong blast waves," Phys. Fluids **7,** No. 8, 1225–1232 (1964).

60. R. Ya. Tugazakov and A. S. Fonarev, "Initial stage of collision of blast waves," Izv. Akad. Nauk SSSR, Mekh. Zhidk. Gaza No. 5, 41–48 (1971).

61. R. Courant and K. O. Friedrichs, *Supersonic Flow and Shock Waves* (Wiley, New York, 1948).

62. *Effect of Atomic Weapons* [Russian translation] (Voenizdat, 1960).

63. M. M. Vasil'ev, "Reflection of a spherical shock wave from a plane," in Vychisl. Mat. Akad. Nauk SSSR No. 6, 87–89 (1960).

64. K. E. Gubkin, "Shock-wave investigation using half-shadow photographs," in Fiz. Vzryva, Akad. Nauk SSSR No. 3, 73–86 (1955).

65. O. S. Ryzhov and S. A. Khristianovich, "Nonlinear reflection of weak shock waves," Prikl. Mat. Mekh. **22,** 586–599 (1958).

66. T. V. Bazhenova and L. G. Gvozdeva, *Nonstationary Interactions of Shock Waves* [in Russian] (Nauka, Moscow, 1977).

67. A. S. Fonarev and V. V. Podlubnyi, (a) "Interaction of a spherical blast wave with a plane surface," (b) "Parameters of an air blast wave reflected from a plane

surface," in *Interaction of a Blast Wave with a Solid Surface,* Tr. TsAGI No. 170, pp. 3–16 and 26–148 (1975).

68. S. K. Godunov, A. V. Zabrodin, M. Ya. Ivanov, A. N. Kraiko, and G. P. Prokopov, *Numerical Solution of Multidimensional Gasdynamics Problems* [in Russian] (Nauka, Moscow, 1976).

Chapter 5

1. V. P. Korobeinikov, "Exact solution of a nonlinear blast problem in a gas with variable initial density," Dokl. Akad. Nauk SSSR **117,** 947–948 (1957).

2. L. I. Sedov, "Integration of the equations of one-dimensional motion of a gas," Dokl. Akad. Nauk SSSR **90,** No. 5, 735 (1953).

3. V. P. Korobeinikov and E. V. Ryazanov, "Exact discontinuous solutions of the equations of one-dimensional gasdynamics and their applications," Prikl. Mat. Mekh. **22,** 265–268 (1958).

4. A. N. Golubyatnikov, "Estimates of the motion of shock waves in one-dimensional nonstationary gasdynamics problems," Dokl. Akad. Nauk SSSR **237,** 800–803 (1977).

5. R. I. Nigmatulin, "Planar blast on the interface of two ideal calorically perfect gases," Vestnik MGU Mat. Mekh. No. 1, 83–87 (1965).

6. A. Sakurai, "Blast wave from a plane source at an interface," J. Phys. Soc. Jpn. **36,** No. 2, 610 (1974).

7. V. P. Korobeinikov and G. A. Ostroumov, "More on cavitation damage," Sov. Phys. Acoust. **12,** No. 4, 397–401 (1965).

8. G. A. Ostroumov, "Mechanism of cavitation damage," Sov. Phys. Acoust. **9,** No. 2 (1963).

9. B. N. Rumyantsev, "One limiting case of strong blast-wave propagation in an inhomogeneous medium," Zh. Prikl. Mekh. Tekh. Fiz. No. 1, 127–129 (1963).

10. B. N. Rumyantsev, "Contribution to the theory of a blast in an inhomogeneous medium at an adiabatic exponent close to unity," Zh. Prikl. Mekh. Tekh. Fiz. No. 4, 118–120 (1963).

11. L. V. Shurvalov, "Strong blast on the boundary of a half-space filled with a perfect gas," Prikl. Mat. Mekh. **33,** 358–363 (1969).

12. L. I. Sedov, *Planar Problems of Hydrodynamics* [in Russian] (Nauka, Moscow, 1980).

13. L. A. Galin, "Impact of a body on the surface of a compressible liquid," in *Abstracts of the First All-Union Congress on Theoretical and Applied Mechanics* (Akad. Nauk SSSR, Moscow, 1960).

14. E. P. Borisova, P. P. Koryavov, and N. N. Moiseev, "Planar and axisymmetric self-similar problem of immersion and collision of jets," Prikl. Mat. Mekh. **23,** No. 2 (1959).

15. A. Ya. Sagomonyan, *3D Problems of Nonstationary Motion of a Compressible Liquid* [in Russian] (Moscow University, Moscow, 1962).

16. A. A. Deribas and S. I. Pokhozhaev, "Formulation of the problem of a strong blast on a liquid surface," Sov. Phys.-Dokl. **7,** 383 (1962).

17. V. F. Minin, "Blast on a liquid surface," Prikl. Mekh. Tekh. Fiz. No. 3 (1964).

18. S. S. Grigoryan and M. M. Martirosyan, "Ground waves stimulated by surface blast," Astronaut. Acta **15,** No. 5 (1970).

19. V. P. Korobeinikov, P. I. Shushkin, and K. V. Sharovatova, *Tables of Gasdynamic Functions of the Initial Stage of a Point Blast,* Soobshch. Vychisl. Mat. VTs Akad. Nauk SSSR No. 2 (1963).

20. V. P. Korobeinikov, P. I. Shushkin, and K. V. Sharovatova, in *Point Blast Gasdynamic Functions* (Computation Center, USSR Academy of Sciences, Moscow, 1969).

21. V. D. Alekseenko, "Experimental investigation of the energy distribution in a contact blast," Izv. Akad. Nauk SSSR Fiz. Gor. Vzryva **3**, No. 1, 152–155 (1967).

22. K. P. Babenko, *et al.* "Methods of solving certain 2D problems," in *Problems of Computational Mathematics and Computation Techniques* [in Russian] (Mashgiz, 1963).

23. V. V. Rusanov and E. E. Shnol', "Difference methods in 3D gasdynamic problems," in *Proceedings of the 4th All-Union Mathematical Congress* (Akad. Nauk SSSR, Leningrad, 1963).

24. V. V. Rusanov, "Calculation and Investigation of Multidimensional Gas Flows by the Finite-Difference Method," (Authors' abstract of doctoral dissertation) (Institute of Mechanics Problems, USSR Academy of Sciences, Moscow, 1968).

25. Kh. S. Kestenboim, F. D. Turetskaya, and L. A. Chudov, "Point blast in an inhomogeneous atmosphere," Prikl. Mekh. Tekh. Fiz. No. 5 (1969).

26. V. P. Korobeinikov and V. P. Karlikov, "Determination of the shape and parameters of the shock wave front of a blast in an inhomogeneous medium," Sov. Phys.-Dokl. **8**, 137 (1963).

27. V. P. Korobeinikov, "Gasdynamics of explosion," Annu. Rev. Fluid Mech. **3**, 317-346 (1971).

28. L. V. Ovsyannikov, "New solution of hydrodynamics equations," Dokl. Akad. SSSR **111**, 47–49 (1956).

29. V. P. Karlikov, "Solution of a linearized axisymmetric point-blast problem in a variable-density medium," Dokl. Akad. SSSR **101**, 1009–1012 (1955).

30. V. P. Karlikov, "Linearized problem of strong-blast propagation in an inhomogeneous atmosphere," Vestnik MGU Ser. Mat. Mekh. Astron. Fiz. Khim. No. 4, 60–65 (1959).

31. V. P. Karlikov, "Linearized solution of a strong-blast problem in a medium with linear density distribution," Vestnik MGU Ser. Mat. Mekh. No. 1 (1960).

32. E. I. Andriankin, "Perturbation method for the point-blast problem," Izv. Akad. Nauk SSSR OTN No. 12, 6–14 (1958).

33. A. S. Kompaneets, "Point blast in an inhomogeneous atmosphere," Sov. Phys.-Dokl. **5**, 46 (1960).

34. E. I. Andriankin, A. S. Kompaneets, and V. P. Krainov, "Strong-blast propagation in an inhomogeneous atmosphere," Prikl. Mat. Tekh. Fiz. No. 6, 3–7 (1962).

35. D. D. Laumbach and R. F. Probstein, "A point explosion in a cold exponential atmosphere," Fluid Mech. **35**, pt. 1 (1969).

36. K. E. Gubkin, "Propagation of discontinuities in sound waves," Prikl. Mat. Mekh. **22**, No. 4 (1958).

37. K. E. Gubkin, "Nonlinear geometric acoustics and its applications," in *Some Mathematics and Mechanics Problems* [in Russian] (USSR Academy of Sciences, Novosibirsk, 1961).

38. O. S. Ryzhov, "Damping of shock waves in inhomogeneous media," Prikl. Mekh. Tekh. Fiz. No. 2, 15–25 (1961).

39. O. S. Ryzhov and G. M. Shefter, "Energy of sound waves propagating in moving media," Prikl. Mat. Mekh. **26**, No. 5 (1962).

40. Kh. S. Kestenboim, G. S. Roslyakov, and L. A. Chudov, *Point Blast Calculation Methods. Tables* [in Russian] (Nauka, Moscow, 1974).

41. Kh. S. Kestenboim and G. S. Roslyakov, *Numerical Solution of One-dimensional Blast Problems* [in Russian] (Moscow University, Moscow, 1971).

42. V. A. Bronshten, "Propagation of spherical and cylindrical blast waves in an inhomogeneous atmosphere with allowance for backpressure," Zh. Prikl. Mekh. Tekh. Fiz. No. 3, 84–90 (1972).

43. M. Lutzky and D. L. Lehot, "Shock propagation in spherically symmetric exponential atmospheres," Phys. Fluids **11**, No. 7, 1466–1472 (1968).

44. L. V. Ovsyannikov, "Approximate method of recalculating the law of propagation of one-dimensional shock waves," Zh. Prikl. Mekh. Tekh. Fiz. No. 1, 55–57 (1972).

45. L. V. Shurvalov, "Calculation of shock waves propagating in an inhomogeneous atmosphere," Dokl. Akad. Nauk SSSR **230**, No. 4, 803–806 (1976).

46. V. P. Korobeinikov, P. I. Chushkin, and L. V. Shurshalov, "Allowance for the inhomogeneity of the atmosphere in the calculation of the Tunguska meteorite blast," Zh. Vychisl. Mat. Mat. Fiz. **17**, 737–752 (1977).

47. L. V. Shurshalov, "Two-dimensional calculations of an explosion in the real atmosphere," Arch. Mech. **30**, Nos. 4–5, 629–643 (1978).

48. V. P. Korobeinikov, B. V. Putjatin, P. I. Chushkin, and L. V. Shurshalov, "On computational modeling of the Tunguska catastrophe," in *Proceedings of the 6th International Conference on Numerical Methods in Fluid Dynamics*, Tbilisi, 1978 (Springer, New York, 1979), pp. 323–331.

49. L. V. Shurvalov, "Allowance for radiation in blast calculations in an inhomogeneous atmosphere," Izv. Akad. Nauk SSSR Mech. Zhidk. Gaza No. 3, 105–112 (1980); No. 6, 124–130 (1982).

50. V. P. Korobeinikov, P. I. Chushkin, and L. V. Shurshalov, "Interaction between large cosmic bodies and the atmosphere," Preprint IAF-81, No. 407, 23rd Congress of the IAF, Rome, September 6–12, 1981.

51. L. I. Sedov, "Similarity methods in nonlinear mechanics of a continuous medium," *Proceedings of the 3rd All-Union Mathematical Congress*, Vol. 3 (USSR Academy of Sciences, Moscow, 1958).

52. V. P. Korobeinikov and P. I. Chushkin, "Planar, cylindrical, and spherical blast in a gas with backpressure," Tr. Mineral. Inst., Akad. Nauk SSSR **87**, 4–34 (1966).

53. V. P. Korobeinikov, P. I. Chushkin, and L. V. Shurshalov, "Gasdynamics of the flight and explosion of meteorites," Astronaut. Acta **17**, Nos. 4–5, 339–348 (1972).

54. V. P. Korobeinikov, P. I. Chushkin, and P. I. Shurshalov, "Hydrodynamic effects in flight and explosion of large meteorites in the earth's atmosphere," Meteoritika, No. 32, 73–89 (1973).

55. E. D. Terent'ev, "Perturbations due to the finite momentum in the strong point-blast problem," Prikl. Mat. Mekh. **43**, No. 1, 51–56 (1979).

56. O. S. Ryzhov and E. D. Terent'ev, "Contribution to a general theory of nonstationary nearly self-similar flows." Prikl. Mat. Mekh. **37**, No. 1, 65–74 (1973).

Chapter 6

1. L. I. Sedov, *Similarity and Dimensionality Methods in Mechanics*, 9th Russian ed. (Nauka, Moscow, 1981) [English translation of the 11th edition (Mir, Moscow, 1985)].

2. V. P. Korobeinikov, N. S. Mel'nikova, and E. V. Ryazanov, *Point-Blast Theory* [in Russian] (Fizmatgiz, Moscow, 1961).

3. Ya. B. Zel'dovich and A. S. Kompaneets, *Detonation Theory* [in Russian] (Gostekhizdat, Moscow, 1955).

4. K. I. Shelkin, "Two cases of unstable combustion," Sov. Phys.-JETP **9**, 416–420 (1960).

5. R. M. Zaidel' and Ya. B. Zel'dovich, "One-dimensional instability and detonation damping," Prikl. Mekh. Tekh. Fiz. No. 6, 59–65 (1963).

6. R. I. Soloukhin, "Detonation in a gas heated by a shock wave," Prikl. Mekh. Tekh. Fiz. No. 4, 42–48 (1964).

7. R. B. Gilbert and R. A. Strelow, "Theory of detonation initiation behind reflected shock waves," AIAA J. **4**, No. 10 (1966).

8. V. V. Pukhnachev, "Stability of Chapman–Jouguet detonation," Prikl. Mekh. Tekh. Fiz. No. 6, 66–73 (1963).

9. R. I. Soloukhin, "Exothermal reaction zone in a one-dimensional shock wave in a gas," Fiz. Gor. Vzryva No. 3, 12–18 (1966).

10. V. P. Korobeinikov, "Point blast in a detonating gas," Sov. Phys.-Dokl. **12,** 1003 (1967).

11. V. P. Korobeinikov, "Problem of point blast in a detonating gas" (International Colloquium on Blast Gasdynamics, Brussels, 1967), Astronaut. Acta **14,** No. 5 (1969).

12. V. P. Korobeinikov and V. V. Markov, "On the propagation of combustion and detonation," Arch. Thermodyn. Splan. Polska **8,** 101–118 (1977).

13. I. S. Shikin, "Exact solutions of one-dimensional gasdynamics equations with shock and detonation waves," Sov. Phys.-Dokl. **3,** 915 (1958).

14. V. A. Levin, "Approximate solution of the strong-blast problem in a combustible mixture," Izv. Akad. Nauk SSSR Mekh. Zhidk. Gaza No. 1, 122–124 (1967).

15. V. P. Korobeinikov and P. I. Chushkin, "Planar, cylindrical, and spherical blast in a gas with backpressure," Tr. Mineral. Inst., Akad. Nauk SSSR **37,** 4–34 (1966).

16. G. G. Chernyi, "Asymptotic law of propagation of a planar detonation wave," Sov. Phys.-Dokl. **12,** 21 (1967).

17. E. Bishimov, "Numerical solution of the strong point-blast problem in a detonating gas," in *Differential Equations and Their Applications* [in Russian] (Nauka, Moscow, 1969), pp. 94–103.

18. E. Bishimov, "Planar, cylindrical and spherical blast with backpressure in a detonating gas," Izv. Akad. Nauk Kaz. SSSR Ser. Fiz.-Mat. No. 1 (1969).

19. E. Bishimov, V. P. Korobeinikov, V. A. Levin, and G. G. Chernyi, "One-dimensional nonstationary flows of a combustible gas mixture with allowance for the finite rate of the chemical reactions," Izv. Akad. Nauk SSSR Mekh. Zhidk. Gaza No. 6 (1968).

20. V. P. Korobeinikov and V. A. Levin, "Strong blast in a combustible gas mixture," Izv. Akad. Nauk SSSR Mekh. Zhidk. Gaza No. 6, 48–51 (1969).

21. V. P. Korobeinikov, "On simple theoretical models of two-phase flows associated with combustion," Astronaut. Acta **6,** 931–941 (1979).

22. M. P. Samozvantsev, "Stabilization of detonation waves with the aid of poorly streamlined bodies," Prikl. Mekh. Tekh. Fiz. No. 4, 126–129 (1964).

23. G. G. Chernyi, "Supersonic flow around bodies with formation of detonation and combustion fronts," in *Problems of Hydrodynamics and Mechanics of a Continuous Medium* [in Russian] (Nauka, Moscow, 1969), pp. 561–578.

24. R. I. Soloukhin, J. H. Lee, and A. K. Oppenheim, "Problems of detonation in gases" (International Colloquium on Blast Gasdynamics, Brussels), Astronaut. Acta **14,** 562–582 (1969).

25. L. G. Gvozdeva, "Experimental investigation of the diffraction of detonation waves in a stoichiometric mixture of methane with oxygen," Prikl. Mekh. Tekh. Fiz. No. 5, 53–56 (1961).

26. J. H. Lee, R. Knystautas, and G. G. Bach, *Theory of Explosions.* MERL Report (McGill University, Montreal, 1969), p. 384.

27. V. P. Korobeinikov, "Gasdynamics of explosion," Annu. Rev. Fluid Mech. **3,** 317–346 (1971).

28. V. P. Korobeinikov, V. A. Levin, V. V. Markov, and G. G. Chernyi, "Propagation of blast waves in a combustible gas," Astronaut. Acta **17,** Nos. 5–6, 529–537 (1972).

29. Ya. B. Zel'dovich and Yu. P. Raizer, *Elements of Gas Dynamics and the Classical Theory of Shock Waves* (Academic, New York, 1968).

30. R. I. Soloukhin, "Nonstationary phenomena in gas detonation," 12th International Symposium on Combustion, Poitiers, 1968 (The Combustion Institute, Pittsburgh, 1969), pp. 671–679.

31. E. A. Lundstrom and A. K. Oppenheim, "On the influence of non-steadiness on the thickness of the detonation wave," Proc. Roy. Soc. London Ser. A **310**, 463–478 (1969).

32. J. Brossard, "Experimental study of divergent detonations," Doctoral thesis, University of Poitiers, 1970.

33. G. G. Bach, R. Knystautas, and J. H. Lee, "Direct initiation of spherical detonations in gaseous explosives," 12th International Symposium on Combustion, Poitiers, 1968 (Combustion Institute, Pittsburgh, 1969).

34. L. N. Busurina, V. Ya. Gol'din, N. N. Kalitkin, *et al.*, "Numerical detonation calculation," Vychisl. Mat. Mat. Fiz. **10**, 239–243 (1970).

35. N. N. Semenov, *Development of the Theory of Chain Reactions and Thermal Conflagration* [in Russian] (Znanie, 1969).

36. V. A. Levin and V. V. Markov, "Investigation of the onset of detonation by concentrated energy supply," Fiz. Gor. Vzryva **2**, 623–629 (1975).

37. V. V. Markov, "Numerical simulation of formation of multifront detonation-wave structure," Dokl. Akad. Nauk SSSR **258**, 314–317 (1981).

38. S. K. Godunov, A. V. Zabrodin, and G. P. Prokopov, "Difference scheme for two-dimensional nonstationary gasdynamics problems and calculation of flow with departing shock wave," Vychisl. Mat. Mat. Fiz. **1**, No. 6, 1020–1050 (1961).

39. B. Lewis and G. von Elbe, *Combustion, Flames, and Explosions of Gases* (Academic, New York, 1961).

40. A. Ferri, "Mixing-controlled supersonic combustion," Annu. Rev. Fluid Mech. **5**, 303 (1973).

41. R. I. Soloukhin, *Measurement Methods and Principal Results of Shock-Tube Experiments* [in Russian] (Akad. Nauk SSSR, Novosibirsk, 1969).

42. V. F. Klimkin, R. I. Soloukhin, and P. Wolansky, "Initial stage of a spherical detonation directly initiated by a laser spark," Combust. Flame **21**, No. 2, 111–117 (1973).

43. R. I. Soloukhin, *Shock Waves and Detonation in Gases* [in Russian] (Fizmatgiz, Moscow, 1963).

44. S. Taki and T. Fujiwara, "Numerical analysis of two-dimensional nonsteady detonation," AIAA J. **16**, No. 1, 73–77 (1978).

45. A. L. Fuller and R. A. Gross, "Thermonuclear detonation wave structure," Phys. Fluids **11**, No. 3, 534–544 (1968).

46. M. S. Chu, "Thermonuclear reaction waves at high densities," Phys. Fluids **15**, 413–422 (1972).

47. N. G. Basov and A. N. Oraevskii, Sov. Phys.-JETP **17**, No. 5, 1171–1173 (1963).

48. B. F. Gordiets, N. N. Sobolev, and L. A. Shelepin, "Kinetics of physical processes in CO_2 lasers," Sov. Phys.-JETP **26**, No. 5, 1822 (1968).

49. J. D. Anderson, Jr. and M. T. Madden, "Population inversion behind shock discontinuities" [Russian translation], Raket. Tekh. Kosmonvatika **9**, No. 8, 256–258 (1971).

50. J. D. Anderson, Jr., "A time-dependent analysis of population inversions in an expanding gas," Phys. Fluids **13**, 1983–1989 (1970).

51. S. A. Losev, *Gasdynamic Lasers* [in Russian] (Nauka, Moscow, 1977).

52. N. S. Zakharov and V. P. Korobeinikov, "The piston problem in a relaxing gas," Inzh. Fiz. Zh. **39**, No. 3, 482–485 (1980).

53. J. D. Anderson, *Gas Dynamic Lasers—An Introduction* (Academic, New York, 1976).

54. J. H. Lee, T. D. Bui, and T. D. Knystautas, "Population inversion in blast waves," Appl. Phys. Lett. **22**, No. 9 (1973).

55. V. A. Levin and Yu. V. Tunik, "Optical gain behind an axisymmetric shock wave," Fiz. Gor. Vzryva **13**, No. 3, 447–454 (1977).

56. Yu. V. Tunik, "Flow of a relaxing gas behind a blast wave," in *Research into Mechanics of Liquids and Solids* [in Russian] (Moscow University, Moscow, 1977), pp. 65–71.

57. M. I. Podduev, "Method for obtaining a vibrational population inversion in a CO_2–N_2–He(H_2O) gas mixture," Sov. J. Quantum Mech. **9**, No. 2, 225–227 (1979).

58. V. P. Korobeinikov and M. I. Podduev, "On the inverse population and laser effect in gas flows due to explosion or detonation. Combustion in reactive system," in *Progress in Astronautics and Aeronautics*, Vol. 76 (AIAA, New York, 1981), pp. 89–105.

59. V. P. Korobeinikov and I. S. Mel'nikov, "Mathematical modeling of blast-wave propagation in coal mines," in *Abstracts, 8th International Colloquium on Gasdynamics of Explosions and Reactive Systems*, Minsk, Aug. 24–28, 1981, p. 103.

60. V. V. Lunev, *Hypersonic Aerodynamics* [in Russian] (Mashinostroenie, 1975).

Chapter 7

1. A. G. Kulikovskii and G. A. Lyubimov, "Gas-ionizing magnetohydrodynamic shock waves," Sov. Phys.-Dokl. **4**, 1185 (1959).

2. V. P. Korobeinikov, "Calculation of one-dimensional flows for cylindrical and planar blasts in an ideally conducting gas with allowance for backpressure and for the magnetic field," Sov. Phys.-Dokl. **10**, 1131 (1965).

3. J. D. Murray, "Strong cylindrical shock waves in magnetogasdynamics," Mathematica **8**, pt. 2 (1961).

4. V. P. Korobeinikov and E. V. Ryazanov, "Some solutions of the equations of one-dimensional magnetohydrodynamics," Prikl. Mat. Mekh. **24**, 111–120 (1960).

5. V. P. Korobeinikov, "One-dimensional gas motion accompanied in a magnetic field by shock waves," Prikl. Mekh. Tekh. Fiz. No. 2, 47–53 (1969).

6. A. Sakurai, "Interaction of cylindrical shock waves with magnetic field parallel to the axis," J. Phys. Soc. Jpn. **17**, No. 10 (1962).

7. V. P. Korobeinikov, "Damping of weak magnetohydrodynamic shock waves," Magn. Gidrodin. No. 2, 25–30 (1967).

8. V. P. Korobeinikov, "One-dimensional self-similar motions of a conducting gas in a magnetic field," Sov. Phys.-Dokl. **3**, 739 (1958).

9. V. P. Korobeinikov, "Cylindrical blast and rectilinear discharge in an electrically conducting medium with allowance for the magnetic field," in *Problems of Magnetohydrodynamics and Plasma Dynamics* [in Russian] (Latvian Academy of Sciences, Riga, 1962), pp. 196–206.

10. E. Kamke, *Differential Equations, Solution Methods, and Solutions. Vol. 1, Ordinary Differential Equations* [in German] (Dover, New York, 1971).

11. C. Greifinger and J. D. Cole, "Similarity solution for cylindrical magnetohydrodynamic blast waves," Phys. Fluids **5**, No. 12, 1597–1607 (1962).

12. V. P. Korobeinikov, "Planar blast in an electrically conducting gas in an oblique magnetic field," Sov. Phys.-Dokl. **11**, 929 (1966).

13. A. G. Kulikovskii and G. A. Lyubimov, *Magnetic Hydrodynamics* [in Russian] (Fizmatgiz, Moscow, 1962).

14. L. D. Landau and E. M. Lifshitz, *Fluid Mechanics* (Pergamon, New York, 1959).

15. V. P. Korobeinikov and V. P. Karlikov, "Interaction of strong blast waves with an electromagnetic field," Sov. Phys.-Dokl. **5**, 679 (1960).

16. V. P. Korobeinikov and E. V. Ryazanov, "Propagation of MHD shock waves produced in blasts," in *Problems of Magnetic Hydrodynamics* [in Russian] (Latvian Academy of Sciences, Riga, 1964), pp. 33–34.

17. V. P. Karlikov, "Linearized problem of strong-blast propagation in an inhomogeneous atmosphere" (Abstract of candidate's dissertation) (Moscow State University, Moscow, 1958).

18. N. Ness, I. B. Fanucci, and L. J. Kijewski, "Nonuniform expansion of a piston into an ionized medium with a weak magnetic field," Phys. Fluids **6**, 1241–1249 (1963).

19. V. P. Korobeinikov, "Interaction of shock waves in an ideally conducting gas with weak magnetic fields," Prikl. Mekh. Tekh. Fiz. No. 4, 113–114 (1964).

20. J. M. Burgers, "Penetration of a shock wave in a magnetic field," Russian translation in the collection *Magnetohydrodynamics* [in Russian] (Atomizdat, Moscow, 1958).

21. A. Sakurai, "A blast-wave theory," in *Basic Developments in Fluid Dynamics*, Vol. 1 (Academic, New York, 1965), pp. 309–375.

22. V. P. Korobeinikov and E. V. Ryazanov, "Effect of a magnetic field on the propagation of plane and cylindrical shock waves," Prikl. Mat. Tekh. Fiz. No. 4, 47–51 (1962).

23. P. S. Lykoudis, "Magnetofluidmechanic blast waves in a medium with finite electrical conductivity," Phys. Fluids **7**, No. 8 (1972).

24. L. G. Linhart, *Plasma Physics*, 2nd Ed. (North-Holland, Amsterdam, 1961).

25. G. C. Vlases and D. L. Jones, "Experimental study of cylindrical magnetohydrodynamic blast waves," Phys. Fluids **11**, No. 5, 987–992 (1968).

26. H. P. Greenspan, "Similarity solution for cylindrical shock," Phys. Fluids **5**, No. 3, 255–258 (1962).

27. F. A. Baum, S. A. Kaplan, and K. P. Stanyukovich, *Introduction to Cosmic Gasdynamics* [in Russian] (Fizmatgiz, Moscow, 1958).

28. P. P. Volosevich and V. S. Sokolov, "Self-similar problem of the expansion of an electrically conducting gas into a medium with specified axial magnetic field," Magn. Gidrodin. No. 1 (1967).

29. *Nuclear Explosion in Space, on Land, and Underground* [coll. of Russian translation] (Voenizdat, 1974).

30. A. A. Brish, M. S. Tarasov, and V. A. Tsukerman, Sov. Phys.-JETP **10**, No. 6, 1095–1101 (1960).

31. V. P. Karlikov and V. P. Korobeinikov, "Disturbance of an electromagnetic field by shock waves in the presence of a conduction discontinuity," Prikl. Mat. Mekh. **25**, No. 3, 554–556 (1961).

32. A. A. Barmin and A. G. Kulikovskii, "Shock waves ionizing a gas in an electromagnetic field," Sov. Phys.-Dokl. **13**, 4 (1968).

33. V. P. Korobeinikov, V. V. Markov, and B. V. Putyatin, "Propagation of cylindrical blast waves with allowance for radiation and for the magnetic field," Izv. Akad. Nauk SSSR, Mekh. Zhidk. Gaza No. 4, 133–138 (1977).

34. R. D. Richtmeyer and K. W. Morton, *Difference Methods for Initial Value Problems* (Interscience, New York, 1967).

35. H. L. Brode, "Gasdynamic motion with radiation: a general numerical method," Astronaut. Acta **14**, Nos. 5–6 (1969).

36. L. M. Degtyarev and A. P. Favorskii, "Flux variant of the sweeping method for difference problems with strongly varying coefficients," Zh. Vychisl. Mat. Mat. Fiz. **9**, No. 6 (1969).

37. C. L. Mader, *Numerical Modeling of Detonation* (University of California, Berkeley, 1979).

38. N. M. Kuznetsov, *Thermodynamic Functions and Shock Adiabats of Air at High Temperatures* [in Russian] (Mashinostroenie, 1965).

39. N. N. Kalitkin, L. V. Kuz'mina, and V. S. Rogov, *Tables of Thermodynamic*

Functions and Transport Coefficients of Plasma [in Russian]. Inst. Appl. Mech. Preprint, 1972.

40. A. V. Avilova, L. M. Biberman, V. S. Vorob'ev, V. M. Zamalin, G. A. Kobzev, A. N. Lagar'kov, A. Kh. Mnatsakanyan, and G. E. Norman, *Optical Properties of Hot Air* [in Russian] (Nauka, Moscow, 1970).

41. L. I. Sedov, *Similarity and Dimensionality Methods in Mechanics*, 9th Russian Ed. (Nauka, Moscow, 1981) [English translation published by Mir, Moscow, 1985].

42. A. A. Samarskii and Yu. P. Popov, *Difference Methods of Solving Gasdynamic Methods*, 2nd ed. (Nauka, Moscow, 1980).

Chapter 8

1. E. Parker, *Interplanetary Dynamic Processes* (Wiley, New York, 1963).

2. D. Ya. Martynov, *Course of General Astrophysics* [in Russian] (Nauka, New York, 1966).

3. G. Smith and E. Smith, *Solar Flares* [Russian translation] (Mir, Moscow, 1966).

4. D. S. Colburn and C. P. Sonnett, "Discontinuities in the solar wind," Space Sci. Rev. **5**, 449–506 (1966).

5. A. Maxwell, R. J. Defouw, and P. Cummings, "Radio evidence for solar corpuscular emission," Planet. Space Sci. **12**, 435–449 (1964).

6. R. C. Hwang, C. K. Lin, and C. C. Chang, "A viscous model of solar wind," Astrophys. J. **145**, No. 1, 255–269 (1966).

7. G. S. Bisnovatyi-Kogan, "Flow of ideal gas in a spherically symmetric gravitational field with allowance for radiant heat conduction and radiation pressure," Prikl. Mat. Mekh. **31**, No. 4, 762–769 (1967).

8. V. V. Vitkevich and V. I. Vlasov, "Radio-astronomy investigations of the motion and dimensions of small-scale inhomogeneities of interplanetary plasma," Sov. Phys.-Dokl. **13**, 624 (1968).

9. G. F. Chew, M. L. Goldberger, and F. E. Low, "The Boltzmann equation and the one-fluid hydromagnetic equations in the absence of particle collisions," Proc. R. Soc. London Ser. A **236**, 112–116 (1956).

10. V. B. Baranov and K. V. Krasnobaev, *Hydrodynamic Theory of Space Plasma* [in Russian] (Nauka, Moscow, 1977).

11. V. P. Korobeinikov, "Application of dimensionality analysis to the problems of motion on interplanetary gas during solar flares," Sov. Phys.-Dokl. **14**, 308 (1969).

12. V. P. Korobeinikov, "On the gas flow due to solar flares," Solar Phys. **7**, No. 3, 463–470 (1969).

13. L. N. Sedov, *Similarity and Dimensionality Methods in Mechanics*, 9th ed. (Nauka, Moscow, 1981) [English translation published by Mir, Moscow, 1985].

14. V. P. Korobeinikov, N. S. Mel'nkov, and E. V. Ryazanov, *Point-Blast Theory* [in Russian] (Fizmatgiz, Moscow, 1961).

15. H. Mirels and J. F. Mullen, "Aerodynamic blast simulation in hypersonic tunnels," AIAA J. **3**, No. 11 (1965).

16. M. L. Lidov, "Self-similar motions of a gas with spherical symmetry in the field of a gravitating center," Astron. Zh. **34**, No. 4, 603–608 (1957).

17. R. Z. Sagdeev, "Collective processes and shock waves in a tenuous plasma," in *Reviews of Plasma Physics*, Vol. 4 (Consultants Bureau, New York, 1968).

18. Yu. A. Berezin, R. Kh. Kurmullaev, and Yu. E. Nesterikhin, "Collisionless shock waves in a tenuous plasma," Izv. Akad. Nauk SSSR Fiz. Gor. Vzryva **2**, No. 1 (1966).

19. J. T. Cosling, *et al.*, "Measurement of the interplanetary solar wind during the

large geomagnetic storm of April 17–18, 1965," J. Geophys. Res. **72,** No. 7, 1813 (1967).

20. J. T. Cosling, *et al.*, "Satellite observations of interplanetary shock waves," J. Geophys. Res. **73,** No. 1 (1968).

21. T. Gold, "Discussion of shock waves and rarefied gases," in *Gas Dynamics of Cosmic Clouds,* edited by H. C. van Hulst and J. M. Burgers (North–Holland, Amsterdam, 1955), p. 103.

22. G. G. Chernyi, *Gas Flow at High Supersonic Velocity* [in Russian] (Fizmatgiz, Moscow, 1959).

23. S. S. Grigoryan, G. V. Marchenko, and Yu. L. Yakimov, "Unsteady motion of gas in shock tubes of variable cross section," Prikl. Mekh. Tekh. Fiz. No. 4, 109–113 (1961).

24. D. D. Laumbach and R. F. Probstein, "A point explosion in a cold exponential atmosphere," J. Fluid Mech. **35,** pt. 1.

25. V. P. Korobeinikov, "Application of blast theory to problems of shock-wave propagation in solar flares," *Abstract, 3rd All-Union Conference on Theoretical and Applied Mechanics* (Nauka, Moscow, 1968), p. 171.

26. P. S. Coleman, Jr. "Variations in the interplanetary magnetic field, 1." J. Geophys. Res. **71,** No. 23, 5509–5531 (1966).

27. V. P. Korobeinikov and E. V. Ryazanov, "Propagation of magnetohydrodynamic shock waves produced by blasts," in *Problems of Magnetohydrodynamics* [in Russian] (Latvian Academy of Sciences, Riga, 1964), pp. 33–41.

28. V. P. Korobeinikov, "Interaction of shock waves in an ideally conducting gas with weak magnetic fields," Prikl. Mekh. Tekh. Fiz. No. 4, 113–114 (1964).

29. A. G. Kulikovskii and G. A. Lyubimov, *Magnetic Hydrodynamics* [in Russian] (Fizmatgiz, Moscow, 1962).

30. *Solar Wind,* edited by R. J. Mackin and M. Neugebauer (Pergamon, New York, 1966).

31. K. N. Gringauz, V. V. Bezrukikh, and L. I. Musatov, "Solar wind observation on the interplanetary station Venus-3," paper at Symposium on Solar and Terrestrial Physics [in Russian], Belgrade, 1966.

32. S. N. Vernov, *et al.* "Investigation of cosmic rays by automatic satellite stations," *Proceedings of the 5th All-Union Winter School on Cosmophysics,* Apatity, Akad. Nauk SSSR, 1968.

33. G. P. Lyubimov, "Deceleration of shock waves from solar flares in outer space," Astron. Tsirk. No. 488, 4–6 (1968).

34. S. J. Bame, I. R. Asbridge, A. J. Hunhausen, *et al.* "Solar wind and magnetosheath observations during the 13–14 January 1967 geomagnetic storm," J. Geophys. Res. **73,** No. 17 (1968).

35. M. Dryer and D. L. Jones, "Energy deposition in the solar wind by flare-generated shock waves," J. Geophys. Res. **73,** No. 15 (1968).

36. V. G. Gorbatskii, *Cosmic Blasts* [in Russian], 3rd. ed. (Nauka, Moscow, 1979).

37. V. P. Korobeinikov and Yu. M. Nikolaev, "Propagation of disturbances in solar wind from chromospheric flares," Kosm. Issled. No. 6, 891–895 (1969).

38. V. P. Korobeinikov, "Planetary-gas motion due to solar flare," in *Problems of Applied Mathematics and Mechanics* [in Russian] (Nauka, Moscow, 1971), pp. 211–221.

39. V. P. Korobeinikov and Yu. M. Nikolaev, "Shock waves and magnetic-field configurations in interplanetary space," Cosmic Electrodynamics **3,** No. 1, 25–44 (1972).

40. L. V. Shidlovskaya, "The problem of gas motion in tubes of variable cross section and its application to solar-wind disturbances," Izv. Akad. Nauk SSSR Mekh. Zhidk. Gaza No. 3, 84–89 (1976).

41. A. J. Hundhausen, "Interplanetary shock waves and the structure of solar-wind disturbances," Proceedings of the 1971 NASA Conference (NASA, Washington, DC, 1972).

42. A. J. Hundhausen and R. A. Gentry, "Numerical simulation of flare-generated disturbances in solar wind," J. Geophys. Res. **74**, No. 11, 2908–2929 (1969).

43. L. V. Shidlovskaya, "Solar-flare-induced disturbance propagation in an interplanetary plasma," Dokl. Akad. Nauk SSSR **225**, 29 (1975).

44. V. P. Korobeinikov and L. V. Shidlovskaya, "Numerical solution of problems of a blast in a moving gas," in *Numerical Methods of Fluid Mechanics* [in Russian], Vol. 6, No. 4 (Nauka, Novosibirsk, 1975), pp. 56–68.

45. O. M. Belotserkovskii and Yu. M. Davidov, "Nonstationary large-particles method for gasdynamic calculations," Vychisl. Mat. Mat. Fiz. **11**, No. 1, 176–207 (1971).

46. A. J. Hundhausen, *Coronal Expansion and Solar Wind* (Springer, New York, 1972).

47. V. V. Zakaidakov and V. S. Synakh, "Numerical simulation of shock-wave propagation in the sun's atmosphere and in solar wind, in *Mathematical Model of Near Outer Space* [in Russian] (Nauka, Novosibirsk, 1977), pp. 33–41.

48. L. V. Shidlovskaya, "Numerical solution of a two-dimensional problem of shock-wave propagation in outer space," Vychisl. Mat. Mat. Fiz. **17**, No. 1, 196–208 (1977).

49. Yu. M. Davydov and L. V. Shidlovskaya, "Numerical experiments on physical processes in the near outer space using the large-particles method," in *Mathematical Models of the Near Outer Space* [in Russian] (Nauka, Novosibirsk, 1977), pp. 67–88.

50. O. M. Belotserkovskii and Yu. M. Davydov, *Method of Large Particles in Gasdynamics. Computer Experiment* [in Russian] (Nauka, Moscow, 1982).

51. L. V. Sokolovskaya, "Results of numerical modeling of single and dual solar flares," in *Shock Physics and Wave Dynamics in Space and on Earth* [in Russian] (USSR Academy of Sciences, Moscow, 1983), pp. 37–57.

Printed in the United States
By Bookmasters

Printed in the United States
By Bookmasters